RAT DISSECTION MANUAL

Bruce D. Wingerd

Illustrated by Geoffrey Stein, D.V.M.

THE JOHNS HOPKINS UNIVERSITY PRESS
Baltimore & London

© 1988 The Johns Hopkins University Press
All rights reserved
Printed in the United States of America

9 8 7 6 5 4 3 2

The Johns Hopkins University Press
2715 North Charles Street
Baltimore, Maryland 21218-4363
www.press.jhu.edu

The paper used in this publication meets the minimum requirements of American National Standard for Information Sciences — Permanence of Paper for Printed Library Materials, ANSI Z39.48-1984.

ISBN 0-8018-3690-5

Contents

Illustrations   v
Introduction   vii

1. **EXTERNAL ANATOMY & THE SKIN**   1

   External Anatomy   1
   - Head   1
   - Neck   1
   - Trunk   2
   - Tail   2

   The Skin   2

2. **THE SKELETAL SYSTEM**   5

   Bones of the Axial Skeleton   5
   - Skull   6
   - Vertebral Column   9
   - Thorax   10

   Bones of the Appendicular Skeleton   10
   - Pectoral Girdle   10
   - Cranial Appendages   12
   - Pelvic Girdle   13
   - Caudal Appendages   13

3. **PRINCIPLES & TECHNIQUES OF DISSECTION**   16

   Preparing for Dissection   16
   The Skinning Procedure   17

4. **THE MUSCULAR SYSTEM**   18

   Muscle Dissection   19
   - Muscles of the Head   19
   - Ventral Muscles of the Neck   19
   - Ventral Muscles of the Pectoral Girdle   21
   - Dorsal Muscles of the Neck & Pectoral Girdle   21
   - Muscles of the Shoulder & Cranial Appendages   24
     - Muscles of the Shoulder   24
     - Muscles of the Brachium   24
     - Muscles of the Antebrachium   25

Contents

      Muscles of the Trunk   27
          Thoracic Muscles   27
          Abdominal Muscles   27
          Back Muscles   27
      Muscles of the Pelvic Girdle & Hip   28
      Muscles of the Caudal Appendages   28
          Muscles of the Thigh   28
          Muscles of the Shank   30

5. **THE NERVOUS SYSTEM**   32

    The Central Nervous System   33
      Brain   34
      Cranial Nerves   36
      Spinal Cord   38

    The Peripheral Nervous System   38
      Major Plexi   39

    The Autonomic Nervous System   40

6. **THE DIGESTIVE SYSTEM**   41

    Cranial Digestive Structures   41
      Salivary Glands   41
      Oral Cavity   42
      Pharynx   42
      Esophagus   43

    Caudal Digestive Structures   43
      Peritoneum   45
      Liver   45
      Stomach   45
      Small Intestine   45
      Pancreas   45
      Large Intestine   47

7. **THE RESPIRATORY SYSTEM**   48

    Cranial Respiratory Structures   48
      Rostrum   48
      Pharynx   49
      Larynx   49
      Trachea   49

    Caudal Respiratory Structures   49
      Bronchi   51
      Lungs   51

8. **THE CIRCULATORY SYSTEM**   52

    Heart   52
      External Features of the Heart   52
      Blood Vessels of the Heart   53
      Internal Features of the Heart   54

    Blood Vessels Cranial to the Heart   55
      Arteries   55
      Veins   57

    Blood Vessels Caudal to the Heart   59
      Arteries   59
      Veins   60

9. **THE EXCRETORY & REPRODUCTIVE SYSTEMS**   63

    The Excretory System   63
      Kidneys   63
      Ureters   65
      Urinary Bladder   65
      Urethra   65

    The Reproductive System   65
      Male Reproductive Structures   65
          Scrotum   65
          Testes   65
          Epididymus   65
          Ductus Deferens   65
          Urethra   65
          Penis   66

      Female Reproductive Structures   66
          Ovaries   67
          Fallopian tubes   67
          Uterus   67
          Vagina   67

Illustrations

N.1. Descriptive terminology and planes of section   viii
1.1. External anatomy of the rat: lateral view of the male   2
1.2. External genitals of the male   3
1.3. External genitals of the female   3
1.4. Transverse section of the skin, 400× magnification   4
2.1. The articulated skeleton of the rat, lateral view   6
2.2. Lateral view of the skull and mandible   6
2.3. The skull, dorsal view   8
2.4. The skull, ventral view   8
2.5. The axis and atlas, lateral view   9
2.6. Selected disarticulated vertebrae   10
2.7. The thoracic cage, lateral view of the left side   11
2.8. The left scapula, lateral view   11
2.9. The left humerus, cranial and caudal views   12
2.10. The left radius and ulna in their articulated position, lateral view   13
2.11. The left manus, dorsal view   13
2.12. Lateral view of the pelvic girdle and the adjoining lumbar, sacral, and caudal vertebrae   14
2.13. The innominate bones in their articulated position, ventral view   14
2.14. The left femur, cranial view   15
2.15. The left tibiofibula, cranial view   15
2.16. The left pes, dorsal view   15
3.1. Rat profile, ventral view: a cutting guide for the skinning procedure   17
4.1. Superficial muscles of the rat, lateral view   20
4.2. Muscles of the ventral neck region   21
4.3. The ventral muscles   22
4.4. The dorsal muscles   23
4.5. Deep muscles of the shoulder and superficial muscles of the brachium and antebrachium, dorsolateral view   25
4.6. Deep muscles of the shoulder and superficial muscles of the brachium and antebrachium, medial view   26

Illustrations

4.7. Dorsal and lateral muscles of the pelvic region and lower appendages 29
4.8. Ventral and medial muscles of the pelvic region and lower appendages 30
5.1. The brain, dorsal view 35
5.2. Lateral view of the brain 36
5.3. Lateral view of the brain, midsagittal section showing internal structures 36
5.4. Ventral surface of the brain 37
5.5. Cross section of the spinal cord and surrounding structures 38
5.6. The nervous system of the rat 39
6.1. Salivary glands and associated structures 42
6.2. The oral cavity and pharynx 43
6.3. Organs of the abdominal cavity, ventral view 44
6.4. Abdominal organs, ventral view 46
6.5. Schematic view of the digestive system caudal to the diaphragm 47
7.1. The head and neck regions, midsagittal section 49
7.2. The thoracic cavity and its associated structures 50
8.1. The heart, ventral view 53
8.2. The heart, dorsal view 54
8.3. Internal structures of the heart, ventral view 55
8.4. The cranial blood vessels 56
8.5. Arterial circulation scheme 57
8.6. Venous circulation scheme 58
8.7. The caudal blood vessels 60
8.8. The hepatic portal circulatory scheme 61
9.1. The excretory and reproductive systems of the rat, ventral view of the female 64
9.2. The male reproductive organs and associated structures 66
9.3. The male excretory and reproductive systems, schematic view 67
9.4. The female excretory and reproductive systems, schematic view 68

Introduction

THE RECENT HISTORY of the rat is a story of successful survival that parallels the development of human civilizations. The species that we identify as the common rat and its larger cousin, the Norway rat, originated from their natural habitat in Southeast Asia. From there, they emigrated to China, Europe, Africa, and eventually to America. Their means of disbursement was via man, with whom they formed a commensal relationship. Their successful survival depended on a number of factors including their high rate of reproduction, their incredible adaptability to different environments, and their knack for feeding opportunisticly upon the discards of human civilization.

In today's world, the rat has become an established member of the global community that has a significant impact on human beings. In a negative sense, many of the numerous vectors they carry spread disease among human populations, and their feeding activities on agricultural reserves cause a significant reduction that would otherwise be available for human use.

On a more positive note, however, the rat has provided us with important opportunities in furthering scientific progress. Its physiological and anatomical similarity to humans, its high reproductive rate, and its ease of maintenance in a laboratory environment have made it an invaluable research and diagnostic tool. Indeed, it was the first mammal in space and is presently being used extensively in cancer, AIDS, and brain research.

The abundance of the rat in modern times has also made it a convenient specimen for student anatomical dissection. Its use in the anatomy lab usually focuses on providing beginning students with a firsthand look at a typical small mammal. The experience has important merit, as it affords an opportunity for the examination of real structures and the development of surgical skills. If taught properly, with reverence and respect for all life, dissection can open up new dimensions of understanding body structure that drawings, models, and charts cannot fulfill. It is the purpose of this manual to present dissection in a clear, step-by-step manner in order to enhance this learning experience.

Introduction

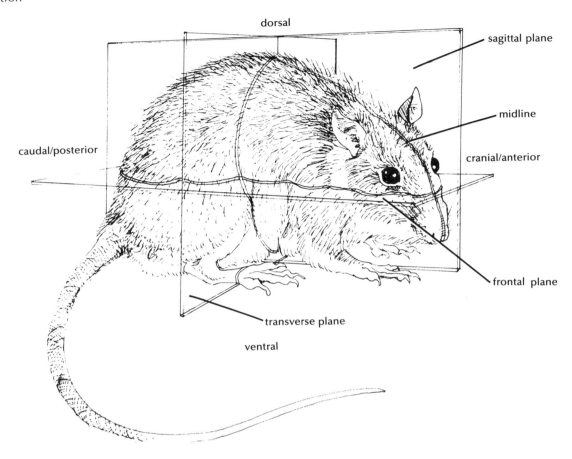

FIGURE N.1. Descriptive terminology and planes of section

The Norway rat, or *Rattus norvegicus*, is the proper name for the white laboratory rat that you will be studying. It is taxonomically classified as follows:

Phylum: Chordata
Subphylum: Vertebrata
Class: Mammalia
Order: Rodentia
Suborder: Myomorpha
Family: Muridae
Genus: *Rattus*
Species: *norvegicus*

Before beginning your study of the rat, study the directional and spatial terms listed below. Note that these terms apply primarily to quadrupeds, or four-legged animals. This introductory step is essential because these terms will be used extensively throughout the text (Fig. N.1).

Directional Terms
Rostral: toward the nose end.
Cranial/anterior[1]: toward the head end.
Caudal/posterior[1]: toward the tail end.
Dorsal: toward the back side.
Ventral: toward the belly side.
Midline: an imaginary plane that bisects the body into right and left halves.
Median: lying in or near the midline.
Medial: lying closer to the midline relative to another structure.
Lateral: lying further from the midline relative to another structure.
Proximal: near a structure's origin or point of attachment to the body.
Distal: away from a structure's origin or point of attachment to the body.
Superficial: toward the body surface.
Deep: away from the body surface.

[1]In the human, **anterior** describes "toward the belly side," and **posterior** describes "toward the back side." Also in the human, **superior** and **inferior** are used to describe "toward the head end" and "away from the head end," respectively. This difference in terminology is due to the bipedal (two-legged) nature of the human posture.

Planes of Section

Transverse (cross): a plane that passes at a right angle to the long axis of a body or body structure, usually resulting in cranial and caudal portions.

Longitudinal: a plane that extends from cranial to caudal along the long axis of the body. The longitudinal plane bisects the transverse plane at a right angle.

Sagittal: a longitudinal plane that divides the body into right and left halves; if this division is into equal halves, it is called **midsagittal**. If it is into unequal halves, it is called **parasagittal**.

Frontal (coronal): a longitudinal plane that extends from cranial to caudal and horizontally from right to left, dividing the body into ventral and dorsal portions.

External Anatomy & the Skin

EXTERNAL ANATOMY

YOUR RAT SPECIMEN comes to you from a biological supply house in a preserved state. The fixative that was initially used during the preservation process is a diluted formalin mixture. This chemical can be an irritant to exposed skin, eyes, and the upper respiratory tract; therefore, wear protective gloves and clothing. Your instructor should be prepared to assure you that the lab is well ventilated. When you are ready to begin your study, remove the rat from the bag in which it arrived and place it on a dissecting tray.

Examine the external anatomy of the rat and compare it with Figures 1.1–1.3. Notice that the body of the rat, like that of all higher vertebrates (reptiles, birds, and mammals), consists of a **head**, **neck**, **trunk**, and **tail**.

HEAD

The features of the head are representative of the Rodentia order. The snout, or **rostrum**, protrudes anteriorly and contains two nostrils, or **external nares** on its ventral side. Posterior to the rostrum is the **mouth**, which is bordered by fleshy **lips**. Notice the cleft in the center of the upper lip. This is called the **philtrum**. On either side of the philtrum are **mystacial pads** which contain long sensory hairs called **vibrissae**. Vibrissae may also be found on the chin, in front of the ears, and below the eyes. The two **eyes** are dorsal to the mouth on both sides of the head and are bounded by **upper** and **lower eyelids**. Make a short incision extending forward from the anterior corner of one eye and pull the upper and lower eyelids apart. The semitransparent membrane between the corner of the eyeball and the lids is the **nictitating membrane**. This membrane is present in most mammals (but is greatly reduced in the human as the conjunctiva) and serves to keep the eyeball free of foreign particles. Dorsal and caudal to the eye on either side of the head is the external ear, or **pinna**, which directs sound waves into the **external auditory meatus**.

NECK

The neck provides support for the head, which it enables to articulate in order to sense the external environment.

External Anatomy & the Skin

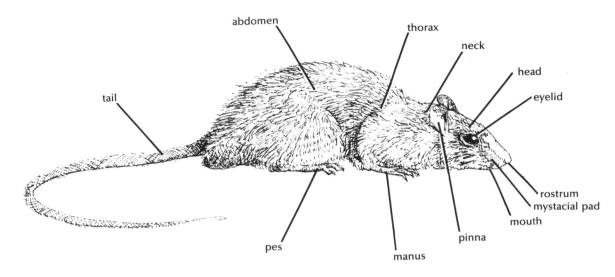

FIGURE 1.1. External anatomy of the rat: lateral view of the male

TRUNK

The trunk of the rat, like that of all vertebrates, consists of a cranial **thorax** and a caudal **abdomen**. The upper appendages are attached to the thorax and the lower appendages are attached to the caudal base of the abdomen. The form of locomotion resulting from the arrangement of the bones in the appendages is called **digitigrade**, which causes the rat to walk on the digits with the remainder of the **manus** (forefoot) and **pes** (hindfoot) elevated (in other words, "on its toes"). For comparison, humans are **plantigrade**, which allows us to walk on the soles of our feet, and horses, cattle, and deer are **unguligrade**, which allows them to walk on their enlarged, flattened nails or hoofs (in other words, "on their tiptoes").

Place your rat on its back to expose its ventral surface and identify the external openings to the **mammary glands**, called **nipples**. The presence of mammary glands, as well as hair during at least some period of its life, identifies the rat as a member of the Class Mammalia. In the rat, there are usually six pairs of mammary glands that are distributed along the ventral thorax and abdomen.

With your specimen remaining on its back, identify the external genitals located caudal to the hind legs (Figs. 1.2, 1.3). If male, these include the small rod-shaped **penis** and the sac of skin, the **scrotum**. The scrotum contains the gonads, the **testes**. If female, you will notice that the urinary and reproductive openings are separated; the ventral opening is the **urethral orifice**, and the dorsal opening is the **vaginal orifice**. In the rat, as in most mammals, the opening to the digestive tract is separate from the urinary and reproductive openings. This is the **anus** and is located near the ventral base of the tail in both sexes.

TAIL

The tail of the rat is quite long, often nearly as long as the trunk. It contains sparse hairs along its length, revealing beneath them the scalelike texture of the epidermal surface.

THE SKIN

The skin, or **integument**, of the rat typifies that of other mammals. Under microscopic examination it can be observed to contain two layers: a thin, superficial **epidermis**, and a thick, deep **dermis** (Fig. 1.4). The epidermis is composed of stratified squamous epithelium that is, in turn, divided into layers. The basement layer, or **stratum basale**, is in a state of continual growth while the superficial layer, the **stratum corneum**, consists of nonliving material. Between the basale and corneum are several layers of cells that are in various stages of keratinization, which progress as they approach the superficial layer. They are, from deep to superficial, the **stratum spinosum**, the **stratum granulosum**, and the thin **stratum lucidum**. From the epidermis, **hair** and **nails** are derived.

The dermis is composed of dense connective tissue. Embedded within it are hair roots, sweat glands, oil glands, blood vessels, and sensory nerve endings. Deep to the dermis is a layer of fat and loose connective tissue connecting the skin and underlying muscle tissue, called the **superficial fascia**.

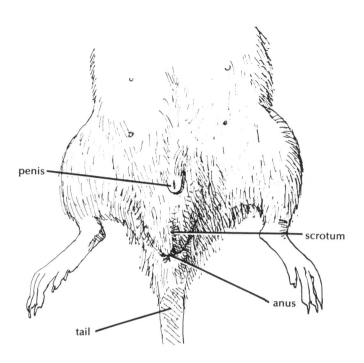

FIGURE 1.2. External genitals of the male

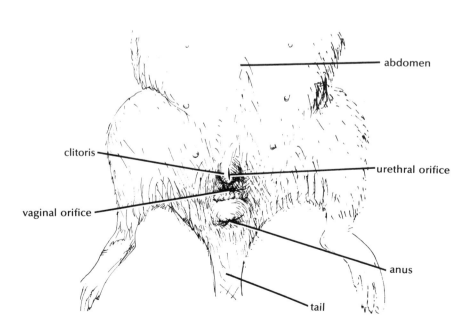

FIGURE 1.3. External genitals of the female

External Anatomy & the Skin

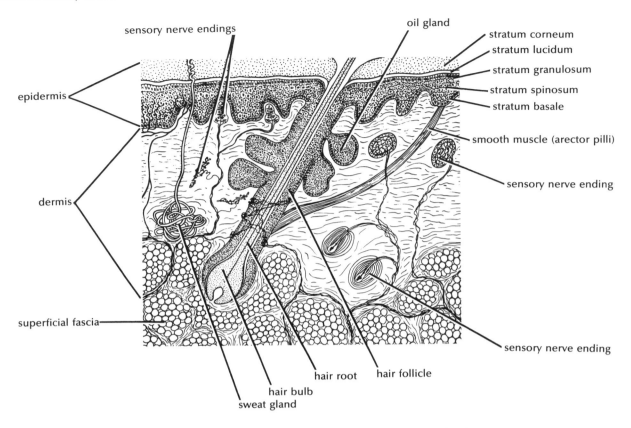

FIGURE 1.4. Transverse section of the skin, 400 × magnification

The Skeletal System

THE SKELETAL SYSTEM is composed of bones and their associated cartilages. The cartilages are mainly located between adjacent bones and form the joints, or **articulations**. As a whole, the skeletal system **supports** soft body tissues, provides a site for muscle attachment to make **movement** possible, **protects** vital organs, **stores** calcium and phosphorus in the form of mineral salts, and **forms blood cells** from within red bone marrow by a process called *hematopoiesis*.

The skeleton of the rat, like the skeleton of all vertebrates, is an **endoskeleton** for it lies within the soft tissues of the body. It is divided into a **visceral skeleton** and a **somatic skeleton**. The visceral skeleton forms the supportive framework for the pharyngeal region. In mammals this consists primarily of the **hyoid bone** attached to the base of the tongue and the three **auditory ossicles** in the tympanic cavity of the ear. The visceral skeleton is more conspicuous in fishes than in mammals, as it forms the jaws and the gill apparatus in these more primitive vertebrates.

The somatic skeleton is composed of two portions: the **axial skeleton**, which contains the bones that lie along the central vertical axis of the body, and the **appendicular skeleton**, which consists of bones that lie lateral to the central axis. The axial skeleton is composed of the bones of the **skull**, the **vertebral column**, and the **sternum**. The appendicular skeleton contains the bones of the **pectoral girdle**, the **cranial appendages**, the **pelvic girdle**, and the **caudal appendages**.

In the following study, the bones of the somatic skeleton and the hyoid bone of the rat are presented. Begin your examination of the skeletal system by studying the complete (articulated) skeleton as shown in Figure 2.1 and a mounted rat skeleton if one is provided by your instructor to orient yourself before proceeding.

BONES OF THE AXIAL SKELETON

Using the following illustrations and descriptions of bones as a guide, identify the bones of the axial skeleton and their major features on the articulated skeleton of the rat and disarticulated bones that may be available in your

The Skeletal System

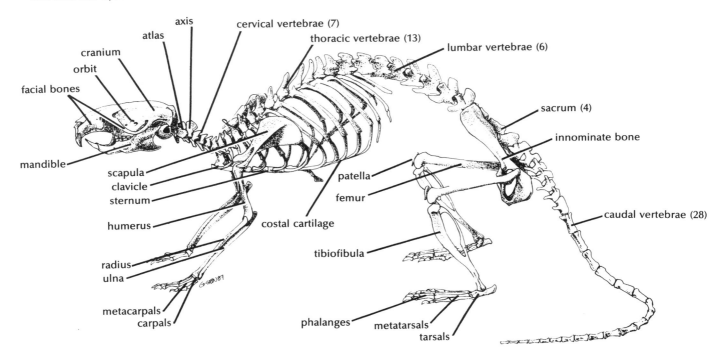

FIGURE 2.1. The articulated skeleton of the rat, lateral view

lab. The number of bones present in each group is given in parentheses following the name of each skeletal component.

SKULL (38)

The skull is divided into two portions: the **cranium**, or **neurocranium**, which encloses and protects the brain, and the **facial bones**, or **splanchnocranium**, which includes most of the bones of the eye orbit, nose, cheek, and jaw.

Bones of the Cranium (11). The cranium occupies the caudal region of the skull (Figs. 2.2–2.4). Its flat, platelike bones are united at immovable fibrous joints called **sutures**.

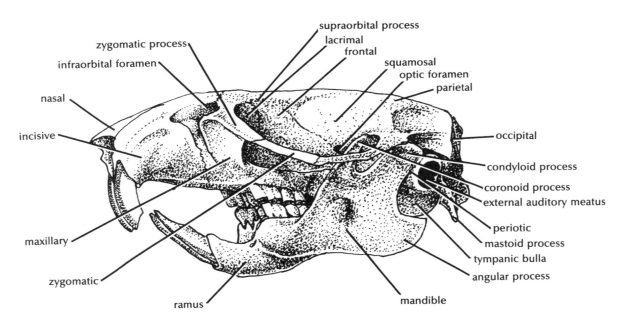

FIGURE 2.2. Lateral view of the skull and mandible

Frontal (2): paired bones that form the dorsal roof of the skull. Each also projects laterally to form the dorsal shelf of the orbit. Along the dorsal margin of the orbit is a ridge called the **supraorbital process**.

Parietal (2): paired bones caudal to the frontals. The parietals also contribute to the formation of the supraorbital process.

Interparietal (1): an unpaired bone caudal to the parietals along the dorsal midline of the skull.

Temporal (2): paired bones that form the lateral walls of the cranium. Each temporal bone consists of the following three segments:

> **Squamosal**: The largest of the three temporal bone segments, it lies ventral to the parietal bone. It contains a large projection, the **zygomatic process**, which forms the caudal end of the zygomatic arch by articulating with the zygomatic bone.
>
> **Tympanic**: Located on the ventrocaudal aspect of the cranium, its main body is occupied by the prominent, round **tympanic bulla** that protects the middle ear ossicles (**malleus**, **incus**, and **stapes**). The ventral projection on the tympanic bulla is the **mastoid process**. The large lateral opening is the **external auditory meatus**, which leads to the middle ear cavity.
>
> **Periotic** (petrosal): a small, irregular segment of the temporal bone located caudal to the tympanic. The periotic encloses the inner ear.

Occipital (1): a composite of four fused bones that form the surface and ventral floor of the cranium: the dorsal **supraoccipital**; the paired, central **exoccipitals**; and the ventral **basioccipital**. Located between the two exoccipitals is a large opening, the **foramen magnum**, which permits passage of the spinal cord. Along its lateral borders are rounded processes that articulate with the first cervical vertebra (the atlas), called the **occipital condyles**. Lateral to each occipital condyle is the union between the basioccipital and the exoccipitals. This union is marked by a ridge, called the **paramastoid process**. Between the occipital condyle and paramastoid process on each side is a small hole through which passes the hypoglossal nerve, called the **hypoglossal canal**. Located at the junction between the basioccipital and the tympanic segment of the temporal bone is a gap called the **jugular foramen**. This important opening allows passage of cerebral veins that later join the jugular vein and the glossopharyngeal, vagus, and spinal accessory cranial nerves.

Basisphenoid (1): a median, unpaired bone rostral to the occipital bone that contributes to the floor of the cranium. It contains lateral wings (**alisphenoids**) that extend to the caudal orbital wall and continue to the palatines, where they form the vertically inclined **pterygoid processes**. Penetrating through the alisphenoids are several prominent pairs of foramina: the lateral **foramen ovale**; the medial **alare canal**; the smaller, medial **carotid canal**, which is located caudal to the alare canal and allows passage for the internal carotid artery; and the large **foramen lacerum**, which borders the tympanic of the temporal bone. The central region of the basisphenoid is occupied internally by a recessed area called the **sella turcica**, which houses the pituitary gland.

Presphenoid (1): a narrow bone located rostral to the basisphenoid along the midventral line. It contains lateral wings (**orbitosphenoids**) that form the caudal ventral wall of the orbit. At the rostral edge of each wing is an opening, called the **optic foramen**, which permits passage of the optic nerve between the eyeball and the brain. Caudal to the optic foramen is a slitlike opening, the **orbital fissure**, through which pass motor nerves that control the muscles of the eyeball and sensory nerves from the region around the eye.

Ethmoid (1): a fragile, well-hidden bone located within the caudal portion of the nasal cavity. Its vertical **perpendicular plate** forms the dorsal part of the nasal septum and its caudal **cribriform plate** forms part of the rostral wall of the cranial cavity. The cribriform plate contains numerous perforations that allow passage of olfactory nerves en route to the brain.

Facial Bones (27). The facial bones form much of the rostral and lateral regions of the skull. Identify the following, using Figures 2.1–2.4 as a guide.

Lacrimal (2): small, paired bones that form the rostral margin of the orbit. Each contains a **nasolacrimal canal**, which allows passage of the nasolacrimal duct extending from the eye to the nasal chamber.

Zygomatic (2): paired bones that form the central portion of the zygomatic arch. Each articulates with the zygomatic processes of the squamosal and maxillary bones.

Maxillary (2): large, paired bones that form the lateral and ventral portions of the rostrum (nasal region) and include much of the hard palate. On their ventral side they contain the upper molar teeth, which articulate with the maxillary at cuplike depressions called **alveolar fossae**. Laterally, the prominent **zygomatic process** projects from its base to articulate with the zygomatic bone. The large opening rostral to the origin of the zygomatic process is the **infraorbital foramen**, which provides a point of attachment for the deep masseter muscle as well as passage for sensory nerves from the face and blood vessels.

Nasal (2): paired bones that form the roof of the nasal cavity.

Incisive (2): Also called the **paramaxillaries**, they are paired bones that form the lateral and ventral walls of the nasal cavity rostral to the maxillaries. The openings into the nasal cavity, the external nares, are bordered by

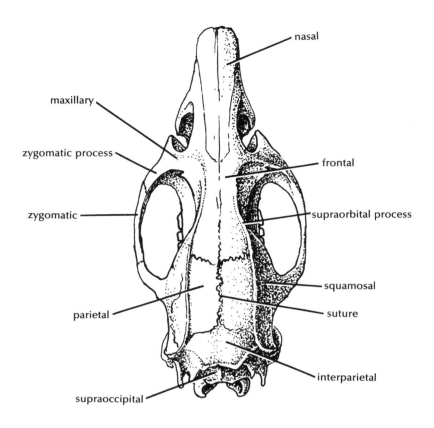

FIGURE 2.3. The skull, dorsal view

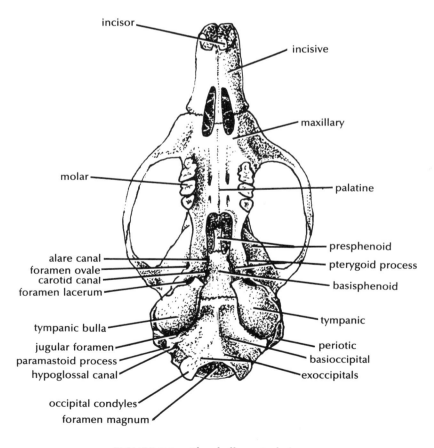

FIGURE 2.4. The skull, ventral view

8

the incisive and nasal bones. Each incisive bone contains the alveolus of one large incisor.

Turbinate (4): located internally on the lateral walls of the nasal cavity, these bones include two pairs of scroll-like bones: the **maxilloturbinals** and the **nasoturbinals**. A third pair of scroll-like bones, the **ethmoturbinals**, are not considered turbinates as they are a part of the ethmoid.

Vomer (1): a single, thin bone that forms the ventral portion of the median nasal septum, which divides the nasal cavity into right and left nasal fossae. The dorsal portion of this partition is formed by the perpendicular plate of the ethmoid bone. Both portions may be viewed by peering through the external nares.

Palatine (2): paired bones that form the caudal portion of the roof of the mouth, called the hard palate. The rostral portion of the hard palate is formed by the maxillary bones.

Mandible (2): Unlike the case in humans, these are paired bones that articulate with each other at their rostral end at a joint called the **mandibular symphysis**. They form the lower jaw. Collectively, they contain six molars caudally and two incisors rostrally which articulate with the mandible at **alveolar fossae**. The caudal dorsal end of each mandible articulates with the squamosal region of the temporal bone to form the **temporomandibular joint**, where a rounded process from the mandible called the **condyloid process** fits into a concave socket at the squamosal called the **mandibular fossa**. Rostral to the condyloid process is a pointed projection called the **coronoid process**, which provides a point of attachment for the temporalis muscle. Ventral to the condyloid process is the **angular process** for attachment of the pterygoid muscles. The **ramus** of the mandible is the region that bears teeth.

Hyoid (6): located on the ventral wall of the pharynx cranial to the larynx and caudal to the mandible, the six hyoid bones are fused together to form a single bone in the adult which does not articulate with other bones. This structure is horseshoe-shaped, and contains a short rostral projection called the **lesser cornu** and a longer caudal projection called the **greater cornu**. It provides attachment to pharyngeal and neck muscles.

VERTEBRAL COLUMN (58)

The vertebral column is a continuous chain of approximately 58 individual or fused bones, called **vertebrae**. Using the illustrations in Figures 2.5 and 2.6, identify the following features that are common among all vertebrae:

The common vertebrae: Each vertebra along the vertebral column contains the following features:
  Body: Also called the **centrum**, it is the central portion.
  Neural arch: A major arch extending dorsally from the body, it surrounds the opening through which the spinal cord passes, called the **neural canal**.
  Pedicle: a constricted portion of the neural arch near its union with the body.
  Spinous process: a prominent dorsal projection.
  Diapophyses: transverse, or lateral, projections.
  Zygapophyses: articular processes at the site of joints. Ventral zygapophyses are termed **prezygapophyses**, and dorsal zygapophyses are called **postzygapophyses**.

As a whole, the vertebral column is divided into five sections that are distinguishable by the structural characteristics of the vertebrae. From cranial to caudal they are **cervical**, **thoracic**, **lumbar**, **sacral**, and **caudal**. On an articulated skeleton of the rat and using Figures 2.1, 2.5–2.7, and 2.12, observe the five sections of the vertebral column and note the following structural differences between the vertebrae within them.

Cervical vertebrae (7): small, lightweight bones that support the neck region. Each contains diapophyses that are penetrated by **transverse foramina**, which are not found in the vertebrae of other sections. The transverse foramina of successive cervical vertebrae combine to form a canal that permits passage of the vertebral arteries and veins. The first two cervical vertebrae, the **atlas** and **axis**, are specialized to permit free movement of the skull. Note the **odontoid process** of the axis, which fits into the **facet** of the atlas for rotational head movement, and the **cranial articular facet** of the atlas, which articulates with the occipital condyles of the occipital bone.

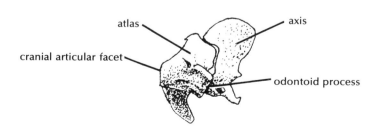

FIGURE 2.5. The axis and atlas, lateral view

The Skeletal System

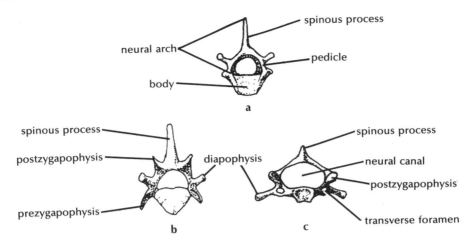

FIGURE 2.6. Selected disarticulated vertebrae, a: thoracic vertebra, b: lumbar vertebra, c: cervical vertebra

Thoracic vertebrae (13): Located caudal to the cervical region, the thoracic vertebrae provide attachment to the ribs. They are characterized by a long, bladelike spinous process, facets on the body, and zygapophyses for articulation with the ribs.

Lumbar vertebrae (6): the largest of the individual vertebrae. They are distinguishable by their robust spinous processes and their large diapophyses that project cranially and ventrally.

Sacral vertebrae (4): In the adult, they have fused to form a single bone, the **sacrum**, which articulates with the pelvic girdle.

Caudal vertebrae (28): The actual number varies according to the length of the tail that they support. They are the smallest of the vertebrae, and their size and complexity decrease as they extend caudally.

THORAX (27)

The thorax, or chest, provides protection for the structures within the thoracic cavity. It is somewhat cone-shaped, with the narrow end, the **apex**, cranial and the broad end, the **base**, caudal. Examine the thorax of the articulated skeleton (Figs. 2.1, 2.7):

Sternum (1): composed of six distinct segments called **sternebrae**. The cranial segment is the **manubrium**, the middle four constitute the **body**, and the caudal segment is the **xiphoid process**, which contains a thin, broad plate of cartilage at its end.

Ribs (26): Ten of the thirteen pairs of ribs contain a ventral extension of hyaline cartilage called **costal cartilage**, which is usually calcified in the adult. The costal cartilages of the first seven pairs articulate directly with the sternum; these ribs are thus called **true ribs**. The costal cartilages of the last six pairs do not unite with the sternum; these ribs are known as **false ribs**. The last three pairs of false ribs are not attached to cartilage and end free in the body wall; they are **floating ribs**. Note the following features of a typical rib (Fig. 2.7):

Head: the proximal end, which articulates with a vertebral body.

Tubercle: a small process that articulates with a vertebral diapophysis.

Neck: a constricted region between the head and tubercle.

Body: the long, distal portion. It is sometimes called the **shaft**.

BONES OF THE APPENDICULAR SKELETON

The appendicular skeleton is divided into four regions: the pectoral girdle, the cranial appendages, the pelvic girdle, and the caudal appendages.

PECTORAL GIRDLE (4)

The bones of the pectoral girdle provide attachment for the cranial appendages to the body trunk and include the clavicles and scapulae (Figs. 2.1, 2.7, & 2.8):

Clavicle (2): small paired bones that connect the scapula to the axial skeleton at the sternum.

Scapula (2): the large, flattened, triangular bones of each shoulder region. Identify the following features:

Spine: a prominent ridge on the lateral side.

Acromion: a process on the tip of the spine.

Coracoid process: a hook-shaped process on the medial surface.

Glenoid fossa: a depression for articulation with the head of the humerus.

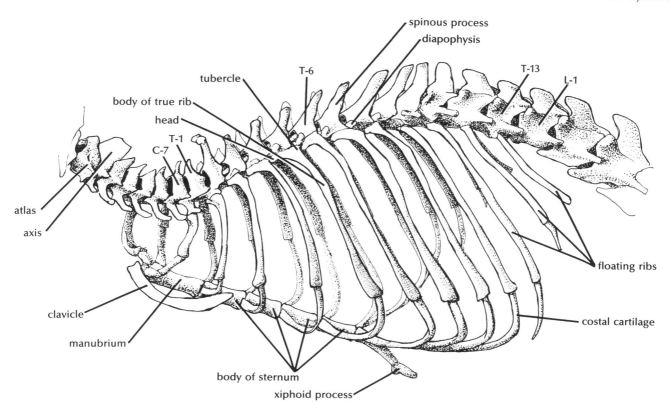

FIGURE 2.7. The thoracic cage, lateral view of the left side

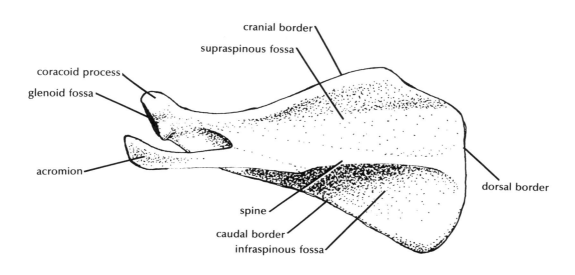

FIGURE 2.8. The left scapula, lateral view

The Skeletal System

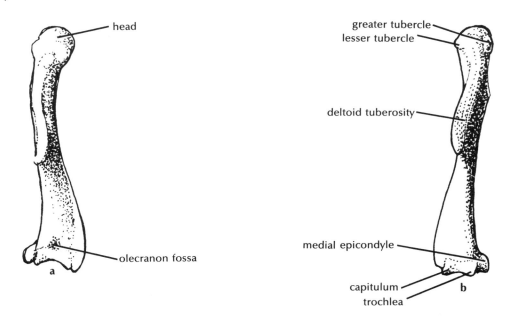

FIGURE 2.9. The left humerus, cranial and caudal views

Supraspinous fossa: a large depression cranial to the spine.
Infraspinous fossa: a large depression caudal to the spine.

CRANIAL APPENDAGES (60)

The bones of the cranial appendages include the humerus in the upper forelimb (**brachium**), the radius and ulna in the lower forelimb (**antebrachium**), the carpals in the wrist, the metacarpals in the forepaw (**manus**), and the phalanges in the five digits.

Humerus (2): the single bone of the brachium. Identify the following features of the humerus (Figs. 2.9, a & b):
  Head: the rounded proximal end that articulates with the scapula.
  Greater tubercle: a large lateral process for muscle attachment.
  Lesser tubercle: a medial process opposite the greater tubercle, also for muscle attachment.
  Trochlea: a medial, pulley-shaped process at the distal end.
  Capitulum: a lateral, rounded process at the distal end.
  Epicondyles: medial and lateral processes bordering the distal articular surfaces.
  Deltoid tuberosity: a prominent ridge on the cranial edge for muscle attachment.
  Olecranon fossa: a depression proximal to the trochlea on the dorsal side.

Ulna (2): the longer, medial bone of the antebrachium. Note its features (Fig. 2.10):
  Olecranon process: a large process at the proximal end that forms the tip of the elbow.
  Trochlear notch: Also called the **semilunar notch**, it is a depression distal to the olecranon that articulates with the trochlea.
  Coronoid process: a projection distal to the trochlear notch.
Radius (2): the shorter lateral bone of the antebrachium. Identify its features (Fig. 2.10):
  Head: the proximal end, which articulates with the capitulum.
  Radial tuberosity: a roughened surface for muscle attachment.
Carpals (16): two rows of small bones in the wrist, arranged with three bones in the proximal row and five in the distal row (Fig. 2.11). From medial to lateral, the proximal bones are the **radiale**, **ulnare**, and **pisiform**. The distal bones are, from medial to lateral, the **trapezium**, **trapezoid**, **centrale**, **capitate**, and **hamate**.
Metacarpals (10): the bones that support the body of the manus. They are numbered according to their position, beginning with the radial side: 1st, 2d, 3d, 4th, and 5th.
Phalanges (28): the bones of the digits. There are fourteen in the digits of each manus: two in the pollex, or digit 1 (proximal and distal), and three each in digits 2–5 (proximal, middle, and distal). Each distal phalanx has a horny claw attached.

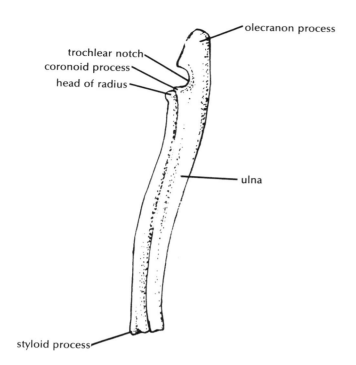

FIGURE 2.10. The left radius and ulna in their articulated position, lateral view

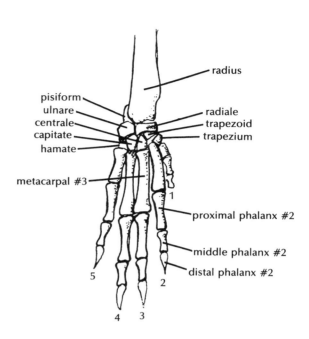

FIGURE 2.11. The left manus, dorsal view

PELVIC GIRDLE (2)

The pelvic girdle provides a point of attachment for the bones of the caudal appendages to the axial skeleton. It consists of the hip or innominate bones and may also include the sacrum.

Innominate bones (2): The two innominate bones (or **os coxae**) unite at the ventral midline at a fibrous joint called the **symphysis pubis**. They articulate dorsomedially with the sacrum at the **sacroiliac** joints. In the young, each innominate bone consists of four closely joined bones: the **ilium**, the **ischium**, the **pubis**, and the **acetabular**. In the adult, these bones fuse completely and can be identified only by their regional location (Figs. 2.12, 2.13):
  Ilium: the cranial region of each innominate. The cranial ridge provides a point for muscle attachment and is called the **iliac crest**. Caudal to the iliac crest are the ventral and dorsal **iliac spines**, also for muscle attachment.
  Ischium: the dorsocaudal region of the innominate. Along its dorsal edge is a ridge, the **ischiadic spine**, that provides a point for muscle attachment.
  Pubis: the narrow, ventrocaudal region of the innominate. It unites with the pubis of the other side at the symphysis pubis.
Acetabulum: the socket for the hip joint. It receives the head of the femur. A small accessory bone, the **acetabu-**

lar bone, forms the ventral portion, the ilium forms the cranial portion, and the ischium forms the caudal portion of this deep socket.
Obturator foramen: a large opening ventrocaudal to the acetabulum and bordered by the pubis and ischium.

CAUDAL APPENDAGES (53)

The caudal appendages consist of the femur in the thigh, the patella of the knee, the tibiofibula in the shank, the tarsals in the ankle, the metatarsal in the hindpaw (**pes**), and the phalanges in the digits.

Femur (2): the single bone of the thigh. Identify its following features (Fig. 2.14):
  Head: the rounded, proximal end, which projects medially. It fits into the acetabulum to form the hip joint.
  Greater trochanter: a roughened, proximal projection for muscle attachment.
  Lesser trochanter: located distal to the head and greater trochanter on the caudal side.
  Trochanteric fossa: a depression between the two trochanters on the caudomedial side.
  Condyles: lateral and medial articular processes at the distal end.
  Intercondylar fossa: a depression between the condyles on the caudal side.

The Skeletal System

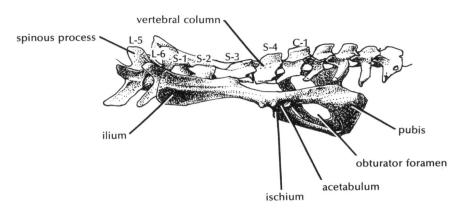

FIGURE 2.12. Lateral view of the pelvic girdle and the adjoining lumbar, sacral, and caudal vertebrae. The left pelvis is depressed ventrally to reveal the major features of the pelvic girdle.

Patellar surface: a shallow depression on the cranial surface at the distal end.
Patella (2): a small sesamoid bone (a bone formed from within a tendon) in the knee. It is formed in the tendon of the quadriceps muscle complex of the thigh.
Tibiofibula (2): the bone of the shank. It is formed by the fusion of the large **tibia** and the smaller **fibula**. The narrow fibula may be viewed free of the tibia along the proximal two-thirds of the tibiofibula. Identify the following features of this fused bone (Fig. 2.15):

Condyles: lateral and medial rounded, proximal surfaces that articulate with the condyles of the femur. They are derived from the tibia only.
Tibial tuberosity: a roughened, projected surface on the cranial side of the tibia for muscle attachment.
Malleoli: lateral and medial processes that project from the distal end of the tibiofibula.
Tarsals (14): the bones of the ankle (Fig. 2.16). From proximal to distal they are the **talus**, which articulates with the tibia at the ankle joint, the **calcaneus**, which forms

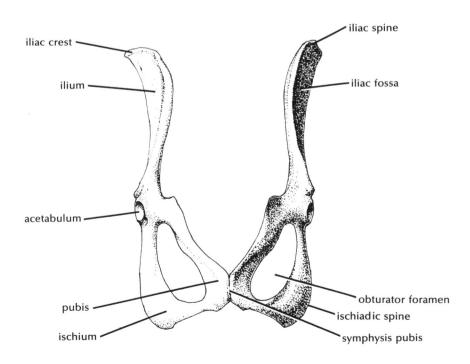

FIGURE 2.13. The innominate bones in their articulated position, ventral view

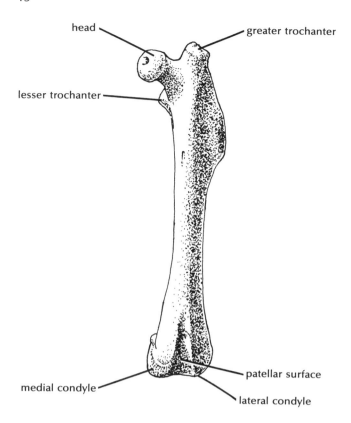

FIGURE 2.14. The left femur, cranial view

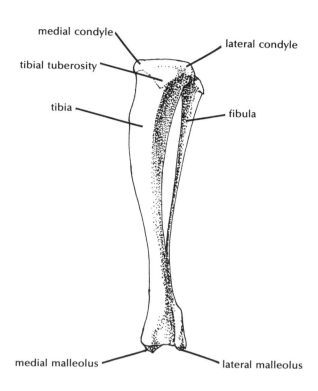

FIGURE 2.15. The left tibiofibula, cranial view

the heel, the **navicular**, which lies between proximal and distal rows, the **cuboid**, which lies on the distal row, and the three **cuneiforms** (lateral, middle, and medial), which also lie on the distal row.

Metatarsals (5): the bones that form the body of each pes. They are numbered 1–5 from medial to lateral.

Phalanges (28): the bones of the digits. Each digit but the first contains three phalanges (proximal, middle, and distal). The first digit contains two phalanges only (proximal and distal).

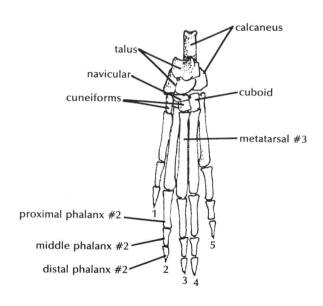

FIGURE 2.16. The left pes, dorsal view

Principles & Techniques of Dissection

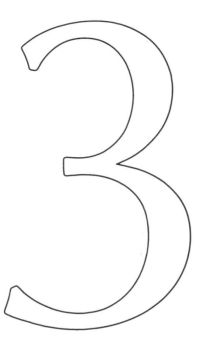

DISSECTION OF preserved specimens is an effective method of studying the structural organization of animals. It is a procedure that enables you to explore by sight and touch real structures, and thereby learn the three-dimensional interrelationships that exist in the body. In this chapter, the techniques of dissection are introduced, and the skinning procedure of the rat is presented as a starting point.

PREPARING FOR DISSECTION

The primary goal of dissection is to bring into view internal structures for the purpose of identification and study. Ideally, the complete dissection of an animal will permit you to study its muscles and internal structures in their proper location in the body. The successful achievement of this goal requires skill, which you can develop if you keep the following points in mind as you proceed through the dissection protocol:

1. *Follow the directions in this manual as they are presented in sequence*, much as you would in a cookbook. This should prevent you from getting lost or otherwise confused.

2. *Exercise caution and patience when cutting through tissues*. Dissection is not merely "cutting up" an animal, but rather a careful process of separating parts from each other. Therefore, it is important to cut only when instructed to and to thoroughly read about the area you are about to cut into or pull apart before doing so. Remember, care must be exercised at all times to avoid damaging structures before they have been identified.

3. *Use the right instrument for the job*. Correct tools are a necessity when precision is desired. Below is a list of common dissection instruments and a brief description of their proper uses:

Blunt probe: a rigid 5-inch steel instrument with a blunt, bent tip. This is useful for gentle manipulation of muscles and internal organs.

Scissors: usually 4–6 inches long with pointed tips. Scissors should be used to cut through skin, muscles, and other large structures.

Scalpel: a rustproof metal handle with replaceable blades. The scalpel should be used to make small incisions. Be very careful in using scalpels and especially in changing their blades, as they are exceedingly sharp.

Needle probe: a 3-inch needle attached to a wooden or plastic handle. This may be used as a pointer or to attach the specimen to the dissecting tray.

Forceps: about 5-inches long, the two pointed ends should contain ridges and should meet together evenly when closed. They are used to grasp small objects.

THE SKINNING PROCEDURE

The following skinning procedure is the first step in dissection. Follow the protocol outlined below when you are ready to begin.

1. Place your rat on its back in the dissecting tray. If it is a preserved, injected specimen, an opening can be found on the midventral surface of the neck. At this opening, insert the point of your scissors between the skin and the underlying muscle and make a midventral incision through the skin extending to the genital region. As you cut through the skin, pull upward away from the underlying muscle. This should prevent you from cutting into the muscle layer, which must be avoided at this point in the procedure.

2. From this cut, make additional incisions through the skin around the genitals and base of the tail, down the lateral surface of each limb, around the wrists and ankles, and around the neck (Fig. 3.1).

3. Beginning from the midventral incision, carefully separate the skin from the underlying muscles. Do this by pulling the skin gently but firmly away from the body with one hand and tearing through the superficial fascia with your fingers or a blunt probe in the other hand. In the chest region you will notice that the skin is attached to a thin, brownish sheet of muscle which is continuous with muscles extending into the armpit, or **axillary**, region. This muscle is the **cutaneous maximus**, which must be removed with the skin. To do this, cut its attachments at the axillary region near its union with the skin.

4. Continue separating the skin from the muscles in the caudal appendages, cranial appendages, and neck region. As you reach the neck you will see fine, parallel muscle fibers attached to the dermis of the skin and extending to the face. This is a muscle of facial expression called the **platysma** and should be removed with the skin. At this point, the skin should be completely removed from the body in a single piece, except for the head, manus and pes, and tail. Retain the skin and wrap it around your specimen when it is to be stored. This

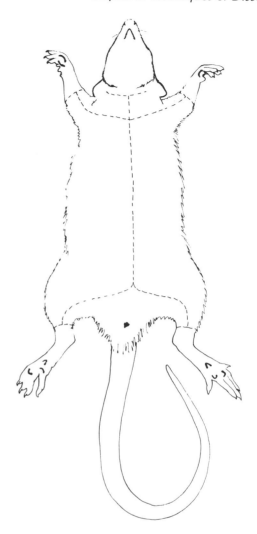

FIGURE 3.1. Rat profile, ventral view: a cutting guide for the skinning procedure

will help prevent the body from drying out and thus will prolong its use.

5. In the head region, make the following separate cuts: on the ventral side of the neck to the mouth; dorsally from the mouth, around the mystacial pad, to the medial corner of the eye on the right side; and caudally from the eye to the back of the neck on the right side. These incisions should expose the ventral neck region and the lateral right side of the head.

6. With a pair of forceps, remove the yellowish fat deposits that remain on the body after skin removal. Also clean the muscle layer by gently picking away with forceps the white superficial fascia until you can see the small cleavage lines that separate adjacent muscles. Some muscles are enveloped by the tough **deep fascia**; allow this to remain for the time being.

The Muscular System

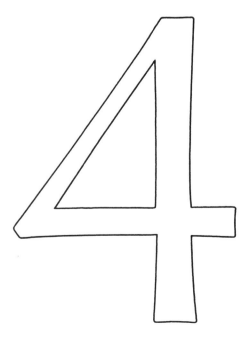

MUSCLE IS AN IMPORTANT type of tissue that is found throughout the body of all vertebrates. There are three basic types of muscle tissue: **smooth** or **visceral muscle**, which forms part of the walls of visceral organs and blood vessels; **cardiac muscle**, which forms the bulk of the heart; and **skeletal** or **somatic muscle**, which lies immediately deep to the skin and is attached to bones. Smooth and cardiac muscle are not considered to be components of the muscular system but are instead part of the organ and organ systems that they help to form.

The muscular system consists only of the type of muscle that is attached to bones, called skeletal muscle, and its associated connective tissue. Skeletal muscle is composed of muscle cells, or **fibers**, that are arranged in parallel bundles to form the muscle unit. Surrounding individual fibers, bundles of fibers called **fasciculi**, and entire muscle units is an extensive network of connective tissue. This connective tissue, or **deep fascia**, continues out of the body of a muscle to form a means of attachment to a bone. A band of deep fascia that extends from a muscle to a bone is generally called a **tendon**. In addition, a broad, thin sheet of deep fascia that attaches a muscle to a bone may also be known as an **aponeurosis**.

During contraction of a muscle, one end remains mostly stationary and the opposite end moves the bone to which it is attached as the muscle shortens. The more stable end is called the **origin** of the muscle, and the mobile end is known as the **insertion**. In the case of limb muscles, the origin is proximal and the insertion is distal.

The primary function of the muscular system is to provide skeletal movement. Movement is produced by the force a contracting muscle exerts on tendons and thus on bone. The effect of this activity is the movement of an articulating bone toward a stationary bone. The various types of movements (**actions**) resulting from skeletal muscle contractions are listed in Table 1.

In this chapter you will study the muscular system of the rat through dissection of your preserved specimen. Each major muscle is listed according to its regional location and identified by a description of its point of origin, point of insertion, and primary action.

The Muscular System

TABLE 1. Movements of Articulating Bones

Action	Description of Movement
Flexion	Decrease in angle between articulating bones
Extension	Increase in angle between articulating bones
Abduction	Movement away from body axis
Adduction	Movement toward body axis
Rotation	Movement around a central axis
Supination	Lateral rotation of the manus upward
Pronation	Medial rotation of the manus downward
Eversion	Rotation of the sole of the pes outward
Inversion	Rotation of the sole of the pes inward
Circumduction	Flexion, extension, abduction, adduction, and rotation

MUSCLE DISSECTION

Muscle dissection involves the careful separation of muscles from each other. This is possible because the individual fibers of each muscle run parallel to each other between attachments. A single muscle is visible and therefore separable when an adjacent muscle contains fibers that travel in a different direction. The following discussion describes the technique you should use when dissecting muscles. Read this through carefully so that you may apply this technique when you begin work on your specimen.

Examine the muscles on your rat closely. Locate the cleavage lines that separate adjacent muscles while comparing your specimen with the diagrams (Figs. 4.1–4.8). If the lines are not visible, they can be accentuated by pulling the muscles apart with your fingers until they appear. To separate individual muscles, use a blunt probe to break the surrounding connective tissue at the cleavage lines by inserting the probe between adjacent muscles and working the instrument forward. Continue this until the probe tip resurfaces at the opposite cleavage line of each muscle.

For each area dissected, you should follow each muscle to its points of origin and insertion as far as is practical. This is done by inserting the probe through both cleavage lines bordering a muscle and sliding the probe along the muscle's length while pulling the muscle slightly outward. The probe will break through the connective tissue between adjacent muscles without causing any damage if this step is done with care.

Once you have separated and identified all the superficial muscles of a region, it will become necessary to **transect** some of them in order to examine the underlying deep muscles. In transection a cut is made across the center of the muscle belly, leaving enough muscular attachment at its opposite ends where it originates and inserts to permit its later identification. These cut superficial muscles may then be **reflected** by pulling the ends backward toward their points of attachment, to reveal the deep muscles. You should dissect for deep muscles in this manner on *one side of the rat only*. This will leave the superficial muscles on the opposite side intact for later study. Proceed now with the following dissection protocol. Keep in mind that it is best to separate the muscles in the sequence in which they are presented below.

MUSCLES OF THE HEAD

The following muscles of the head move the jaws for chewing, a process called **mastication**. They are shown in Figures 4.1 and 4.2. Identify and separate these muscles:

Masseter: a large, round muscle in the cheek region. It is composed of two sections, superficial and deep. These sections are each divided into rostral and caudal portions. Separate the superficial masseter and transect it to reveal the deep masseter. Their origin is from the zygomatic arch and the portion of the maxilla near the infraorbital foramen. Their insertion is upon the lateral surface of the mandible. Their combined action: elevation of the mandible.

Temporalis: a large muscle that occupies the lateral surface of the cranium. It originates from the parietal bone and the squamosal part of the temporal bone and inserts at the coronoid process of the mandible. Action: elevation of the mandible.

Digastricus: a narrow muscle that can be viewed just caudal to the chin on the ventral side. It extends from its origin at the paramastoid process of the occipital bone to its insertion at the angular process of the mandible. A small tendon divides the muscle into caudal and rostral bellies. Action: depression of the mandible.

VENTRAL MUSCLES OF THE NECK

The following muscles on the ventral side of the neck mainly effect movement of the tongue, the larynx (often called the "voicebox"), and the head as a whole. With the rat on its back to expose the ventral side, separate and identify these muscles (Figs. 4.2, 4.3):

Sternomastoid: a long muscle that extends from the manubrium of the sternum to the mastoid process of the temporal bone. When viewed ventrally, the two sternomastoids form a V at the center of the neck. Action: Single contraction turns the head laterally; double contraction depresses the snout.

Cleidomastoid: a long muscle that is immediately deep to the sternomastoid. To observe it, transect and reflect the sternomastoid on one side. The origin of the cleidomastoid is at the clavicle, and its insertion is at the mastoid process. Action: same as the sternomastoid.

Sternohyoid: a long, narrow muscle that lies along the midventral line of the neck. Trace it from its origin at the

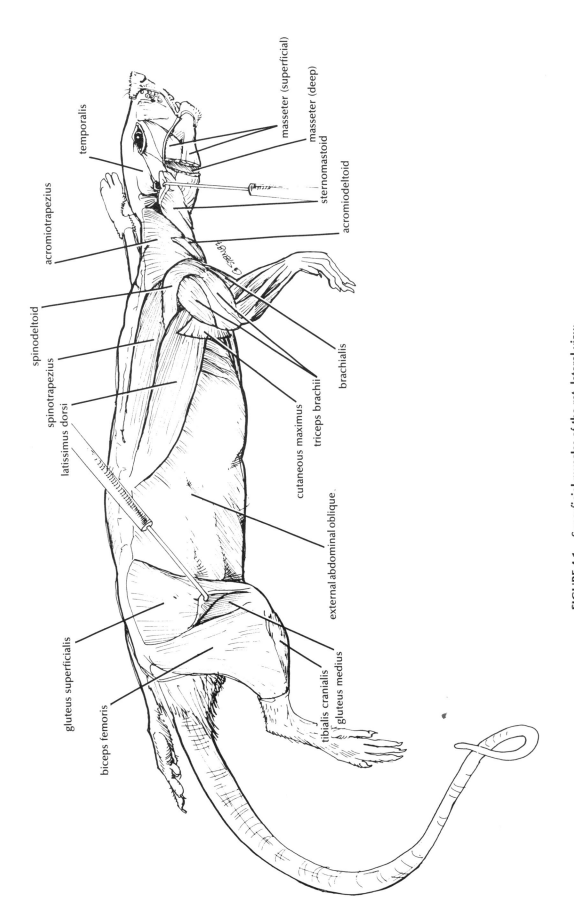

FIGURE 4.1. Superficial muscles of the rat, lateral view

20

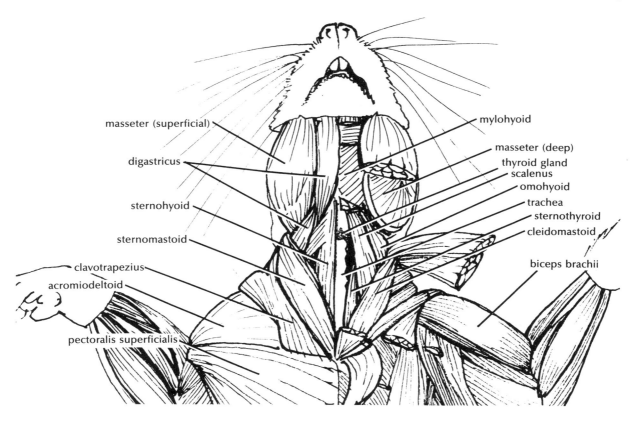

FIGURE 4.2. Muscles of the ventral neck region. Superficial muscles are shown on the right side, deeper muscles on the left.

manubrium of the sternum to its insertion at the hyoid and transect it on one side. Action: draws the hyoid caudally.

Sternothyroid: a narrow muscle lying immediately deep to the sternohyoid; it may be viewed by reflecting its transected ends. It arises from the manubrium and first two costal cartilages and inserts on the thyroid cartilage of the larynx. Note the tiny **thyroid gland** that lies beneath it. Action: draws the larynx caudally.

Omohyoid: a narrow muscle that extends from the rostral border of the scapula to the hyoid. It passes deep to the cleidomastoid and may be viewed by pulling this muscle to the side and reflecting the cut sternohyoid. Action: draws the hyoid caudally and laterally.

Mylohyoid: a broad muscle deep to the digastricus. To view it, make a shallow transverse cut across the digastricus on one side and reflect its ends. The mylohyoid originates from the mandible and inserts at the hyoid. Action: raises the floor of the mouth and draws the hyoid forward.

VENTRAL MUSCLES OF THE PECTORAL GIRDLE

With the rat remaining on its back, examine the following muscles that move the pectoral girdle (Fig. 4.3).

Pectorales: the large, triangular muscle complex covering the chest. The common action is adduction of the humerus. It consists of the following muscles:

- **Pectoralis superficialis**: a superficial muscle that extends from the manubrium and first two sternebrae to the deltoid tuberosity of the humerus.
- **Pectoralis profundus**: a deeper, more extensive muscle that extends from the second to fifth sternebrae and xiphoid process to the coracoid process of the scapula and to the humerus with the pectoralis superficialis.

Serratus ventralis: Transect the pectorales and reflect them and notice the serrated muscle that lies deep. This is the serratus ventralis, which is also partially deep to the latissimus dorsi (below). Its serrated appearance is due to the muscle's origin from the first seven ribs. These fibers merge into a compact muscle that inserts on the vertebral border of the scapula. Action: depresses the scapula.

DORSAL MUSCLES OF THE NECK & PECTORAL GIRDLE

Turn the rat over on its belly to expose its dorsal side. Beginning at the upper back and working caudally, locate and separate the following muscles (Figs. 4.1, 4.4):

The Muscular System

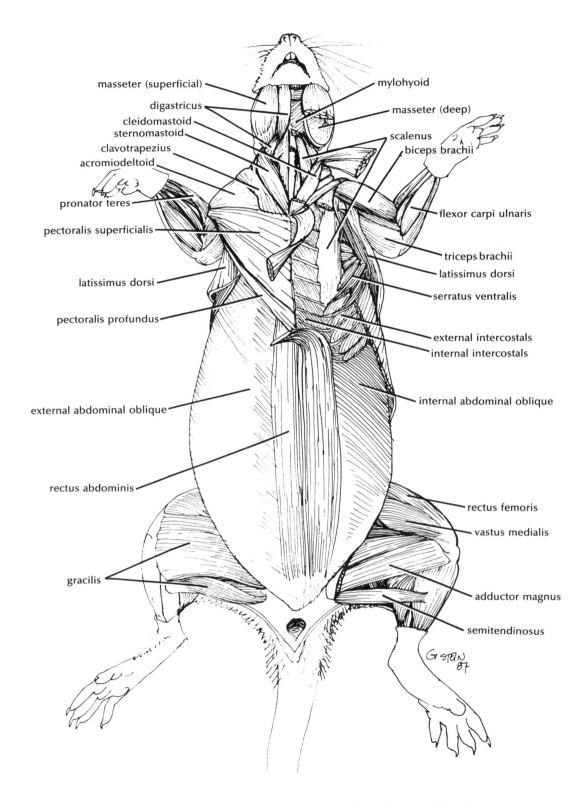

FIGURE 4.3. The ventral muscles. The superficial muscles are reflected on the left side to reveal deeper muscles.

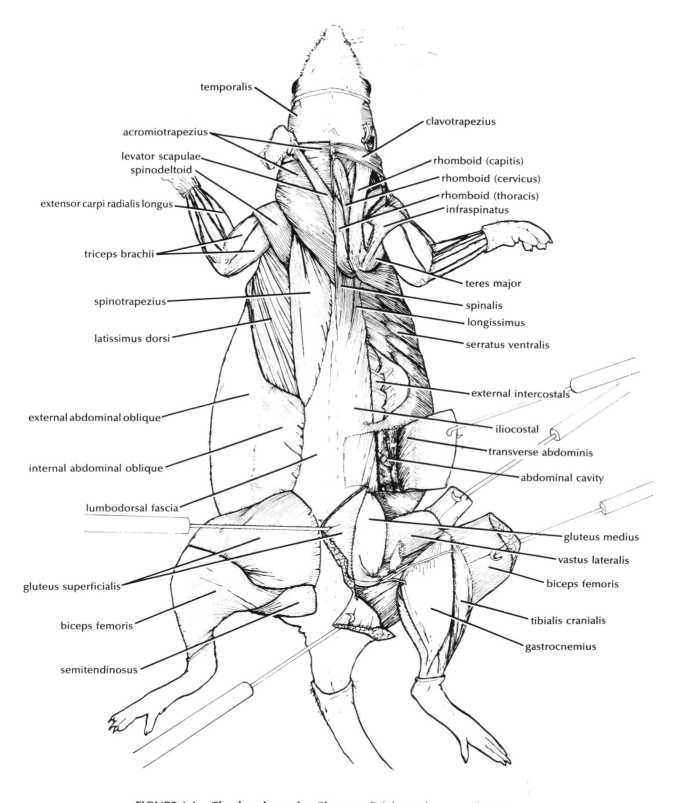

FIGURE 4.4. The dorsal muscles. The superficial muscles are reflected on the left side to reveal deeper muscles.

The Muscular System

Acromiotrapezius: a thin, fragile muscle that extends from the spinous processes of the cervical and first four thoracic vertebrae to the acromion process and spine of the scapula. After you have separated this muscle, lift the limb to loosen it from deeper muscles and transect it on one side. Action: draws the scapula medially.

Levator scapulae: a narrow muscle that extends superficially across the dorsal half of the acromiotrapezius. It originates from the last four cervical vertebrae and inserts at the vertebral border of the scapula. Action: depresses the scapula.

Clavotrapezius: a narrow muscle partially deep to the acromiotrapezius. It arises from the basioccipital and wraps around the lateral side of the neck to insert on the ventral aspect of the clavicle. Action: draws the scapula and clavicle forward.

Spinotrapezius: located caudal to the acromiotrapezius, this thin muscle originates from the spinous processes of the fourth thoracic to the third lumbar vertebrae and inserts at the spine of the scapula. Transect and reflect this muscle on one side. Action: draws the scapula caudally.

Latissimus dorsi: a flat, broad muscle of the middle back that originates from the spinous processes of the eighth and twelfth thoracic vertebrae and from a thick sheet of fascia of the lower back called the **lumbodorsal fascia**. From here it extends to its insertion at the proximal end of the humerus. Transect and reflect this muscle on one side, and clean away the connective tissue and fat from beneath it. Action: extends and adducts the brachium and draws the shoulder backward.

Rhomboids: Lying deep to the acromiotrapezius, the rhomboids are a group of muscles that extend from the vertebral column to the scapula. To view them, draw a limb ventrally under the body and reflect the acromiotrapezius. There are three distinct rhomboid muscles. From cranial to caudal, they are the following:

    Capitis: a narrow muscle that extends from the basioccipital bone to the dorsal border of the scapula. Action: draws the scapula cranially.

    Cervicis: extending from the spinous processes of the first three cervical vertebrae, it inserts at the vertebral border of the scapula. Action: draws the scapula medially.

    Thoracis: the largest rhomboid, its origin is from the spinous processes of No. 4–No. 7 cervical vertebrae, and its insertion is at the dorsal border of the scapula. Action: draws the scapula medially.

MUSCLES OF THE SHOULDER & CRANIAL APPENDAGES

In this section you will dissect the shoulder, brachium, and antebrachium. Follow the protocol below and refer often to Figures 4.1 and 4.3–4.6.

Muscles of the Shoulder. Shoulder muscles extend between the cranial appendage and the pectoral girdle. They stabilize the shoulder joints and move the brachium. Separate and identify the following muscles:

Acromiodeltoid: a superficial muscle that passes from the clavicle and acromion process of the scapula to the deltoid tuberosity of the humerus. It lies on the cranial edge of the shoulder and can be observed most clearly from the ventral side (Fig. 4.3). Action: extends and rotates the humerus medially.

Spinodeltoid: located caudal to the acromiodeltoid, its origin is from the spine of the scapula, and its insertion is also at the deltoid tuberosity. Action: flexes and rotates the humerus laterally.

Supraspinatus: Transect and reflect the acromiodeltoid and spinodeltoid and reflect once again the acromiotrapezius and latissimus dorsi. For the remaining shoulder muscles, keep these superficial muscles in this position (Fig. 4.5). The supraspinatus is located deep and occupies the supraspinous fossa of the scapula, where it originates. From here it passes to its insertion at the greater tubercle of the humerus. Action: extends the humerus.

Infraspinatus: a deep muscle caudal to the supraspinatus, it occupies the infraspinous fossa of the scapula (Fig. 4.5). From this point of origin it extends to the greater tuberosity of the humerus, where it inserts. Action: rotates the humerus caudally and flexes the humerus.

Teres major: arises from the caudal border of the scapula and passes forward to insert at the proximal end of the humerus. You may observe it as it lies ventral to the infraspinatus. Action: rotates the humerus caudolaterally.

Teres minor: a small muscle tucked in between the infraspinatus and the teres major. Its origin is from the caudal border of the scapula, and its insertion is at the greater tubercle of the humerus. Action: rotates the humerus caudolaterally.

Subscapularis: a muscle that occupies the subscapular fossa of the scapula on its medial side. From its origin here it extends to the lesser tubercle of the humerus, where it inserts. To observe it, transect and reflect the rhomboids and tuck the arm beneath the body ventrally (Fig. 4.6). Action: adducts the humerus.

Muscles of the Brachium. The muscles of the brachium provide movement primarily at the elbow joint. Beginning on the lateral side, separate and identify the following (Figs. 4.5, 4.6):

Triceps brachii: the large muscle mass on the lateral side of the brachium. The triceps has three heads, or places of origin, which merge to insert on the olecranon process of the ulna. The **long head** originates from the proximal end of the humerus and the medial border of the

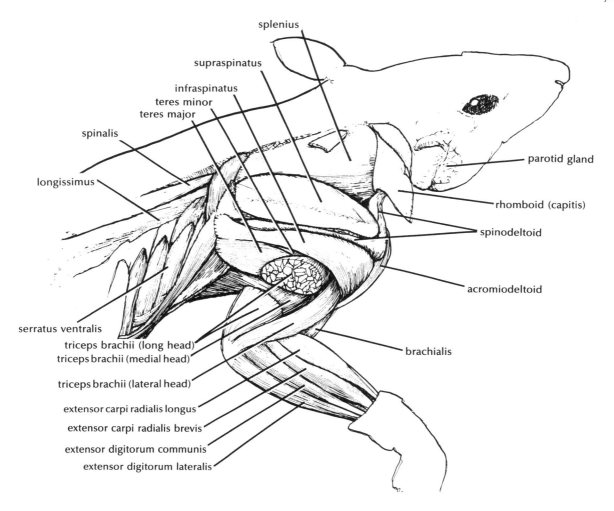

FIGURE 4.5. Deep muscles of the shoulder and superficial muscles of the brachium and antebrachium, dorsolateral view. The trapezius muscle group and latissimus dorsi have been removed, and the spinodeltoid is partially removed in this view.

scapula. The **lateral head**, which is cranial to the long head, arises from the greater tubercle of the humerus. The small, deep **medial head** arises from the dorsal surface of the humerus. In order to observe the medial head you must pull the lateral head away from the plane of the forelimb. Action: all three heads extend the antebrachium.

Brachialis: a muscle located cranial to the lateral head of the triceps and visible from the lateral side. It originates from the greater tubercle of the humerus and inserts at the medial surface of the ulna. Action: flexes the antebrachium.

Biceps brachii: a large muscle on the medial side of the brachium (Fig. 4.6). It arises from two heads of origin: the **long head** from the glenoid fossa, and the **short head** from the coracoid process. The two heads converge to form a prominent belly over the humerus before extending beyond the elbow to insert at the radial tuberosity. Action: Both heads of the biceps flex the antebrachium.

Epitrochlearis: an extremely thin muscle on the medial side adjacent to the biceps. To separate it, tease it free of the triceps long head that lies deep to it. It arises from the latissimus dorsi by a thin, fragile tendon and inserts at the medial epicondyle of the humerus. Action: extends the antebrachium.

Muscles of the Antebrachium. Antebrachium muscles move the manus and the digits. To observe them, you must remove the tough sheath of fascia (**antebrachial fascia**) that surrounds them by making an incision through it and pulling it away. The muscles of the antebrachium are

The Muscular System

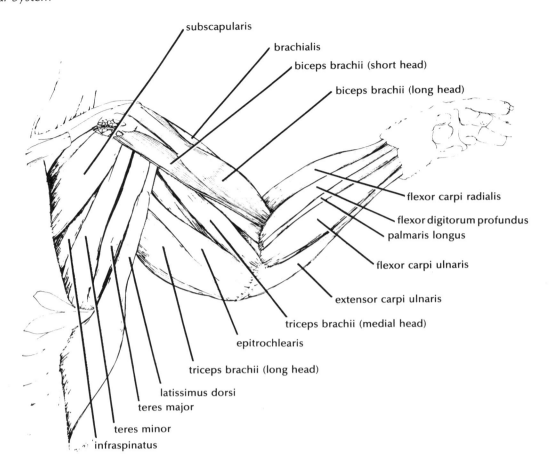

FIGURE 4.6. Deep muscles of the shoulder and superficial muscles of the brachium and antebrachium, medial view. The pectorales muscle group is removed in this view.

divided into an extensor group, located on the craniolateral surface, and a flexor group, located on the caudomedial surface.

Extensor Group: The muscles in this group are basically extensors of the wrist joint. Their common origin is from the lateral epicondyle of the humerus. Examine the lateral side of the antebrachium and separate the following muscles from the radial to the ulnar side (Fig. 4.5):

Extensor carpi radialis longus: a narrow, cranial muscle that extends from the lateral epicondyle to the second metacarpal.

Extensor carpi radialis brevis: Located lateral to the longus, it arises from the lateral epicondyle and inserts at the third metacarpal.

Extensor digitorum communis: originates from the lateral epicondyle and extends to the digits, where it divides into four tendons that insert at the distal phalanx of digits 2–5.

Extensor digitorum lateralis: arises from the lateral epicondyle and extends to digit 4, where its tendon inserts at the distal phalanx.

Extensor carpi ulnaris: Located on the ulnar border of the antebrachium, it arises from the lateral epicondyle and inserts at the base of the fifth metacarpal (shown in Fig. 4.6).

Flexor Group: The muscles in this group are primarily flexors and pronators of the wrist joint and are located on the medial side of the antebrachium. They originate from the medial epicondyle of the humerus by a common flexor tendon. Identify and separate the following muscles from the radial to the ulnar side (Fig. 4.6).

Pronator teres: passes diagonally from its origin at the medial epicondyle to its insertion on the medial surface of the radius (shown in Fig. 4.3).

Flexor carpi radialis: extends from the medial epicondyle to the third metacarpal.

Flexor digitorum profundus: arises from five heads of origin at the medial epicondyle and the radius and ulna to extend into the palm as a powerful flexor plate. From the palm it breaks up into four tendons that insert at the distal phalanx of digits 2–5.

Palmaris longus: originates from the medial epicondyle and inserts superficially into the fascia of the palm.

Flexor carpi ulnaris: Located on the caudal edge of the antebrachium, it arises from two heads of origin: the medial epicondyle by means of the common flexor tendon, and the medial surface of the olecranon process. It inserts into the fascia of the palm.

MUSCLES OF THE TRUNK

The following muscles of the trunk are divided into three regions: thoracic, abdominal, and back. The thoracic and abdominal muscles of the ventral side will be studied first.

Thoracic Muscles. The muscles of the thoracic region are primarily involved in respiration. They include the following (Fig. 4.3):

Scalenus: With the rat lying on its back to expose the ventral side, reflect the pectorales and latissimus dorsi muscles to locate the underlying scalenus. This muscle is actually a group of three muscles: a cranial, a middle, and a caudal. Each of these arises from the last four cervical vertebrae and inserts on the upper ribs. Action: their combined efforts elevates the ribs and expands the thoracic cavity. If the thorax is kept rigid, these muscles also move the head.

External intercostals: thirteen pairs of muscles that form the outermost layer of the thoracic wall between the ribs. The fibers of each muscle extend from the caudal border of one rib downward to the cranial border of the next rib. Action: they move the ribs closer together and thus play a major role in inspiration and expiration.

Internal intercostals: Make a shallow incision through an external intercostal to reveal the underlying muscle. Note that the space between the two muscle layers is called the **intercostal space**. The fibers of the thirteen pairs of internal intercostals run at right angles to the fibers of the external intercostals. Each internal intercostal extends from the caudal border of one rib to the cranial border of the rib below. Action: They move the ribs closer together and thus play a role in inspiration and expiration.

Diaphragm: a circular sheet of muscle that divides the thoracic and abdominal cavities internally. This muscle will be observed when the internal structures of the respiratory and digestive systems are dissected. Action: it pushes downward, which expands the thoracic cavity, causing air to be drawn into the lungs during inspiration.

Abdominal Muscles. Continue in a caudal direction the examination of the ventral trunk muscles. For each of the following muscles, do not attempt to separate to their origin and insertion. Rather, identify them based on their superficial location (Figs. 4.3, 4.4).

Rectus abdominis: a long, flat muscle that extends from the symphysis pubis to the first rib, the clavicle, and the sternum along the ventral midline. Its origin and insertion are dependent on which end of the muscle is fixed during a given action and are therefore reversible. Action: flexes the vertebral column.

External abdominal oblique: an extensive superficial sheet of muscle in the ventrolateral wall of the abdomen. Originating from the fourth through twelfth ribs, its fibers travel in a dorsocaudal direction and insert by means of an aponeurosis at the iliac crest, the pubis, and the linea alba. Action: constricts the abdomen and thorax.

Internal abdominal oblique: Make a shallow incision along the lateral surface of the abdomen from the base of the ribcage to the iliac crest. Take special care to cut through only the most superficial layer, the external oblique. Reflect the cut ends and observe the sheet of muscle that is deep, the internal oblique. This muscle, whose fibers run perpendicular to those of the external oblique, arises from the lumbodorsal fascia, the iliac crest, and a band of fascia in the lower abdomen called the inguinal ligament. It inserts at the linea alba. Action: constricts the abdomen.

Transverse abdominis: Carefully separate the internal oblique from the underlying muscle, which is closely associated with it. Transect a lateral segment of the internal oblique and reflect it. The deep muscle is the transverse abdominis and is the innermost layer of the ventrolateral wall of the abdomen. It arises from the lower seven ribs, the lumbodorsal fascia, and the inguinal ligament and inserts by means of a thin aponeurosis on the linea alba. Note the transverse direction of its fibers. Action: constricts the abdomen.

Back Muscles. In order to observe the muscles of the back, reflect the latissimus dorsi, acromiotrapezius, and spinotrapezius to their insertions. Cut through the lumbodorsal fascia along the midline and peel it laterally. Again, do not attempt to separate the following muscles to their origins and insertions, but instead identify them by their locations (Fig. 4.4):

Erector spinae: A group of muscles that make up most of the dorsal musculature. Their contraction extends the vertebral column. They consist of the following:

  Iliocostal: The lateral component of the erector spinae group, it extends between the ilium, dorsolumbar fascia, and vertebral column to the lumbar vertebrae and ribs cranial to the origin. It consists of lumbar, thoracic, and cervical parts.

  Longissimus: The intermediate erector spinae, it arises from the iliac crest and lumbar and thoracic vertebrae and inserts at the thoracic and cervical vertebrae and ribs. The longissimus consists of lumbar, thoracic, cervical, and capitis parts.

Spinalis: The medial component of the erector spinae, it extends between the thoracic and cranial lumbar vertebrae and the cranial thoracic and cervical vertebrae.

MUSCLES OF THE PELVIC GIRDLE & HIP

The pelvic muscles move the hind limb at the hip joint. They include the gluteal complex and the deeper pelvic muscles on the dorsal side and the iliopsoas complex on the ventral side. Beginning on the dorsal side of the rat, separate and identify the following muscles (Fig. 4.7):

Gluteus superficialis: the large, cranial muscle of the pelvic region. This muscle incorporates the **tensor fascia latae**, **sartorius**, and **gluteus maximus** muscles, which are distinctly separate from each other in larger mammals. It arises from the dorsal border of the ilium and inserts at the greater trochanter of the femur and a wide band of fascia that covers the thigh called the **fascia latae**. Transect this large muscle between the hip and the knee. Action: abducts the thigh and tenses the fascia latae.

Gluteus medius: Reflect the gluteus superficialis to its origin and insertion and observe the deeper gluteus medius. It arises from the ilium and inserts at the greater trochanter of the femur. Action: abducts and extends the thigh.

Gluteus profundus: Also called the **gluteus minimus**, it is located deep to the gluteus medius. To observe it, pull back the medius without transecting it (not shown). The profundus arises from the dorsal border of the ilium and inserts at the greater trochanter. Action: rotates the thigh in a medial direction.

Piriformis: a thin muscle that lies partially deep to the caudal portion of the gluteus medius and dorsal to the gluteus profundus and is closely invested with them. Therefore, you should not attempt to separate it, but rather identify it by its location. It originates from the sacrum and inserts at the greater trochanter. Action: abducts the thigh.

Quadratus femoris: a small triangular muscle that lies partially deep to the semimembranosus and semitendinosus muscles of the thigh (below). It extends between the ischium and the proximal end of the femur. Action: extends and rotates the thigh.

Iliopsoas: The iliopsoas is actually a complex of two muscles that share the same insertion at the lesser trochanter of the femur. Their common action is rotation of the femur in a lateral direction and flexion of the hip joint. The fibers of both muscles travel caudally from the dorsal abdominal wall toward their origins. Because of their deep location against the dorsal wall of the abdomen, they may be observed later when the abdominal cavity is dissected. Their description follows:

Iliacus: a narrow muscle located dorsal to the gluteus medius. It originates from the fifth and sixth lumbar vertebrae.

Psoas major: located dorsal to the iliacus. From its origin at the lumbar vertebrae, it travels parallel with the iliacus to the lesser trochanter.

Psoas minor: a narrow muscle that is medial to the psoas major. It extends between the second through sixth lumbar vertebrae and the ilium (not shown). This muscle may also be observed once the abdominal cavity is dissected. Action: flexes the vertebral column.

MUSCLES OF THE CAUDAL APPENDAGES

The caudal appendages are divided into the cranial portion of the hind leg, called the thigh, and the caudal portion, called the shank. As with the muscles of the cranial appendages, begin separating the caudal appendage muscles at the proximal end and work distally.

Muscles of the Thigh. Thigh muscles move the hind limb at the hip joint and the knee joint. Begin dissecting the thigh region by clearing away as much fascia as necessary to expose the muscle fibers. Starting on the dorsal side of the rat with the hind limbs spread laterally, separate and identify the following muscles (Figs. 4.7, 4.8):

Biceps femoris: The prominent muscle of the thigh on the lateral side (Fig. 4.7), it originates by three heads from the sacral and caudal vertebrae. The fibers from each head fan out to insert by means of a broad aponeurosis on the femur and proximal end of the tibia. Transect this muscle. Action: abducts the thigh and flexes the shank.

Semitendinosus: a narrow muscle located caudal to the biceps femoris. Its origin is from the sacral and caudal vertebrae and its insertion is at the tibial tuberosity. Action: flexes the shank.

Semimembranosus: Reflect the biceps femoris to its origin and insertion and observe the semimembranosus beneath. The semimembranosus arises from the ischium, sacrum, and first caudal vertebra and inserts at the tibia and femur. It consists of two portions, a caudal and a cranial (sometimes called the caudofemoralis). Action: extends and abducts the thigh and flexes the shank.

Quadriceps femoris: Most of the cranial surface of the thigh is occupied by this large muscle complex, which is actually a group of four muscles that share a common point of insertion: a large tendon that extends to the patella, attaches around it, and continues as the **patellar ligament** to insert on the tibial tuberosity. Their common action is extension of the shank. The following muscles are part of the quadriceps femoris complex:

Rectus femoris: Reflect the previously transected gluteus superficialis to observe the underlying rec-

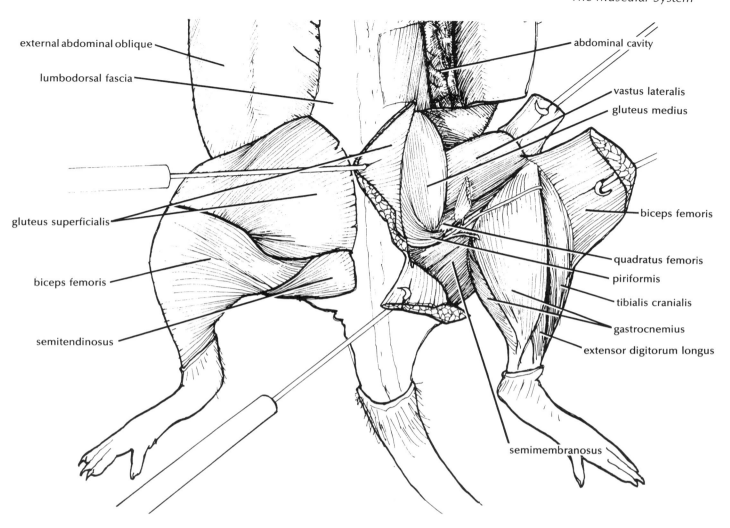

FIGURE 4.7. Dorsal and lateral muscles of the pelvic region and lower appendages. Superficial muscles are intact on the left side and reflected on the right to reveal deeper muscles.

tus femoris on the cranial edge of the thigh (Fig. 4.8). The rectus femoris has two portions due to its two heads of origin: a cranial portion that arises from the ilium and a caudal portion that originates from a tubercle cranial to the acetabulum. Both heads insert at the patellar ligament.

Vastus lateralis: Reflect the biceps femoris on the lateral side of the thigh to expose the deeper vastus lateralis (Fig. 4.7). The vastus lateralis extends from the greater trochanter of the femur to the patellar ligament.

Vastus medialis: Visible from the medial side of the thigh, it lies medial to the rectus femoris (Fig. 4.8).

The vastus medialis arises from the neck and proximal end of the femur.

Vastus intermedius: Pull the vastus medialis away from the femur to view the small vastus intermedius, which lies deep (not shown). It originates from the shaft of the femur.

Gracilis: Visible from the medial side of the thigh, the gracilis is a flat, superficial muscle complex that consists of the following two muscles. Once they have been identified, transect and reflect both muscles:

Gracilis cranialis: It extends between the symphysis pubis and the shaft of the tibia. Action: adducts the thigh.

The Muscular System

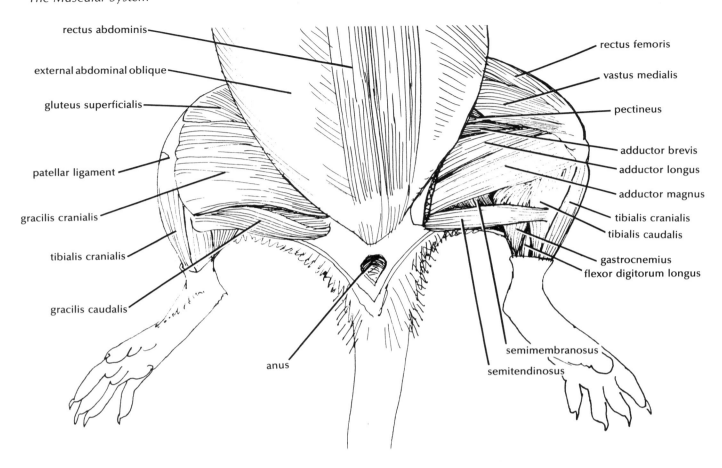

FIGURE 4.8. Ventral and medial muscles of the pelvic region and lower appendages. The gracilis muscle group and gluteus superficialis are removed on the left side to reveal deeper muscles.

Gracilis caudalis: the caudal muscle that extends between the ischium and the tibial tuberosity. Action: adducts the thigh.

Pectineus: a small, triangular muscle on the medial side of the thigh. It arises from the pubic arch and inserts at the shaft of the femur. Action: adducts the thigh.

Adductor brevis: a thin muscle caudal to the pectineus. It arises from the pubis and inserts at the proximal end of the femur. Action: adducts the thigh.

Adductor longus: Somewhat larger than the adductor brevis and located caudal to it, the adductor longus originates from the pubis and inserts at the femur. Action: adducts the thigh.

Adductor magnus: The largest and most caudal of the adductors, it arises from the pubis and symphysis pubis. It inserts at the patellar ligament. Action: adducts the thigh.

Muscles of the Shank. On one side of the rat, prepare the shank for dissection by removing any remaining skin and superficial fascia. Then cut and remove the broad sheet of fascia that encases the shank at the level of the knee. Beginning on the lateral side, separate the following shank muscles (Figs. 4.7, 4.8):

Tibialis cranialis: a narrow muscle that covers the cranial surface of the tibiofibula. It arises from the lateral condyle and tuberosity of the tibia and inserts at the first cuneiform and first metatarsal. Action: flexes the pes.

Extensor digitorum longus: located caudal to the tibialis cranialis and visible from the lateral side. Its point of origin is the lateral epicondyle of the femur, and its tendon of insertion divides into four tendons that pass down digits 2–5 to insert at the distal phalanges. Action: extends the pes and the digits.

Peroneus longus: a long, narrow muscle on the lateral side that is caudal to the extensor digitorum longus (not shown). It arises from the head of the fibula and lateral condyle of the tibia and inserts at the first cuneiform and first metatarsal. Action: flexes the pes.

Gastrocnemius: The prominent muscle on the caudal side, it is the largest muscle of the shank. It has two heads of origin: a lateral head, which arises from the lateral epicondyle of the femur; and a medial head, which arises from the medial epicondyle. Both heads converge to insert at the calcaneus by means of the large **calcaneus tendon**. Transect this superficial muscle on one side. Action: extends the pes.

Soleus: Reflect the gastrocnemius to view the soleus, which lies directly deep to the lateral head (not shown). The soleus originates from the head of the fibula and inserts in common with the gastrocnemius at the calcaneus by means of the calcaneus tendon. Action: extends the pes.

Flexor digitorum longus: With the gastrocnemius reflected, observe this deep flexor muscle from the medial side. Its origin is from the tibiofibula, and its insertion is at the distal phalanx of digits 2–5. Action: flexes the digits.

Tibialis caudalis: A small muscle that occupies the caudomedial surface of the tibiofibula, it can be viewed from the medial side cranial to the flexor digitorum longus. It arises from the proximal end of the tibiofibula and inserts at the sole of the pes on the first cuneiform. Action: extends the pes.

The Nervous System

IN THE RAT as in all vertebrates, the nervous system is divided into a **central nervous system (CNS)** and a **peripheral nervous system (PNS)**. The CNS consists of the **brain** and **spinal cord** and serves as the information input and output control center for the body. The PNS is composed of neurons arranged in bundles, called **nerves**, which extend between the CNS and the peripheral regions of the body. PNS nerves contain sensory and motor neurons that transport impulses to and from the CNS. A third division, the **autonomic nervous system**, contains neurons that often parallel PNS neurons within nerves. Autonomic neurons convey involuntary information to and from the CNS.

The primary function of the nervous system as a whole is to maintain the body in a relatively stable condition, or **homeostasis**, despite changes that may occur in the internal and external environments. This is accomplished by the generation of nerve impulses that travel along nerve conduction pathways. The impulses usually originate from specialized sensory structures called **receptors** and terminate in muscles and glands called **effectors**. This means of body control is rapid and specific to the region that receives the impulse.

The **endocrine system** also plays an important role in homeostasis. In contrast to the rapid and specific effects of nervous control, endocrine control is by way of the secretions of **hormones** that are carried more sluggishly by the bloodstream. Hormonal stimulation thus results in slower effects that may affect different parts of the body. Hormones are secreted by glands that are widely scattered, making dissection of the endocrine system as a whole very difficult. Therefore, the endocrine system will not be studied in one place. Rather, endocrine glands are presented as they are observed in the course of dissection. Their activities are summarized in Table 2.

In this chapter, you will be introduced to the nervous system of the rat through dissection. The brain is the focus of this study, due to its relative importance to the body and the extensive amount of information available on its structure.

TABLE 2. Endocrine Gland Hormones and Their Effects

Gland	Hormone Secreted	Primary Effect
Hypophysis:		
Adenohypophysis	Thyrotrophic hormone	Stimulates the thyroid gland
	Growth hormone	Controls growth and development
	Adenocorticotrophic hormone (ACTH)	Stimulates the adrenal gland (cortex)
	Luteinizing hormone (LH)	Stimulates the secretions of sex hormones by the gonads; also the conversion of ovarian follicles into corpora lutea
	Follicle-stimulating hormone (FSH)	Stimulates the maturation of the testes and the ovarian follicles
	Prolactin	Stimulates milk secretion by the mammary glands
Neurohypophysis	Oxytocin	Stimulates contractions of the uterus and the release of milk by the mammary glands
	Vasopressin (ADH)	Stimulates water reabsorption in the kidneys
Pineal	Melatonin	May inhibit gonad activities
Thyroid	Thyroxin	Controls catabolic metabolism
	Thyrocalcitonin	Helps regulate blood calcium levels
Parathyroid	Parathormone	Regulates calcium-phosphate metabolism (antagonist to thyrocalcitonin)
Thymus	Thymosin	Stimulates the functional maturation of "T" lymphocytes
Pancreas (Islets of Langerhans)	Insulin	Stimulates glucose uptake and carbohydrate catabolism in all body cells
	Glucagon	Stimulates the conversion of glycogen into glucose
Adrenals:		
Cortex	Glucocorticoids	Stimulates glycogen formation and storage, increases body resistance to stress, reduces inflammation of tissues
	Mineralocorticoids	Regulates sodium-potassium metabolism
Medulla	Epinephrine	Prolongs the conditions responsible for the "flight or fight" response: increase in heart rate, blood pressure, etc.
	Norepinephrine	Stimulates the constriction of arteries
Gonads:		
Testes	Testosterone	Stimulates the development of male sex characteristics and spermatozoa production
Ovaries	Estrogen	Stimulates the development of female sex characteristics, the ovarian cycle, and the menstruation cycle
	Progesterone	Maintains the growth of the endometrium of the uterus

THE CENTRAL NERVOUS SYSTEM

The central nervous system integrates, regulates, and controls the body's activities and relays impulses between the brain and peripheral nerves. It is a hollow structure; through it runs a **central canal** that expands in certain regions of the brain to form **ventricles**. Within these cavities and also surrounding the spinal cord and brain is **cerebrospinal fluid**, a slowly circulating colorless fluid that provides nutritive support for active cells and a protective liquid cushion for the central nervous system.

Before study of the brain is possible, you must remove it from the cranial cavity. To do this, follow the protocol outlined below.

1. Clean the dorsal surface of the skull and neck completely of skin, muscle, and connective tissue.
2. Using bone shears or heavy gauge scissors, cut the third, second, and first cervical vertebrae and remove the bone fragments with forceps. Sever the spinal cord as far caudal as possible.

3. Insert the pointed half of your bone shears slightly through the foramen magnum at the base of the skull—don't go too deep! Close the scissors and apply pressure. This will crack the occipital bone, which is the thickest bone of the skull.

4. From this point, cut through the cranium along the dorsal midline to the orbits with scissors. Pick away remaining pieces of bone from the dorsal cranium with forceps until the brain is completely exposed. Note the tough **dura mater** that covers the brain. It is the outermost of a series of membranes that surround the brain and spinal cord called **meninges**. Carefully remove the dura mater to expose the deeper meningeal layers. As you do so, note the extension of the dura mater between the two cerebral hemispheres, called the **falx cerebri**, and the extension between the cerebrum and the cerebellum, called the **tentorium cerebelli**. These membranes help to anchor the brain within the cranial cavity. Once you have removed the dura mater, identify the middle meningeal membrane, the **arachnoid**. Its sparse arrangement of fibers passes over the shallow grooves (**sulci**) of the brain. Also note the deep **pia mater**. This thin meningeal membrane closely invests the outer surface of the brain and is separated from the arachnoid by a cavity called the **subarachnoid space**. In life, **cerebrospinal fluid** circulates within this cavity.

5. To remove the brain from the cranial cavity, lift the severed spinal cord from the floor of the neural canal as a first step. Working caudally, lift the brain from the floor of the cranial cavity, severing the cranial nerves as you proceed. Make your cuts midway between the brain and their points of departure from the skull to facilitate later identification of these nerves (Table 3). As you raise the brain from the floor of the cranium, try to avoid tearing the hypophysis on the ventral surface and the lateral extensions of the cerebellum by clearing all connective tissues near their bony attachments.

6. To observe certain structures of the brain, it will be necessary to section it along the midsagittal plane. Do this only after you have examined the structures of the brain that are visible from the dorsal and ventral aspects. Perform this cut with a large scalpel or single-edged razor in a single motion.

BRAIN

The brain of the rat is similar to that of other vertebrates in that it consists of three primary regions: the forebrain, or **prosencephalon**; the midbrain, or **mesencephalon**; and the hindbrain, or **rhombencephalon**. Identify these regions and their components using dorsal, ventral, lateral, and midsagittal views of the brain (Figs. 5.1–5.4).

Prosencephalon: The largest region of the brain, it occupies the anterior and middle areas. It consists of the **telencephalon** and the **diencephalon**.

Telencephalon: It includes the largest single structure of the mammalian brain, the **cerebrum**.

Cerebrum: Although smooth-textured in the rat, in larger mammals its convolutions contain many folds (**gyri**) and grooves (**sulci**). It is divided into **right** and **left hemispheres** by a deep, middorsal furrow called the **longitudinal fissure**. A second prominent furrow, the **transverse fissure**, separates the cerebrum from the cerebellum on the posterior side of the brain. At its anterior end, the cerebrum contains a pair of swellings called the **olfactory bulbs**, which receive sensory neurons from the nasal cavity. From the bulb, the **olfactory tract** passes in a posterior direction to merge with the cerebrum on its ventral surface (Fig. 5.4). The outer region of the cerebrum is the **cortex** and is composed of gray matter. The inner region is composed of white matter and includes the **corpus callosum**, which bridges the two hemispheres together. The corpus callosum can be viewed dorsally at the floor of the longitudinal fissure. Observe the sagittal section (Fig. 5.3) and identify the **genu**, the expanded rostral end, the **trunk** in the center, and the **splenium** at the posterior end. Also identify the **fornix**, which is a ventral continuation of the corpus callosum. Embedded within the inner region of the cerebrum are clusters of gray matter, called **basal ganglia**. Immediately ventral to the corpus callosum lies a pair of cavities, the **lateral ventricles**, which are also visible in the sagittal section. In life they are filled with cerebrospinal fluid, which is produced from a capillary network called **choroid plexi**. The choroid plexi are associated with a thin membrane dividing the two lateral ventricles, called the **septum pellucidum**. Together, the choroid plexi and septum pellucidum constitute the **tela choroidea**. The thin tela choroidea covers each of the four ventricles.

Diencephalon: Located in the center of the brain ventral to the telencephalon, it contains the following structures:

Thalamus: masses of gray matter that form the lateral walls of the diencephalon. It is divided into right and left portions, which communicate by way of the **interthalamic adhesion**. Because of the internal location of the thalamus, its approximate location can be observed in the midsagittal section (Fig. 5.3). The cavity located between the thalami and ventral to the interthalamic adhesion is the **third ventricle**. It communicates with the lateral ventricles via a short channel called the **foramen of Monro**.

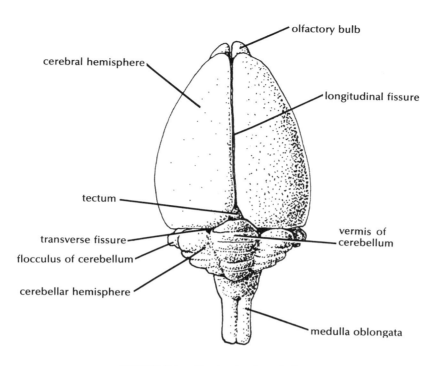

FIGURE 5.1. The brain, dorsal view

Hypothalamus: Located ventral to the thalamus, it is a small region that is the center of control for the autonomic nervous system.

Hypophysis: Also called the **pituitary gland**, it is an important endocrine gland that is attached to the hypothalamus via a short stalk, the **infundibulum**. From the ventral aspect (Fig. 5.4), it can be observed posterior to the hypothalamus.

Epithalamus: Dorsal to the thalamus, it is a small area that forms the roof of the diencephalon. Occupying most of the epithalamus is a small endocrine gland, the **pineal body**.

Optic chiasma: located on the ventral surface of the brain anterior to the hypophysis. It is the region where the optic nerves (cranial nerve II) cross.

Mesencephalon: A small region located posterior to the prosencephalon. It contains the following structures:

Tectum: forms the dorsal roof of the mesencephalon and can be located in the intact brain by spreading the cerebrum and cerebellum apart. From the sagittal aspect it can be observed posterior to the cerebrum. The tectum consists of four large swellings called the **corpora quadrigemina**. The larger, rostral pair is called the **rostral colliculi**, and the smaller, caudal pair is the **caudal colliculi**. The rostral colliculi contain optic reflex centers, and the caudal colliculi contain auditory reflex centers.

Cerebral peduncles: a pair of fiber bundles that form the ventrolateral surface of the mesencephalon. They provide a connection between the cerebrum and the mesencephalon.

Rhombencephalon: the posterior portion of the brain. It consists of two parts, anterior and posterior:

Metencephalon: the anterior portion of the rhombencephalon. It contains the following structures:

Cerebellum: the dorsal portion of the metencephalon. Its surface area is vastly increased by numerous folds (**folia**), which are separated from each other by shallow grooves, or **sulci**. The central portion of the cerebellum, the **vermis**, is composed of white matter. As can be seen in the sagittal aspect, numerous branches extend from it toward the outer surface. This arrangement is called the **arbor vitae**. Lateral to the vermis are the **right** and **left cerebellar hemispheres** and their lateral extensions, the **flocculi**. The gray matter of the cerebellum occupies the external surface of the folia and is called the **cerebellar cortex**. The cerebellum coordinates motor activities and maintains muscle tone.

Pons varolii: The ventral portion of the metencephalon. The pons is connected to the cerebellum by transverse fibers, which also connect the two cerebellar hemispheres.

Myelencephalon: the posterior portion of the rhombencephalon. It contains the following structures:

Medulla oblongata: the transitional region between the anterior parts of the brain and the spinal cord. The medulla oblongata, pons varo-

The Nervous System

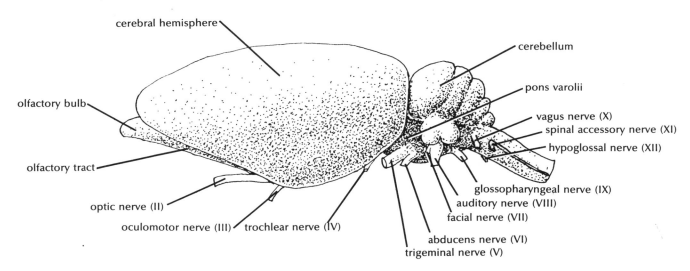

FIGURE 5.2. Lateral view of the brain

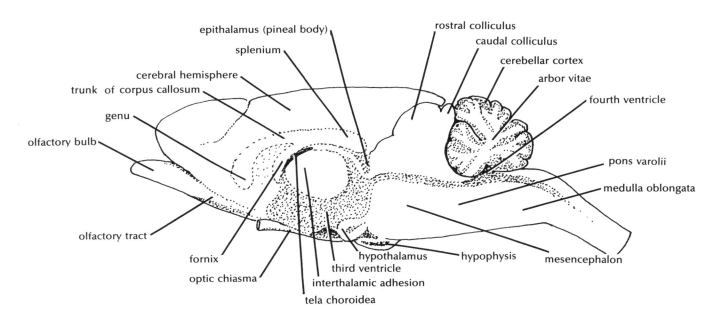

FIGURE 5.3. Lateral view of the brain, midsagittal section showing internal structures

lii, and mesencephalon form the **brain stem**. The medulla contains reflex centers that regulate respiration, blood pressure, and heart rate. It also regulates sensory impulses entering the brain.

Fourth ventricle: the posteriormost of the four brain cavities; it may be observed in the sagittal aspect between the cerebellum and the medulla. Cerebrospinal fluid passes between it and the third ventricle by way of a narrow canal, the **cerebral aqueduct**, which runs through the mesencephalon.

CRANIAL NERVES

Cranial nerves are nerves that originate from the brain. Mammals have twelve pairs, each with its own number (based on its position in human beings) and its own name. During the removal of the brain, you have severed them near their origin to leave their stumps for identification. Their names and corresponding numbers, as well as their points of union with the brain, innervations, and sensations or actions are listed in Table 3. Identify the cranial nerves of the rat by studying this list and the illustrations in Figures 5.2 and 5.4.

The Nervous System

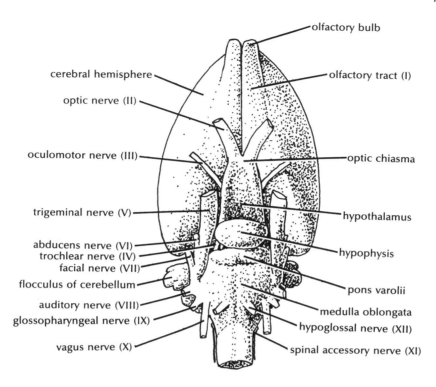

FIGURE 5.4. Ventral surface of the brain

TABLE 3. The Cranial Nerves

Cranial Nerve	Point of Union with Brain	Innervations	Sensation or Action
Olfactory (I)	Olfactory bulb	Sensory endings in nasal epithelium	Sense of smell
Optic (II)	Rostral colliculus and thalamus	Sensory cells in retina of each eye	Sense of sight
Oculomotor (III)	Cerebral peduncles of midbrain	Dorsal, ventral, medial recti and ventral oblique extrinsic eye muscle; iris of eye	Movement of eye; control of light into eye
Trochlear (IV)	Roof of midbrain	Dorsal oblique extrinsic eye muscle	Movement of eye
Trigeminal (V) 1. Ophthalmic 2. Maxillary 3. Mandibular	Lateral portions of pons	Skin of skull, external eyeball Vibrissae and upper teeth Jaw muscles, teeth of lower jaw, mouth, and tongue	Sensations of face region Sensations of rostrum Movement of jaw; sensations of jaw regions
Abducens (VI)	Rostral end of medulla	Lateral rectus; extrinsic eye muscle	Movement of eye
Facial (VII)	Lateral edges of medulla	Jaw and face muscles	Mastication and facial expression
Vestibulocochlear (VIII)	Lateral edge of medulla caudal to VII	Sensory cells within cochlea and vestibule of inner ear	Sensation of hearing and equilibrium
Glossopharyngeal (IX)	Lateral edge of medulla near X	Pharynx, parotid gland, tongue	Sensation of taste and throat region
Vagus (X)	Lateral edge of medulla caudal to IX	Muscles of the pharynx, larynx, thoracic viscera, abdominal viscera; sensory cells of larynx, and thoracic and abdominal viscera	Contraction of innervated muscles; sensation of innervated regions
Spinal Accessory (XI)	Lateral edge of medulla near its caudal end	Muscles of the neck and pharyngeal region	Movement of the neck and pharynx
Hypoglossal (XII)	Ventral surface of medulla	Tongue and hyoid muscles	Movement of the tongue

The Nervous System

SPINAL CORD

The spinal cord is a semicylindrical mass of gray and white matter surrounded by the meninges. Beginning cranially as a caudal continuation of the medulla oblongata, it passes through the neural canal and terminates as a slender filament, the **filum terminale**, in the base of the tail. In the cervical and lumbar regions, it contains swellings that give rise to the nerves that innervate the appendages.

A pair of **spinal nerves** arise from each segment and exit the vertebral column through intervertebral foramina. Each spinal nerve is formed by the union of a **ventral root** and a **dorsal root**, which merge with the spinal cord. The spinal nerves are named for the region from which they arise: cervical, thoracic, lumbar, sacral, and caudal.

You should not attempt to remove the entire spinal cord of your rat. Instead, obtain a cross section of the cord by cutting off the caudal tip of the region below the brain stem. With a magnifying lens, locate the following structures from your cross section (Fig. 5.5):

Meninges: Continuous with the meninges of the brain, they share the same names: the outermost **dura mater**, the middle **arachnoid**, and the deep **pia mater**. As in the brain, the arachnoid and pia mater are separated by the **subarachnoid space**, in which cerebrospinal fluid circulates.

Gray matter: the H-shaped mass of nervous tissue at the center of the spinal cord. Its two ventral arms are called the **anterior gray horns**, and the two dorsal arms are the **posterior gray horns**. The smaller, lateral projections are the **lateral gray horns**. The central bar is the **gray commissure**, which contains a small cavity that runs through its center called the **central canal**. Gray matter consists largely of neuron cell bodies and supportive tissue. Functionally, the ventral side is motor, and the dorsal side is sensory.

White matter: the nervous tissue of the spinal cord that surrounds the central gray matter. It is composed of **ascending tracts** of sensory fibers and **descending tracts** of motor fibers. From the cross-sectioned spinal cord, you should observe its distribution into four columns, or **funiculi**: one ventral, one dorsal, and two lateral to the gray matter.

THE PERIPHERAL NERVOUS SYSTEM

The peripheral nervous system (PNS) consists of **motor** (efferent) and **sensory** (afferent) nerves that carry impulses between the CNS and receptors and effectors. To dissect these nerves, they must be traced between the spinal cord and the appendages as they pass through the musculature. As you proceed, refer often to Figure 5.6.

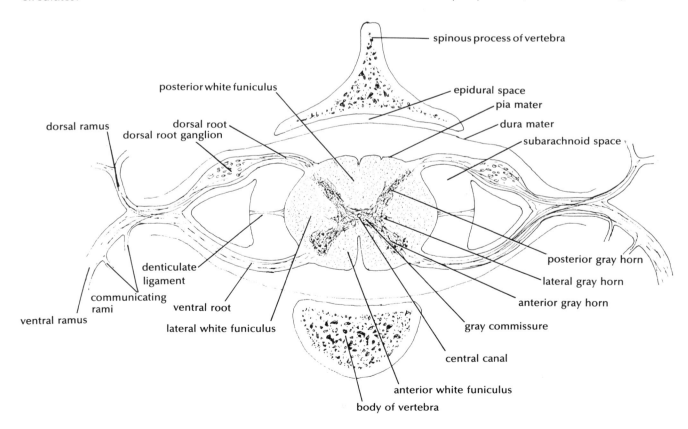

FIGURE 5.5. Cross section of the spinal cord and surrounding structures

MAJOR PLEXI

Plexi are networks of nerves located lateral to the spinal cord. They are formed from branches of the ventral rami of certain spinal nerves. In the musculature of the rat, locate the following plexi and peripheral nerves (Fig. 5.6):

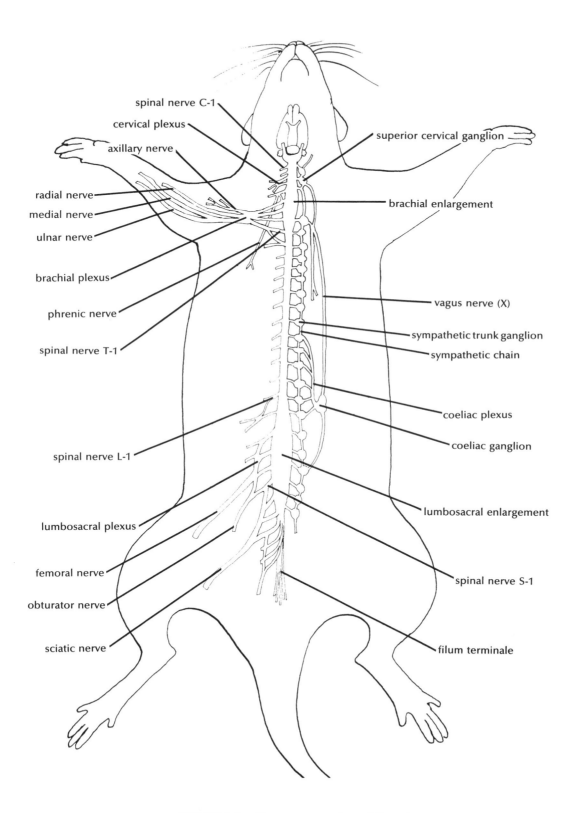

FIGURE 5.6. The nervous system of the rat

Cervical plexus: The smallest of the three plexi, it is formed by the first three and part of the fourth cervical nerves. It can be located by tracing the ventral rami of the neck region from the dorsal side as they emerge from the vertebral column. The cervical plexus supplies the musculature and skin of the neck region.

Brachial plexus: Lying medial to the shoulder and cranial to the first rib, this extensive network can be viewed from the ventral side deep to the pectorales muscles. The brachial plexus is formed by the union of part of the fourth and the fifth to eighth cervical nerves, as well as the first thoracic nerve. From the brachial plexus you should trace the following major nerves:

- **Axillary nerve**: a large nerve that passes along the caudal side of the humerus ventral to the long head of the triceps. It is formed by the union of the seventh and eighth cervical and first thoracic nerves.
- **Radial nerve**: a large, deep nerve located caudal to the axillary. It is formed by the union of the seventh and eighth cervical and first thoracic nerves and can be traced as it passes down the cranial side of the forelimb.
- **Ulnar nerve**: a prominent nerve that passes down the caudal side of the forelimb.
- **Median nerve**: a large nerve that crosses the rostral surface of the elbow and extends down the forelimb along the median surface.

Lumbosacral plexus: Formed by ventral rami of seven spinal nerves (fourth lumbar to third sacral), it supplies the skin and muscles of the pelvis and hind limbs. It lies deep within the abdominal and pelvic cavities and should not be dissected at this time. However, identify the following peripheral nerves that arise from this plexus in the hip and thigh musculature:

- **Femoral nerve**: Arising primarily from the fifth and sixth lumbar nerves, it passes through the abdominal wall to extend down the medial side of the thigh and shank. It can be observed as it extends along the medial edge of the vastus medialis.
- **Obturator nerve**: a smaller nerve that passes through the obturator foramen to the gluteus complex and deeper muscles of the hip.
- **Sciatic nerve**: a prominent nerve that extends through the deep caudal muscles of the thigh. It can be observed by reflecting the gluteus superficialis and biceps femoris muscles on the lateral side of the thigh.

THE AUTONOMIC NERVOUS SYSTEM

The autonomic nervous system consists of plexi, ganglia, and nerves that are associated with involuntary body activities. Due to the difficulty of dissection, do not try to locate the following structures in your specimen. Rather, identify the following in Figure 5.6:

Sympathetic trunk ganglion: ganglia, or clusters of neuron cell bodies, that lie in a vertical row on both sides of the vertebral column from the base of the skull to the coccyx. Collectively, the vertical rows of ganglia are referred to as the **sympathetic chain**.

Superior cervical ganglion: a large swelling that represents the union of the first four cervical sympathetic trunk ganglia. From it extend efferent fibers that supply the iris of the eye and the salivary glands.

Coeliac ganglion: a large swelling in the abdominal viscera that receives numerous autonomic nerves. The network of nerves in this region is called the **coeliac plexus**. Efferent fibers of the coeliac plexus supply visceral organs of the abdominal cavity.

Vagus nerve: cranial nerve X, which arises from the brain stem. It is a large nerve that supplies major thoracic and abdominal organs.

Phrenic nerve: a large nerve that arises from the ventral rami of the fifth and sixth cervical nerves. From here, it passes caudally through the thoracic cavity to innervate the diaphragm.

The Digestive System

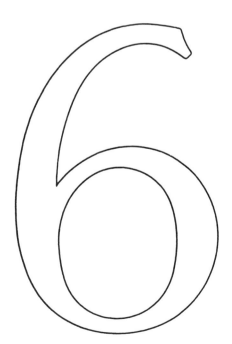

THE DIGESTIVE SYSTEM is composed of a series of organs that form one long, continuous, compartmentalized tube extending through the ventral body cavity from the mouth to the anus. Known as the **gastrointestinal (G.I.) tract**, it is here that the functions of mechanical digestion, chemical digestion, absorption of nutrients and water, and the storage and release of solid waste material take place. Each of these processes occurs within a compartment of the tract that is specialized to accommodate it. The specialized compartments, or organs, of the G.I. tract are the mouth, pharynx, esophagus, stomach, small intestine, and large intestine.

Also included in the digestion process are a number of structures that are closely associated with the G.I. tract. These structures, which are located within G.I. tract organs or communicate with them by means of a duct, are called **accessory organs**. They include the teeth, tongue, salivary glands, liver, and pancreas.

The organs and associated structures of the digestive system will be discussed sequentially from the salivary glands around the mouth to the anus. The system is divided into a cranial portion and a caudal portion, with the diaphragm serving as the line of division between the two.

CRANIAL DIGESTIVE STRUCTURES

The digestive structures that lie cranial to the diaphragm include the salivary glands, oral cavity or mouth, pharynx, and esophagus.

SALIVARY GLANDS

If possible, study the salivary glands on the side of the head that was used for dissection of the muscles. If these glands are damaged, turn your rat over and dissect the salivary glands on the intact side of the head. In the rat there are three pairs of salivary glands (Fig. 6.1):

Parotid gland: a large, lobular gland lying just below the ear along the caudal border of the mandible. Emerging from its rostral edge is the **parotid duct**, which crosses

The Digestive System

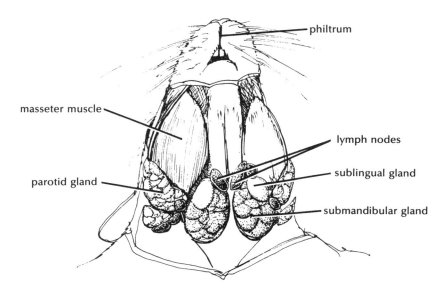

FIGURE 6.1. Salivary glands and associated structures of the ventral neck region

over the surface of the masseter muscle and enters the oral cavity through the cheek wall.

Submandibular gland: a lobular structure located ventral to the parotid gland and caudal to the angle of the mandible.

Sublingual gland: a pale, oval structure located rostral and immediately adjacent to the submandibular gland and medial to the mandible. A small tube, called Wharton's duct, empties its secretions and those from the submandibular gland into the oral cavity near the incisors. The sublingual gland is sometimes mistaken for the nearby **lymph nodes**, which are also oval (but somewhat smaller) structures in the ventral neck region.

ORAL CAVITY

In order to expose the oral cavity for study, cut through the muscles and connective tissue suspending the jaw on one side of your rat. Using bone shears, cut the condyloid portion of the jaw on the same side, pry it open, and locate the following structures (Fig. 6.2):

Vestibule: a narrow space between the teeth and cheeks.
Teeth: four incisors and twelve molars.
Hard palate: the rostral portion of the roof of the oral cavity. It is formed by the incisive, maxillary, and palatine bones and is lined with mucus membrane.
Soft palate: Caudal to the hard palate, it is a muscular partition between two chambers of the pharynx and is lined with mucus membrane.
Tongue: The large muscular structure that makes up the floor of the oral cavity, the tongue is attached to the floor of the mouth by a ventral fold of mucus membrane called the **lingual frenulum**. In order to examine the caudal portion of the tongue, sever all remaining points of attachment of the lower jaw on the side you have previously cut. Continue this cut down the lateral side of the throat and reflect the jaw to one side. Do not cut the pharynx or trachea. Locate the following features on the dorsal side of the tongue (Fig. 6.2):

  Lingual torus: an elevated region on the caudal part of the tongue.
  Median lingual sulcus: a central groove that superficially divides the tongue into right and left halves.
  Papillae: elevations of mucus membrane on the surface of the tongue, which contain sensory structures called **taste buds**. They include the numerous spinelike **filiform papillae** on the dorsal surface, several round **fungiform papillae** on the rostral surface, and a single round **circumvallate papilla** on the caudal surface of the tongue.
  Fauces: the opening at the extreme caudal end of the oral cavity. The fauces leads into the pharynx.

PHARYNX

A chamber lying caudal to the fauces, the pharynx serves as a muscular passageway for air traveling to and from the lungs and for food traveling from the mouth to the esophagus. It is commonly divided into the following sections (see Fig. 7.1):

Nasopharynx: the chamber dorsal to the soft palate. To view the nasopharynx, make a longitudinal incision along the midline of the soft palate. Spread the two cut ends apart as far as possible and shine a light into the exposed space. Note the two slitlike openings on the laterodorsal wall of the nasopharynx. These open into

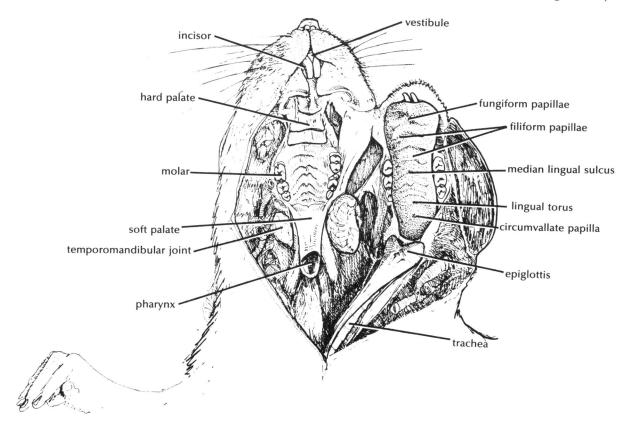

FIGURE 6.2. The oral cavity and pharynx, shown with the jaw cut on the right side and the floor of the mouth and pharynx swung laterally

the **internal auditory canals** (Eustachian tubes), which communicate with the middle ear. The two openings on the rostrodorsal wall are the **internal nares**, which open into the nasal cavity.

Oropharynx: the chamber located caudal to the fauces. Embedded within its laterodorsal walls are a pair of **palatine tonsils**, which lie partially recessed in a depression called the **tonsilar fossa**. Tonsils are lymphatic structures that play a role in immunity.

Laryngopharynx: the caudal-most portion of the pharynx. It lies dorsal and immediately cranial to the larynx. It communicates caudally with the esophagus and ventrally with the larynx. Its slitlike opening to the larynx, called the **glottis**, is protected from passing food particles by a movable fold, or diverter valve, called the **epiglottis**.

ESOPHAGUS

The esophagus is a long, muscular tube that transports material from the pharynx to the stomach. It lies dorsal to the trachea and travels the length of the thoracic cavity. At its caudal end it perforates the diaphragm to unite with the stomach in the abdominal cavity. Because it is located in the thoracic cavity, the esophagus will not be dissected at this time. Rather, you should examine it when other thoracic cavity structures are considered in Chapter 7.

CAUDAL DIGESTIVE STRUCTURES

Most of the digestive organs lie caudal to the diaphragm within the abdominal cavity. To observe these structures, cut through the abdominal wall by making a longitudinal incision from the symphysis pubis to a point caudal to the sternum. Do not continue this incision cranially, for you want to avoid damaging the large, circular sheet of muscle dividing the thoracic and abdominal cavities, the **diaphragm**. Do not insert your scissors too deeply while cutting, or you will damage the internal organs. From this midventral incision, make another incision on each side of the caudal border of the diaphragm to the dorsal body wall. Again from the midventral incision, make a pair of incisions at the level of the iliac crest at the caudal end of the abdomen in opposite directions to the dorsal body wall.

Pull back the flaps of the abdominal wall and examine the **abdominal cavity** (Fig. 6.3). This closed body cavity

The Digestive System

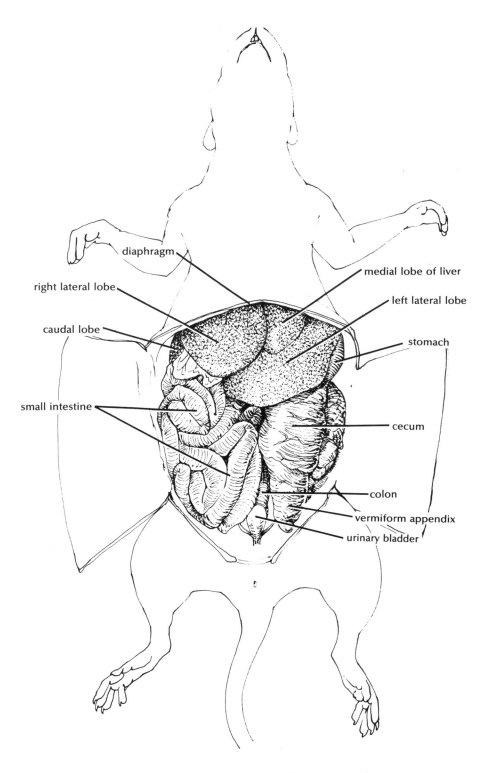

FIGURE 6.3. Organs of the abdominal cavity, ventral view

extends from the diaphragm to the pelvis. Examine the following structures within it.

PERITONEUM

The peritoneum is an extensive serous membrane that lines the abdominal cavity and covers many visceral organs. The portion of the peritoneum that lines the internal surface of the abdominal wall is called the **parietal peritoneum**, and the portion that covers the organs is the **visceral peritoneum**. The potential space between the two membranes is called the **peritoneal cavity**. It normally contains nothing more than a small amount of fluid that reduces friction between adjacent visceral organs.

In addition to the peritoneal membranes, there are **peritoneal folds**. These extensions of the peritoneum arise during embryonic development as the **dorsal** and **ventral mesenteries**. These mesenteries differentiate into the following peritoneal folds that are present in the adult rat (Figs. 6.3–6.5):

Ventral mesentery: folds of the peritoneum that originate from the ventral abdominal wall.
 Falciform ligament: attaches the liver to the ventral abdominal wall and the diaphragm.
 Lesser omentum: extends between the lesser curvature of the stomach and the liver.
Dorsal mesentery: folds of the peritoneum that anchor associated organs to the dorsal wall of the abdomen.
 Mesogaster: anchors the stomach.
 Mesentery proper: anchors the small intestine.
 Mesocolon: anchors the large intestine.
 Mesorchium: anchors the testes in the male.
 Mesovarium: anchors the ovaries in the female.
 Greater omentum: a double layer of peritoneum that extends from the greater curvature of the stomach partially over the intestines.

LIVER

The liver is the large, dark brown organ lying caudal to the diaphragm (Figs. 6.3–6.5). It is divided into four lobes: a large **medial lobe**, a **right lateral lobe**, a **left lateral lobe**, and a small **caudal lobe**. The liver receives blood from the G.I. tract and processes it by way of nutrient modification and detoxification. It also produces **bile** for the digestion of fats; bile passes into the duodenum via the **central bile duct** (ductus choledochus). There is no gallbladder for bile storage as there is in most mammals.

STOMACH

The stomach is located directly caudal to the diaphragm on the left side of the abdominal cavity (Figs. 6.3–6.5). Its lateral border forms a rounded, convex surface called the **greater curvature**, and its medial border forms a concave angle called the **lesser curvature**. Note the flat, elongate **spleen** near the dorsolateral surface of the stomach. This organ is part of the lymphatic system, and it serves as a storage area for red blood cells and white blood cells. The stomach is divided into the following regions:

Cardia: The forestomach, it is the portion of the stomach that receives the esophagus. Its internal lining, or **mucosa**, is similar to that of the esophagus. It serves as a holding chamber for food prior to digestion.
Fundus: the central portion of the stomach. Its mucosa contains secretory cells (**gastric glands**) that release a protein-digesting enzyme (**pepsin**), hydrochloric acid, and mucus.
Pylorus: the caudal region of the stomach. The secretory cells within its mucosa release only mucus. At its caudal end it forms the **pyloric valve**, a thick sphincter muscle that regulates the movement of material into the bordering small intestine.

SMALL INTESTINE

The small intestine is a long, winding tube (It is roughly six times the length of the animal from snout to anus!) that extends from the pyloric valve to the **ileocecal valve** of the large intestine (Figs. 6.3–6.5). It chemically digests material entering from the stomach and absorbs nutrients across the walls of its specialized mucosa. The small intestine is divided into three segments:

Duodenum: the cranial segment of the small intestine. It extends from the pyloric valve to its union with the jejunum. The duodenum receives digestive enzymes from the liver via the **central bile duct** and the pancreas via the **pancreatic duct**.
Jejunum: the middle segment of the small intestine. It is the greatest in length but not easily distinguishable from the other two segments.
Ileum: the caudal segment. It extends from its union with the jejunum to the ileocecal valve, where the large intestine originates. The ileum is characterized by the presence of large lymph nodules, which appear as small bumps on its surface.

PANCREAS

The pancreas is a diffuse mass of tissue concealed by mesentery near the duodenum and passing dorsal to the stomach (Figures 6.4–6.5). It secretes a variety of digestive enzymes, or **pancreatic juice**, which pass through the **pancreatic duct** en route to the duodenum. It also secretes two hormones, glucagon and insulin, from clusters of endocrine cells within its tissue called **Islets of Langerhans** (see Table 2).

The Digestive System

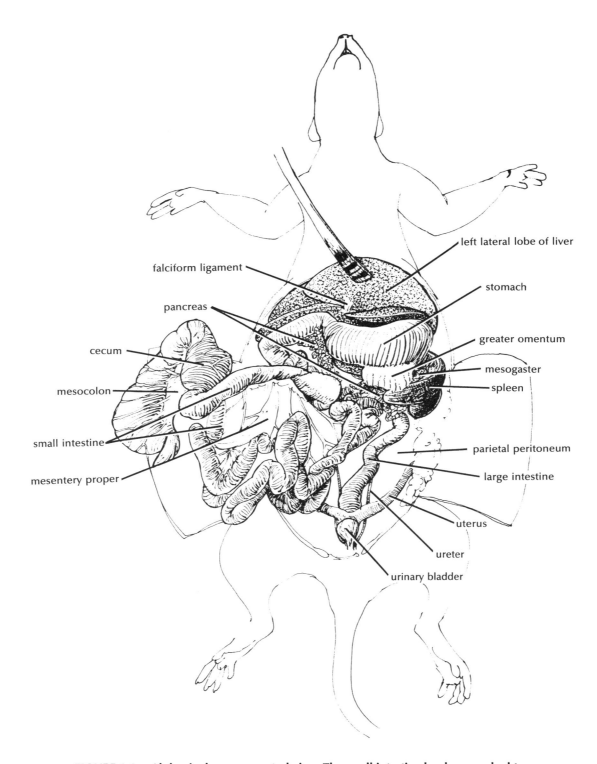

FIGURE 6.4. Abdominal organs, ventral view. The small intestine has been pushed to the right side and the liver moved cranially in this female specimen.

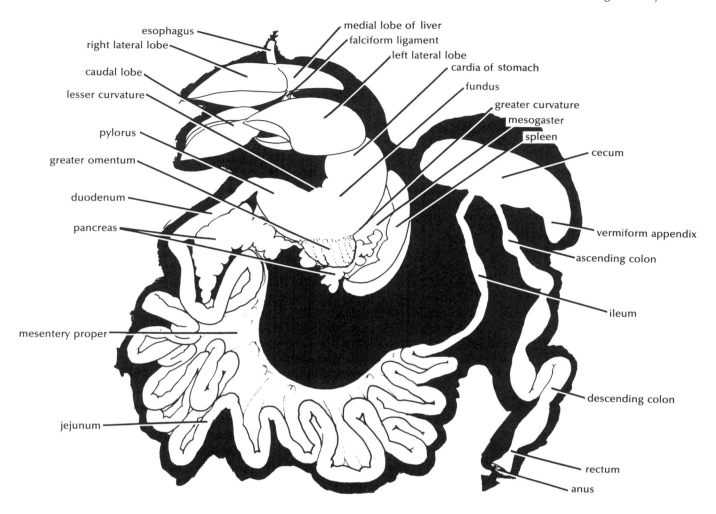

FIGURE 6.5. Schematic view of the digestive system caudal to the diaphragm. The arrows indicate the normal direction of food passage.

LARGE INTESTINE

The caudal portion of the G.I. tract, the large intestine extends from the ileocecal valve to the anus (Figs. 6.3–6.5). It functions in water absorption, fecal formation, and elimination. It is larger in diameter than the small intestine and is divided into the following segments:

Cecum: the first segment of the large intestine. It communicates with the ileum via the ileocecal valve. The cecum is a large, blind sac that projects caudally and can be identified by its thin walls and large diameter (which exceeds that of other segments of the large intestine). At its terminal end, a narrow, thicker-walled projection called the **vermiform appendix** extends.

Colon: a long, wide tube that extends from the cecum near the ileocecal valve to the rectum. Initially, it can be traced extending cranially to the stomach (the **ascending colon**). From the level of the stomach it passes caudally (the **descending colon**) to the rectum.

Rectum: a short, straight tube that passes from the colon to the exterior. The terminal portion of the rectum is called the **anal canal** and is bordered at its external opening, or **anus**, by external and internal sphincter muscles.

The Respiratory System

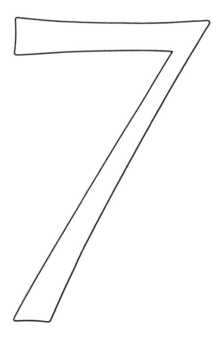

THE PRIMARY FUNCTIONS of the respiratory system are to provide the body with a continuous supply of oxygen and to remove the metabolic waste product, carbon dioxide. Structurally, it consists of a thoroughfare of chambers and tubes that conduct air between the external environment and the microscopic, thin-walled sacs within the lungs.

In this chapter you will study the organs and associated structures of the respiratory system. For convenience of discussion, they are divided into cranial respiratory structures, which are located cranial to the thoracic cavity, and caudal respiratory structures, which lie within the thoracic cavity.

CRANIAL RESPIRATORY STRUCTURES

The respiratory organs that are located cranial to the thoracic cavity include the rostrum, pharynx, larynx, and trachea.

ROSTRUM

The rostrum contains the initial warming and humidifying chamber for incoming air, the **nasal cavity**. This internal cavity lies between the **external nares** and the openings to the nasopharynx, the **internal nares** (choanae). It contains the following structures (Fig. 7.1):

Nasal septum: a vertical partition that divides the nasal cavity into right and left chambers, or **fossae**. It is formed by the vomer, the perpendicular plate of the ethmoid, and cartilage, which are overlain with mucus membrane.

Nasal meati: channels within the nasal fossae that serve to warm and moisten air as it passes through. The two meati within each fossa are formed by the turbinate bones.

Nasal epithelium: mucus membrane surrounding the caudal end of the fossae where the special sensory organs for smell are located. Their receptors are dendrites to sensory neurons that lie embedded within the epithelium.

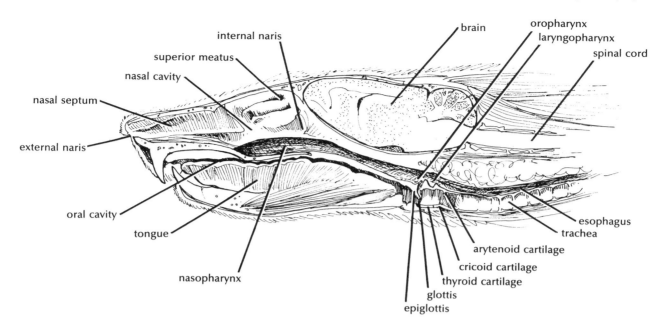

FIGURE 7.1. The head and neck regions, midsagittal section

PHARYNX

For a description of the pharynx, see Chapter 6.

LARYNX

The larynx is a boxlike structure composed of five cartilages that support the entrance from the pharynx into the trachea. Identify the following in your specimen (Figs. 7.1, 7.2):

Glottis: the opening between the laryngopharynx and the larynx.
Epiglottis: a leaf-shaped, single cartilage that lies immediately cranial to the glottis. During swallowing, it drops to form a lid over the glottis to prevent the passage of food or water into the trachea. At its base, the epiglottis is attached to the thyroid cartilage.
Thyroid cartilage: a single, large cartilage that forms the ventral and lateral walls of the larynx. It lies caudal to the epiglottis.
Thyroid gland: an endocrine gland that lies over the caudal part of the larynx. To view it, remove the sternohyoid muscles. Note that it is immediately caudal to the thyroid cartilage. Embedded within the thyroid gland are four to five small, oval **parathyroid glands** that are also endocrine structures.
Cricoid cartilage: Remove the thyroid gland and observe the underlying, single cricoid cartilage. It is a small ring that forms the dorsal wall of the larynx. At its caudal ridge, it is attached to the first tracheal cartilage.
Arytenoid cartilage: Located dorsal to the thyroid cartilage at the margin of the glottis, this paired cartilage serves as the point of origin of the vocal cords.
Vocal cords: To observe these, cut open the larynx along its middorsal line. The pair of whitish, lateral folds that extend from the arytenoids to the thyroid cartilage are the vocal cords.

TRACHEA

The trachea is a tubular air passageway that extends from the larynx to the level of the fourth or fifth thoracic vertebral segment, where it divides. It lies ventral to the esophagus, and its walls are formed by smooth muscle and connective tissue encircled by a series of incomplete horizontal rings of cartilage. The openings in the cartilage rings face the esophagus. Locate the cranial portion of the trachea in the neck region of your specimen before proceeding into the thoracic cavity (Fig. 7.2).

CAUDAL RESPIRATORY STRUCTURES

The remainder of the respiratory organs lie within the thoracic cavity. In order to examine them, first make a longitudinal incision slightly to one side of the midventral line from the diaphragm to the cranial margin of the sternum. You will be cutting through the ribs, muscles of the thoracic wall, and the clavicle, so use a strong pair of scissors. To prevent damage to the internal organs, do not in-

The Respiratory System

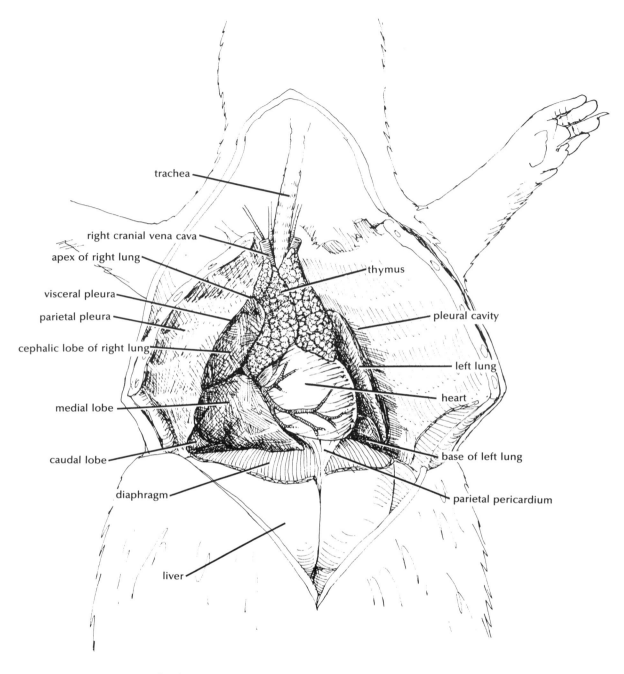

FIGURE 7.2. The thoracic cavity and its associated structures. The parietal pericardium has been partially removed in this specimen.

sert your scissors too deeply. Then make a transverse incision immediately cranial to the diaphragm, extending it laterally and dorsally to the back on both sides. Follow the contour margin of the diaphragm as you make this cut. Finally, spread the thoracic walls outward, breaking the ribs near their point of attachment to the vertebral column.

The internal walls of the thoracic cavity are largely lined with a serous membrane similar to the peritoneum, called the **parietal pleura**. The **pleural cavity** is a potential space that surrounds each lung and is located between the parietal pleura and a **visceral pleura**, which covers the outer surface of each lung. The potential space between the medial walls of the two visceral pleurae is called the **mediastinum**. It is nearly filled with structures, including an endocrine gland called the **thymus**, the **parietal pericardium** (pericardial sac) that encloses the **heart**, and major blood vessels of the heart. Identify these cavities and structures as shown in Figure 7.2 before proceeding.

The respiratory organs that lie within the thoracic cavity are the bronchial tubes and the lungs.

BRONCHI

The trachea terminates in the thoracic cavity on the dorsal side of the heart by dividing into the **right** and **left primary bronchi**, which lead to the lungs. Both bronchi contain incomplete rings of cartilage like those of the trachea. Locate the primary bronchi by gently pushing aside the heart.

Once inside each lung, the primary bronchus divides into **secondary** (lobar) **bronchi**, which conduct air to and from individual lobes of the lung. Because the left lung contains only one lobe, it receives a single secondary bronchus. The secondary bronchi branch further to form smaller **tertiary** (segmental) **bronchi**, which divide into yet smaller **bronchioles**. The microscopic bronchioles terminate in **vestibules**, which contain **alveoli**. At this stage in the dissection procedure, the subdivisions of the bronchi within the lung cannot be traced.

LUNGS

The lungs are large, spongy structures lying lateral to the centrally located heart. As mentioned above, the left lung consists of a single lobe or divisible section. The right lung is divided into four lobes: **cephalic**, **medial**, **caudal**, and **postcaval** (Fig. 7.2; the postcaval lobe is not visible here because it lies dorsal).

The exterior surface of the lungs can be described as follows. The cranial tapering of each lung into a somewhat conical point is called the **apex**, and the caudal, concave portion that touches the diaphragm is the **base**. The **mediastinal** surface is the surface area facing the mediastinum, and the **costal** surface faces the ribcage.

Make an incision through one of the lungs, and view its internal features with a magnifying hand lens. Note the cut subdivisions of the bronchi, the numerous blood vessels, and, if possible, the microscopic sacs called **alveoli**. In a preserved specimen the alveoli are filled with fluid, but in life they are partially filled with air for gas exchange. Alveoli, with their associated connective tissue and blood vessel elements, make up the bulk of the lung.

The Circulatory System

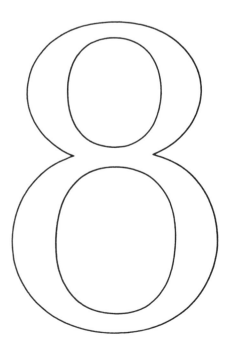

THE CIRCULATORY SYSTEM provides a means of internal transportation for a large variety of substances, including oxygen, carbon dioxide, nutrients, hormones, and nitrogenous wastes. This function is accomplished by a vast network of **blood vessels**, which carry these substances in a dissolved or suspended state within the **blood**. The vessels that transport blood away from the heart toward the microscopic, thin-walled **capillaries** are called **arteries**, and those that transport blood toward the heart are called **veins**.

A separate series of vessels that parallels the course of major veins carries a fluid called **lymph**. These are the **lymphatic vessels**, which form a part of the lymphatic system with the **spleen**, **tonsils**, and **nodes**. The lymphatic system filters body fluids of dead or dying cells and invading microbes.

The vessels of the circulatory and lymphatic systems constitute an extremely extensive network that invests most organs and tissues. Therefore, it is necessary to limit the following dissection protocol to a study of the heart, the major arteries and veins, and their primary tributaries. Lymphatic vessels and structures that have not yet been identified will be briefly mentioned when they become observable during the course of the dissection procedure.

HEART

The heart is the central pumping organ of the circulatory system. In the following study, you will examine its external features, the major blood vessels that emerge from it or terminate in it, and its internal structure.

EXTERNAL FEATURES OF THE HEART

The heart is located in the thoracic cavity between the two lungs. Partially obscuring the cranial end of the heart is an endocrine gland known as the **thymus**. Remove this gland and locate the following external features of the heart (Figs. 7.2 and 8.1):

The Circulatory System

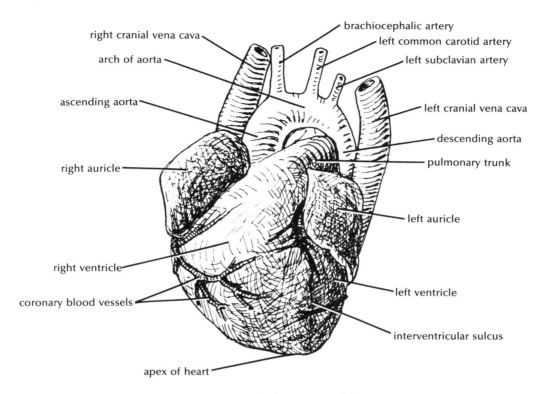

FIGURE 8.1. The heart, ventral view

Pericardium: the two protective layers of serous membrane that surround the heart. The outermost membrane, the **parietal pericardium** (pericardial sac), looks like a sac that wraps around the heart. Lift this membrane away from the heart surface and carefully make an incision through it. Continue this incision along the vertical length of the heart and peel the pericardium away. The space immediately deep to the parietal pericardium is the **pericardial cavity**, which is filled with fluid in the living rat. Immediately deep to the pericardial cavity is the second serous membrane layer, the **visceral pericardium**. This membrane is directly attached to the underlying **cardiac muscle** (myocardium), which forms the walls of the heart.

With the parietal pericardium peeled away, examine the general shape of the heart. The pointed, caudal end is called the **apex**, and the somewhat flattened, cranial end is the **base**. The two major divisions of the heart can now be identified externally:

- **Ventricles**: occupy most of the heart. They are completely separated internally into a **right** and **left ventricle**. This separation is represented externally by a shallow groove called the **interventricular sulcus**, which may not be visible if your specimen was young. **Coronary blood vessels** can be seen within this sulcus.
- **Atria**: the two small sacks on either side of the cranial portion of the heart. The **right** and **left atria**, also completely separated internally, are divided from each other externally by the two large vessels that leave the heart. The externally attached extension of each atrium is the **auricle**.

BLOOD VESSELS OF THE HEART

The following prominent blood vessels either originate or terminate at the heart. To aid in their identification, they and their tributaries have had colored material (usually latex) injected after death. This color-coding distinguishes between arteries and veins and makes the blood vessels tougher and more elastic. Doubly injected specimens have a red material in their arteries and a blue material in their veins. This coding is reversed in the pulmonary circulation. In triply injected specimens, a yellow material is injected into the hepatic portal system. With this coding in mind, examine the large vessels on the ventral side of the heart.

Pulmonary trunk: It arises from the right ventricle and extends dorsally to the lungs. If injected, this large artery will be blue, even though it carries blood away from the heart. Frequently, however, it is not injected. Observe that it emerges between the left and right atria.

- **Pulmonary arteries** (paired): originate from a bifurcation of the pulmonary trunk. As right and left branches of the trunk, these arteries carry deoxygenated blood to the respective lungs.

The Circulatory System

Aorta: Located adjacent to the pulmonary trunk, the aorta ascends from its origin at the left ventricle (**ascending aorta**), forms a prominent arch (**arch of the aorta**), and descends behind the heart (**descending aorta**) to continue to its bifurcation at the caudal end of the abdominal cavity. Before it begins its descent, this thick-walled artery gives off the following major branches:

Coronary arteries (paired): vessels that originate from the base of the aorta and pass to the heart wall. Running parallel to these arteries are the **coronary veins** that drain the heart wall into the right atrium.

Brachiocephalic artery: a single vessel that is the first branch to arise from the arch of the aorta. Also called the innominate artery, it passes cranially until it divides into the **right common carotid artery** and the **right subclavian artery**. These vessels will be traced later.

Left common carotid artery: a single artery arising from the arch of the aorta at a point lateral to the brachiocephalic artery origin. It supplies the left side of the neck and head via its tributaries.

Left subclavian artery: the third branch from the arch of the aorta. It passes to the left shoulder deep to the clavicle. It supplies the left forelimb via its tributaries.

Vena cavae: large veins that drain blood from the systemic circulatory network into the right atrium. Push the heart of your rat to the left side of the thorax and note that the vena cavae enter the right atrium at the dorsal side. Do not attempt to view the entire dorsal side of the heart at this time; you will examine it after the heart has been removed from the chest cavity.

Cranial vena cavae (paired): large veins that drain the body cranial to the heart (excluding the lungs). They enter the cranial portion of the right atrium.

Caudal vena cava: a single vein that drains the body caudal to the heart. Arising from the union of the **common iliac veins** near the caudal end of the abdominal cavity, it enters the caudal portion of the right atrium.

INTERNAL FEATURES OF THE HEART

Cut the large vessels associated with the heart no closer than half an inch from their points of union with the heart, as shown in Figure 8.2. As you cut, note particularly the **pulmonary veins**, which were not visible ventrally. Passing from the lungs, these veins enter the left atrium. The pulmonary trunk, arteries, veins, and their smaller tributaries constitute the **pulmonary circuit**. All other vessels of the body make up the **systemic circuit**. Remove the heart from the thoracic cavity and again identify the heart chambers and associated vessels on the

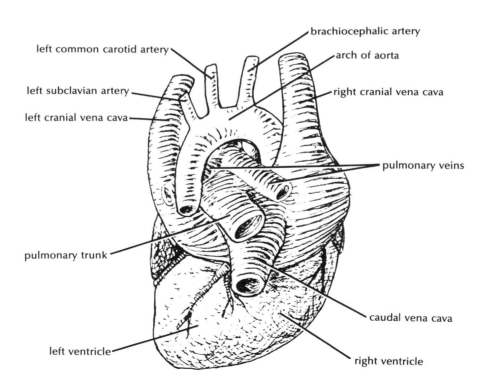

FIGURE 8.2. The heart, dorsal view

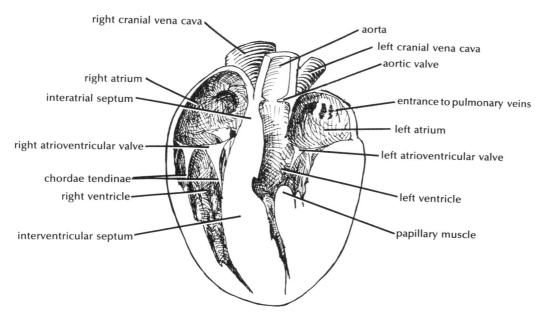

FIGURE 8.3. Internal structures of the heart, ventral view

ventral side. Now turn the heart over and identify the chambers and vessels on the dorsal side as well.

To study the internal anatomy of the heart, section the heart completely across the frontal plane. Carefully separate the ventral and dorsal halves and remove any dried blood or latex that may occupy the ventricles. Locate each of the following structures and identify their associated features (Fig. 8.3):

Right atrium: the thin-walled chamber formed by the union of the vena cavae. It also receives the coronary veins. Note the smooth internal surface of this chamber.

Left atrium: the thin-walled chamber that receives the four pulmonary veins. It is separated from the right atrium by a thin, muscular partition called the **interatrial septum**.

Atrioventricular valves (paired): one-way valves located between each atrium and ventricle (right and left). Each consists of two flaps that point downward into the ventricle. Note the conspicuous white cords, called **chordae tendinae**, which connect the lower margins of the flaps to the wall of the ventricle. Many of these cords are anchored to bulblike extensions of the ventricular myocardium, the **papillary muscles**. These structures prevent the valves from everting into the atria during ventricular contraction.

Right ventricle: a thick-walled, caudal chamber that receives deoxygenated blood from the right atrium through the right atrioventricular valve. It pushes blood out through the pulmonary trunk to the lungs for oxygenation.

Left ventricle: a large, thick-walled caudal chamber. The left ventricle is larger in area and wall thickness than the right ventricle. This size difference is a consequence of its requirement to push newly oxygenated blood it has received from the left atrium through the vast systemic circuit. It is separated internally from the right ventricle by a thick-walled partition, the **interventricular septum**.

Semilunar valves (paired): one-way valves located between the ventricles and the two major arteries. Each semilunar valve consists of three pocket-shaped cusps located at the base of each vessel. The valve in the pulmonary trunk is the **pulmonary valve**, and that in the aorta is the **aortic valve**.

BLOOD VESSELS CRANIAL TO THE HEART

As you dissect the blood vessels, you should keep in mind that they are subject to considerable variation and may therefore be somewhat different in each specimen. Arteries and veins also tend to share the same pathways through the body. Consequently, they commonly share a name, which usually corresponds to the region they supply or pass through.

ARTERIES

Cranial arteries are tributaries of the two large vessels that exit from the ventricles: the pulmonary trunk and the aorta. Since the pulmonary trunk tributaries have been identified, locate only the aorta once again and trace its following tributaries (Figs. 8.4, 8.5):

Brachiocephalic artery: Trace this short vessel once more and recall that it divides into the **right common carotid**

The Circulatory System

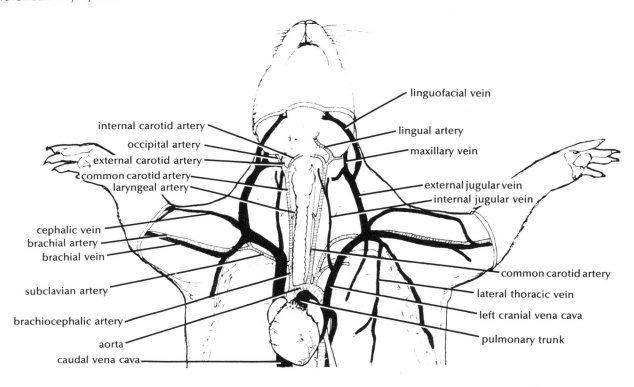

FIGURE 8.4. The cranial blood vessels. The thymus and lungs have been removed from the thoracic cavity.

and **right subclavian arteries**. Identify these vessels and trace their following branches:

Right common carotid artery: passes cranially along the right side of the trachea. Near the larynx, it divides into the following:

Right external carotid artery: passes upward through the parotid salivary gland to supply structures of the head and neck. Its major tributaries include the **laryngeal artery** (to the larynx), the **thyroid artery** (to the thyroid gland), the **lingual artery** (to the tongue), the **parotid artery** (to the parotid gland), and the **occipital artery** (to the occipital muscles).

Right internal carotid artery: extends cranially through the ventral neck muscles to supply the brain. It enters the cranial cavity through the carotid canal at the base of the skull.

Right subclavian artery: extends from its origin at the brachiocephalic artery to the right shoulder, where it becomes the **right axillary artery**. Along its length, branch the **vertebral artery**, which passes to the brain (where it unites with the vertebral artery of the opposite side to form the **basilar artery**); the **internal thoracic artery**, which supplies the ventral chest wall; the **superficial cervical artery**, which supplies the dorsal muscles and structures of the neck and chest; and the **cervical trunk**, which sup-

plies neck and shoulder muscles, salivary glands, and the skin of the neck region.

Right axillary artery: A continuation of the right subclavian in the shoulder region, it passes through the armpit (axilla) where it gives off minor branches to the shoulder and chest muscles before proceeding into the arm as the **right brachial artery**.

Right brachial artery: A continuation of the axillary artery as it enters the brachium, it passes downward between the triceps and biceps muscles. Along its length, branch numerous smaller vessels that supply the muscles of the brachium. At the level of the elbow, it divides into vessels that supply the muscles of the antebrachium and manus.

Left common carotid artery: Arising directly from the arch of the aorta, recall that it supplies the left side of the head and neck. Its tributaries, the **left external** and **left internal carotid arteries** and their branches, correspond to those sharing the same name on the right side.

Left subclavian artery: Recall that this artery also originates directly from the arch of the aorta. Along its length, it gives rise to vessels that correspond to branches of the right subclavian artery. Its continuation into the left shoulder is called the **left axillary**

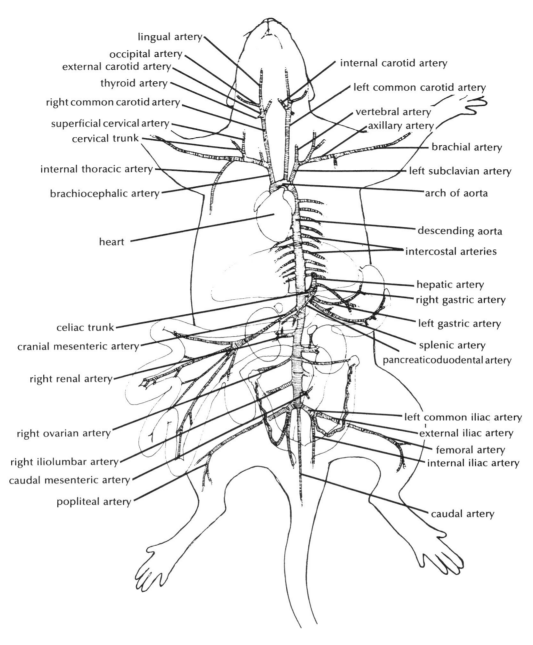

FIGURE 8.5. Arterial circulation scheme

artery, which then continues down the arm as the **left brachial artery**.

VEINS

Most peripheral veins closely parallel the arteries with which they share a common name, but there are some notable exceptions as the veins approach the heart. Beginning from the heart, locate the following major cranial veins (Figs. 8.4 and 8.6):

Cranial vena cava (paired): Recall that the **right and left cranial vena cavae** drain the body cranial to the heart into the right atrium. Near their union with the heart, a number of smaller veins enter. These veins include, from caudal point of entry to cranial on each side, the

The Circulatory System

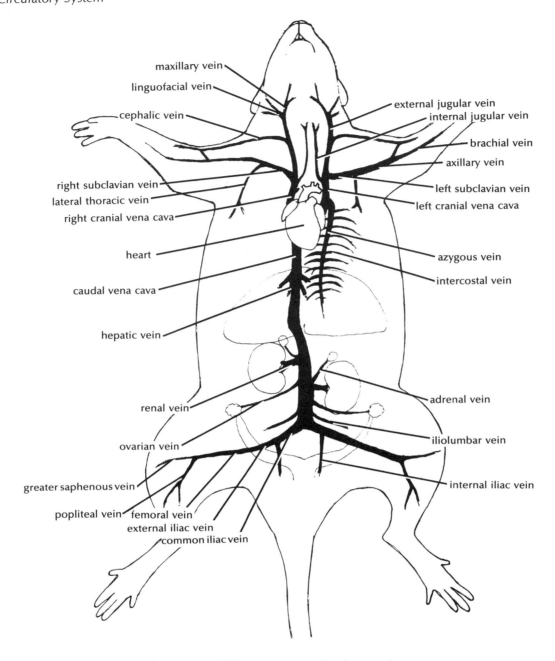

FIGURE 8.6. Venous circulation scheme

unpaired **azygos vein**, which in turn receives the segmental **intercostal veins** from the thoracic wall; the paired **internal thoracic vein**, and the paired **vertebral vein**. In addition, the right and left cranial vena cavae each receive the following prominent veins:

Subclavian vein: right and left. Uniting with the cranial vena cava at the level of the 1st rib, each is a continuation of the **axillary vein** from the armpit re-

gion of each side. The right and left subclavians each receive the **lateral thoracic vein**, which drains the pectorales muscles.

Axillary vein: right and left. A cranial continuation of the **brachial vein** in the armpit region, which drains the forelimb.

Brachial vein: right and left. It originates at the angle of the elbow as a cranial continuation of the **median vein**, which drains the antebrachium and manus. It parallels the brachial artery and median nerve along its route.

External jugular vein: right and left. A large, superficial vein on each side that extends from the angle of the jaw to the cranial vena cava. From caudal to cranial, each receives the following veins: the **accessory cephalic vein**, which is a small, superficial vessel that drains the shoulder where it emerges from the cephalic vein; the larger **cephalic vein**, which originates from the lateral surface of the brachium and drains the antebrachium; the **linguofacial vein**, which drains the salivary glands, tongue, and facial region; and the **maxillary vein**, which drains the eye orbit, the palate, the ear, and the jaw muscles.

Internal jugular vein: right and left. Formed by the union of veins that drain structures lying within the cranial cavity, the small internal jugulars exit the cranial cavity through the jugular foramina and descend to the lower neck region of each side. Each internal jugular passes medial to the external jugular and lateral to the common carotid artery of the same side and unites with the cranial vena cava dorsal to the junction of the external jugular.

BLOOD VESSELS CAUDAL TO THE HEART

Locate the following caudal arteries and veins in your specimen. Refer often to Figures 8.5 through 8.8.

ARTERIES

All arteries caudal to the heart are tributaries of the **descending aorta**, the section of the aorta that passes caudally through the thoracic and abdominal cavities (where it is called the **thoracic aorta** and **abdominal aorta**, respectively). Trace the path of the vessel from its origin at the aortic arch, along the dorsal wall of the thorax, through the **aortic hiatus** in the diaphragm, and along the dorsal abdominal wall. Note the numerous, segmental **intercostal arteries** that extend from the thoracic aorta to supply the intercostal muscles of the ribcage. Also note that the abdominal aorta is located behind the peritoneum. Now identify the following tributaries of the abdominal aorta (Figs. 8.5, 8.7).

Celiac trunk: a large, single vessel that emerges from the abdominal aorta at a point slightly below the diaphragm. In order to locate it and its tributaries, push the liver upward and the G.I. track organs to one side. The celiac supplies the abdominal organs at the cranial end of the abdomen via its branches. They include the **left gastric artery**, which passes to the lesser curvature of the stomach; the **hepatic artery**, which supplies the liver by its main trunk and the greater curvature of the stomach, pancreas, and duodenum by its tributaries (via the **right gastric**, **pancreaticoduodenal**, and **gastroepiploic arteries**, respectively); and the **splenic artery**, which supplies the spleen and part of the pancreas and omentum.

Cranial mesenteric artery: a large, unpaired vessel that arises from the abdominal aorta just caudal to the origin of the celiac trunk. It supplies the small intestine, large intestine, and mesentery via smaller branches that extend through the corresponding peritoneal folds.

Renal arteries: paired vessels that supply the right and left kidneys. The right renal artery normally arises from the abdominal aorta cranial to the left renal artery. The renal arteries are located dorsal to the parietal peritoneum.

Gonadal arteries: paired vessels that arise from the abdominal aorta caudal to the renal arteries. In some specimens they may alternatively emerge from the renal arteries. In the male, they are the **spermatic arteries**, which pass to the scrotum and testes. In the female, they are the **ovarian arteries**, which pass to the ovaries.

Iliolumbar arteries: paired vessels that pass to the dorsal body wall. They originate from the abdominal aorta caudal to the gonadal arteries.

Lumbar arteries: small, segmental vessels extending from the abdominal aorta to the dorsal body wall that they supply (not shown).

Caudal mesenteric artery: an unpaired vessel that arises from the aorta cranial to its bifurcation into the common iliacs (described below). The caudal mesenteric extends through the mesocolon and supplies the colon and rectum.

Common iliac arteries: paired vessels, right and left, that emerge from the bifurcation of the abdominal aorta at the caudal end of the cavity. At this bifurcation emerges a smaller vessel as well, the **caudal artery**, which passes to the tail. In some specimens, each common iliac soon divides into an **internal iliac artery**, which passes into the pelvic cavity along its dorsal wall to supply structures of the pelvic and gluteal regions, and an **external iliac artery**, which passes along the lateral body wall and exits the pelvic cavity to continue down the thigh as the **femoral artery**. In many specimens, however, this division of the common iliacs does not occur. In these rats the common iliac of each side gives off numerous tributaries that supply the pelvic and gluteal regions directly before continuing out of the pelvic cavity to the thigh.

Femoral artery: paired vessels, right and left, that continue from the common iliacs or external iliacs down the medial side of the thigh. At the level of the knee, the femoral continues as the **popliteal artery** through the popliteal fossa before it divides into the

The Circulatory System

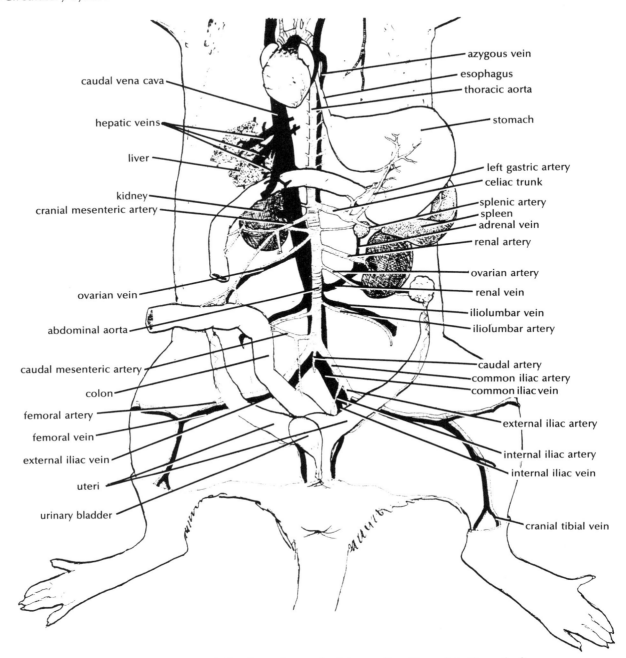

FIGURE 8.7. The caudal blood vessels. Much of the small and large intestines, the liver, the pancreas, the lungs, the thymus, and the diaphragm have been removed from this female specimen.

ventral and **dorsal tibial arteries**, which extend through the shank.

VEINS

All veins caudal to the heart drain deoxygenated blood into the **caudal vena cava**, the large vein that travels parallel to the aorta before uniting with the right atrium. Locate this vessel and trace its path through the thoracic and abdominal cavities (Fig. 8.7). Note the small, uninjected vessel parallel to it and the aorta against the dorsal thoracic wall. This is the **thoracic duct**, which is the main collecting trunk for the lymphatic system. It collects lymph from lymphatic vessels at its caudal end, called the **cysterna chyli**, and recirculates it into the bloodstream by uniting with the left subclavian vein. Now locate the following veins in your specimen (Figs. 8.6, 8.7, and 8.8):

Hepatic veins: On the medial surface of the liver, scrape away tissue to locate the points of origin of these large

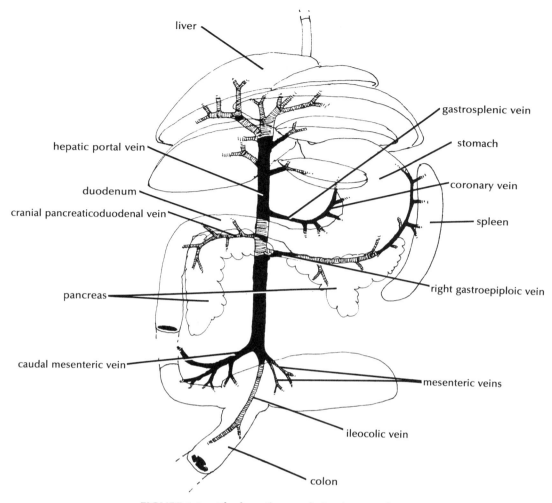

FIGURE 8.8. The hepatic portal circulatory scheme

veins. There are usually four in the rat, each of which drains blood into the caudal vena cava from a lobe of the liver.

Renal veins: Corresponding to the renal arteries, these paired vessels drain the kidneys and unite with the caudal vena cava directly. Normally, the right renal vein is cranial to the left renal vein. Merging with each renal vein is the **adrenal vein**, which drains the adrenal gland at the cranial border of each kidney.

Gonadal veins: Although the paired gonadal veins sometimes correspond to the gonadal arteries by draining directly into the caudal vena cava, the left gonadal vein usually drains into the left renal vein instead. As in the arteries, they are **spermatic veins** in the male and **ovarian veins** in the female.

Lumbar veins: segmental veins that parallel the lumbar arteries. They drain the muscles of the lumbar region (not shown).

Iliolumbar veins: paired vessels that drain the dorsal muscles of the lumbar region and enter the caudal vena cava near its origin.

Common iliac veins: the large, paired vessels that drain the pelvic and hind limb regions. Their convergence at the caudal end of the abdominal cavity forms the caudal vena cava. Each receives the following veins:

   External iliac veins: paired veins that continue from the **femoral vein** of each thigh into the pelvic cavity to unite with the common iliacs.

   Femoral veins: paired veins that parallel the artery of the same name through the medial side of the thigh. A superficial, medial vein that unites with each femoral vein is the **greater saphenous vein**. Each femoral vein originates at the level of the knee as a cranial continuation of the **popliteal vein**, which in turn arises from the **cranial** and **caudal tibial veins** of the shank.

   Internal iliac veins: paired veins that drain the pelvic region. They merge directly with the common iliacs at their origin.

Hepatic portal vein: a prominent, unpaired vein that receives blood from the G.I. tract organs, the pancreas, and the spleen and transports it to the liver for modification, detoxification, and nutrient storage. Trace the hepatic portal along its length from its cranial branches, which extend into the liver, to its caudal origin. It lies in the center of the abdominal cavity medial to the visceral organs that communicate with it. Identify the following vessels that drain into it (Fig. 8.8):

Gastrosplenic vein: a cranial vein that receives the **splenic vein** from the spleen and pancreas and the **coronary vein** from the lesser curvature of the stomach before it unites with the hepatic portal vein.

Cranial pancreaticoduodenal vein: unites with the hepatic portal vein near the union of the gastrosplenic vein. It drains the duodenum and the pancreas.

Right gastroepiploic vein: drains the greater curvature of the stomach. It unites with the hepatic portal at a point caudal to the union of the cranial pancreaticoduodenal vein.

Caudal mesenteric vein: unites with the hepatic portal vein caudal to the union of the right gastroepiploic vein. It drains the large intestine.

Mesenteric veins: numerous veins that originate from the small intestine. They unite with the hepatic portal along much of its caudal length.

Ileocolic vein: arises from the area surrounding the ileocecal valve. Its cranial end is continuous with the caudal end of the hepatic portal vein.

The Excretory & Reproductive Systems

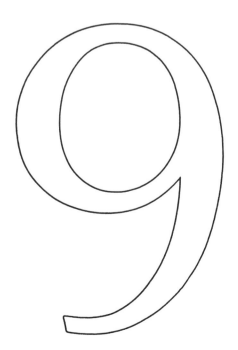

DISCUSSION OF THE EXCRETORY and reproductive systems is combined in this chapter because of the common location and close association of a number of their organs. As you will discover, several excretory organs lie in direct contact with reproductive structures, and in some cases the functions of both systems take place within one organ. The overall functions of the two systems are quite distinct, however.

The excretory system performs several important functions, each of which is primarily accomplished by the kidneys. These functions include the removal of nitrogen-containing waste products from the bloodstream and their transport out of the body in the form of liquid urine, maintenance of the osmotic balance of fluids and electrolytes, blood pressure regulation, and the control of red blood cell formation in red bone marrow.

In sharp contrast, the reproductive system has the sole function of procreation. The system is unique in that the male and female structures are very different from each other—a condition called **sexual dimorphism**. In the female, the reproductive organs are the site for gamete (**ova**) production, internal fertilization, internal incubation of the developing embryo, and the birth process (**parturition**). In the male, the reproductive organs are the site of gamete (**spermatozoa**) production and transfer.

THE EXCRETORY SYSTEM

The excretory, or urinary, organs include the kidneys, the ureters, the urinary bladder, and the urethra.

KIDNEYS

The paired kidneys are bean-shaped organs that lie partially embedded in fat against the dorsal body wall. To locate them, you must first push the visceral digestive structures completely to one side. Note that the kidneys are not suspended within the abdominal cavity by mesentery as are the visceral organs but lie dorsal to it. Only their ventral surface is covered with parietal peritoneum. This positioning is called **retroperitoneal**. Also note the position of the **adrenal glands**, small endocrine glands that lie cranial and slightly medial to each kidney.

The Excretory & Reproductive Systems

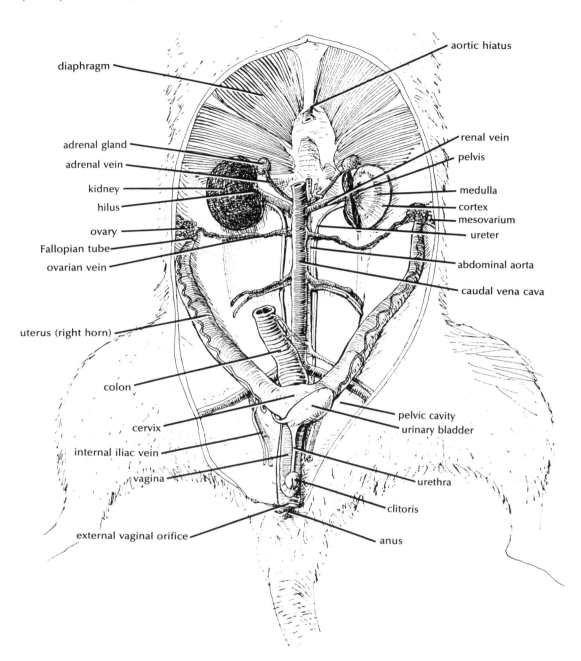

FIGURE 9.1. The excretory and reproductive systems of the rat, ventral view of the female. The digestive organs have been removed to expose the abdominal cavity and retroperitoneal structures.

Carefully remove the fat from one of the kidneys, section it across the frontal plane, and locate the following features (Fig. 9.1):

Hilus: a depression near the center of each kidney's concave (medial) border through which the renal artery and vein, lymphatic vessels, and nerves enter and exit.

Pelvis: a membrane-enclosed region in the center of each kidney. Within it is a cavity, called the **renal sinus**. Along its peripheral edges are cuplike extensions called **calyces**, which collect newly formed urine and channel it into the sinus. The renal pelvis is actually the cranial, expanded end of the ureter.

Cortex: the outer region of each kidney. Its smooth texture is due to the presence of many **renal corpuscles**, which perform the function of blood filtration.

Medulla: the inner region of each kidney. It contains from six to eighteen triangular, striated structures called **renal**

pyramids. The striations are caused by the presence of **renal tubules**, which transport filtrate from the renal corpuscles to the calyces. Along this route, water is reabsorbed from the tubules into the bloodstream. A renal corpuscle and its associated renal tubules constitute a single **nephron**. The bases of the pyramids face the cortex, and the apices, or **renal papillae**, point to the pelvis.

URETERS

The paired ureters are conducting tubes that transport urine from their union with the kidneys to the urinary bladder. They can be distinguished from other structures as small, white tubes that lie against the dorsal body wall.

URINARY BLADDER

Follow the path of the ureters to their caudal ends. Here they unite with a single saclike structure, the urinary bladder, on its dorsal side. Section the bladder frontally and note the wrinkled texture of the internal surface. These "wrinkles" are called **rugae**. Also note the points of entry of the ureters and the exit orifice of the urethra. These three openings form a triangle called the **trigone**.

URETHRA

The urethra is a small tube that transports urine from the urinary bladder to the exterior. It exhibits sexual dimorphism in that its length varies between males and females. In the female it is a short duct that serves only excretory functions, although it passes through a small structure that is homologous to the male penis, the **clitoris**, before opening to the exterior. In the male the urethra is comparatively long. It extends from the urinary bladder to the tip of the penis and carries both excretory and reproductive fluids.

THE REPRODUCTIVE SYSTEM

The following treatment of the reproductive system is divided into the male and female systems. On the basis of the descriptions that follow, determine the sex of your specimen and follow the appropriate dissection protocol. Then locate in your lab section a specimen of the opposite sex and also examine it.

MALE REPRODUCTIVE STRUCTURES

Locate the following components of the male reproductive system (Figs. 9.2, 9.3):

Scrotum. The scrotum is a large, exterior sac located ventral to the anus that contains the male gonads called the **testes**. During breeding periods it is greatly enlarged and clearly visible, but between periods it is retracted into the abdominal cavity. Cut open the scrotum and note its layers that are, from superficial to deep, the **integument**, or **skin**; the **scrotal fascia**, which contains the **cremaster muscle**; and an inner membrane called the **tunica vaginalis**.

Testes. The paired testes are the organs that produce the male gametes (**spermatozoa**) and the male sex hormone (**testosterone**). They lie within the tunica vaginalis layer of the scrotum and are externally lined with a white layer of connective tissue, the **tunica albuginea**. With a sharp scalpel, cut one testis in half and note the coiled tubules within. These are called **seminiferous tubules** and are the site where the spermatozoa are produced. Between the tubules are **interstitial cells**, which secrete testosterone.

Epididymus. The epididymus is a tightly coiled tubule that lies on the surface of each testis. Locate the epididymus on the unsectioned testis of your rat by carefully making a slit through and peeling away the layers of connective tissue that enclose both structures. The epididymus forms a band that curves around the testis, originating on the cranial surface (where it is called the **head**), and extending laterally (the **body**) and caudally (the **tail**). At its caudal end, it is continuous with the **ductus deferens**.

Ductus Deferens. The ductus, or vas, deferens is a narrow tube that extends from each epididymus to the urethra, where it joins the ductus deferens of the opposite side. Extend the midventral incision that was made earlier to the anus (but bypass the penis and scrotum) and peel back the wall of the abdomen. Trace one ductus deferens from its origin to its point of entry into the body cavity. Note that it passes through an opening called the **inguinal canal**. Once within the body cavity, it passes ventrally to the ureter and turns medial to a position behind (dorsal to) the urinary bladder. Near the neck of the bladder the ductus deferens passes with its counterpart from the opposite side through the large, lobulated **prostate glands** before uniting with the urethra.

Urethra. The male urethra extends from the urinary bladder to the tip of the penis. Due to its deep location in the floor of the abdomen and subsequent difficulty of dissection, rely on Figures 9.2 and 9.3 for your study. The male urethra is divided into three sections:

Prostatic urethra: the cranial portion, which extends from its emergence from the urinary bladder to the floor of the pelvic cavity. Near its union with the ductus deferens, it receives secretions from the **prostate glands**; the **vesicular glands**, which are large, paired structures that extend in a cranial direction to lie ventral to the ureters; the paired **coagulating glands**, which lie against the medial border of the vesicular glands; and the four small, oval **ampullary glands**, which surround the point

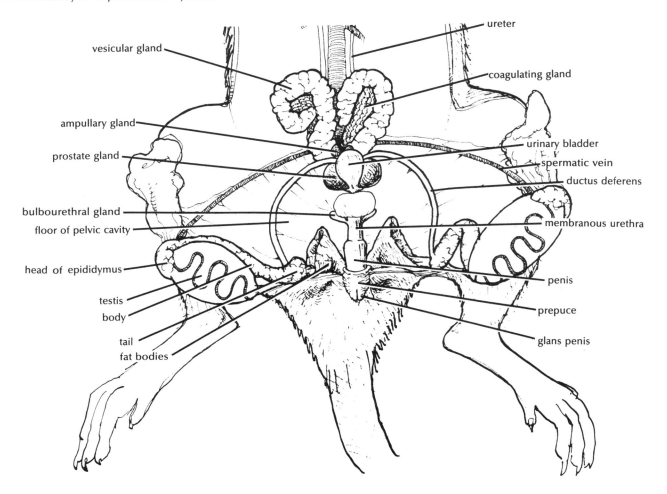

FIGURE 9.2. The male reproductive organs and associated structures. The scrotum has been incised to expose both testes in this semidiagrammatic view.

of union between the ductus deferens and the prostatic urethra.

Membranous urethra: the middle segment. It extends from the floor of the pelvic cavity to the base of the penis. The membranous urethra receives secretions from the paired **bulbourethral glands**, which unite with it via ducts near the base of the penis.

Spongy urethra: the caudal portion of the urethra. It extends through the length of the penis. Its opening to the exterior is called the **urogenital orifice**. Near this opening it receives ducts from a pair of glands located in the **prepuce** of the penis, called the **preputial glands**.

Penis. The penis encloses the spongy urethra. Its free end, called the **glans penis**, lies in a pocket of skin called the **prepuce**. Cut open the prepuce to reveal completely the glans. Note the preputial glands in this region. Now make a complete transverse section through the glans and note the three cylindrical bodies that make up the internal substance of the penis. These are the two dorsal **corpora cavernosa penis** and the single ventral **corpus spongiosum urethra**. Near the internal base of the penis, these bodies diverge from each other to form the **crura**, which are surrounded by muscle that help maintain the state of erection. Note that the spongy urethra passes through the corpus cavernosum urethra. Also note the presence of a small bone, the **baculum** or **os penis**, in the transverse section.

FEMALE REPRODUCTIVE STRUCTURES

Using Figures 9.1 and 9.4 as guides, locate the following components of the female reproductive system.

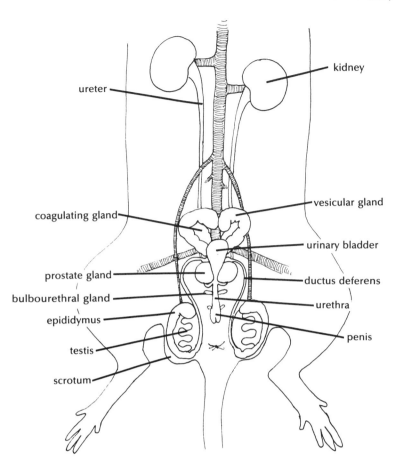

FIGURE 9.3. The male excretory and reproductive systems, schematic view

Ovaries. The paired ovaries are the small, oval organs that lie slightly caudal to the kidneys. They may be partially obscured by fat, and if so, this should be removed. Deep to the fat layer is a capsule that encloses each ovary, which should not be removed. This is the **ovarian bursa** and will be described below. The ovaries are the female gonads. As such, they produce the female gametes, or **ova**, and the hormone, **estrogen**. Note that each ovary is supported by a mesentery called the **mesovarium**. An additional mesentery, the **ovarian ligament**, attaches each ovary to the lateral edge of the uterus.

Fallopian Tubes. The paired Fallopian tubes are small, coiled ducts that transport the ova from the ovaries to the **uteri**. In rats, the cranial end of each tube forms a capsule that completely encloses the ovary. Called the **ovarian bursa**, it directs the movement of the ovum as it emerges out of the ovary during **ovulation** immediately into the Fallopian tube. In other mammals (including humans), the ovum must pass briefly through the peritoneal cavity before it enters the Fallopian tube. Note the mesentery that supports the Fallopian tubes, called the **mesosalpinx**. It also suspends the ovaries on the lateral edge of the uterus.

Uterus. The uterus is a paired structure in the rat. The right and left uteri, or **horns**, are caudal continuations of the Fallopian tubes but with thicker, more muscular walls. At their caudal end they unite to form the **vagina**, with a single, thickened **cervix** marking the junction. The arrangement of a single cervix serving two uteri is called **bipartite**. Cut open one of the uterine horns and examine the interior. If your specimen was pregnant, embryos should be visibly attached to its internal surface.

Vagina. The vagina is an unpaired organ located caudal to the uteri. It extends through the floor of the pelvic cavity between the urethra and the rectum to open to the exterior via the **external vaginal orifice**. Note that the urethra, which lies ventral, opens to the exterior separate of the vaginal orifice. Also recall that the **clitoris** is associated with the urethra in the rat, which is not the case in most other mammals.

The Excretory & Reproductive Systems

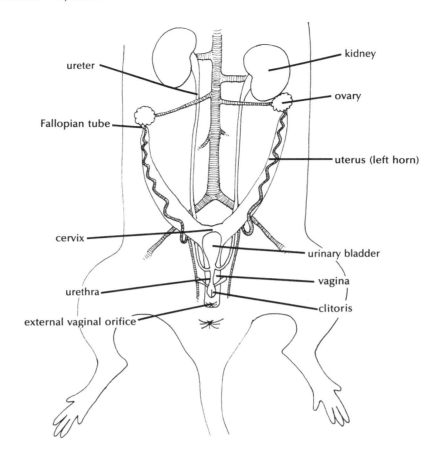

FIGURE 9.4. The female excretory and reproductive systems, schematic view

also by Bruce D. Wingerd and available from Johns Hopkins:

Human Anatomy and Rabbit Dissection
Rabbit Dissection Manual
Dogfish Dissection Manual
Frog Dissection Manual

The Johns Hopkins University Press

RAT DISSECTION MANUAL

This book was composed in Optima (Oracle) type by Brushwood Graphics, Inc., from a design by Susan P. Fillion. It was printed by Thomson-Shore, Inc., on 60-lb. Spring Forge Offset.

Microsoft® Word 2002 Exploring

Microsoft® Exploring Word 2002

Robert T. Grauer
University of Miami

Maryann Barber
University of Miami

PRENTICE HALL   *Upper Saddle River, New Jersey 07458*

Senior Acquisitions Editor: David Alexander
VP/Publisher: Natalie Anderson
Managing Editor: Melissa Whitaker
Assistant Editor: Kerri Limpert
Editorial Assistant: Maryann Broadnax
Technical Editor: Cecil Yarbrough
Media Project Manager: Cathleen Profitko
Marketing Assistant: Jason Smith
Production Manager: Gail Steier de Acevedo
Project Manager: Lynne Breitfeller
Production Editor: Greg Hubit
Associate Director, Manufacturing: Vincent Scelta
Manufacturing Buyer: Lynne Breitfeller
Design Manager: Pat Smythe
Interior Design: Jill Yutkowitz
Cover Design: Blair Brown
Cover Illustration: Marjorie Dressler
Composition: GTS
Printer/Binder: Banta Menasha

Microsoft and the Microsoft Office User Specialist logo are trademarks or registered trademarks of Microsoft Corporation in the United States and/or other countries. Prentice Hall is independent from Microsoft Corporation, and not affiliated with Microsoft in any manner. This publication may be used in assisting students to prepare for a Microsoft Office User Specialist Exam. Neither Microsoft Corporation, its designated review company, nor Prentice Hall warrants that use of this publication will ensure passing the relevant Exam.

Use of the Microsoft Office User Specialist Approved Courseware Logo on this product signifies that it has been independently reviewed and approved in complying with the following standards:

Acceptable coverage of all content related to the expert level Microsoft Office Exam entitled, "Word 2002," and sufficient performance-based exercises that relate closely to all required content, based on sampling of text.

Copyright © 2002 by Pearson Education, Inc., Upper Saddle River, New Jersey, 07458. All rights reserved. Printed in the United States of America. This publication is protected by copyright and permission should be obtained from the publisher prior to any prohibited reproduction, storage in a retrieval system, or transmission in any form or by any means, electronic, mechanical, photocopying, recording, or likewise. For information regarding permission(s), write to: Rights and Permissions Department.

10 9 8 7 6 5 4 3 2 1
ISBN 0-13-092444-X

To Marion —
my wife, my lover, and my best friend

Robert Grauer

To Frank —
for giving me the encouragement, love, and the space

Maryann Barber

APPROVED COURSEWARE

What does this logo mean?

It means this courseware has been approved by the Microsoft® Office User Specialist Program to be among the finest available for learning **Word 2002**. It also means that upon completion of this courseware, you may be prepared to become a Microsoft Office User Specialist.

What is a Microsoft Office User Specialist?

A Microsoft Office User Specialist is an individual who has certified his or her skills in one or more of the Microsoft Office desktop applications of Microsoft Word, Microsoft Excel, Microsoft PowerPoint®, Microsoft Outlook® or Microsoft Access, or in Microsoft Project. The Microsoft Office User Specialist Program typically offers certification exams at the "Core" and "Expert" skill levels.[*] The Microsoft Office User Specialist Program is the only Microsoft approved program in the world for certifying proficiency in Microsoft Office desktop applications and Microsoft Project. This certification can be a valuable asset in any job search or career advancement.

More Information:

To learn more about becoming a Microsoft Office User Specialist, visit www.mous.net

To purchase a Microsoft Office User Specialist certification exam, visit www.DesktopIQ.com

To learn about other Microsoft Office User Specialist approved courseware from Prentice Hall, visit http://www.prenhall.com/phit/mous_frame.html

---

[*]The availability of Microsoft Office User Specialist certification exams varies by application, application version and language. Visit www.mous.net for exam availability.

Microsoft, the Microsoft Office User Specialist Logo, PowerPoint and Outlook are either registered trademarks or trademarks of Microsoft Corporation in the United States and/or other countries.

CONTENTS

PREFACE XI

EXPLORING MICROSOFT® WORD 2002

1
MICROSOFT WORD 2002: WHAT WILL WORD PROCESSING DO FOR ME? 1

OBJECTIVES 1
OVERVIEW 1
The Basics of Word Processing 2
   The Insertion Point 2
   Word Wrap 2
   Toggle Switches 3
   Insert versus Overtype 4
   Deleting Text 5
Introduction to Microsoft Word 5
The File Menu 8
Learning by Doing 9
HANDS-ON EXERCISE 1: MY FIRST DOCUMENT 10
Troubleshooting 17
HANDS-ON EXERCISE 2: MODIFYING AN EXISTING DOCUMENT 19
The Spell Check 24
   AutoCorrect and AutoText 26
Thesaurus 27
Grammar Check 28
Save Command 30
   Backup Options 30
HANDS-ON EXERCISE 3: THE SPELL CHECK, THESAURUS, AND GRAMMAR CHECK 32
Summary 39
Key Terms 39
Multiple Choice 40
Practice with Word 42
On Your Own 47

2
GAINING PROFICIENCY: EDITING AND FORMATTING 49

OBJECTIVES 49
OVERVIEW 49
Select-Then-Do 50
Moving and Copying Text 50
Find, Replace, and Go To Commands 51
Scrolling 53
View Menu 55
HANDS-ON EXERCISE 1: EDITING A DOCUMENT 56
Typography 64
   Type Size 65
   Format Font Command 65
Page Setup Command 67
   Page Breaks 69
An Exercise in Design 69
HANDS-ON EXERCISE 2: CHARACTER FORMATTING 70
Paragraph Formatting 77
   Alignment 77
   Indents 77
   Tabs 80
   Hyphenation 81
   Line Spacing 81
Format Paragraph Command 81
   Borders and Shading 83
Column Formatting 84
HANDS-ON EXERCISE 3: PARAGRAPH FORMATTING 85
Summary 92
Key Terms 92
Multiple Choice 93
Practice with Word 95
On Your Own 101

3

ENHANCING A DOCUMENT: THE WEB AND OTHER RESOURCES 103

OBJECTIVES 103
OVERVIEW 103
Enhancing a Document 104
   The Media Gallery 105
   The Insert Symbol Command 106
   Microsoft WordArt 107
The Drawing Toolbar 108
HANDS-ON EXERCISE 1: CLIP ART AND WORDART 109
Microsoft Word and the Internet 116
   Copyright Protection 117
HANDS-ON EXERCISE 2: MICROSOFT WORD AND THE WEB 118
Wizards and Templates 127
Mail Merge 130
HANDS-ON EXERCISE 3: MAIL MERGE 133
Summary 142
Key Terms 142
Multiple Choice 143
Practice with Word 145
On Your Own 152

4

ADVANCED FEATURES: OUTLINES, TABLES, STYLES, AND SELECTIONS 153

OBJECTIVES 153
OVERVIEW 153
Bullets and List 154
Creating an Outline 154
HANDS-ON EXERCISE 1: BULLETS, LISTS, AND OUTLINES 156
Tables 162
HANDS-ON EXERCISE 2: TABLES 164
Styles 162
The Outline View 173
   The AutoFormat Command 174
HANDS-ON EXERCISE 3: STYLES 175
Working in Long Documents 183
   Page Numbers 183
   Headers and Footers 183
   Sections 184
   Table of Contents 184
   Creating an Index 185
   The Go To Command 185
HANDS-ON EXERCISE 4: WORKING IN LONG DOCUMENTS 187
Summary 198

Key Terms 198
Multiple Choice 199
Practice with Word 201
On Your Own 207

5

DESKTOP PUBLISHING: CREATING A NEWSLETTER AND OTHER DOCUMENTS 209

OBJECTIVES 209
OVERVIEW 209
The Newsletter 210
   Typography 212
   The Columns Command 213
HANDS-ON EXERCISE 1: NEWSPAPER COLUMNS 214
Elements of Graphic Design 222
   The Grid 222
   Emphasis 224
   Clip Art 224
The Drawing Toolbar 226
HANDS-ON EXERCISE 2: COMPLETE THE NEWSLETTER 227
Object Linking and Embedding 237
HANDS-ON EXERCISE 3: OBJECT LINKING AND EMBEDDING 239
Summary 247
Key Terms 247
Multiple Choice 248
Practice with Word 250
On Your Own 256

6

INTRODUCTION TO HTML: CREATING A HOME PAGE AND WEB SITE 257

OBJECTIVES 257
OVERVIEW 257
Introduction to HTML 258
   Microsoft Word 260
HANDS-ON EXERCISE 1: INTRODUCTION TO HTML 261
Publish Your Home Page 267
HANDS-ON EXERCISE 2: PUBLISHING YOUR HOME PAGE 268
Creating a Web Site 274
HANDS-ON EXERCISE 3: CREATING A WEB SITE 277
Summary 285
Key Terms 285
Multiple Choice 286
Practice with Word 288
On Your Own 294

7

THE EXPERT USER: WORKGROUPS, FORMS, MASTER DOCUMENTS, AND MACROS 295

OBJECTIVES 295
OVERVIEW 295
Workgroups 296
    Versions 297
Forms 298
HANDS-ON EXERCISE 1: WORKGROUPS AND FORMS 299
Table Math 307
HANDS-ON EXERCISE 2: TABLE MATH 309
Master Documents 316
HANDS-ON EXERCISE 3: MASTER DOCUMENTS 318
Introduction to Macros 324
    The Visual Basic Editor 325
HANDS-ON EXERCISE 4: INTRODUCTION TO MACROS 328
Summary 337
Key Terms 337
Multiple Choice 338
Practice with Word 340
On Your Own 345

APPENDIX A: TOOLBARS 347

A VBA PRIMER: EXTENDING MICROSOFT® OFFICE XP

OBJECTIVES 1
OVERVIEW 2
Introduction to VBA 2
The Msgbox Statement 3
The Input Function 4
    Declaring Variables 5
The VBA Editor 6
HANDS-ON EXERCISE 1: INTRODUCTION TO VBA 7
If...Then...Else Statement 16
Case Statement 18
Custom Toolbar 19
HANDS-ON EXERCISE 2: DECISION MAKING 20
For...Next Statement 28
Do Loops 29
Debugging 30
HANDS-ON EXERCISE 3: LOOPS AND DEBUGGING 32
Putting VBA to Work (Microsoft Excel) 41
HANDS-ON EXERCISE 4: EVENT-DRIVEN PROGRAMMING (MICROSOFT EXCEL) 43
Putting VBA to Work (Microsoft Access) 52
HANDS-ON EXERCISE 5: EVENT-DRIVEN PROGRAMMING (MICROSOFT ACCESS) 54
Summary 62
Key Terms 62
Multiple Choice 63
Practice with VBA 65

ESSENTIALS OF MICROSOFT® WINDOWS®

OBJECTIVES 1
OVERVIEW 1
The Desktop 2
The Common User Interface 5
    Moving and Sizing a Window 7
    Pull-Down Menus 7
    Dialog Boxes 8
The Mouse 10
    The Mouse versus the Keyboard 10
The Help Command 11
Formatting a Floppy Disk 11
HANDS-ON EXERCISE 1: WELCOME TO WINDOWS 2000 13
Files and Folders 20
My Computer 22
The Exploring Windows Practice Files 22
HANDS-ON EXERCISE 2: THE PRACTICE FILES VIA THE WEB 24
Windows Explorer 33
    Expanding and Collapsing a Drive or Folder 35

HANDS-ON EXERCISE 3: THE PRACTICE FILES VIA A LOCAL AREA NETWORK 36
The Basics of File Management 41
    Moving and Copying a File 41
    Deleting a File 41
    Renaming a File 42
    Backup 42
    Write Protection 42
    Our Next Exercise 42
HANDS-ON EXERCISE 4: FILE MANAGEMENT 43
Summary 51
Key Terms 52
Multiple Choice 52
Practice with Windows 54
On Your Own 62

INDEX

PREFACE

ABOUT THIS SERIES........

Continuing a tradition of excellence, Prentice Hall is proud to announce the latest update in Microsoft Office texts: the new Exploring Microsoft Office XP series by Robert T. Grauer and Maryann Barber.

The hands-on approach and conceptual framework of this comprehensive series helps students master all aspects of the Microsoft Office XP software, while providing the background necessary to transfer and use these skills in their personal and professional lives.

WHAT'S NEW IN THE EXPLORING OFFICE SERIES FOR XP

The entire Exploring Office series has been revised to include the new features found in the Office XP Suite, which contains Word 2002, Excel 2002, Access 2002, PowerPoint 2002, Publisher 2000, FrontPage 2002, and Outlook 2002.

In addition, this revision includes fully revised end-of-chapter material that provides an extensive review of concepts and techniques discussed in the chapter. Many of these exercises feature the World Wide Web and application integration.

Building on the success of the Web site provided for previous editions of this series, Exploring Office XP will introduce the MyPHLIP Companion Web site, a site customized for each instructor that includes on-line, interactive study guides, data file downloads, current news feeds, additional case studies and exercises, and other helpful information. Start out at www.prenhall.com/grauer to explore these resources!

Organization of the Exploring Office Series for XP

The new Exploring Microsoft Office XP series includes four combined Office XP texts from which to choose:

- ***Volume I*** is MOUS certified in each of the major applications in the Office suite (Word, Excel, Access, and PowerPoint). Three additional modules (Essential Computer Concepts, Essentials of Windows, and Essentials of the Internet) are also included.

- ***Volume II*** picks up where Volume I left off, covering the advanced topics for the individual applications. A VBA primer has been added.

- The ***Brief Microsoft Office XP*** edition provides less coverage of the individual applications than Volume I (a total of 8 chapters as opposed to 14). The supplementary modules (Windows, Internet, and Concepts) are not included.

- A new volume, ***Getting Started with Office XP***, contains the first chapter from each application (Word, Excel, Access, and PowerPoint), plus three additional modules: Essentials of Windows, Essentials of the Internet, and Essential Computer Concepts.

Individual texts for Word 2002, Excel 2002, Access 2002, and PowerPoint 2002 provide complete coverage of the application and are MOUS certified. For shorter courses, we have created brief versions of the Exploring texts that give students a four-chapter introduction to each application. Each of these volumes is MOUS certified at the Core level.

To complete the full coverage of this series, custom modules on Microsoft Outlook 2002, Microsoft FrontPage 2002, Microsoft Publisher 2002, and a generic introduction to Microsoft Windows are also available.

This book has been approved by Microsoft to be used in preparation for Microsoft Office User Specialist exams.

APPROVED COURSEWARE

The Microsoft Office User Specialist (MOUS) program is globally recognized as the standard for demonstrating desktop skills with the Microsoft Office suite of business productivity applications (Microsoft Word, Microsoft Excel, Microsoft PowerPoint, Microsoft Access, and Microsoft Outlook). With a MOUS certification, thousands of people have demonstrated increased productivity and have proved their ability to utilize the advanced functionality of these Microsoft applications.

By encouraging individuals to develop advanced skills with Microsoft's leading business desktop software, the MOUS program helps fill the demand for qualified, knowledgeable people in the modern workplace. At the same time, MOUS helps satisfy an organization's need for a qualitative assessment of employee skills.

Customize the Exploring Office Series with Prentice Hall's Right PHit Binding Program

The Exploring Office XP series is part of the Right PHit Custom Binding Program, enabling instructors to create their own texts by selecting modules from Office XP Volume I, Volume II, Outlook, FrontPage, and Publisher to suit the needs of a specific course. An instructor could, for example, create a custom text consisting of the core modules in Word and Excel, coupled with the brief modules for Access and PowerPoint, and a brief introduction to computer concepts.

Instructors can also take advantage of Prentice Hall's Value Pack program to shrinkwrap multiple texts together at substantial savings to the student. A value pack is ideal in courses that require complete coverage of multiple applications.

INSTRUCTOR AND STUDENT RESOURCES

The **Instructor's CD** that accompanies the Exploring Office series contains:

- Student data disks
- Solutions to all exercises and problems
- PowerPoint lectures
- Instructor's manuals in Word format enable the instructor to annotate portions of the instructor manual for distribution to the class
- A Windows-based test manager and the associated test bank in Word format

Prentice Hall's New MyPHLIP Companion Web site at www.prenhall.com/grauer offers current events, exercises, and downloadable supplements. This site also includes an on-line study guide containing true/false, multiple-choice, and essay questions.

WebCT www.prenhall.com/webct

GOLD LEVEL CUSTOMER SUPPORT available exclusively to adopters of Prentice Hall courses is provided free-of-charge upon adoption and provides you with priority assistance, training discounts, and dedicated technical support.

Blackboard www.prenhall.com/blackboard

Prentice Hall's abundant on-line content, combined with Blackboard's popular tools and interface, result in robust Web-based courses that are easy to implement, manage, and use—taking your courses to new heights in student interaction and learning.

CourseCompass www.coursecompass.com

CourseCompass is a dynamic, interactive on-line course management tool powered by Blackboard. This exciting product allows you to teach with marketing-leading Pearson Education content in an easy-to-use customizable format.

ABOUT THE BOOK

Exploring Microsoft Office XP assumes no prior knowledge of the operating system. A 64-page section introduces the reader to the Essentials of Windows and provides an overview of the operating system. Students are shown the necessary file-management operations to use Microsoft Office successfully.

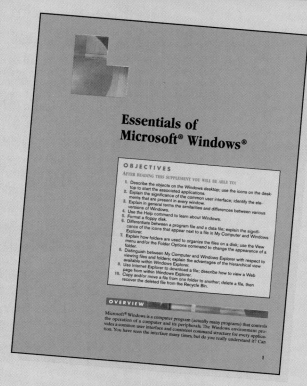

In-depth tutorials throughout all the Office XP applications enhance the conceptual introduction to each task and guide the student at the computer. Every step in every exercise has a full-color screen shot to illustrate the specific commands. Boxed tips provide alternative techniques and shortcuts and/or anticipate errors that students may make.

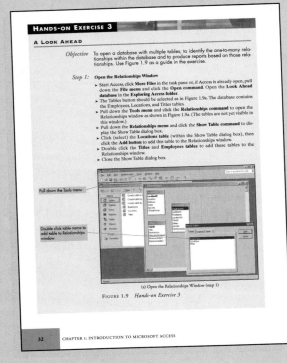

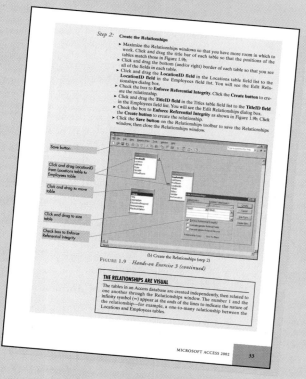

The authors have created an entirely new set of end-of-chapter exercises for every chapter in all of the applications. These new exercises have been written to provide the utmost in flexibility, variety, and difficulty.

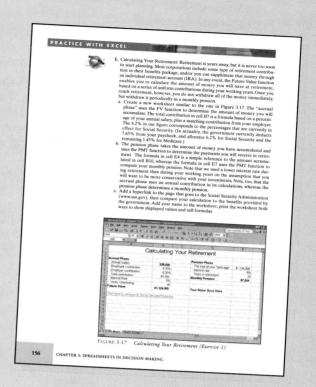

 Web-based Practice Exercises and On Your Own Exercises are marked by an icon in the margin and allow further exploration and practice via the World Wide Web.

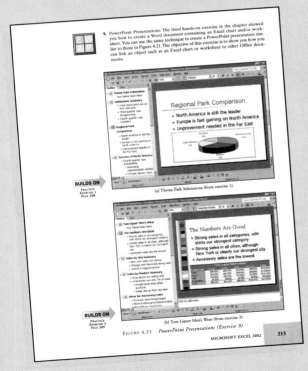

 Integration Exercises are marked by an icon in the margin. These exercises take advantage of the Microsoft Office Suite's power to use multiple applications in one document, spreadsheet, or presentation.

BUILDS ON ***Builds On Exercises*** require students to use selected application files as the starting point in later exercises, thereby introducing new information to students only as needed.

The end-of-chapter material includes multiple-choice questions for self-evaluation plus additional "on your own" exercises to encourage the reader to further explore the application.

ACKNOWLEDGMENTS

We want to thank the many individuals who have helped to bring this project to fruition. David Alexander, senior editor at Prentice Hall, has provided new leadership in extending the series to Office XP. Cathi Profitko did an absolutely incredible job on our Web site. Melissa Whitaker coordinated the myriad details of production and the certification process. Greg Christofferson was instrumental in the acquisition of supporting software. Lynne Breitfeller was the project manager and manufacturing buyer. Greg Hubit has been masterful as the external production editor for every book in the series. Cecil Yarbrough did an outstanding job in checking the manuscript for technical accuracy. Chuck Cox did his usual fine work as copyeditor. Kerri Limpert was the supplements editor. Cindy Stevens, Tom McKenzie, and Michael Olmstead wrote the instructor manuals. Patricia Smythe developed the innovative and attractive design. We also want to acknowledge our reviewers who, through their comments and constructive criticism, greatly improved the series.

Lynne Band, Middlesex Community College
Don Belle, Central Piedmont Community College
Stuart P. Brian, Holy Family College
Carl M. Briggs, Indiana University School of Business
Kimberly Chambers, Scottsdale Community College
Alok Charturvedi, Purdue University
Jerry Chin, Southwest Missouri State University
Dean Combellick, Scottsdale Community College
Cody Copeland, Johnson County Community College
Larry S. Corman, Fort Lewis College
Janis Cox, Tri-County Technical College
Martin Crossland, Southwest Missouri State University
Paul E. Daurelle, Western Piedmont Community College
Carolyn DiLeo, Westchester Community College
Judy Dolan, Palomar College
David Douglas, University of Arkansas
Carlotta Eaton, Radford University
Judith M. Fitspatrick, Gulf Coast Community College
James Franck, College of St. Scholastica
Raymond Frost, Central Connecticut State University
Midge Gerber, Southwestern Oklahoma State University
James Gips, Boston College
Vernon Griffin, Austin Community College
Ranette Halverson, Midwestern State University
Michael Hassett, Fort Hays State University
Mike Hearn, Community College of Philadelphia
Wanda D. Heller, Seminole Community College
Bonnie Homan, San Francisco State University
Ernie Ivey, Polk Community College
Mike Kelly, Community College of Rhode Island
Jane King, Everett Community College

Rose M. Laird, Northern Virginia Community College
John Lesson, University of Central Florida
David B. Meinert, Southwest Missouri State University
Alan Moltz, Naugatuck Valley Technical Community College
Kim Montney, Kellogg Community College
Bill Morse, DeVry Institute of Technology
Kevin Pauli, University of Nebraska
Mary McKenry Percival, University of Miami
Delores Pusins, Hillsborough Community College
Gale E. Rand, College Misericordia
Judith Rice, Santa Fe Community College
David Rinehard, Lansing Community College
Marilyn Salas, Scottsdale Community College
John Shepherd, Duquesne University
Barbara Sherman, Buffalo State College
Robert Spear, Prince George's Community College
Michael Stewardson, San Jacinto College—North
Helen Stoloff, Hudson Valley Community College
Margaret Thomas, Ohio University
Mike Thomas, Indiana University School of Business
Suzanne Tomlinson, Iowa State University
Karen Tracey, Central Connecticut State University
Antonio Vargas, El Paso Community College
Sally Visci, Lorain County Community College
David Weiner, University of San Francisco
Connie Wells, Georgia State University
Wallace John Whistance-Smith, Ryerson Polytechnic University
Jack Zeller, Kirkwood Community College

A final word of thanks to the unnamed students at the University of Miami, who make it all worthwhile. Most of all, thanks to you, our readers, for choosing this book. Please feel free to contact us with any comments and suggestions.

Robert T. Grauer
rgrauer@miami.edu
www.bus.miami.edu/~rgrauer
www.prenhall.com/grauer

Maryann Barber
mbarber@miami.edu
www.bus.miami.edu/~mbarber

CHAPTER 1

Microsoft® Word 2002: What Will Word Processing Do for Me?

OBJECTIVES

AFTER READING THIS CHAPTER YOU WILL BE ABLE TO:

1. Define word wrap; differentiate between a hard and a soft return.
2. Distinguish between the insert and overtype modes.
3. Describe the elements on the Microsoft Word screen.
4. Create, save, retrieve, edit, and print a simple document.
5. Check a document for spelling; describe the function of the custom dictionary.
6. Describe the AutoCorrect and AutoText features; explain how either feature can be used to create a personal shorthand.
7. Use the thesaurus to look up synonyms and antonyms.
8. Explain the objectives and limitations of the grammar check; customize the grammar check for business or casual writing.
9. Differentiate between the Save and Save As commands; describe various backup options that can be selected.

OVERVIEW

Have you ever produced what you thought was the perfect term paper only to discover that you omitted a sentence or misspelled a word, or that the paper was three pages too short or one page too long? Wouldn't it be nice to make the necessary changes, and then be able to reprint the entire paper with the touch of a key? Welcome to the world of word processing, where you are no longer stuck with having to retype anything. Instead, you retrieve your work from disk, display it on the monitor and revise it as necessary, then print it at any time, in draft or final form.

This chapter provides a broad-based introduction to word processing in general and Microsoft Word in particular. We begin by presenting (or perhaps reviewing) the essential concepts of a word processor, then show you how these concepts are implemented in Word. We show you how to create a document, how to save it on disk, then retrieve the document you just created. We introduce you to the spell check and thesaurus, two essential tools in any word processor. We also present the grammar check as a convenient way of finding a variety of errors but remind you there is no substitute for carefully proofreading the final document.

THE BASICS OF WORD PROCESSING

All word processors adhere to certain basic concepts that must be understood if you are to use the programs effectively. The next several pages introduce ideas that are applicable to any word processor (and which you may already know). We follow the conceptual material with a hands-on exercise that enables you to apply what you have learned.

The Insertion Point

The ***insertion point*** is a flashing vertical line that marks the place where text will be entered. The insertion point is always at the beginning of a new document, but it can be moved anywhere within an existing document. If, for example, you wanted to add text to the end of a document, you would move the insertion point to the end of the document, then begin typing.

Word Wrap

A newcomer to word processing has one major transition to make from a typewriter, and it is an absolutely critical adjustment. Whereas a typist returns the carriage at the end of every line, just the opposite is true of a word processor. One types continually *without* pressing the enter key at the end of a line because the word processor automatically wraps text from one line to the next. This concept is known as ***word wrap*** and is illustrated in Figure 1.1.

The word *primitive* does not fit on the current line in Figure 1.1a, and is automatically shifted to the next line, *without* the user having to press the enter key. The user continues to enter the document, with additional words being wrapped to subsequent lines as necessary. The only time you use the enter key is at the end of a paragraph, or when you want the insertion point to move to the next line and the end of the current line doesn't reach the right margin.

Word wrap is closely associated with another concept, that of hard and soft returns. A ***hard return*** is created by the user when he or she presses the enter key at the end of a paragraph; a ***soft return*** is created by the word processor as it wraps text from one line to the next. The locations of the soft returns change automatically as a document is edited (e.g., as text is inserted or deleted, or as margins or fonts are changed). The locations of the hard returns can be changed only by the user, who must intentionally insert or delete each hard return.

There are two hard returns in Figure 1.1b, one at the end of each paragraph. There are also six soft returns in the first paragraph (one at the end of every line except the last) and three soft returns in the second paragraph. Now suppose the margins in the document are made smaller (that is, the line is made longer) as shown in Figure 1.1c. The number of soft returns drops to four and two (in the first and second paragraphs, respectively) as more text fits on a line and fewer lines are needed. The revised document still contains the two original hard returns, one at the end of each paragraph.

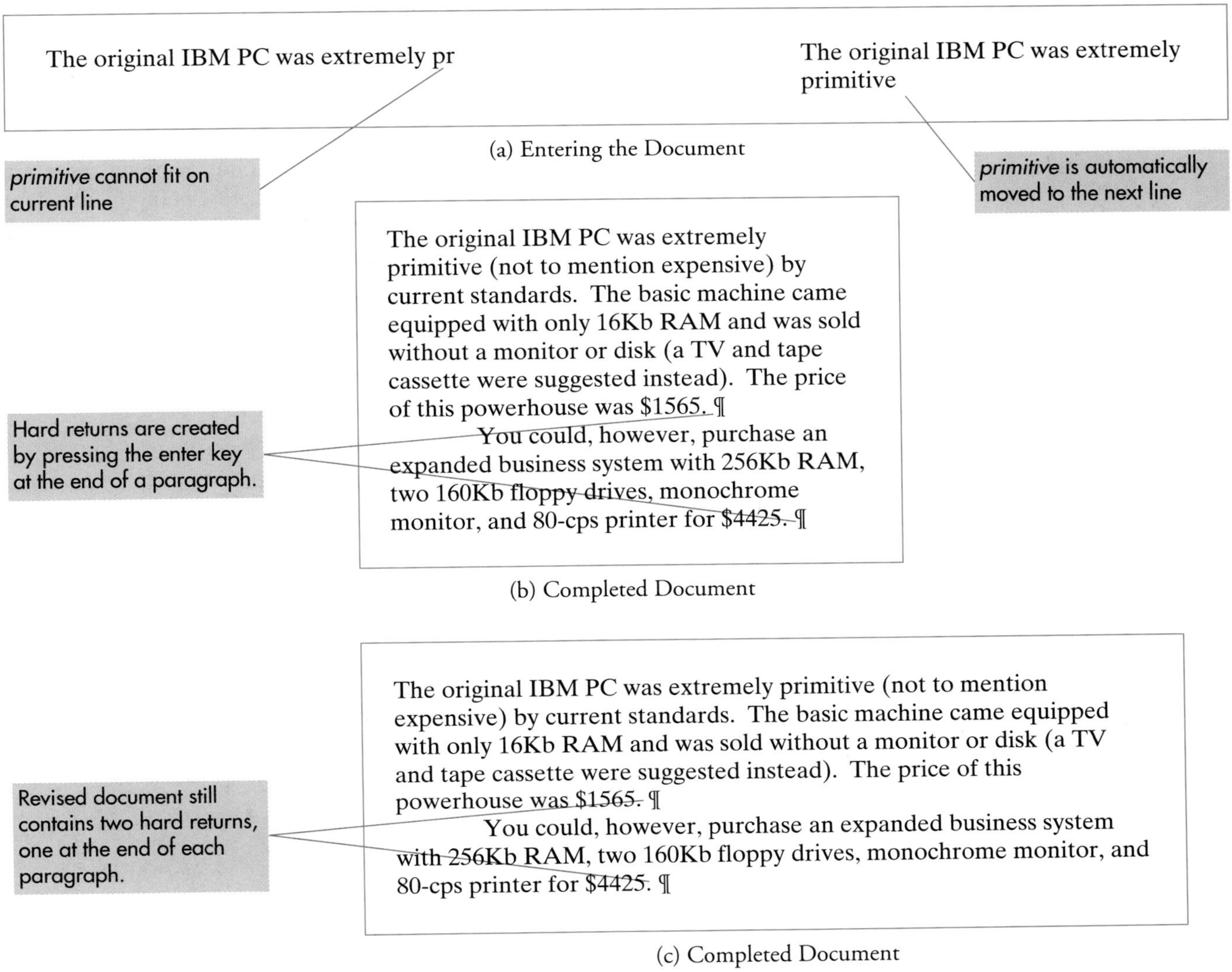

The original IBM PC was extremely pr

The original IBM PC was extremely primitive

primitive cannot fit on current line

(a) Entering the Document

primitive is automatically moved to the next line

The original IBM PC was extremely primitive (not to mention expensive) by current standards. The basic machine came equipped with only 16Kb RAM and was sold without a monitor or disk (a TV and tape cassette were suggested instead). The price of this powerhouse was $1565. ¶
You could, however, purchase an expanded business system with 256Kb RAM, two 160Kb floppy drives, monochrome monitor, and 80-cps printer for $4425. ¶

Hard returns are created by pressing the enter key at the end of a paragraph.

(b) Completed Document

The original IBM PC was extremely primitive (not to mention expensive) by current standards. The basic machine came equipped with only 16Kb RAM and was sold without a monitor or disk (a TV and tape cassette were suggested instead). The price of this powerhouse was $1565. ¶
You could, however, purchase an expanded business system with 256Kb RAM, two 160Kb floppy drives, monochrome monitor, and 80-cps printer for $4425. ¶

Revised document still contains two hard returns, one at the end of each paragraph.

(c) Completed Document

FIGURE 1.1 *Word Wrap*

Toggle Switches

Suppose you sat down at the keyboard and typed an entire sentence without pressing the Shift key; the sentence would be in all lowercase letters. Then you pressed the Caps Lock key and retyped the sentence, again without pressing the Shift key. This time the sentence would be in all uppercase letters. You could repeat the process as often as you like. Each time you pressed the Caps Lock key, the sentence would switch from lowercase to uppercase and vice versa.

The point of this exercise is to introduce the concept of a ***toggle switch***, a device that causes the computer to alternate between two states. The Caps Lock key is an example of a toggle switch. Each time you press it, newly typed text will change from uppercase to lowercase and back again. We will see several other examples of toggle switches as we proceed in our discussion of word processing.

Insert versus Overtype

Microsoft Word is always in one of two modes, **insert** or **overtype**, and uses a toggle switch (the Ins key) to alternate between the two. Press the Ins key once and you switch from insert to overtype. Press the Ins key a second time and you go from overtype back to insert. Text that is entered into a document during the insert mode moves existing text to the right to accommodate the characters being added. Text entered from the overtype mode replaces (overtypes) existing text. Regardless of which mode you are in, text is always entered or replaced immediately to the right of the insertion point.

The insert mode is best when you enter text for the first time, but either mode can be used to make corrections. The insert mode is the better choice when the correction requires you to add new text; the overtype mode is easier when you are substituting one or more character(s) for another. The difference is illustrated in Figure 1.2.

Figure 1.2a displays the text as it was originally entered, with two misspellings. The letters *se* have been omitted from the word *insert,* and an *x* has been erroneously typed instead of an *r* in the word *overtype*. The insert mode is used in Figure 1.2b to add the missing letters, which in turn moves the rest of the line to the right. The overtype mode is used in Figure 1.2c to replace the *x* with an *r*.

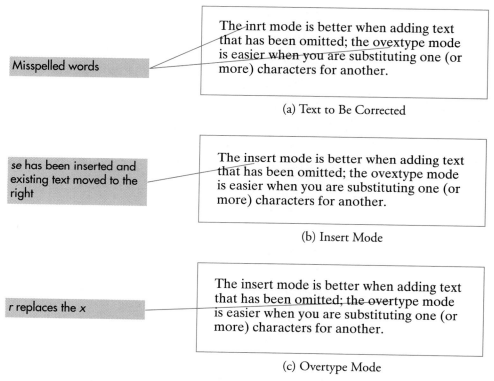

FIGURE 1.2   *Insert and Overtype Modes*

Deleting Text

The Backspace and Del keys delete one character immediately to the left or right of the insertion point, respectively. The choice between them depends on when you need to erase a character(s). The Backspace key is easier if you want to delete a character immediately after typing it. The Del key is preferable during subsequent editing.

You can delete several characters at one time by selecting (dragging the mouse over) the characters to be deleted, then pressing the Del key. And finally, you can delete and replace text in one operation by selecting the text to be replaced and then typing the new text in its place.

LEARN TO TYPE

The ultimate limitation of any word processor is the speed at which you enter data; hence the ability to type quickly is invaluable. Learning how to type is easy, especially with the availability of computer-based typing programs. As little as a half hour a day for a couple of weeks will have you up to speed, and if you do any significant amount of writing at all, the investment will pay off many times.

INTRODUCTION TO MICROSOFT WORD

We used Microsoft Word to write this book, as can be inferred from the screen in Figure 1.3. Your screen will be different from ours in many ways. You will not have the same document nor is it likely that you will customize Word in exactly the same way. You should, however, be able to recognize the basic elements that are found in the Microsoft Word window that is open on the desktop.

There are actually two open windows in Figure 1.3—an application window for Microsoft Word and a document window for the specific document on which you are working. The application window has its own Minimize, Maximize (or Restore) and Close buttons. The document window has only a Close button. There is, however, only one title bar that appears at the top of the application window and it reflects the application (Microsoft Word) as well as the document name (Word Chapter 1). A menu bar appears immediately below the title bar. Vertical and horizontal scroll bars appear at the right and bottom of the document window. The Windows taskbar appears at the bottom of the screen and shows the open applications.

Microsoft Word is also part of the Microsoft Office suite of applications, and thus shares additional features with Excel, Access, and PowerPoint, that are also part of the Office suite. *Toolbars* provide immediate access to common commands and appear immediately below the menu bar. The toolbars can be displayed or hidden using the Toolbars command in the *View menu*.

The *Standard toolbar* contains buttons corresponding to the most basic commands in Word—for example, opening a file or printing a document. The icon on the button is intended to be indicative of its function (e.g., a printer to indicate the Print command). You can also point to the button to display a *ScreenTip* showing the name of the button. The *Formatting toolbar* appears under the Standard toolbar and provides access to such common formatting operations as boldface, italics, or underlining.

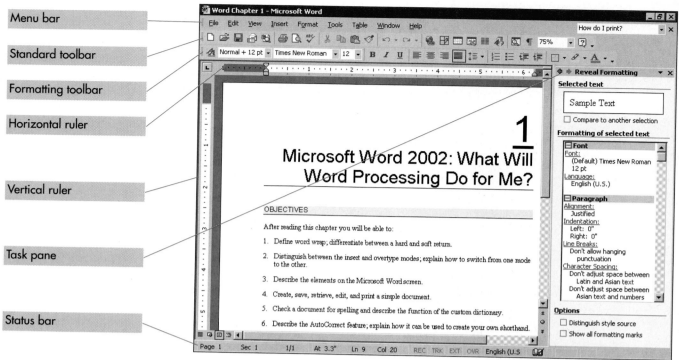

FIGURE 1.3 *Microsoft Word*

The toolbars may appear overwhelming at first, but there is absolutely no need to memorize what the individual buttons do. That will come with time. We suggest, however, that you will have a better appreciation for the various buttons if you consider them in groups, according to their general function, as shown in Figure 1.4a. Note, too, that many of the commands in the pull-down menus are displayed with an image that corresponds to a button on a toolbar.

The ***horizontal ruler*** is displayed underneath the toolbars and enables you to change margins, tabs, and/or indents for all or part of a document. A ***vertical ruler*** shows the vertical position of text on the page and can be used to change the top or bottom margins.

The ***status bar*** at the bottom of the document window displays the location of the insertion point (or information about the command being executed). The status bar also shows the status (settings) of various indicators—for example, OVR to show that Word is in the overtype, as opposed to the insert, mode.

THE TASK PANE

All applications in Office XP provide access to a ***task pane*** that facilitates the execution of subsequent commands. (Microsoft refers to the suite as Office XP, but designates the individual applications as version 2002.) The task pane serves many functions. It can be used to display the formatting properties of selected text, open an existing document, or search for appropriate clip art. The task pane will open automatically in response to certain commands. It can also be toggled open or closed through the Task pane command in the View menu. The task pane is discussed in more detail throughout the chapter.

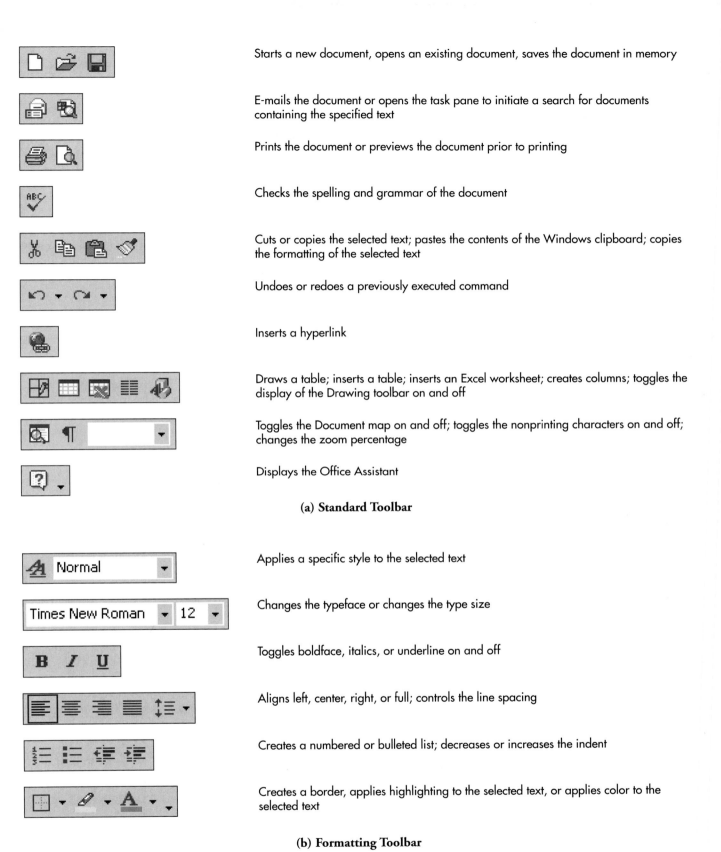

FIGURE 1.4 *Toolbars*

THE FILE MENU

The ***File Menu*** is a critically important menu in virtually every Windows application. It contains the Save and Open commands to save a document on disk, then subsequently retrieve (open) that document at a later time. The File Menu also contains the ***Print command*** to print a document, the ***Close command*** to close the current document but continue working in the application, and the ***Exit command*** to quit the application altogether.

The ***Save command*** copies the document that you are working on (i.e., the document that is currently in memory) to disk. The command functions differently the first time it is executed for a new document, in that it displays the Save As dialog box as shown in Figure 1.5a. The dialog box requires you to specify the name of the document, the drive (and an optional folder) in which the document is stored, and its file type. All subsequent executions of the command will save the document under the assigned name, each time replacing the previously saved version with the new version.

The ***file name*** (e.g., My First Document) can contain up to 255 characters including spaces, commas, and/or periods. (Periods are discouraged, however, since they are too easily confused with DOS extensions.) The Save In list box is used to select the drive (which is not visible in Figure 1.5a) and the optional folder (e.g., Exploring Word). The ***Places bar*** provides a shortcut to any of its folders without having to search through the Save In list box. Click the Desktop icon, for example, and the file is saved automatically on the Windows desktop. The ***file type*** defaults to a Word 2002 document. You can, however, choose a different format such as Word 95 to maintain compatibility with earlier versions of Microsoft Word. You can also save any Word document as a Web page (or HTML document).

The ***Open command*** is the opposite of the Save command as it brings a copy of an existing document into memory, enabling you to work with that document. The Open command displays the Open dialog box in which you specify the file name, the drive (and optionally the folder) that contains the file, and the file type. Microsoft Word will then list all files of that type on the designated drive (and folder), enabling you to open the file you want. The Save and Open commands work in conjunction with one another. The Save As dialog box in Figure 1.5a, for example, saves the file My First Document in the Exploring Word folder. The Open dialog box in Figure 1.5b loads that file into memory so that you can work with the file, after which you can save the revised file for use at a later time.

The toolbars in the Save As and Open dialog boxes have several buttons in common that facilitate the execution of either command. The Views button lets you display the files in either dialog box in one of four different views. The Details view (in Figure 1.5a) shows the file size as well as the date and time a file was last modified. The Preview view (in Figure 1.5b) shows the beginning of a document, without having to open the document. The List view displays only the file names, and thus lets you see more files at one time. The Properties view shows information about the document, including the date of creation and number of revisions.

SORT BY NAME, DATE, OR FILE SIZE

The files in the Save As and Open dialog boxes can be displayed in ascending or descending sequence by name, date modified, or size. Change to the Details view, then click the heading of the desired column; for example, click the Modified column to list the files according to the date they were last changed. Click the column heading a second time to reverse the sequence.

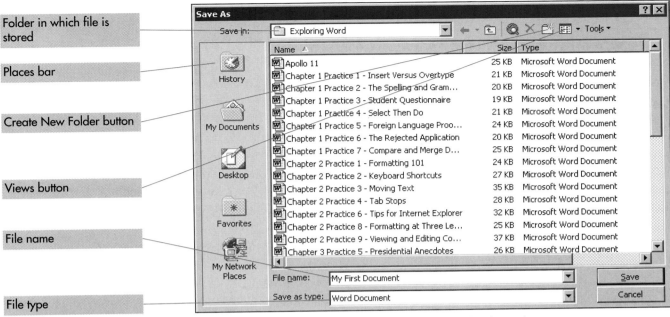

(a) Save As Dialog Box (details view)

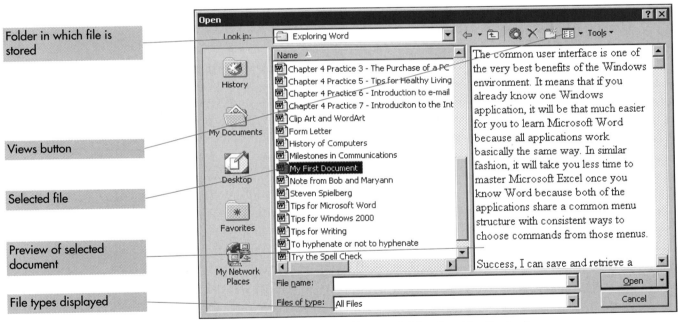

(b) Open Dialog Box (details view)

FIGURE 1.5 *The Save and Open Commands*

LEARNING BY DOING

Every chapter contains a series of hands-on exercises that enable you to apply what you learn at the computer. The exercises in this chapter are linked to one another in that you create a simple document in exercise one, then open and edit that document in exercise two. The ability to save and open a document is critical, and you do not want to spend an inordinate amount of time entering text unless you are confident in your ability to retrieve it later.

HANDS-ON EXERCISE 1

MY FIRST DOCUMENT

Objective   To start Microsoft Word in order to create, save, and print a simple document; to execute commands via the toolbar or from pull-down menus. Use Figure 1.6 as a guide in doing the exercise.

Step 1:   **The Windows Desktop**

> ➤ Turn on the computer and all of its peripherals. The floppy drive should be empty prior to starting your machine. This ensures that the system starts from the hard disk, which contains the Windows files, as opposed to a floppy disk, which does not.
> ➤ Your system will take a minute or so to get started, after which you should see the Windows desktop in Figure 1.6a. Do not be concerned if the appearance of your desktop is different from ours.
> ➤ You may see a Welcome to Windows dialog box with command buttons to take a tour of the operating system. If so, click the appropriate button(s) or close the dialog box.
> ➤ You should be familiar with basic file management and very comfortable moving and copying files from one folder to another. If not, you may want to review this material.

(a) The Windows Desktop (step 1)

FIGURE 1.6   *Hands-on Exercise 1*

Step 2: **Obtain the Practice Files**

➤ We have created a series of practice files (also called a "data disk") for you to use throughout the text. Your instructor will make these files available to you in a variety of ways:
- The files may be on a network drive, in which case you use Windows Explorer to copy the files from the network to a floppy disk.
- There may be an actual "data disk" that you are to check out from the lab in order to use the Copy Disk command to duplicate the disk.

➤ You can also download the files from our Web site provided you have an Internet connection. Start Internet Explorer, then go to the Exploring Windows home page at **www.prenhall.com/grauer**.
- Click the book for **Office XP**, which takes you to the Office XP home page. Click the **Student Resources tab** (at the top of the window) to go to the Student Resources page as shown in Figure 1.6b.
- Click the link to **Student Data Disk** (in the left frame), then scroll down the page until you can select Word 2002. Click the link to download the student data disk.
- You will see the File Download dialog box asking what you want to do. The option button to save this program to disk is selected. Click **OK**. The Save As dialog box appears.
- Click the ***down arrow*** in the Save In list box to enter the drive and folder where you want to save the file. It's best to save the file to the Windows desktop or to a temporary folder on drive C.
- Double click the file after it has been downloaded to your PC, then follow the onscreen instructions.

➤ Check with your instructor for additional information.

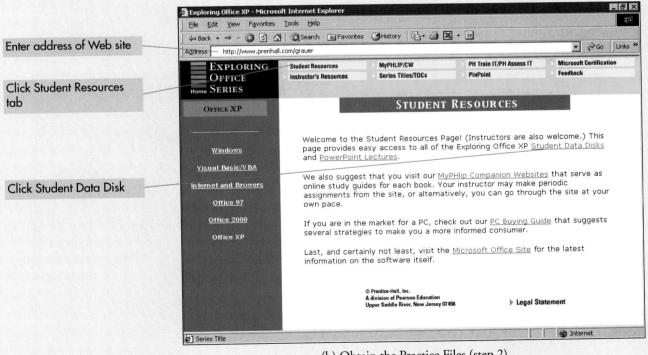

(b) Obtain the Practice Files (step 2)

FIGURE 1.6 *Hands-on Exercise 1 (continued)*

MICROSOFT WORD 2002   11

Step 3: **Start Microsoft Word**

➤ Click the **Start button** to display the Start menu. Click (or point to) the Programs menu, then click **Microsoft Word 2002** to start the program.
➤ You should see a blank document within the Word application window. (Click the **New Blank document** button on the Standard toolbar if you do not see a document.) Close the task pane if it is open.
➤ Click and drag the Office Assistant out of the way. (The Assistant is illustrated in step six of this exercise.)
➤ Do not be concerned if your screen is different from ours as we include a troubleshooting section immediately following the exercise.

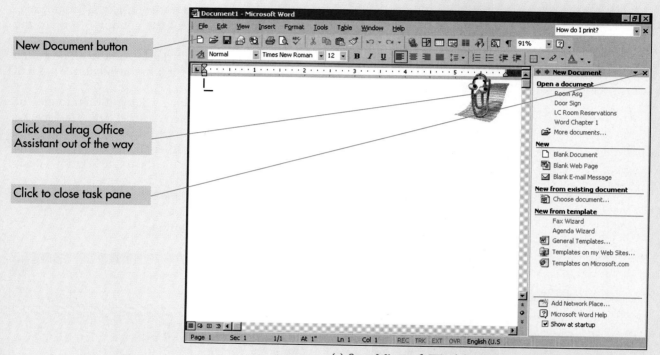

(c) Start Microsoft Word (step 3)

FIGURE 1.6 *Hands-on Exercise 1 (continued)*

ASK A QUESTION

Click in the "Ask a Question" list box that appears at the right of the document window, enter the text of a question such as "How do I save a document?", press enter, and Word returns a list of potential help topics. Click any topic that appears promising to open the Help window with detailed information. You can ask multiple questions during a Word session, then click the down arrow in the list box to return to an earlier question, which will return you to the help topics. You can also access help through the Help menu.

Step 4: **Create the Document**

- Create the document in Figure 1.6d. Type just as you would on a typewriter with one exception; do *not* press the enter key at the end of a line because Word will automatically wrap text from one line to the next.
- Press the **enter key** at the end of the paragraph.
- You may see a red or green wavy line to indicate spelling or grammatical errors, respectively. Both features are discussed later in the chapter.
- Point to the red wavy line (if any), click the **right mouse button** to display a list of suggested corrections, then click (select) the appropriate substitution.
- Ignore the green wavy line (if any).

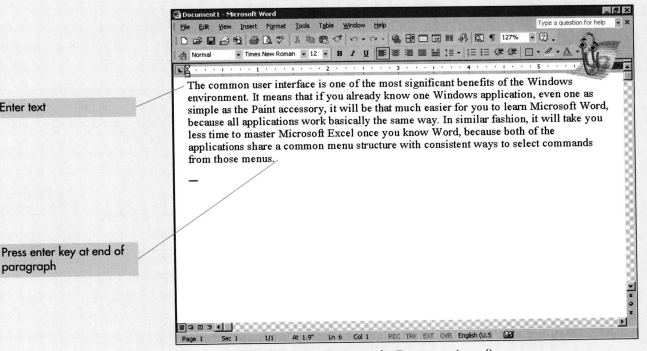

(d) Create the Document (step 4)

FIGURE 1.6 *Hands-on Exercise 1 (continued)*

SEPARATE THE TOOLBARS

You may see the Standard and Formatting toolbars displayed on one row to save space within the application window. If so, we suggest that you separate the toolbars, so that you see all of the buttons on each. Click the Toolbar Options down arrow that appears at the end of any visible toolbar to display toolbar options, then click the option to show the buttons on two rows. Click the down arrow a second time to show the buttons on one row if you want to return to the other configuration.

Step 5: **Save the Document**

> ➤ Pull down the **File menu** and click **Save** (or click the **Save button** on the Standard toolbar). You should see the Save As dialog box in Figure 1.6e.
> ➤ If necessary, click the **drop-down arrow** on the View button and select the **Details view**, so that the display on your monitor matches our figure. (The title bar shows Document1 because the file has not yet been saved.)
> ➤ To save the file:
> - Click the **drop-down arrow** on the Save In list box.
> - Click the appropriate drive, e.g., drive C or drive A, depending on whether or not you installed the data disk on your hard drive.
> - Double click the **Exploring Word folder**, to make it the active folder (the folder in which you will save the document).
> - Click and drag over the default entry in the File name text box. Type **My First Document** as the name of your document. (A DOC extension will be added automatically when the file is saved to indicate that this is a Word document.)
> - Click **Save** or press the **enter key**. The title bar changes to reflect the new document name (My First Document).
> ➤ Add your name at the end of the document, then click the **Save button** on the Standard toolbar to save the document with the revision.
> ➤ This time the Save As dialog box does not appear, since Word already knows the name of the document.

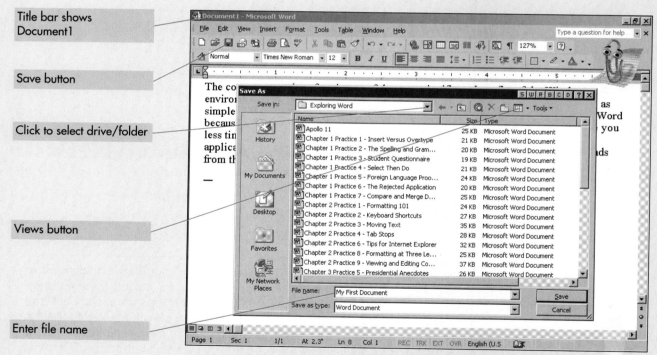

(e) Save the Document (step 5)

FIGURE 1.6 *Hands-on Exercise 1 (continued)*

Step 6: **The Office Assistant**

➤ If necessary, pull down the **Help menu** and click the command to **Show the Office Assistant**. You may see a different character than the one we have selected.
➤ Click the **Assistant**, enter the question, **How do I print?** as shown in Figure 1.6f, then click the **Search button** to look for the answer. The size of the Assistant's balloon expands as the Assistant suggests several topics that may be appropriate.
➤ Click the topic, **Print a document**, which in turn displays a Help window that contains links to various topics, each with detailed information. Click the **Office Assistant** to hide the balloon (or drag the Assistant out of the way).
➤ Click any of the links in the Help window to read the information. You can print the contents of any topic by clicking the **Print button** in the Help window. Close the Help window when you are finished.

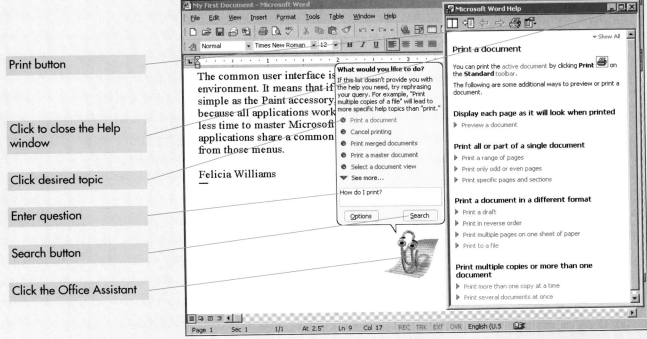

(f) The Office Assistant (step 6)

FIGURE 1.6   *Hands-on Exercise 1 (continued)*

TIP OF THE DAY

You can set the Office Assistant to greet you with a "tip of the day" each time you start Word. Click the Microsoft Word Help button (or press the F1 key) to display the Assistant, then click the Options button to display the Office Assistant dialog box. Click the Options tab, then check the Show the Tip of the Day at Startup box and click OK. The next time you start Microsoft Word, you will be greeted by the Assistant, who will offer you the tip of the day.

Step 7: **Print the Document**

- ➤ You can print the document in one of two ways:
  - Pull down the **File menu**. Click **Print** to display the dialog box of Figure 1.6g. Click the **OK command button** to print the document.
  - Click the **Print button** on the Standard toolbar to print the document immediately without displaying the Print dialog box.
- ➤ Submit this document to your instructor.

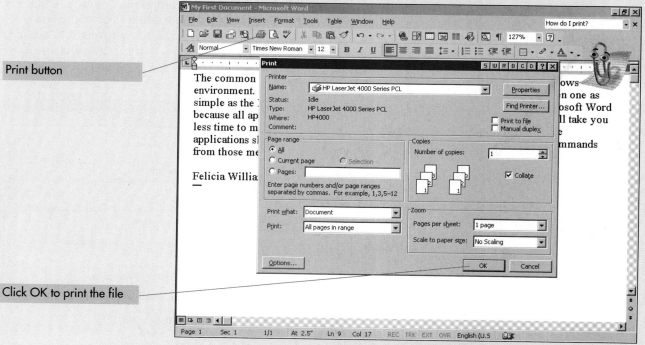

(g) Print the Document (step 7)

FIGURE 1.6 *Hands-on Exercise 1 (continued)*

ABOUT MICROSOFT WORD

Pull down the Help menu and click About Microsoft Word to display the specific version number and other licensing information, including the product ID. This help screen also contains two very useful command buttons, System Information and Technical Support. The first button displays information about the hardware installed on your system, including the amount of memory and available space on the hard drive. The Technical Support button describes various ways to obtain technical assistance.

Step 8: **Close the Document**

- ➤ Pull down the **File menu**. Click **Close** to close this document but remain in Word. (If you don't see the Close command, click the **double arrow** at the bottom of the menu.) Click **Yes** if prompted to save the document.
- ➤ Pull down the **File menu** a second time. Click **Exit** to close Word if you do not want to continue with the next exercise at this time.

TROUBLESHOOTING

We trust that you completed the hands-on exercise without difficulty, and that you were able to create, save, and print the document in the exercise. There is, however, considerable flexibility in the way you do the exercise in that you can display different toolbars and menus, and/or execute commands in a variety of ways. This section describes various ways in which you can customize Microsoft Word, and in so doing, will help you to troubleshoot future exercises.

Figure 1.7 displays two different views of the same document. Your screen may not match either figure, and indeed, there is no requirement that it should. You should, however, be aware of different options so that you can develop preferences of your own. Consider:

- Figure 1.7a uses short menus (note the double arrow at the bottom of the menu to display additional commands) and a shared row for the Standard and Formatting toolbars. Figure 1.7b displays the full menu and displays the toolbars on separate rows. We prefer the latter settings, which are set through the Customize command in the Tools menu.
- Figure 1.7a shows the Office Assistant (but drags it out of the way), whereas Figure 1.7b hides it. We find the Assistant distracting, and display it only when necessary by pressing the F1 key. You can also use the appropriate option in the Help menu to hide or show the Assistant, and/or you can right click the Assistant to hide it.
- Figure 1.7a shows the document with the task pane open, whereas the task pane is closed in Figure 1.7b. The task pane serves a variety of functions as you will see throughout the text. These include opening a document, inserting clip art, creating a mail merge, or displaying the formatting properties of selected text.
- Figure 1.7a displays the document in the ***Normal view*** whereas Figure 1.7b uses the ***Print Layout view***. The Normal view is simpler, but the Print Layout view more closely resembles the printed page as it displays top and bottom margins, headers and footers, graphic elements in their exact position, a vertical ruler, and other elements not seen in the Normal view.
- Figure 1.7a displays the ¶ and other nonprinting symbols, whereas they are hidden in Figure 1.7b. We prefer the cleaner screen without the symbols, but on occasion display the symbols if there is a problem in formatting a document. The ***Show/Hide ¶ button*** toggles the symbols on or off.
- Figure 1.7b displays an additional toolbar, the Drawing toolbar, at the bottom of the screen. Microsoft Word has more than 20 toolbars that are suppressed or displayed through the Toolbars command in the View menu. Note, too, that you can change the position of any visible toolbar by dragging its move handle (the parallel lines) at the left of the toolbar.

THE MOUSE VERSUS THE KEYBOARD

Almost every command in Microsoft Office can be executed in different ways, using either the mouse or the keyboard. Most people start with the mouse and add keyboard shortcuts as they become more proficient. There is no right or wrong technique, just different techniques, and the one you choose depends entirely on personal preference in a specific situation. If, for example, your hands are already on the keyboard, it is faster to use the keyboard. Other times, your hand will be on the mouse and that will be the fastest way.

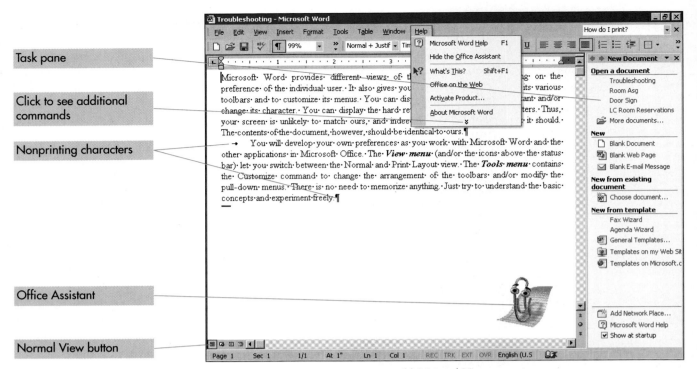

(a) Normal View

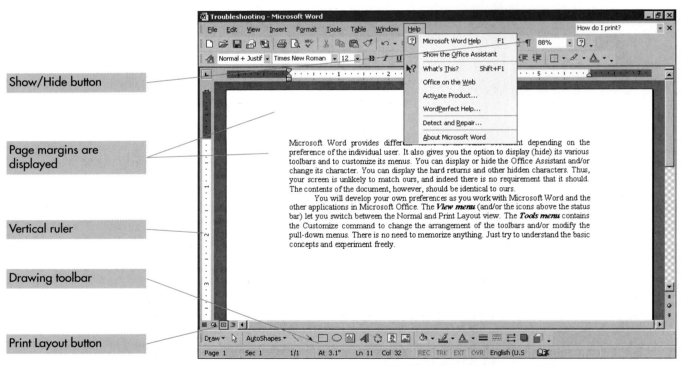

(b) Print Layout View

FIGURE 1.7 *Troubleshooting*

Hands-on Exercise 2

Modifying an Existing Document

Objective To open an existing document, revise it, and save the revision; to use the Undo, Redo, and Help commands. Use Figure 1.8 as a guide in doing the exercise.

Step 1: **Open an Existing Document**

➤ Start Word. Click and drag the Assistant out of the way if it appears. Close the task pane if it is open because we want you to practice locating a document within the Open dialog box.

➤ Pull down the **File menu** and click the **Open command** (or click the **Open button** on the Standard toolbar). You should see an Open dialog box similar to Figure 1.8a.

➤ If necessary, click the **drop-down arrow** on the Views button and change to the Details view. Click and drag the vertical border between columns to increase (decrease) the size of a column.

➤ Click the **drop-down arrow** on the Look in list box. Select (click) the drive that contains the Exploring Windows folder. Double click the folder to open it.

➤ Click the **down arrow** on the vertical scroll bar until you can select **My First Document** from the previous exercise.

➤ Double click the document (or click the **Open button** within the dialog box). Your document should appear on the screen.

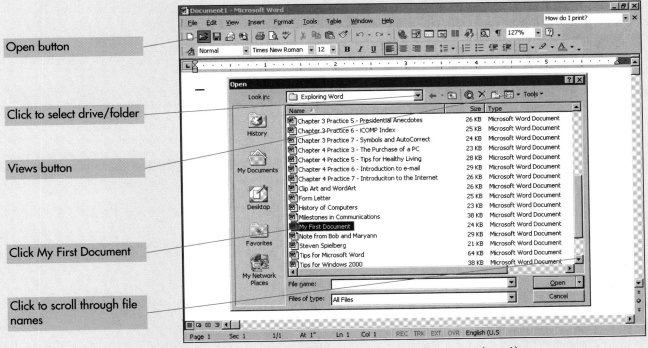

(a) Open an Existing Document (step 1)

FIGURE 1.8 *Hands-on Exercise 2*

Step 2: **Troubleshooting**

➤ Modify the settings within Word so that the document on your screen matches Figure 1.8b.
- To separate the Standard and Formatting toolbars, pull down the **Tools menu**, click **Customize**, click the **Options tab**, then check the box that indicates the Standard and Formatting toolbars should be displayed on two rows. Click the **Close button**.
- To display the complete menus, pull down the **Tools menu**, click **Customize**, click the **Options tab**, then check the box to always show full menus. Click the **Close Button**.
- To change to the Normal view, pull down the **View menu** and click **Normal** (or click the **Normal View button** at the bottom of the window).
- To change the amount of text that is visible on the screen, click the **drop-down arrow** on the **Zoom box** on the Standard toolbar and select **Page Width**.
- To display (hide) the ruler, pull down the **View menu** and toggle the **Ruler command** on or off. End with the ruler on. (If you don't see the Ruler command, click the **double arrow** at the bottom of the menu, or use the **Options command** in the Tools menu to display the complete menus.)
- To show or hide the Office Assistant, pull down the **Help menu** and click the appropriate command.
- Pull down the **View menu** and click the **Toolbars command** to display or hide additional toolbars.

➤ Click the **Show/Hide ¶ button** to display or hide the hard returns as you see fit. The button functions as a toggle switch.

➤ There may still be subtle differences between your screen and ours, depending on the resolution of your monitor. These variations, if any, need not concern you as long as you are able to complete the exercise.

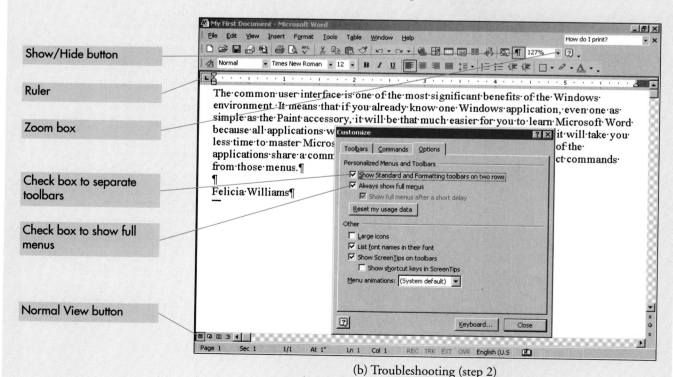

(b) Troubleshooting (step 2)

FIGURE 1.8 *Hands-on Exercise 2 (continued)*

Step 3: **Modify the Document**

➤ Press **Ctrl+End** to move to the end of the document. Press the **up arrow key** once or twice until the insertion point is on a blank line above your name. If necessary, press the **enter key** once (or twice) to add additional blank line(s).

➤ Add the sentence, **Success, I can save and retrieve a document!**, as shown in Figure 1.8c.

➤ Make the following additional modifications to practice editing:
- Change the phrase *most significant* to **very best**.
- Change *Paint accessory* to **game of Solitaire**.
- Change the word *select* to **choose**.

➤ Use the **Ins key** to switch between insert and overtype modes as necessary. (You can also double click the **OVR indicator** on the status bar to toggle between the insert and overtype modes.)

➤ Pull down the **File menu** and click **Save**, or click the **Save button**.

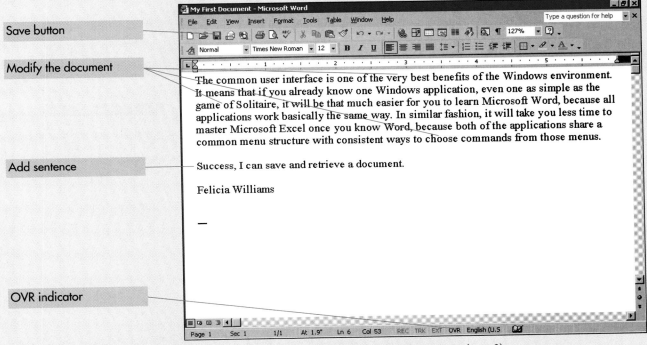

(c) Modify the Document (step 3)

FIGURE 1.8 *Hands-on Exercise 2 (continued)*

MOVING WITHIN A DOCUMENT

Press Ctrl+Home and Ctrl+End to move to the beginning and end of a document, respectively. You can also press the Home or End key to move to the beginning or end of a line. These shortcuts work not just in Word, but in any other Office application, and are worth remembering as they allow your hands to remain on the keyboard as you type.

Step 4: **Deleting Text**

- Press and hold the left mouse button as you drag the mouse over the phrase, **even one as simple as the game of Solitaire**, as shown in Figure 1.8d.
- Press the **Del** key to delete the selected text from the document. Pull down the **Edit menu** and click the **Undo command** (or click the **Undo button** on the Standard toolbar) to reverse (undo) the last command. The deleted text should be returned to your document.
- Pull down the **Edit menu** a second time and click the **Redo command** (or click the **Redo button**) to repeat the Delete command.
- Try this simple experiment. Click the **Undo button** repeatedly to undo the commands one at a time, until you have effectively canceled the entire session. Now click the **Redo command** repeatedly, one command at a time, until you have put the entire document back together.
- Click the **Save button** on the Standard toolbar to save the revised document a final time.

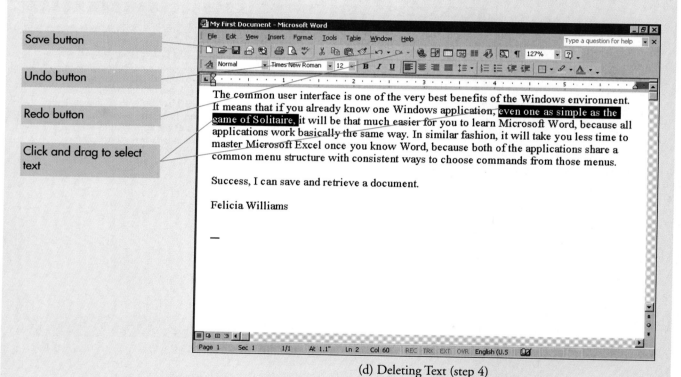

(d) Deleting Text (step 4)

FIGURE 1.8 *Hands-on Exercise 2 (continued)*

THE UNDO AND REDO COMMANDS

Click the drop-down arrow next to the Undo button to display a list of your previous actions, then click the action you want to undo, which also undoes all of the preceding commands. Undoing the fifth command in the list, for example, will also undo the preceding four commands. The Redo command works in reverse and cancels the last Undo command.

Step 5: **E-mail Your Document**

- You should check with your professor before attempting this step.
- Click the **E-mail button** on the Standard toolbar to display a screen similar to Figure 1.8e. The text of your document is entered automatically into the body of the e-mail message.
- Enter your professor's e-mail address in the To text box. The document title is automatically entered in the Subject line. Press the **Tab key** to move to the Introduction line. Type a short note above the inserted document to your professor, then click the **Send a Copy button** to mail the message.
- The e-mail window closes and you are back in Microsoft Word. The introductory text has been added to the document. Pull down the **File menu**. Click **Close** to close the document (there is no need to save the document).
- Pull down the **File menu**. Click **Exit** if you do not want to continue with the next exercise at this time.

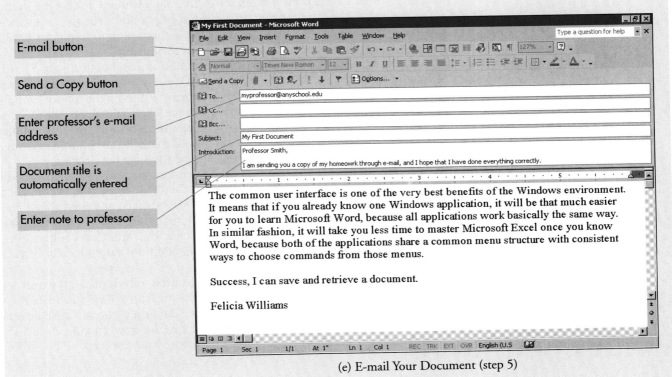

(e) E-mail Your Document (step 5)

FIGURE 1.8 *Hands-on Exercise 2 (continued)*

DOCUMENT PROPERTIES

Prove to your instructor how hard you've worked by printing various statistics about your document, including the number of revisions and the total editing time. Pull down the File menu, click the Print command to display the Print dialog box, click the drop-down arrow in the Print What list box, select Document properties, then click OK. You can view the information (without printing) by pulling down the File menu, clicking the Properties command, then selecting the Statistics tab.

THE SPELL CHECK

There is simply no excuse to misspell a word, since the ***spell check*** is an integral part of Microsoft Word. (The spell check is also available for every other application in the Microsoft Office.) Spelling errors make your work look sloppy and discourage the reader before he or she has read what you had to say. They can cost you a job, a grade, a lucrative contract, or an award you deserve.

The spell check can be set to automatically check a document as text is entered, or it can be called explicitly by clicking the Spelling and Grammar button on the Standard toolbar. The spell check compares each word in a document to the entries in a built-in dictionary, then flags any word that is in the document, but not in the built-in dictionary, as an error.

The dictionary included with Microsoft Office is limited to standard English and does not include many proper names, acronyms, abbreviations, or specialized terms, and hence, the use of any such item is considered a misspelling. You can, however, add such words to a ***custom dictionary*** so that they will not be flagged in the future. The spell check will inform you of repeated words and irregular capitalization. It cannot, however, flag properly spelled words that are used improperly, and thus cannot tell you that *Two be or knot to be* is not the answer.

The capabilities of the spell check are illustrated in conjunction with Figure 1.9a. Microsoft Word will indicate the errors as you type by underlining them in red. Alternatively, you can click the Spelling and Grammar button on the Standard toolbar at any time to move through the entire document. The spell check will then go through the document and return the errors one at a time, offering several options for each mistake. You can change the misspelled word to one of the alternatives suggested by Word, leave the word as is, or add the word to a custom dictionary.

The first error is the word *embarassing,* with Word's suggestion(s) for correction displayed in the list box in Figure 1.9b. To accept the highlighted suggestion, click the Change command button, and the substitution will be made automatically in the document. To accept an alternative suggestion, click the desired word, then click the Change command button. Alternatively, you can click the AutoCorrect button to correct the mistake in the current document, and, in addition, automatically correct the same mistake in any future document.

The spell check detects both irregular capitalization and duplicated words, as shown in Figures 1.9c and 1.9d, respectively. The last error, *Grauer,* is not a misspelling per se, but a proper noun not found in the standard dictionary. No correction is required, and the appropriate action is to ignore the word (taking no further action)—or better yet, add it to the custom dictionary so that it will not be flagged in future sessions.

A spell check will catch embarassing mistakes, iRregular capitalization, and duplicate words words. It will also flag proper nouns, for example Robert Grauer, but you can add these terms to a custom dictionary. It will not notice properly spelled words that are used incorrectly; for example, too bee or knot to be is not the answer.

(a) The Text

FIGURE 1.9   *The Spell Check*

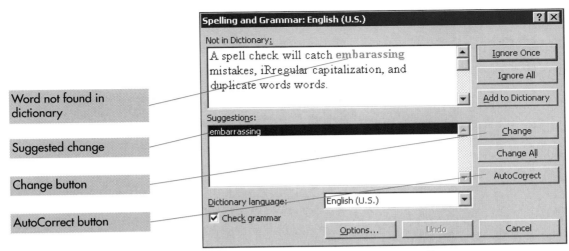

(b) Ordinary Misspelling

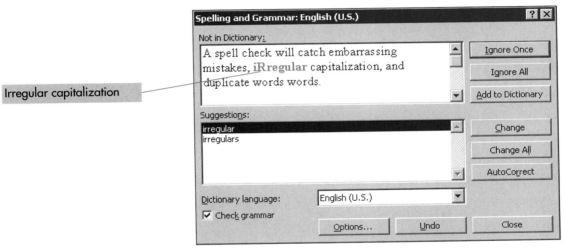

(c) Irregular Capitalization

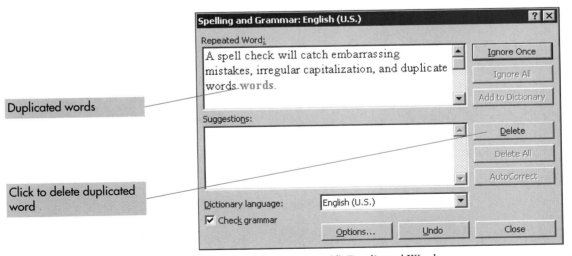

(d) Duplicated Word

FIGURE 1.9 *The Spell Check (continued)*

AutoCorrect and AutoText

The ***AutoCorrect*** feature corrects mistakes as they are made without any effort on your part. It makes you a better typist. If, for example, you typed *teh* instead of *the*, Word would change the spelling without even telling you. Word will also change *adn* to *and, i* to *I*, and *occurence* to *occurrence*. All of this is accomplished through a predefined table of common mistakes that Word uses to make substitutions whenever it encounters an entry in the table. You can add additional items to the table to include the frequent errors you make. You can also use the feature to define your own shorthand—for example, cis for Computer Information Systems as shown in Figure 1.10a.

The AutoCorrect feature will also correct mistakes in capitalization; for example, it will capitalize the first letter in a sentence, recognize that MIami should be Miami, and capitalize the days of the week. It's even smart enough to correct the accidental use of the Caps Lock key, and it will toggle the key off!

The ***AutoText*** feature is similar in concept to AutoCorrect in that both substitute a predefined item for a specific character string. The difference is that the substitution occurs automatically with the AutoCorrect entry, whereas you have to take deliberate action for the AutoText substitution to take place. AutoText entries can also include significantly more text, formatting, and even clip art.

Microsoft Word includes a host of predefined AutoText entries. And as with the AutoCorrect feature, you can define additional entries of your own. (You may, however, not be able to do this in a computer lab environment.) The entry in Figure 1.10b is named "signature" and once created, it is available to all Word documents. To insert an AutoText entry into a new document, just type the first several letters in the AutoText name (signature in our example), then press the enter key when Word displays a ScreenTip containing the text of the entry.

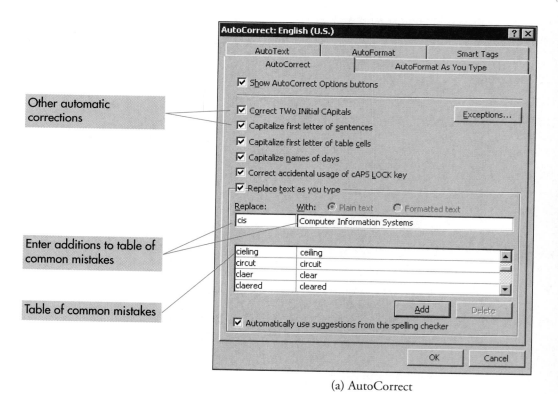

(a) AutoCorrect

FIGURE 1.10 *AutoCorrect and AutoText*

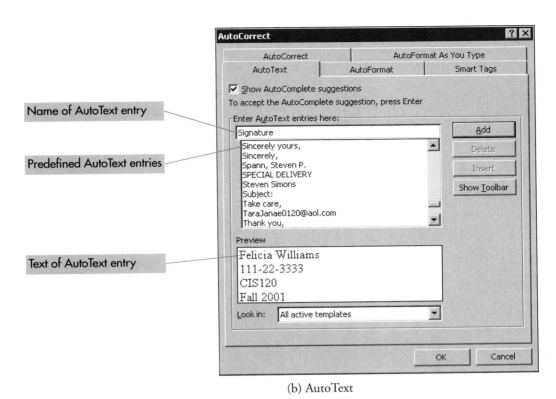

(b) AutoText

FIGURE 1.10 *AutoCorrect and AutoText (continued)*

THESAURUS

The ***thesaurus*** helps you to avoid repetition and polish your writing. The thesaurus is called from the Language command in the Tools menu. You position the cursor at the appropriate word within the document, then invoke the thesaurus and follow your instincts. The thesaurus recognizes multiple meanings and forms of a word (for example, adjective, noun, and verb) as in Figure 1.11a. Click a meaning, then double click a synonym to produce additional choices as in Figure 1.11b. You can explore further alternatives by selecting a synonym or antonym and clicking the Look Up button. We show antonyms in Figure 1.11c.

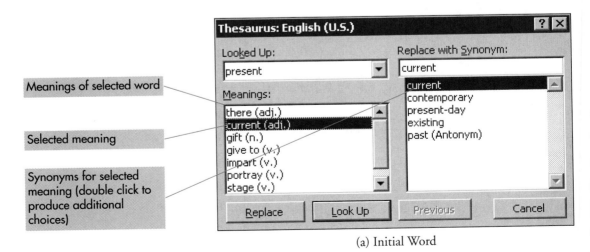

(a) Initial Word

FIGURE 1.11 *The Thesaurus*

MICROSOFT WORD 2002

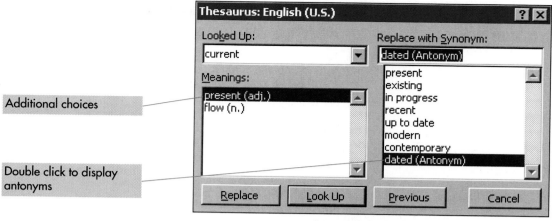

(b) Additional Choices

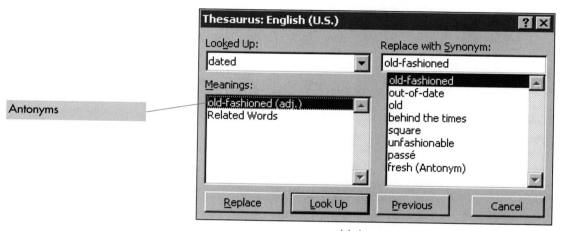

(c) Antonyms

FIGURE 1.11   *The Thesaurus (continued)*

GRAMMAR CHECK

The ***grammar check*** attempts to catch mistakes in punctuation, writing style, and word usage by comparing strings of text within a document to a series of predefined rules. As with the spell check, errors are brought to the screen, where you can accept the suggested correction and make the replacement automatically, or more often, edit the selected text and make your own changes.

You can also ask the grammar check to explain the rule it is attempting to enforce. Unlike the spell check, the grammar check is subjective, and what seems appropriate to you may be objectionable to someone else. Indeed, the grammar check is quite flexible, and can be set to check for different writing styles; that is, you can implement one set of rules to check a business letter and a different set of rules for casual writing. Many times, however, you will find that the English language is just too complex for the grammar check to detect every error, although it will find many errors.

The grammar check caught the inconsistency between subject and verb in Figure 1.12a and suggested the appropriate correction (am instead of are). In Figure 1.12b, it suggested the elimination of the superfluous comma. These examples show the grammar check at its best, but it is often more subjective and less capable. It missed the error "no perfect" in Figure 1.12c (although it did catch "to" instead of "too"). Suffice it to say, that there is no substitute for carefully proofreading every document.

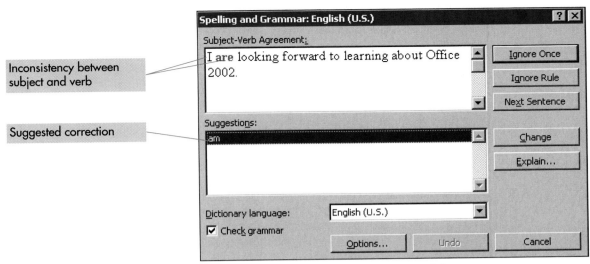

(a) Inconsistent Verb

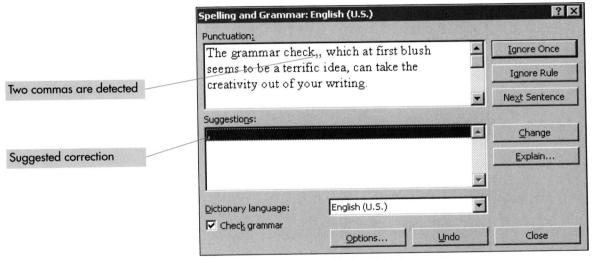

(b) Doubled Punctuation

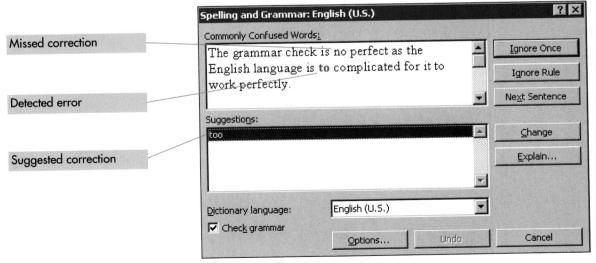

(c) Limitations

FIGURE 1.12 *The Grammar Check*

SAVE COMMAND

The Save command was used in the first two exercises. The Save As command will be introduced in the next exercise as a very useful alternative. We also introduce you to different backup options. We believe that now, when you are first starting to learn about word processing, is the time to develop good working habits.

You already know that the Save command copies the document currently being edited (the document in memory) to disk. The initial execution of the command requires you to assign a file name and to specify the drive and folder in which the file is to be stored. All subsequent executions of the Save command save the document under the original name, replacing the previously saved version with the new one.

The ***Save As command*** saves another copy of a document under a different name (and/or a different file type), and is useful when you want to retain a copy of the original document. The Save As command provides you with two copies of a document. The original document is kept on disk under its original name. A copy of the document is saved on disk under a new name and remains in memory. All subsequent editing is done on the new document.

We cannot overemphasize the importance of periodically saving a document, so that if something does go wrong, you won't lose all of your work. Nothing is more frustrating than to lose two hours of effort, due to an unexpected program crash or to a temporary loss of power. Save your work frequently, at least once every 15 minutes. Pull down the File menu and click Save, or click the Save button on the Standard toolbar. Do it!

Backup Options

Microsoft Word offers several different ***backup*** options. We believe the two most important options are to create a backup copy in conjunction with every save command, and to periodically (and automatically) save a document. Both options are implemented in step 3 in the next hands-on exercise.

Figure 1.13 illustrates the option to create a backup copy of the document every time a Save command is executed. Assume, for example, that you have created the simple document, *The fox jumped over the fence* and saved it under the name "Fox". Assume further that you edit the document to read, *The quick brown fox jumped over the fence,* and that you saved it a second time. The second save command changes the name of the original document from "Fox" to "Backup of Fox", then saves the current contents of memory as "Fox". In other words, the disk now contains two versions of the document: the current version "Fox" and the most recent previous version "Backup of Fox".

The cycle goes on indefinitely, with "Fox" always containing the current version, and "Backup of Fox" the most recent previous version. Thus if you revise and save the document a third time, "Fox" will contain the latest revision while "Backup of Fox" would contain the previous version alluding to the quick brown fox. The original (first) version of the document disappears entirely since only two versions are kept.

The contents of "Fox" and "Backup of Fox" are different, but the existence of the latter enables you to retrieve the previous version if you inadvertently edit beyond repair or accidentally erase the current "Fox" version. Should this occur (and it will), you can always retrieve its predecessor and at least salvage your work prior to the last save operation.

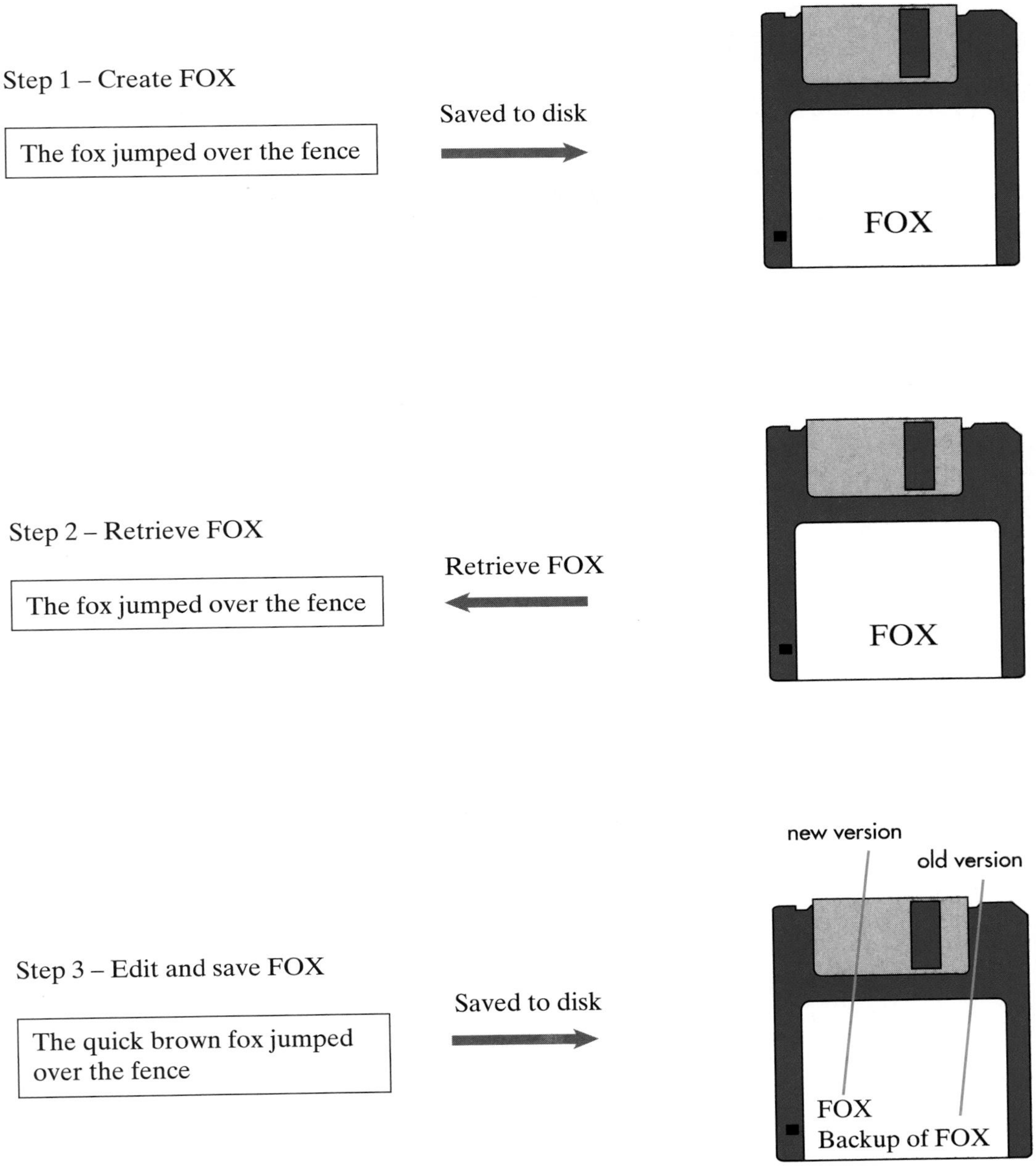

Step 1 – Create FOX

| The fox jumped over the fence |

Saved to disk →

FOX

Step 2 – Retrieve FOX

| The fox jumped over the fence |

← Retrieve FOX

FOX

Step 3 – Edit and save FOX

| The quick brown fox jumped over the fence |

Saved to disk →

new version
old version

FOX
Backup of FOX

FIGURE 1.13   *Backup Procedures*

COMPARE AND MERGE DOCUMENTS

The Compare and Merge Documents command lets you compare the content of two documents to one another in order to see the differences between those documents. It is very useful if you have lost track of different versions of a document and/or if you are working with others. You can also use the command to see how well you did the hands-on exercises. The command highlights the differences in the documents and then it gives you the option to accept or reject changes. See exercise 7 at the end of the chapter.

HANDS-ON EXERCISE 3

THE SPELL CHECK, THESAURUS, AND GRAMMAR CHECK

Objective   To open an existing document, check it for spelling, then use the Save As command to save the document under a different file name. Use Figure 1.14 as a guide in the exercise.

Step 1:   **Preview a Document**

➤ Start Microsoft Word. Pull down the **Help menu**. Click the command to **Hide the Office Assistant**.
➤ If necessary, pull down the **View menu** and click the command to open the **Task pane**. Click the link to **More documents** in the task pane or pull down the **File menu** and click **Open** (or click the **Open button** on the Standard toolbar). You should see a dialog box similar to the one in Figure 1.14a.
➤ Select the appropriate drive, drive C or drive A, depending on the location of your data. Double click the **Exploring Word folder** to make it the active folder (the folder from which you will open the document).
➤ Scroll in the Name list box until you can select (click) the **Try the Spell Check** document. Click the **drop-down arrow** on the **Views button** and click **Preview** to preview the document as shown in Figure 1.14a.
➤ Click the **Open command button** to open the file. Your document should appear on the screen.

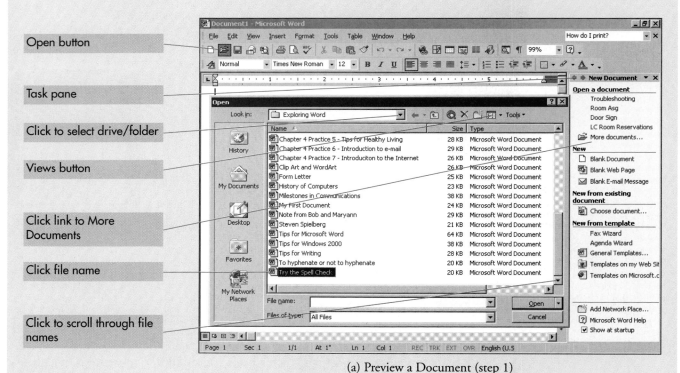

(a) Preview a Document (step 1)

FIGURE 1.14   *Hands-on Exercise 3*

Step 2: **The Save As Command**

- Pull down the **File menu**. Click **Save As** to produce the dialog box in Figure 1.14b.
- Enter **Modified Spell Check** as the name of the new document. (A file name may contain up to 255 characters, and blanks are permitted.) Click the **Save command button**.
- There are now two identical copies of the file on disk: Try the Spell Check, which we supplied, and Modified Spell Check, which you just created.
- The title bar shows the latter name (Modified Spell Check) as it is the document in memory. All subsequent changes will be made to this document.

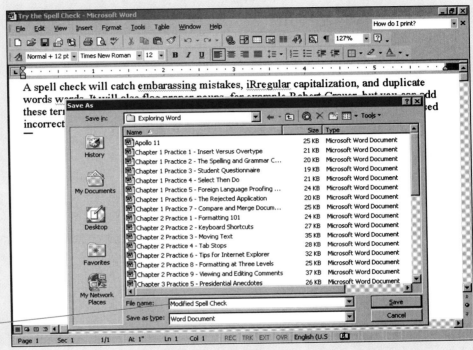

(b) The Save As Command (step 2)

FIGURE 1.14 *Hands-on Exercise 3 (continued)*

THE WORD COUNT TOOLBAR

How close are you to completing the 500-word paper that your professor assigned? Pull down the Tools menu and click the Word Count command to display a dialog box that shows the number of pages, words, paragraphs, and characters in your document. There is also a command button to display the Word Count toolbar so that it remains on the screen throughout the session. Click the Recount button on the toolbar at any time to see the current statistics for your document.

Step 3: **The Spell Check**

- ➤ If necessary, press **Ctrl+Home** to move to the beginning of the document. Click the **Spelling and Grammar button** on the Standard toolbar to check the document.
- ➤ "Embarassing" is flagged as the first misspelling as shown in Figure 1.14c. Click the **Change button** to accept the suggested spelling.
- ➤ "iRregular" is flagged as an example of irregular capitalization. Click the **Change button** to accept the suggested correction.
- ➤ Continue checking the document, which displays misspellings and other irregularities one at a time. Click the appropriate command button as each mistake is found.
  - Click the **Delete button** to remove the duplicated word.
  - Click the **Ignore Once button** to accept Grauer (or click the **Add button** to add Grauer to the custom dictionary).
- ➤ The last sentence is flagged because of a grammatical error and is discussed in the next step.

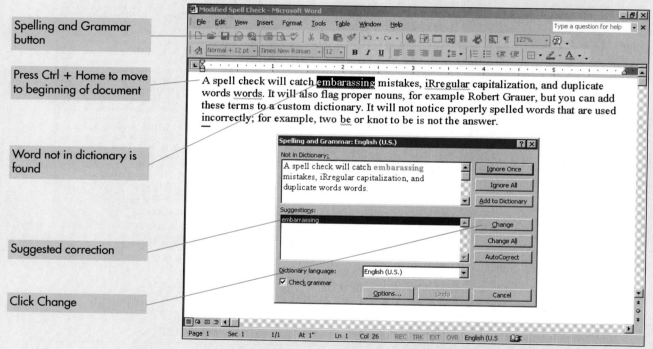

(c) The Spell Check (step 3)

FIGURE 1.14 *Hands-on Exercise 3 (continued)*

AUTOMATIC SPELLING AND GRAMMAR CHECKING

Red and green wavy lines may appear throughout a document to indicate spelling and grammatical errors, respectively. Point to any underlined word, then click the right mouse button to display a context-sensitive help menu with suggested corrections. To enable (disable) these options, pull down the Tools menu, click the Options command, click the Spelling and Grammar tab, and check (clear) the options to check spelling (or grammar) as you type.

Step 4: **The Grammar Check**

➤ The last phrase, "Two be or knot to be is not the answer", should be flagged as an error, as shown in Figure 1.14d. If this is not the case:
- Pull down the **Tools menu**, click **Options**, then click the **Spelling and Grammar tab**.
- Check the box to **Check Grammar with Spelling**, then click the button to **Recheck document**. Click **Yes** when told that the spelling and grammar check will be reset, then click **OK** to close the Options dialog box.
- Press **Ctrl+Home** to return to the beginning of the document, then click the **Spelling and Grammar button** to recheck the document.

➤ The Grammar Check suggests substituting "are" for "be", which is not what you want. Click in the preview box and make the necessary corrections. Change "two" to "to" and "knot" to "not". Click **Change**.

➤ Click **OK** when you see the dialog box indicating that the spelling and grammar check is complete. Enter any additional grammatical changes manually. Save the document.

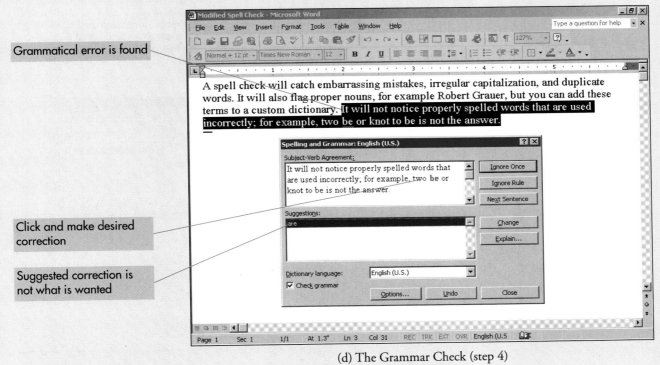

(d) The Grammar Check (step 4)

FIGURE 1.14 *Hands-on Exercise 3 (continued)*

CHECK SPELLING ONLY

The grammar check is invoked by default in conjunction with the spell check. You can, however, check the spelling of a document without checking its grammar. Pull down the Tools menu, click Options to display the Options dialog box, then click the Spelling and Grammar tab. Clear the box to check grammar with spelling, then click OK to accept the change and close the dialog box.

Step 5: **The Thesaurus**

- Select the word **flag**, which appears toward the beginning of your document.
- Pull down the **Tools menu**, click **Language**, then click **Thesaurus** to display the associated dialog box as shown in Figure 1.14e.
- Choose the proper form of the word; that is, you want to find synonyms for the word "flag" when it is used as a verb as opposed to its use as a noun.
- Select **identify** from the list of synonyms, then click the **Replace button** to make the change automatically.
- Right click the word **incorrectly** (which appears in the last sentence) to display a context-sensitive menu, click **synonyms**, then choose **inaccurately** to make the substitution into the document.
- Save the document.

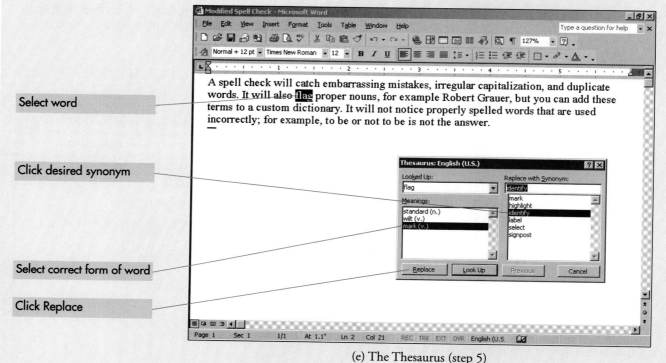

(e) The Thesaurus (step 5)

FIGURE 1.14 *Hands-on Exercise 3 (continued)*

FOREIGN LANGUAGE PROOFING TOOLS

The English version of Microsoft Word supports the spelling, grammar, and thesaurus features in more than 80 foreign languages. Support for Spanish and French is built in at no additional cost, whereas you will have to pay an additional fee for other languages. Just pull down the Tools menu and click the Select Language command to change to a different language. You can even check multiple languages within the same document. See practice exercise 5 at the end of the chapter.

Step 6: **AutoCorrect**

- Press **Ctrl+End** to move to the end of the document. Press the **enter key** twice.
- Type the *misspelled* phrase, **Teh AutoCorrect feature corrects common spelling mistakes**. Word will automatically change "Teh" to "The".
- Press the **Home key** to return to the beginning of the line, where you will notice a blue line, under the "T", indicating that an automatic correction has taken place. Point to the blue line, then click the **down arrow** to display the AutoCorrect options.
- Click the command to **Control AutoCorrect options**, which in turn displays the dialog box in Figure 1.14f. Click the **AutoCorrect tab**, then click the **down arrow** on the scroll bar to view the list of corrections. Close the dialog box.
- Add the sentence, **The feature also changes special symbols such as :) to ☺ to indicate I understand my work**. (You will have to use the AutoCorrect options to change the first ☺ back to the :) within the sentence.)

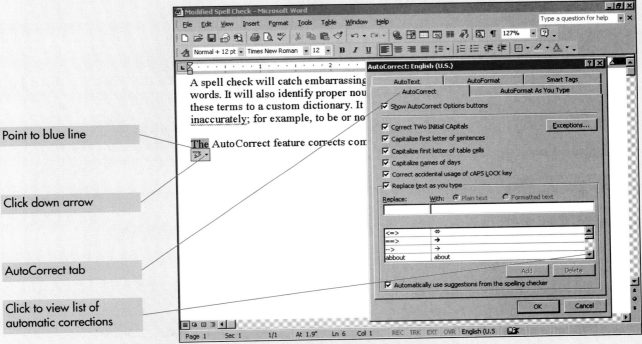

(f) AutoCorrect (step 6)

FIGURE 1.14   *Hands-on Exercise 3 (continued)*

CREATE YOUR OWN SHORTHAND

Use AutoCorrect to expand abbreviations such as "usa" for United States of America. Pull down the Tools menu, click AutoCorrect Options, type the abbreviation in the Replace text box and the expanded entry in the With text box. Click the Add command button, then click OK to exit the dialog box and return to the document. The next time you type usa in a document, it will automatically be expanded to United States of America.

Step 7: **Create an AutoText Entry**

- Press **Ctrl+End** to move to the end of the document. Press the **enter key** twice. Enter your name and class.
- Click and drag to select the information you just entered. Pull down the **Insert menu**, select the **AutoText command**, then select **AutoText** to display the AutoCorrect dialog box in Figure 1.14g.
- Your name (Felicia Williams in our example) is suggested automatically as the name of the AutoText entry. Click the **Add button**.
- To test the entry, you can delete your name and other information, then use the AutoText feature. Your name and other information should still be highlighted. Press the **Del key** to delete the information.
- Type the first few letters of your name and watch the screen as you do. You should see a ScreenTip containing your name and other information. Press the **enter key** or the **F3 key** when you see the ScreenTip.
- Save the document. Print the document for your instructor. Exit Word.

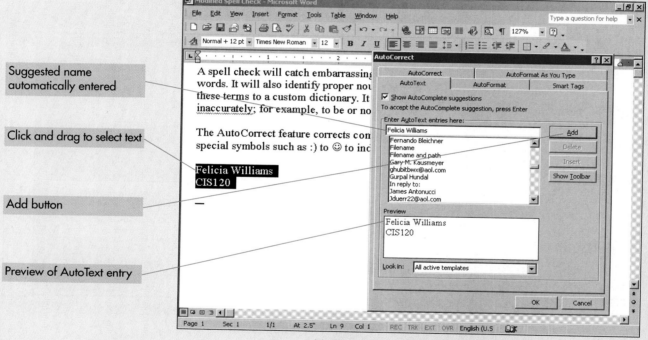

(g) Create an AutoText Entry (step 7)

FIGURE 1.14 *Hands-on Exercise 3 (continued)*

THE AUTOTEXT TOOLBAR

Point to any visible toolbar, click the right mouse button to display a context-sensitive menu, then click AutoText to display the AutoText toolbar. The AutoText toolbar groups the various AutoText entries into categories, making it easier to select the proper entry. Click the down arrow on the All Entries button to display the various categories, click a category, then select the entry you want to insert into the document.

SUMMARY

The chapter provided a broad-based introduction to word processing in general and to Microsoft Word in particular. Help is available from many sources. You can use the Help menu or the Office Assistant as you can in any Office application. You can also go to the Microsoft Web site to obtain more recent, and often more detailed, information.

Microsoft Word is always in one of two modes, insert or overtype; the choice between the two depends on the desired editing. The insertion point marks the place within a document where text is added or replaced.

The enter key is pressed at the end of a paragraph, but not at the end of a line because Word automatically wraps text from one line to the next. A hard return is created by the user when he or she presses the enter key; a soft return is created by Word as it wraps text and begins a new line.

The Save and Open commands work in conjunction with one another. The Save command copies the document in memory to disk under its existing name. The Open command retrieves a previously saved document. The Save As command saves the document under a different name and is useful when you want to retain a copy of the current document prior to all changes.

A spell check compares the words in a document to those in a standard and/or custom dictionary and offers suggestions to correct the mistakes it finds. It will detect misspellings, duplicated phrases, and/or irregular capitalization, but will not flag properly spelled words that are used incorrectly. Foreign-language proofing tools for French and Spanish are built into the English version of Microsoft Word 2002.

The AutoCorrect feature corrects predefined spelling errors and/or mistakes in capitalization, automatically, as the words are entered. The AutoText feature is similar in concept except that it can contain longer entries that include formatting and clip art. Either feature can be used to create a personal shorthand to expand abbreviations as they are typed.

The thesaurus suggests synonyms and/or antonyms. It can also recognize multiple forms of a word (noun, verb, and adjective) and offer suggestions for each. The grammar check searches for mistakes in punctuation, writing style, and word usage by comparing strings of text within a document to a series of predefined rules.

KEY TERMS

AutoCorrect (p. 26)
AutoText (p. 26)
Backup (p. 30)
Close command (p. 8)
Custom dictionary (p. 24)
Exit command (p. 8)
File menu (p. 8)
File name (p. 8)
File type (p. 8)
Formatting toolbar (p. 5)
Grammar check (p. 28)
Hard return (p. 2)
Horizontal ruler (p. 6)
Insert mode (p. 4)
Insertion point (p. 2)
Normal view (p. 17)
Office Assistant (p. 12)
Open command (p. 8)
Overtype mode (p. 4)
Places Bar (p. 8)
Print command (p. 8)
Print Layout view (p. 17)
Save As command (p. 30)
Save command (p. 8)
ScreenTip (p. 5)
Show/Hide ¶ button (p. 17)
Soft return (p. 2)
Spell check (p. 24)
Standard toolbar (p. 5)
Status bar (p. 6)
Task pane (p. 6)
Thesaurus (p. 27)
Toggle switch (p. 3)
Toolbar (p. 5)
Undo command (p. 22)
Vertical ruler (p. 6)
View menu (p. 20)
Word wrap (p. 2)

MULTIPLE CHOICE

1. When entering text within a document, the enter key is normally pressed at the end of every:
   (a) Line
   (b) Sentence
   (c) Paragraph
   (d) All of the above

2. Which menu contains the commands to save the current document, or to open a previously saved document?
   (a) The Tools menu
   (b) The File menu
   (c) The View menu
   (d) The Edit menu

3. How do you execute the Print command?
   (a) Click the Print button on the standard toolbar
   (b) Pull down the File menu, then click the Print command
   (c) Use the appropriate keyboard shortcut
   (d) All of the above

4. The Open command:
   (a) Brings a document from disk into memory
   (b) Brings a document from disk into memory, then erases the document on disk
   (c) Stores the document in memory on disk
   (d) Stores the document in memory on disk, then erases the document from memory

5. The Save command:
   (a) Brings a document from disk into memory
   (b) Brings a document from disk into memory, then erases the document on disk
   (c) Stores the document in memory on disk
   (d) Stores the document in memory on disk, then erases the document from memory

6. What is the easiest way to change the phrase, *revenues, profits, gross margin,* to read *revenues, profits, and gross margin*?
   (a) Use the insert mode, position the cursor before the *g* in *gross,* then type the word *and* followed by a space
   (b) Use the insert mode, position the cursor after the *g* in *gross,* then type the word *and* followed by a space
   (c) Use the overtype mode, position the cursor before the *g* in *gross,* then type the word *and* followed by a space
   (d) Use the overtype mode, position the cursor after the *g* in *gross,* then type the word *and* followed by a space

7. A document has been entered into Word with a given set of margins, which are subsequently changed. What can you say about the number of hard and soft returns before and after the change in margins?
   (a) The number of hard returns is the same, but the number and/or position of the soft returns is different
   (b) The number of soft returns is the same, but the number and/or position of the hard returns is different
   (c) The number and position of both hard and soft returns is unchanged
   (d) The number and position of both hard and soft returns is different

8. Which of the following will be detected by the spell check?
   (a) Duplicate words
   (b) Irregular capitalization
   (c) Both (a) and (b)
   (d) Neither (a) nor (b)

9. Which of the following is likely to be found in a custom dictionary?
   (a) Proper names
   (b) Words related to the user's particular application
   (c) Acronyms created by the user for his or her application
   (d) All of the above

10. Ted and Sally both use Word. Both have written a letter to Dr. Joel Stutz and have run a spell check on their respective documents. Ted's program flags *Stutz* as a misspelling, whereas Sally's accepts it as written. Why?
    (a) The situation is impossible; that is, if they use identical word processing programs they should get identical results
    (b) Ted has added *Stutz* to his custom dictionary
    (c) Sally has added *Stutz* to her custom dictionary
    (d) All of the above reasons are equally likely as a cause of the problem

11. The spell check will do all of the following *except:*
    (a) Flag properly spelled words used incorrectly
    (b) Identify misspelled words
    (c) Accept (as correctly spelled) words found in the custom dictionary
    (d) Suggest alternatives to misspellings it identifies

12. The AutoCorrect feature will:
    (a) Correct errors in capitalization as they occur during typing
    (b) Expand user-defined abbreviations as the entries are typed
    (c) Both (a) and (b)
    (d) Neither (a) nor (b)

13. When does the Save As dialog box appear?
    (a) The first time a file is saved using either the Save or Save As commands
    (b) Every time a file is saved
    (c) Both (a) and (b)
    (d) Neither (a) nor (b)

14. Which of the following is true about the thesaurus?
    (a) It recognizes different forms of a word; for example, a noun and a verb
    (b) It provides antonyms as well as synonyms
    (c) Both (a) and (b)
    (d) Neither (a) nor (b)

15. The grammar check:
    (a) Implements different rules for casual and business writing
    (b) Will detect all subtleties in the English language
    (c) Is always run in conjunction with a spell check
    (d) All of the above

ANSWERS

1. c	6. a	11. a
2. b	7. a	12. c
3. d	8. c	13. a
4. a	9. d	14. c
5. c	10. c	15. a

PRACTICE WITH WORD

1. **Insert versus Overtype:** Open the *Chapter 1 Practice 1* document that is shown in Figure 1.15 and make the following changes.
   a. Enter your instructor's name and your name in the To and From lines, respectively.
   b. Change "better" to "preferable" in the third line of the first paragraph.
   c. Delete the word "then" from the last line in the first paragraph.
   d. Click at the end of the first paragraph, and add the sentence, "The insert mode adds characters at the insertion point while moving existing text to the right in order to make room for the new text."
   e. Delete the last paragraph, which describes how to delete text. Create a new paragraph in its place with the following text: "There are two other keys that function as toggle switches of which you should be aware. The Caps Lock key toggles between upper- and lowercase letters. The Num Lock key alternates between typing numbers and using the arrow keys."
   f. Print the revised document for your instructor.
   g. Create a cover sheet for the assignment with your name, your instructor's name, today's date, and the assignment number.

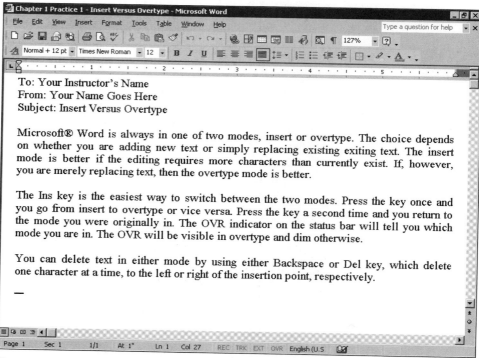

FIGURE 1.15 *Insert versus Overtype (Exercise 1)*

2. **The Spelling and Grammar Check:** Open the *Chapter 1 Practice 2* document that is displayed in Figure 1.16, then run the spelling and grammar check to correct the various errors that are contained in the original document. Print this version of the corrected document for your instructor.
   Read the corrected document carefully and make any other necessary corrections. You should find several additional errors because the English language is very complicated and it is virtually impossible to correct every error automatically. Print this version of the document as well. Add a cover page and submit both versions of the corrected document to your instructor.

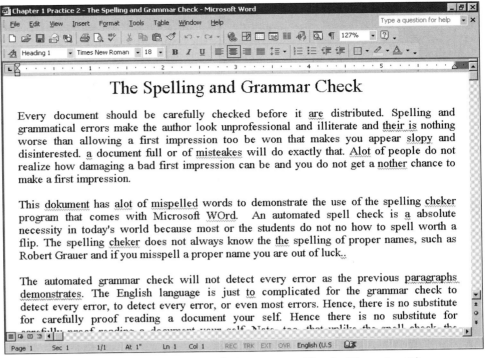

FIGURE 1.16   *The Spelling and Grammar Check (Exercise 2)*

3. Student Questionnaire: Use the partially completed document in *Chapter 1 Practice 3* to describe your background. If there is time in the class, your instructor can have you exchange assignments with another student. There are many variations on this "icebreaker," and the assignment will let you gain practice with Microsoft Word.

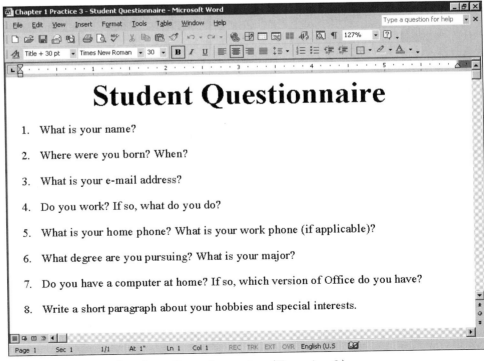

FIGURE 1.17   *Student Questionnaire (Exercise 3)*

4. **Select-Then-Do:** Formatting is not covered in this chapter, but it is very easy to apply basic formatting to a document, especially if you have used another application in Microsoft Office. Many formatting operations are implemented in the context of "select-then-do" as described in the document in Figure 1.18. You select the text that you want to format, then you execute the appropriate formatting command, most easily by clicking the appropriate button on the Formatting toolbar. The function of each button should be apparent from its icon, but you can simply point to a button to display a ScreenTip that is indicative of its function.

   An unformatted version of the document in Figure 1.18 is found in the *Chapter 1 Practice 4* document in the Exploring Word folder. Open the document, then format it to match the completed document in the figure. The title of the document is centered in 22-point Arial, whereas the rest of the document is set in 12-point Times New Roman. Boldface, italicize, and highlight the text as indicated in the actual document. A color font is also indicated, but do not be concerned if you do not have a color printer. Add your name to the bottom of the completed document, then print the document for your instructor.

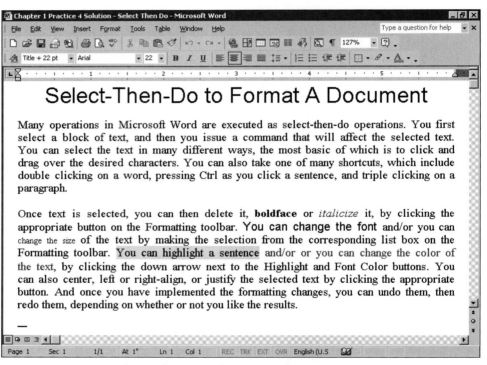

FIGURE 1.18 *Select-Then-Do (Exercise 4)*

5. **Foreign Language Proofing Tools:** Use the document in Figure 1.19 to practice with the foreign language proofing tools for French and Spanish that are built into Microsoft Word. We have entered the text of the document for you, but it is your responsibility to select the appropriate proofing tool for the different parts of the document. Open the document in *Chapter 1 Practice 5*, which will indicate multiple misspellings because the document is using the English spell check.

   English is the default language. To switch to a different language, select the phrase, pull down the Tools menu, click the Language command, and then click the Set Language command to set (or change) the language in effect. Add your name to the completed document and print it for your instructor.

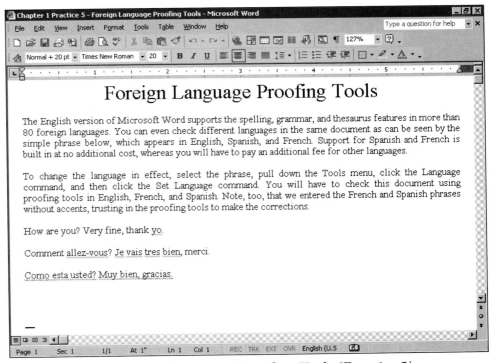

FIGURE 1.19 *Foreign Language Proofing Tools (Exercise 5)*

6. The Rejected Job Applicant: The individual who wrote the letter in Figure 1.20 has been rejected for every job for which he has applied. He is a good friend and you want to help. Open the document in *Chapter 1 Practice 6* and correct the obvious spelling errors. Read the document carefully and make any additional corrections that you think appropriate. Sign your name and print the corrected letter.

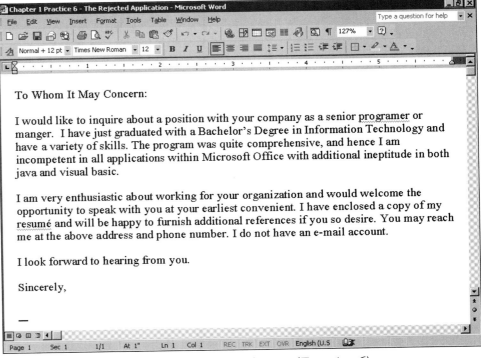

FIGURE 1.20 *The Rejected Job Applicant (Exercise 6)*

BUILDS ON

HANDS-ON
EXERCISE 2,
PAGES 19–23

7. **Compare and Merge Documents:** The Compare and Merge Documents command lets you compare two documents in order to see the changes between those documents. You can use the command to see how well you completed the first two hands-on exercises. Proceed as follows:
   a. Open the completed *Chapter 1 Practice 7* document in the Exploring Word folder. This document contains a paragraph from the hands-on exercise followed by a paragraph that describes the Compare and Merge Documents command.
   b. Pull down the Tools menu and click the Compare and Merge Documents command to display the associated dialog box and select *My First Document* (the document you created).
   c. Check the box for Legal Blackline (to display a thin black line in the left margin showing where the changes occur), then click the Compare button. The two documents are merged together as shown in Figure 1.21. Click the Print button to print the merged documents for your instructor.
   d. Accept or reject the changes as you see fit. Use the Help command to learn more about merging documents and tracking changes.
   e. Summarize your thoughts about these commands in a short note to your instructor. Compare your findings to those of your classmates.
   f. Learn how to insert comments into a document that contain suggestions for improving the document. What happens if different users insert comments into the same document?

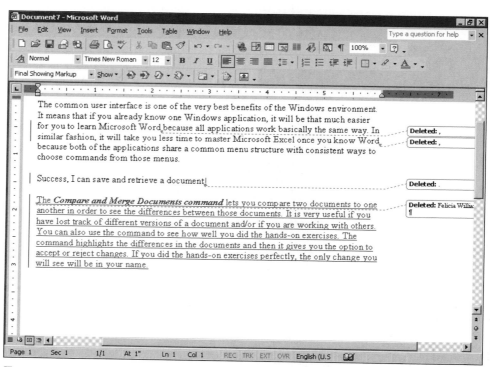

FIGURE 1.21   *Compare and Merge Documents (Exercise 7)*

ON YOUR OWN

Acronym Finder

Do you know what the acronym, PCMCIA stands for? Some might say it stands for "People Can't Memorize Computer Industry Acronyms", although the real meaning is "Personal Computer Memory Card International Association", which refers to the PC cards that are used with notebook computers. Use your favorite Internet search engine to locate a site that publishes lists of acronyms. Select five computer-related terms and create a short document with the acronym and its meaning. Select a second list of any five acronyms that appeal to you. Print the document and submit it to your instructor.

The Reference Desk

The Reference Desk, at www.refdesk.com, contains a treasure trove of information for the writer. You will find access to several online dictionaries, encyclopedias, and other references. Go to the site and select five links that you think will be of interest to you as a writer. Create a short document that contains the name of the site, its Web address, and a brief description of the information that is found at the site. Print the document and submit it to your instructor.

Planning for Disaster

Do you have a backup strategy? Do you even know what a backup strategy is? You should learn, because sooner or later you will wish you had one. You will erase a file, be unable to read from a floppy disk, or worse yet suffer a hardware failure in which you are unable to access the hard drive. The problem always seems to occur the night before an assignment is due. The ultimate disaster is the disappearance of your computer, by theft or natural disaster (e.g., Hurricane Andrew). Describe in 250 words or less the backup strategy you plan to implement in conjunction with your work in this class.

A Letter Home

You really like this course and want very much to have your own computer, but you're strapped for cash and have decided to ask your parents for help. Write a one-page letter describing the advantages of having your own system and how it will help you in school. Tell your parents what the system will cost, and that you can save money by buying through the mail. Describe the configuration you intend to buy (don't forget to include the price of software) and then provide prices from at least three different companies. Cut out the advertisements and include them in your letter. Bring your material to class and compare your research with that of your classmates.

Computer Magazines

A subscription to a computer magazine should be given serious consideration if you intend to stay abreast in a rapidly changing field. The reviews on new products are especially helpful, and you will appreciate the advertisements should you need to buy. Go to the library or a newsstand and obtain a magazine that appeals to you, then write a brief review of the magazine for class. Devote at least one paragraph to an article or other item you found useful.

A Junior Year Abroad

How lucky can you get? You are spending the second half of your junior year in Paris. The problem is you will have to submit your work in French, and the English version of Microsoft Word won't do. Is there a foreign-language version available? What about the dictionary and thesaurus? How do you enter the accented characters, which occur so frequently? You are leaving in two months, so you'd better get busy. What are your options? *Bon voyage!*

Changing Menus and Toolbars

Office XP enables you to display a series of short menus that contain only basic commands. The additional commands are made visible by clicking the double arrow that appears at the bottom of the menu. New commands are added to the menu as they are used, and conversely, other commands are removed if they are not used. A similar strategy is followed for the Standard and Formatting toolbars that are displayed on a single row, and thus do not show all of the buttons at one time. The intent is to simplify Office XP for the new user by limiting the number of commands that are visible. The consequence, however, is that the individual is not exposed to new commands, and hence may not use Office to its full potential. Which set of menus do you prefer? How do you switch from one set to the other?

CHAPTER 2

Gaining Proficiency: Editing and Formatting

OBJECTIVES

AFTER READING THIS CHAPTER YOU WILL BE ABLE TO:

1. Define the select-then-do methodology; describe several shortcuts with the mouse and/or the keyboard to select text.
2. Move and copy text within a document; distinguish between the Windows clipboard and the Office clipboard.
3. Use the Find, Replace, and Go To commands to substitute one character string for another.
4. Define scrolling; scroll to the beginning and end of a document.
5. Distinguish between the Normal and Print Layout views; state how to change the view and/or magnification of a document.
6. Define typography; distinguish between a serif and a sans serif typeface; use the Format Font command to change the font and/or type size.
7. Use the Format Paragraph command to change line spacing, alignment, tabs, and indents, and to control pagination.
8. Use the Borders and Shading command to box and shade text.
9. Describe the Undo and Redo commands and how they are related to one another.
10. Use the Page Setup command to change the margins and/or orientation; differentiate between a soft and a hard page break.
11. Enter and edit text in columns; change the column structure of a document through section formatting.

OVERVIEW

The previous chapter taught you the basics of Microsoft Word and enabled you to create and print a simple document. The present chapter significantly extends your capabilities, by presenting a variety of commands to change the contents and appearance of a document. These operations are known as editing and formatting, respectively.

You will learn how to move and copy text within a document and how to find and replace one character string with another. You will also learn the basics of typography and how to switch between different fonts. You will be able to change alignment, indentation, line spacing, margins, and page orientation. All of these commands are used in three hands-on exercises, which require your participation at the computer, and which are the very essence of the chapter.

SELECT-THEN-DO

Many operations in Word take place within the context of a *select-then-do* methodology; that is, you select a block of text, then you execute the command to operate on that text. The most basic way to select text is by dragging the mouse; that is, click at the beginning of the selection, press and hold the left mouse button as you move to the end of the selection, then release the mouse.

Selected text is affected by any subsequent operation; for example, clicking the Bold or Italic button changes the selected text to boldface or italics, respectively. You can also drag the selected text to a new location, press the Del key to erase the selected text, or execute any other editing or formatting command. The text continues to be selected until you click elsewhere in the document.

MOVING AND COPYING TEXT

The ability to move and/or copy text is essential in order to develop any degree of proficiency in editing. A move operation removes the text from its current location and places it elsewhere in the same (or even a different) document; a copy operation retains the text in its present location and places a duplicate elsewhere. Either operation can be accomplished using the Windows clipboard and a combination of the ***Cut***, ***Copy***, and ***Paste commands***.

The ***Windows clipboard*** is a temporary storage area available to any Windows application. Selected text is cut or copied from a document and placed onto the clipboard from where it can be pasted to a new location(s). A move requires that you select the text and execute a Cut command to remove the text from the document and place it on the clipboard. You then move the insertion point to the new location and paste the text from the clipboard into that location. A copy operation necessitates the same steps except that a Copy command is executed rather than a cut, leaving the selected text in its original location as well as placing a copy on the clipboard. (The ***Paste Special command*** can be used instead of the Paste command to paste the text without the associated formatting.)

The Cut, Copy, and Paste commands are found in the Edit menu, or alternatively, can be executed by clicking the appropriate buttons on the Standard toolbar. The contents of the Windows clipboard are replaced by each subsequent Cut or Copy command, but are unaffected by the Paste command. The contents of the clipboard can be pasted into multiple locations in the same or different documents.

Microsoft Office has its own clipboard that enables you to collect and paste multiple items. The ***Office clipboard*** differs from the Windows clipboard in that the contents of each successive Copy command are added to the clipboard. Thus, you could copy the first paragraph of a document to the Office clipboard, then copy (add) a bulleted list in the middle of the document to the Office clipboard, and finally copy (add) the last paragraph (three items in all) to the Office clipboard. You could then go to another place in the document or to a different document altogether, and paste the contents of the Office clipboard (three separate items) with a single command.

Selected text is copied automatically to the Office clipboard regardless of whether you use the Copy command in the Edit menu, the Copy button on the Standard toolbar, or the Ctrl+C shortcut. The Office clipboard is accessed through the Edit menu and/or the task pane.

FIND, REPLACE, AND GO TO COMMANDS

The Find, Replace, and Go To commands share a common dialog box with different tabs for each command as shown in Figure 2.1. The ***Find command*** locates one or more occurrences of specific text (e.g., a word or phrase). The ***Replace command*** goes one step further in that it locates the text, and then enables you to optionally replace (one or more occurrences of) that text with different text. The ***Go To command*** goes directly to a specific place (e.g., a specific page) in the document.

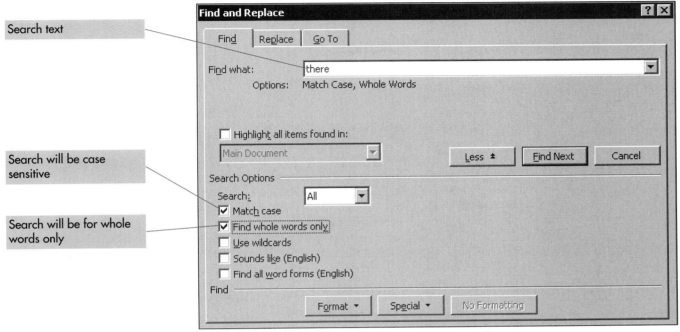

(a) Find Command

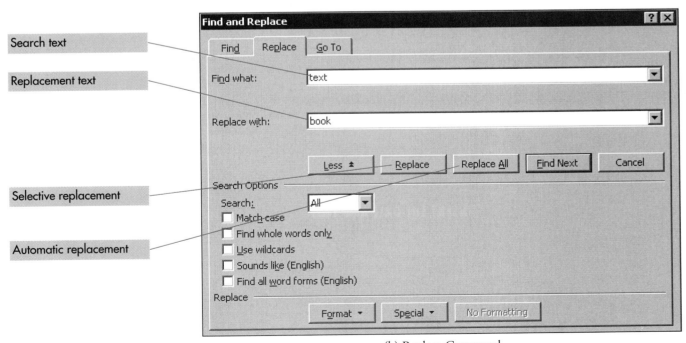

(b) Replace Command

FIGURE 2.1 *The Find, Replace, and Go To Commands*

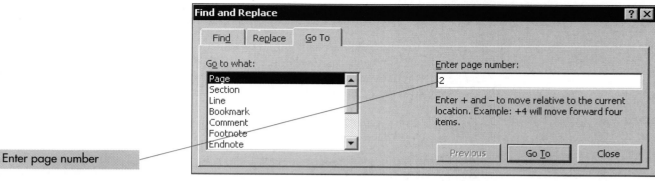

(c) Go To Command

FIGURE 2.1 *The Find, Replace, and Go To Commands (continued)*

The search in both the Find and Replace commands is case sensitive or case insensitive. A ***case-sensitive search*** (where Match Case is selected as in Figure 2.1a) matches not only the text, but also the use of upper- and lowercase letters. Thus, *There* is different from *there,* and a search on one will not identify the other. A ***case-insensitive search*** (where Match Case is *not* selected as in Figure 2.1b) is just the opposite and finds both *There* and *there.* A search may also specify ***whole words only*** to identify *there,* but not *therefore* or *thereby.* And finally, the search and replacement text can also specify different numbers of characters; for example, you could replace *16* with *sixteen.*

The Replace command in Figure 2.1b implements either ***selective replacement***, which lets you examine each occurrence of the character string in context and decide whether to replace it, or ***automatic replacement***, where the substitution is made automatically. Selective replacement is implemented by clicking the Find Next command button, then clicking (or not clicking) the Replace button to make the substitution. Automatic replacement (through the entire document) is implemented by clicking the Replace All button. This often produces unintended consequences and is not recommended; for example, if you substitute the word *text* for *book,* the word *textbook* would become *texttext,* which is not what you had in mind.

The Find and Replace commands can include formatting and/or special characters. You can, for example, change all italicized text to boldface, or you can change five consecutive spaces to a tab character. You can also use special characters in the character string such as the "any character" (consisting of ^?). For example, to find all four-letter words that begin with "f" and end with "l" (such as *fall, fill,* or *fail*), search for f^?^?l. (The question mark stands for any character, just like a ***wild card*** in a card game.) You can also search for all forms of a word; for example, if you specify *am,* it will also find *is* and *are.* You can even search for a word based on how it sounds. When searching for *Marion,* for example, check the Sounds Like check box, and the search will find both *Marion* and *Marian.*

INSERT THE DATE AND TIME

Most documents include the date and time they were created. Pull down the Insert menu, select the Date and Time command to display the Date and Time dialog box, then choose a format. Check the box to update the date automatically if you want your document to reflect the date on which it is opened, or clear the box to retain the date on which the document was created. See practice exercise 5 at the end of the chapter.

SCROLLING

Scrolling occurs when a document is too large to be seen in its entirety. Figure 2.2a displays a large printed document, only part of which is visible on the screen as illustrated in Figure 2.2b. In order to see a different portion of the document, you need to scroll, whereby new lines will be brought into view as the old lines disappear.

To: Our Students
From: Robert Grauer and Mary Ann Barber

Welcome to the wonderful world of word processing. Over the next several chapters we will build a foundation in the basics of Microsoft Word, and then teach you to format specialized documents, create professional looking tables and charts, publish well-designed newsletters, and create Web pages. Before you know it, you will be a word processing and desktop publishing wizard!

The first chapter presented the basics of word processing and showed you how to create a simple document. You learned how to insert, replace, and/or delete text. This chapter will teach you about fonts and special effects (such as **boldfacing** and *italicizing*) and how to use them effectively — how too little is better than too much.

You will go on to experiment with margins, tab stops, line spacing, and justification, learning first to format simple documents and then going on to longer, more complex ones. It is with the latter that we explore headers and footers, page numbering, widows and orphans (yes, we really did mean widows and orphans). It is here that we bring in graphics, working with newspaper-type columns, and the elements of a good page design. And without question, we will introduce the tools that make life so much easier (and your writing so much more impressive) — the Spell Check, Grammar Check, Thesaurus, and Styles.

If you are wondering what all these things are, read on in the text and proceed with the hands-on exercises. We will show you how to create a simple newsletter, and then improve it by adding graphics, fonts, and WordArt. You will create a simple calendar using the Tables feature, and then create more intricate forms that will rival anything you have seen. You will learn how to create a résumé with your beginner's skills, and then make it look like so much more with your intermediate (even advanced) skills. You will learn how to download resources from the Internet and how to create your own Web page. Last, but not least, run a mail merge to produce the cover letters that will accompany your résumé as it is mailed to companies across the United States (and even the world).

It is up to you to practice, for it is only through working at the computer, that you will learn what you need to know. Experiment and don't be afraid to make mistakes. Practice and practice some more.

Our goal is for you to learn and to enjoy what you are learning. We have great confidence in you, and in our ability to help you discover what you can do. Visit the home page for the Exploring Windows series. You can also send us e-mail. Bob's address is rgrauer@miami.edu. Mary Ann's address is mbarber@miami.edu. As you read the last sentence, notice that Microsoft Word is Web-enabled and that the Internet and e-mail references appear as hyperlinks in this document. Thus, you can click the address of our home page from within this document and then you can view the page immediately, provided you have an Internet connection. You can also click the e-mail address to open your mail program, provided it has been configured correctly.

We look forward to hearing from you and hope that you will like our textbook. You are about to embark on a wonderful journey toward computer literacy. Be patient and inquisitive.

(a) Printed Document

FIGURE 2.2 *Scrolling*

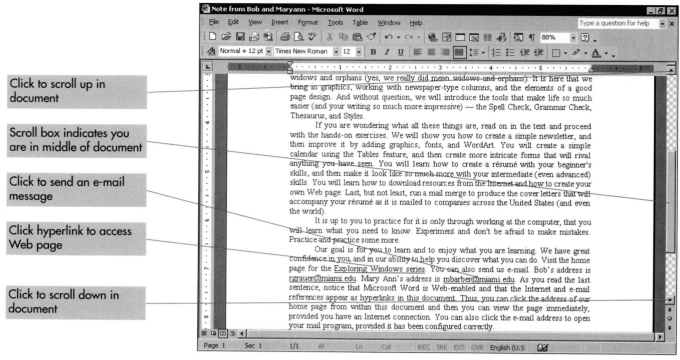

(b) Screen Display

FIGURE 2.2 *Scrolling (continued)*

Scrolling comes about automatically as you reach the bottom of the screen. Entering a new line of text, clicking on the down arrow within the scroll bar, or pressing the down arrow key brings a new line into view at the bottom of the screen and simultaneously removes a line at the top. (The process is reversed at the top of the screen.)

Scrolling can be done with either the mouse or the keyboard. Scrolling with the mouse (e.g., clicking the down arrow in the scroll bar) changes what is displayed on the screen, but does not move the insertion point, so that you must click the mouse after scrolling prior to entering the text at the new location. Scrolling with the keyboard, however (e.g., pressing Ctrl+Home or Ctrl+End to move to the beginning or end of a document, respectively), changes what is displayed on the screen as well as the location of the insertion point, and you can begin typing immediately.

Scrolling occurs most often in a vertical direction as shown in Figure 2.2. It can also occur horizontally, when the length of a line in a document exceeds the number of characters that can be displayed horizontally on the screen.

WRITE NOW, EDIT LATER

You write a sentence, then change it, and change it again, and one hour later you've produced a single paragraph. It happens to every writer—you stare at a blank screen and flashing cursor and are unable to write. The best solution is to brainstorm and write down anything that pops into your head, and to keep on writing. Don't worry about typos or spelling errors because you can fix them later. Above all, resist the temptation to continually edit the few words you've written because overediting will drain the life out of what you are writing. The important thing is to get your ideas on paper.

VIEW MENU

The ***View menu*** provides different views of a document. Each view can be displayed at different magnifications, which in turn determine the amount of scrolling necessary to see remote parts of a document.

The ***Normal view*** is the default view and it provides the fastest way to enter text. The ***Print Layout view*** more closely resembles the printed document and displays the top and bottom margins, headers and footers, page numbers, graphics, and other features that do not appear in the Normal view. The Normal view tends to be faster because Word spends less time formatting the display.

The ***Zoom command*** displays the document on the screen at different magnifications; for example, 75%, 100%, or 200%. (The Zoom command does not affect the size of the text on the printed page.) A Zoom percentage (magnification) of 100% displays the document in the approximate size of the text on the printed page. You can increase the percentage to 200% to make the characters appear larger. You can also decrease the magnification to 75% to see more of the document at one time.

Word will automatically determine the magnification if you select one of four additional Zoom options—Page Width, Text Width, Whole Page, or Many Pages (Whole Page and Many Pages are available only in the Print Layout view). Figure 2.3, for example, displays a two-page document in Print Layout view. The 40% magnification is determined automatically once you specify the number of pages.

The View menu also provides access to two additional views—the Outline view and the Web Layout view. The Outline view does not display a conventional outline, but rather a structural view of a document that can be collapsed or expanded as necessary. The Web Layout view is used when you are creating a Web page. Both views are discussed in later chapters.

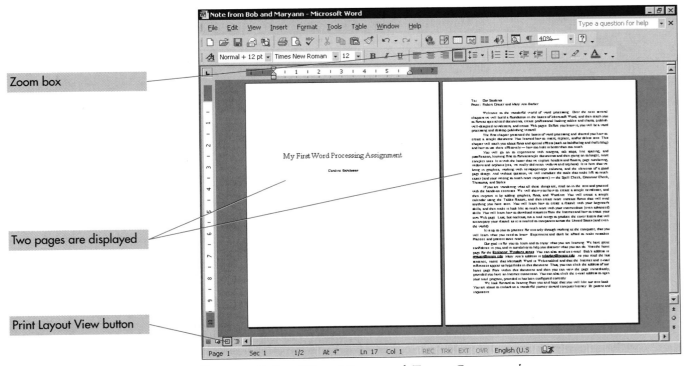

FIGURE 2.3 *View Menu and Zoom Command*

Hands-on Exercise 1

Editing a Document

Objective To edit an existing document; to use the Find and Replace commands; to move and copy text using the clipboard and the drag-and-drop facility. Use Figure 2.4 as a guide in the exercise.

Step 1: **The View Menu**

➤ Start Word as described in the hands-on exercises from Chapter 1. Pull down the **File menu** and click **Open** (or click the **Open button** on the toolbar).
  • Click the **drop-down arrow** on the Look In list box. Click the appropriate drive, drive C or drive A, depending on the location of your data.
  • Double click the **Exploring Word folder** to make it the active folder (the folder in which you will save the document).
  • Scroll in the Name list box (if necessary) until you can click the **Note from Bob and Maryann** to select this document. Double click the **document icon** or click the **Open command button** to open the file.
➤ The document should appear on the screen as shown in Figure 2.4a.
➤ Change to the Print Layout view at Page Width magnification:
  • Pull down the **View menu** and click **Print Layout** (or click the **Print Layout View button** above the status bar) as shown in Figure 2.4a.
  • Click the **down arrow** in the Zoom box to change to **Page Width**.
➤ Click and drag the mouse to select the phrase **Our Students**, which appears at the beginning of the document. Type your name to replace the selected text.
➤ Pull down the **File menu**, click the **Save As command**, then save the document as **Modified Note**. (This creates a second copy of the document.)

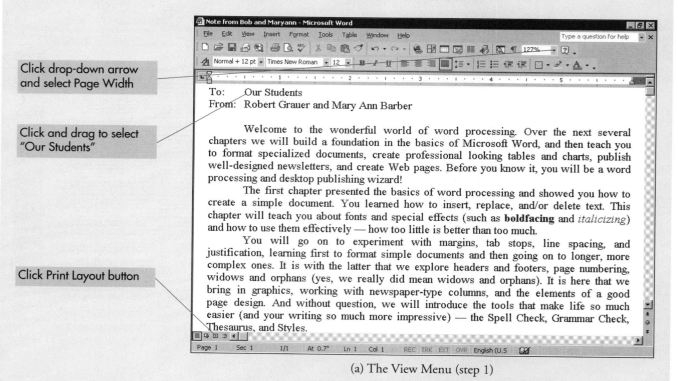

(a) The View Menu (step 1)

Figure 2.4 *Hands-on Exercise 1*

Step 2: **Scrolling**

> ➤ Click and drag the **scroll box** within the vertical scroll bar to scroll to the end of the document as shown in Figure 2.4b. Click immediately before the period at the end of the last sentence.
> ➤ Type a **comma** and a space, then insert the phrase **but most of all, enjoy**.
> ➤ Drag the **scroll box** to the top of the scroll bar to get back to the beginning of the document.
> ➤ Click immediately before the period ending the first sentence, press the **space bar**, then add the phrase **and desktop publishing**.
> ➤ Use the keyboard to practice scrolling shortcuts. Press **Ctrl+Home** and **Ctrl+End** to move to the beginning and end of a document, respectively. Press **PgUp** or **PgDn** to scroll one screen in the indicated direction.
> ➤ Save the document.

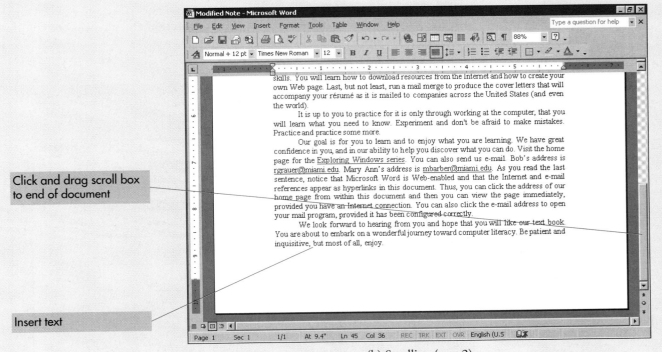

(b) Scrolling (step 2)

FIGURE 2.4   *Hands-on Exercise 1 (continued)*

THE MOUSE AND THE SCROLL BAR

Scroll quickly through a document by clicking above or below the scroll box to scroll up or down an entire screen. Move to the top, bottom, or an approximate position within a document by dragging the scroll box to the corresponding position in the scroll bar; for example, dragging the scroll box to the middle of the bar moves the mouse pointer to the middle of the document. Scrolling with the mouse does not change the location of the insertion point, however, and thus you must click the mouse at the new location prior to entering text at that location.

Step 3: **The Replace Command**

➤ Press **Ctrl+Home** to move to the beginning of the document. Pull down the **Edit menu**. Click **Replace** to produce the dialog box of Figure 2.4c. Click the **More button** to display the available options. Clear the check boxes.

➤ Type **text** in the Find what text box. Press the **Tab key**. Type **book** in the Replace with text box.

➤ Click the **Find Next button** to find the first occurrence of the word *text*. The dialog box remains on the screen and the first occurrence of *text* is selected. This is *not* an appropriate substitution.

➤ Click the **Find Next button** to move to the next occurrence without making the replacement. This time the substitution is appropriate.

➤ Click **Replace** to make the change and automatically move to the next occurrence where the substitution is again inappropriate. Click **Find Next** a final time. Word will indicate that it has finished searching the document. Click **OK**.

➤ Change the Find and Replace strings to **Mary Ann** and **Maryann**, respectively. Click the **Replace All button** to make the substitution globally without confirmation. Word will indicate that two replacements were made. Click **OK**.

➤ Close the dialog box. Save the document.

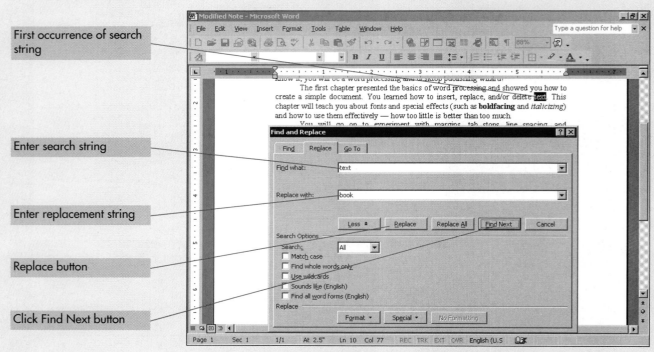

(c) The Replace Command (step 3)

FIGURE 2.4 *Hands-on Exercise 1 (continued)*

SEARCH FOR SPECIAL CHARACTERS

Use the Find and Replace commands to search for special characters such as tabs or paragraph marks. Click the More button in either dialog box, then click the Special command button that appears in the expanded dialog box to search for the additional characters. You could, for example, replace erroneous paragraph marks with a simple space, or replace five consecutive spaces with a Tab character.

Step 4: **The Windows Clipboard**

➤ Press **PgDn** to scroll toward the end of the document until you come to the paragraph beginning **It is up to you**. Select the sentence **Practice and practice some more** by dragging the mouse. (Be sure to include the period.)

➤ Pull down the **Edit menu** and click the **Copy command** or click the **Copy button**.

➤ Press **Ctrl+End** to scroll to the end of the document. Press the **space bar**. Pull down the **Edit menu** and click the **Paste command** (or click the **Paste button**).

➤ Click the **Paste Options button** if it appears as shown in Figure 2.4d to see the available options, then press **Esc** to suppress the context-sensitive menu.

➤ Click and drag to select the sentence asking you to visit our home page, which includes a hyperlink (underlined blue text). Click the **Copy button**.

➤ Press **Ctrl+End** to move to the end of the document. Pull down the **Edit menu**, click the **Paste Special command** to display the Paste Special dialog box. Select **Unformatted text** and click **OK**.

➤ The sentence appears at the end of the document, but without the hyperlink formatting. Click the **Undo button** since we do not want the sentence. You have, however, seen the effect of the Paste Special command.

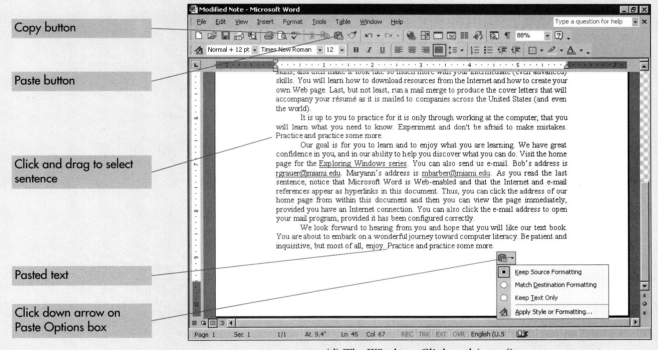

(d) The Windows Clipboard (step 4)

FIGURE 2.4 *Hands-on Exercise 1 (continued)*

PASTE OPTIONS

Text can be copied with or without the associated formatting according to the selected option in the Paste Options button. (The button appears automatically whenever the source and destination paragraphs have different formatting.) The default is to keep the source formatting (the formatting of the copied object). The button disappears as soon as you begin typing.

Step 5: **The Office Clipboard**

➤ Pull down the **Edit menu** and click the **Office Clipboard command** to open the task pane as shown in Figure 2.4e. The contents of your clipboard will differ.
➤ Right click the first entry in the task pane that asks you to visit our home page, then click the **Delete command**. Delete all other items except the one urging you to practice what was copied in the last hands-on exercise.
➤ Click and drag to select the three sentences that indicate you can send us e-mail, and that contain our e-mail addresses. Click the **Copy button** to copy these sentences to the Office clipboard, which now contains two icons.
➤ Press **Ctrl+End** to move to the end of the document, press **enter** to begin a new paragraph, and press the **Tab key** to indent the paragraph. Click the **Paste All button** on the Office clipboard to paste both items at the end of the document. (You may have to add a space between the two sentences.)
➤ Close the task pane.

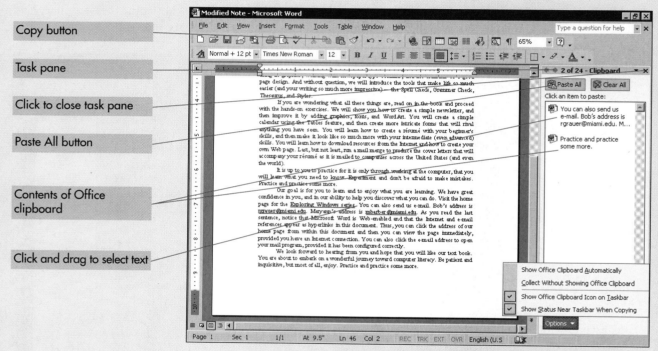

(e) The Office Clipboard (step 5)

FIGURE 2.4 *Hands-on Exercise 1 (continued)*

THE OFFICE CLIPBOARD

The Office clipboard is different from the Windows clipboard. Each successive Cut or Copy command (in any Office application) adds an object to the Office clipboard (up to a maximum of 24), whereas it replaces the contents of the Windows clipboard. You may, however, have to set the option to automatically copy to the Office clipboard for this to take place. Pull down the Edit menu, click the Office Clipboard command to open the task pane, and click the Options button at the bottom of the task pane. Check the option to always copy to the Office clipboard.

Step 6: **Undo and Redo Commands**

- Click the **drop-down arrow** next to the Undo button to display the previously executed actions as in Figure 2.4f. The list of actions corresponds to the editing commands you have issued since the start of the exercise.
- Click **Paste** (the first command on the list) to undo the last editing command; the sentence asking you to send us e-mail disappears from the last paragraph.
- Click the **Undo button** a second time and the sentence, Practice and practice some more, disappears from the end of the last paragraph.
- Click the remaining steps on the undo list to retrace your steps through the exercise one command at a time. Alternatively, you can scroll to the bottom of the list and click the last command.
- Either way, when the undo list is empty, you will have the document as it existed at the start of the exercise. Click the **drop-down arrow** for the Redo command to display the list of commands you have undone.
- Click each command in sequence (or click the command at the bottom of the list) and you will restore the document.
- Save the document.

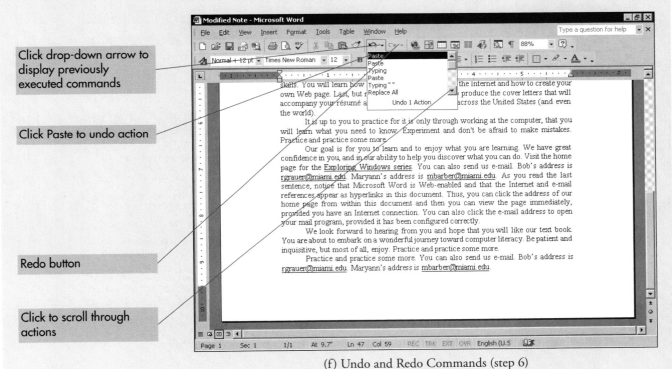

(f) Undo and Redo Commands (step 6)

FIGURE 2.4 *Hands-on Exercise 1 (continued)*

KEYBOARD SHORTCUTS—CUT, COPY AND PASTE

Ctrl+X, Ctrl+C, and Ctrl+V are keyboard shortcuts to cut, copy, and paste, respectively. The "X" is supposed to remind you of a pair of scissors. The shortcuts are easier to remember when you realize that the operative letters, X, C, and V, are next to each other on the keyboard. The shortcuts work in virtually any Windows application. See practice exercise 2 at the end of the chapter.

Step 7: **Drag and Drop**

➤ Scroll to the top of the document. Click and drag to select the phrase **format specialized documents** (including the comma and space) as shown in Figure 2.4g, then drag the phrase to its new location immediately before the word *and*. (A dotted vertical bar appears as you drag the text, to indicate its new location.)

➤ Release the mouse button to complete the move. Click the **drop-down arrow** for the Undo command; click **Move** to undo the move.

➤ To copy the selected text to the same location (instead of moving it), press and hold the **Ctrl key** as you drag the text to its new location. (A plus sign appears as you drag the text, to indicate it is being copied rather than moved.)

➤ Practice the drag-and-drop procedure several times until you are confident you can move and copy with precision.

➤ Click anywhere in the document to deselect the text. Save the document.

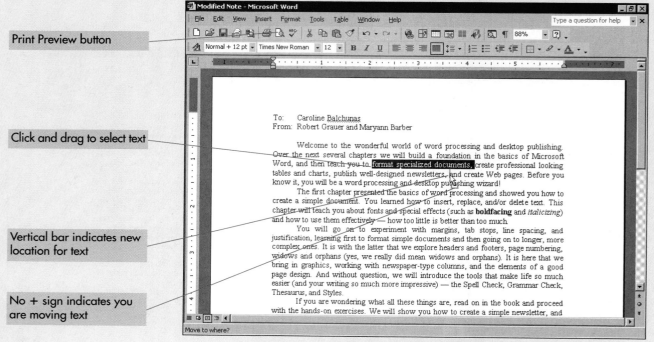

(g) Drag and Drop (step 7)

FIGURE 2.4 *Hands-on Exercise 1 (continued)*

SELECTING TEXT

The selection bar, a blank column at the far left of the document window, makes it easy to select a line, paragraph, or the entire document. To select a line, move the mouse pointer to the selection bar, point to the line and click the left mouse button. To select a paragraph, move the mouse pointer to the selection bar, point to any line in the paragraph, and double click the mouse. To select the entire document, move the mouse pointer to the selection bar and press the Ctrl key while you click the mouse.

Step 8: **The Print Preview Command**

- Pull down the **File menu** and click **Print Preview** (or click the **Print Preview button** on the Standard toolbar). You should see your entire document as shown in Figure 2.4h.
- Check that the entire document fits on one page—that is, check that you can see the last paragraph. If not, click the **Shrink to Fit button** on the toolbar to automatically change the font size in the document to force it onto one page.
- Click the **Print button** to print the document so that you can submit it to your instructor. Click the **Close button** to exit Print Preview and return to your document.
- Close the document. Exit Word if you do not want to continue with the next exercise at this time.

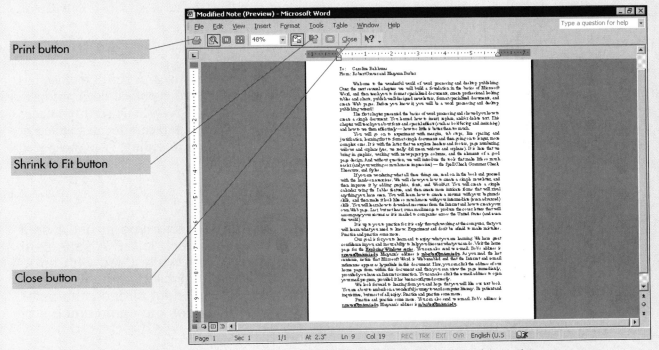

(h) The Print Preview Command (step 8)

FIGURE 2.4 *Hands-on Exercise 1 (continued)*

INSERT COMMENTS INTO A DOCUMENT

Share your thoughts electronically with colleagues and other students by inserting comments into a document. Click in the document where you want the comment to go, pull down the Insert menu, click the Comment command, and enter the text of the comment. All comments appear on the screen in the right margin of the document. The comments can be printed or suppressed according to the option selected in the Print command. See exercise 9 at the end of the chapter.

TYPOGRAPHY

Typography is the process of selecting typefaces, type styles, and type sizes. The importance of these decisions is obvious, for the ultimate success of any document depends greatly on its appearance. Type should reinforce the message without calling attention to itself and should be consistent with the information you want to convey.

A ***typeface*** or ***font*** is a complete set of characters (upper- and lowercase letters, numbers, punctuation marks, and special symbols). Figure 2.5 illustrates three typefaces—***Times New Roman***, ***Arial***, and ***Courier New***—that are accessible from any Windows application.

A definitive characteristic of any typeface is the presence or absence of tiny cross lines that end the main strokes of each letter. A ***serif typeface*** has these lines. A ***sans serif typeface*** (*sans* from the French for *without*) does not. Times New Roman and Courier New are examples of a serif typeface. Arial is a sans serif typeface.

Typography is the process of selecting typefaces, type styles, and type sizes. A serif typeface has tiny cross strokes that end the main strokes of each letter. A sans serif typeface does not have these strokes. Serif typefaces are typically used with large amounts of text. Sans serif typefaces are used for headings and limited amounts of text. A proportional typeface allocates space in accordance with the width of each character and is what you are used to seeing. A monospaced typeface uses the same amount of space for every character.

(a) Times New Roman (serif and proportional)

Typography is the process of selecting typefaces, type styles, and type sizes. A serif typeface has tiny cross strokes that end the main strokes of each letter. A sans serif typeface does not have these strokes. Serif typefaces are typically used with large amounts of text. Sans serif typefaces are used for headings and limited amounts of text. A proportional typeface allocates space in accordance with the width of each character and is what you are used to seeing. A monospaced typeface uses the same amount of space for every character.

(b) Arial (sans serif and proportional)

```
Typography is the process of selecting typefaces, type styles,
and type sizes. A serif typeface has tiny cross strokes that end
the main strokes of each letter. A sans serif typeface does not
have these strokes. Serif typefaces are typically used with large
amounts of text. Sans serif typefaces are used for headings and
limited amounts of text. A proportional typeface allocates space
in accordance with the width of each character and is what you
are used to seeing. A monospaced typeface uses the same amount of
space for every character.
```

Courier New (serif and monospaced)

FIGURE 2.5 *Typefaces*

Serifs help the eye to connect one letter with the next and are generally used with large amounts of text. This book, for example, is set in a serif typeface. A sans serif typeface is more effective with smaller amounts of text and appears in headlines, corporate logos, airport signs, and so on.

A second characteristic of a typeface is whether it is monospaced or proportional. A ***monospaced typeface*** (e.g., Courier New) uses the same amount of space for every character regardless of its width. A ***proportional typeface*** (e.g., Times New Roman or Arial) allocates space according to the width of the character. Monospaced fonts are used in tables and financial projections where text must be precisely lined up, one character underneath the other. Proportional typefaces create a more professional appearance and are appropriate for most documents. Any typeface can be set in different ***type styles*** (such as regular, **bold**, or *italic*).

Type Size

Type size is a vertical measurement and is specified in points. One ***point*** is equal to $1/72$ of an inch; that is, there are 72 points to the inch. The measurement is made from the top of the tallest letter in a character set (for example, an uppercase T) to the bottom of the lowest letter (for example, a lowercase y). Most documents are set in 10 or 12 point type. Newspaper columns may be set as small as 8 point type, but that is the smallest type size you should consider. Conversely, type sizes of 14 points or higher are ineffective for large amounts of text.

Figure 2.6 shows the same phrase set in varying type sizes. Some typefaces appear larger (smaller) than others even though they may be set in the same point size. The type in Figure 2.6a, for example, looks smaller than the corresponding type in Figure 2.6b even though both are set in the same point size. Note, too, that you can vary the type size of a specific font within a document for emphasis. The eye needs at least two points to distinguish between different type sizes.

Format Font Command

The ***Format Font command*** gives you complete control over the typeface, size, and style of the text in a document. Executing the command before entering text will set the format of the text you type from that point on. You can also use the command to change the font of existing text by selecting the text, then executing the command. Either way, you will see the dialog box in Figure 2.7, in which you specify the font (typeface), style, and point size.

You can choose any of the special effects, such as SMALL CAPS, superscripts, or subscripts. You can also change the underline options (whether or not spaces are to be underlined). You can even change the color of the text on the monitor, but you need a color printer for the printed document. (The Character Spacing and Text Effects tabs produce different sets of options in which you control the spacing and appearance of the characters and are beyond the scope of our discussion.)

TYPOGRAPHY TIP—USE RESTRAINT

More is not better, especially in the case of too many typefaces and styles, which produce cluttered documents that impress no one. Try to limit yourself to a maximum of two typefaces per document, but choose multiple sizes and/or styles within those typefaces. Use boldface or italics for emphasis; but do so in moderation, because if you emphasize too many elements, the effect is lost.

This is Arial 8 point type

This is Arial 10 point type

This is Arial 12 point type

This is Arial 18 point type

This is Arial 24 point type

This is Arial 30 point type

(a) Sans Serif Typeface

This is Times New Roman 8 point type

This is Times New Roman 10 point type

This is Times New Roman 12 point type

This is Times New Roman 18 point type

This is Times New Roman 24 point type

This is Times New Roman 30 point

(b) Serif Typeface

FIGURE 2.6 *Type Size*

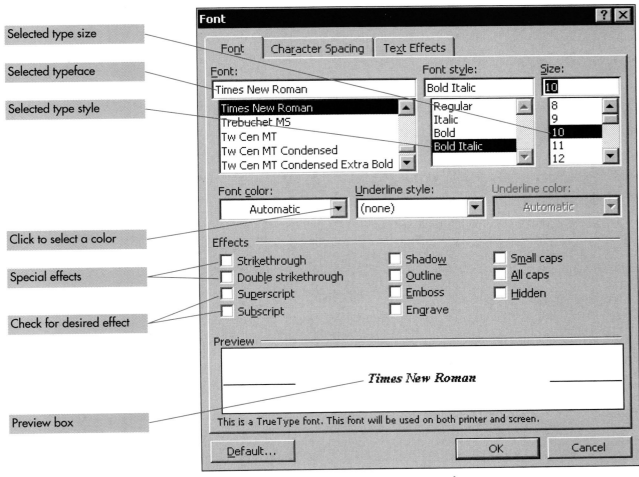

FIGURE 2.7 *Format Font Command*

The Preview box shows the text as it will appear in the document. The message at the bottom of the dialog box indicates that Times New Roman is a TrueType font and that the same font will be used on both the screen and the printer. TrueType fonts ensure that your document is truly WYSIWYG (What You See Is What You Get) because the fonts you see on the monitor will be identical to those in the printed document.

PAGE SETUP COMMAND

The ***Page Setup command*** in the File menu lets you change margins, paper size, orientation, paper source, and/or layout. All parameters are accessed from the dialog box in Figure 2.8 by clicking the appropriate tab within the dialog box.

The default margins are indicated in Figure 2.8a and are one inch on the top and bottom of the page, and one and a quarter inches on the left and right. You can change any (or all) of these settings by entering a new value in the appropriate text box, either by typing it explicitly or clicking the up/down arrow. All of the settings in the Page Setup command apply to the whole document regardless of the position of the insertion point. (Different settings for any option in the Page Setup dialog box can be established for different parts of a document by creating sections. Sections also affect column formatting, as discussed later in the chapter.)

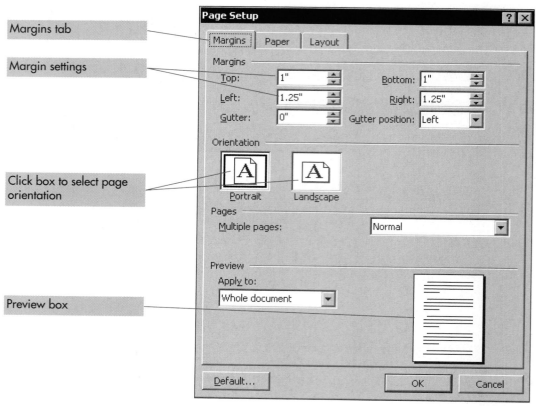

(a) Margins Tab

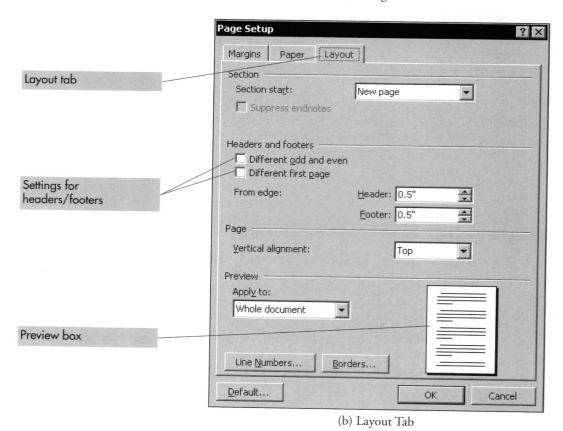

(b) Layout Tab

FIGURE 2.8 *Page Setup Command*

The ***Margins tab*** also enables you to change the orientation of a page as shown in Figure 2.8b. ***Portrait orientation*** is the default. ***Landscape orientation*** flips the page 90 degrees so that its dimensions are 11 × 8½ rather than the other way around. Note, too, the Preview area in both Figures 2.8a and 2.8b, which shows how the document will appear with the selected parameters.

The Paper tab (not shown in Figure 2.8) is used to specify which tray should be used on printers with multiple trays, and is helpful when you want to load different types of paper simultaneously. The Layout tab in Figure 2.8b is used to specify options for headers and footers (text that appears at the top or bottom of each page in a document), and/or to change the vertical alignment of text on the page.

Page Breaks

One of the first concepts you learned was that of word wrap, whereby Word inserts a soft return at the end of a line in order to begin a new line. The number and/or location of the soft returns change automatically as you add or delete text within a document. Soft returns are very different from the hard returns inserted by the user, whose number and location remain constant.

In much the same way, Word creates a ***soft page break*** to go to the top of a new page when text no longer fits on the current page. And just as you can insert a hard return to start a new paragraph, you can insert a ***hard page break*** to force any part of a document to begin on a new page. A hard page break is inserted into a document using the Break command in the Insert menu or more easily through the Ctrl+enter keyboard shortcut. (You can prevent the occurrence of awkward page breaks through the Format Paragraph command as described later in the chapter.)

AN EXERCISE IN DESIGN

The following exercise has you retrieve an existing document from the set of practice files, then experiment with various typefaces, type styles, and point sizes. The original document uses a monospaced (typewriter style) font, without boldface or italics, and you are asked to improve its appearance. The first step directs you to save the document under a new name so that you can always return to the original if necessary.

There is no right and wrong with respect to design, and you are free to choose any combination of fonts that appeals to you. The exercise takes you through various formatting options but lets you make the final decision. It does, however, ask you to print the final document and submit it to your instructor. Experiment freely and print multiple versions with different designs.

> **IMPOSE A TIME LIMIT**
>
> A word processor is supposed to save time and make you more productive. It will do exactly that, provided you use the word processor for its primary purpose—writing and editing. It is all too easy, however, to lose sight of that objective and spend too much time formatting the document. Concentrate on the content of your document rather than its appearance. Impose a time limit on the amount of time you will spend on formatting. End the session when the limit is reached.

Hands-on Exercise 2

Character Formatting

Objective To experiment with character formatting; to change fonts and to use boldface and italics; to copy formatting with the format painter; to insert a page break and see different views of a document. Use Figure 2.9 as a guide in the exercise.

Step 1: **Open the Existing Document**

➤ Start Word. Pull down the **File menu** and click **Open** (or click the **Open button** on the toolbar). To open a file:
  • Click the **drop-down arrow** on the Look In list box. Click the appropriate drive, drive C or drive A, depending on the location of your data.
  • Double click the **Exploring Word folder** to make it the active folder (the folder in which you will open and save the document).
  • Scroll in the **Open list box** (if necessary) until you can click **Tips for Writing** to select this document.
➤ Double click the **document icon** or click the **Open command button** to open the document shown in Figure 2.9a.
➤ Pull down the **File menu**. Click the **Save As command** to save the document as **Modified Tips**. The new document name appears on the title bar.
➤ Pull down the **View menu** and click **Normal** (or click the **Normal View button** above the status bar). Set the magnification (zoom) to **Page Width**.

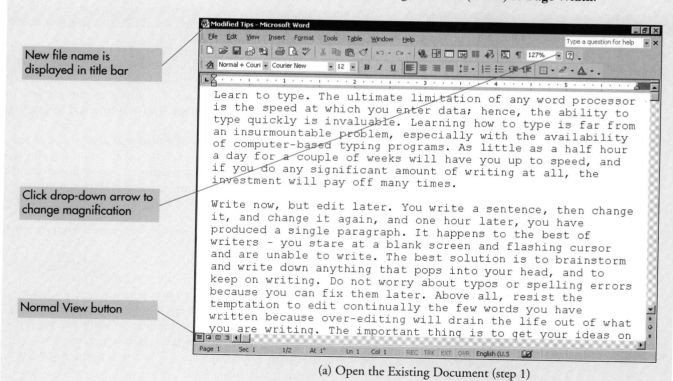

(a) Open the Existing Document (step 1)

FIGURE 2.9 *Hands-on Exercise 2*

Step 2: **Change the Font**

- Pull down the **Edit menu** and click the **Select All command** (or press **Ctrl+A**) to select the entire document as shown in Figure 2.9b.
- Click the **down arrow** on the Font List box and choose a different font. We selected **Times New Roman**. Click the **down arrow** on the Font Size list box and choose a different type size.
- Pull down the **Format menu** and select the **Font command** to display the Font dialog box, where you can also change the font and/or font size.
- Experiment with different fonts and font sizes until you are satisfied. We ended with 12 point Times New Roman.
- Save the document.

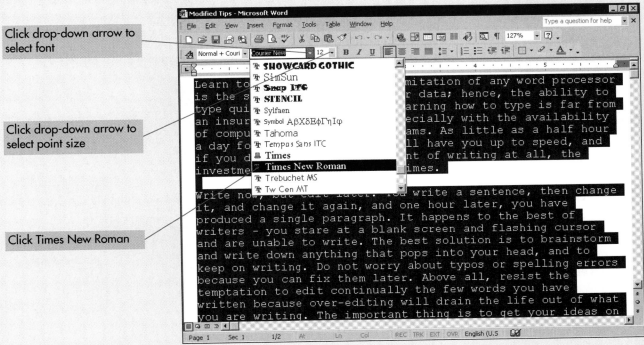

(b) Change the Font (step 2)

FIGURE 2.9 *Hands-on Exercise 2 (continued)*

FIND AND REPLACE FORMATTING

The Replace command enables you to replace formatting as well as text. To replace any text set in bold with the same text in italics, pull down the Edit menu, and click the Replace command. Click the Find what text box, but do *not* enter any text. Click the More button to expand the dialog box. Click the Format command button, click Font, click Bold in the Font Style list, and click OK. Click the Replace with text box and again do *not* enter any text. Click the Format command button, click Font, click Italic in the Font Style list, and click OK. Click the Find Next or Replace All command button to do selective or automatic replacement. Use a similar technique to replace one font with another.

Step 3: **Boldface and Italics**

- Select the sentence **Learn to type** at the beginning of the document.
- Click the **Italic button** on the Formatting toolbar to italicize the selected phrase, which will remain selected after the italics take effect.
- Click the **Bold button** to boldface the selected text. The text is now in bold italic.
- Pull down the **View menu** and open the task pane. Click the **down arrow** in the task pane and select **Reveal Formatting** as shown in Figure 2.9c.
- Click anywhere in the heading, **Learn to Type**, to display its formatting properties. This type of information can be invaluable if you are unsure of the formatting in effect. Close the task pane.
- Experiment with different styles (bold, italics, underlining, bold italics) until you are satisfied. Each button functions as a toggle switch to turn the selected effect on or off.

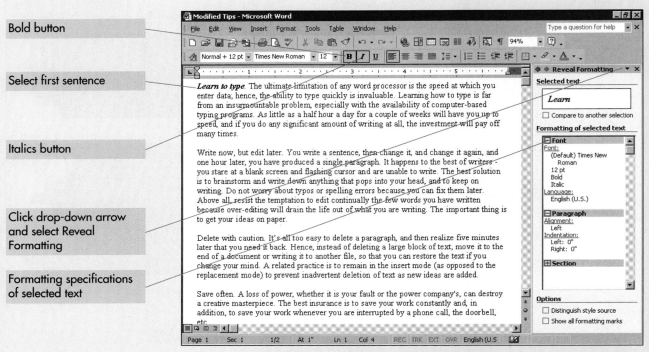

(c) Boldface and Italics (step 3)

FIGURE 2.9 *Hands-on Exercise 2 (continued)*

UNDERLINING TEXT

Underlining is less popular than it was, but Word provides a complete range of underlining options. Select the text to underline, pull down the Format menu, click Font to display the Font dialog box, and click the Font tab if necessary. Click the down arrow on the Underline Style list box to choose the type of underlining you want. You can choose whether to underline the words only (i.e., the underline does not appear in the space between words). You can also choose the type of line you want—solid, dashed, thick, or thin.

Step 4: **The Format Painter**

- Click anywhere within the sentence Learn to Type. **Double click** the **Format Painter button** on the Standard toolbar. The mouse pointer changes to a paintbrush as shown in Figure 2.9d.
- Drag the mouse pointer over the next title, **Write now, but edit later**, and release the mouse. The formatting from the original sentence (bold italic) has been applied to this sentence as well.
- Drag the mouse pointer (in the shape of a paintbrush) over the remaining titles (the first sentence in each paragraph) to copy the formatting. You can click the down arrow on the vertical scroll bar to bring more of the document into view.
- Click the **Format Painter button** after you have painted the title of the last tip to turn the feature off. (Note that clicking the Format Painter button, rather than double clicking, will paint only one item.)
- Save the document.

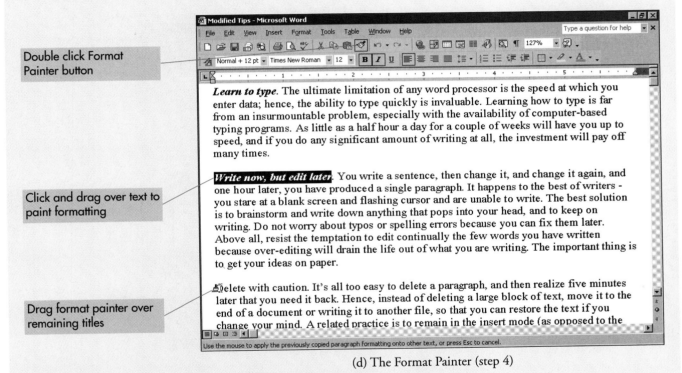

(d) The Format Painter (step 4)

FIGURE 2.9 *Hands-on Exercise 2 (continued)*

HIGHLIGHTING TEXT

You will love the Highlight tool, especially if you are in the habit of highlighting text with a pen. Click the down arrow next to the tool to select a color (yellow is the default) to change the mouse pointer to a pen, then click and drag to highlight the desired text. Continue dragging the mouse to highlight as many selections as you like. Click the Highlight tool a second time to turn off the feature. See practice exercise 2 at the end of the chapter.

Step 5: **Change Margins**

- Press **Ctrl+End** to move to the end of the document as shown in Figure 2.9e. You will see a dotted line indicating a soft page break. (If you do not see the page break, it means that your document fits on one page because you used a different font and/or a smaller point size. We used 12 point Times New Roman.)
- Pull down the **File menu**. Click **Page Setup**. Click the **Margins tab** if necessary. Change the bottom margin to **.75** inch.
- Check that these settings apply to the **Whole Document**. Click **OK**. Save the document.
- The page break disappears because more text fits on the page.

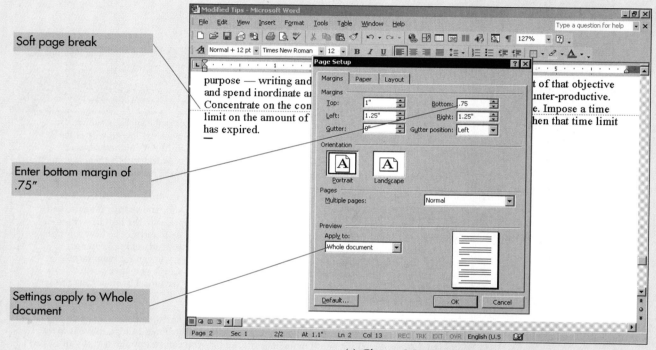

(e) Change Margins (step 5)

FIGURE 2.9 *Hands-on Exercise 2 (continued)*

DIALOG BOX SHORTCUTS

You can use keyboard shortcuts to select options in a dialog box. Press Tab (Shift+Tab) to move forward (backward) from one field or command button to the next. Press Alt plus the underlined letter to move directly to a field or command button. Press enter to activate the selected command button. Press Esc to exit the dialog box without taking action. Press the space bar to toggle check boxes on or off. Press the down arrow to open a drop-down list box once the list has been accessed, then press the up or down arrow to move between options in a list box. These are uniform shortcuts that apply to any Windows application.

Step 6: **Create the Title Page**

- Press **Ctrl+Home** to move to the beginning of the document. Press **enter** three or four times to add a few blank lines.
- Press **Ctrl+enter** to insert a hard page break. You will see the words "Page Break" in the middle of a dotted line as shown in Figure 2.9f.
- Press the **up arrow key** three times. Enter the title **Tips for Writing**. Select the title, and format it in a larger point size, such as 24 points.
- Press **enter** to move to a new line. Type your name and format it in a different point size, such as 14 points.
- Select both the title and your name as shown in the figure. Click the **Center button** on the Formatting toolbar.
- Save the document.

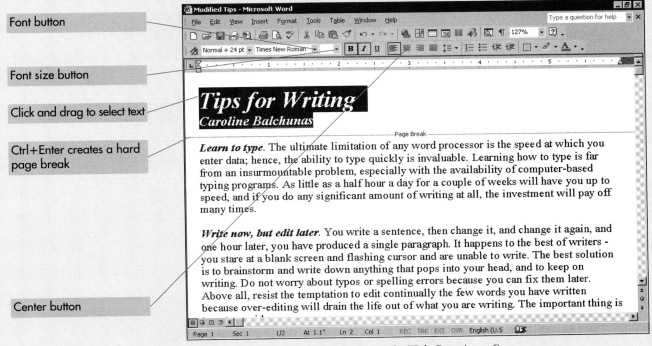

(f) Create the Title Page (step 6)

FIGURE 2.9 *Hands-on Exercise 2 (continued)*

DOUBLE CLICK AND TYPE

Creating a title page is a breeze if you take advantage of the (double) click and type feature. Pull down the View menu and change to the Print Layout view. Double click anywhere on the page and you can begin typing immediately at that location, without having to type several blank lines, or set tabs. The feature does not work in the Normal view or in a document that has columns. To enable (disable) the feature, pull down the Tools menu, click the Options command, click the Edit tab, then check (clear) the Enable Click and Type check box.

Step 7: **The Completed Document**

- Pull down the **View menu** and click **Print Layout** (or click the **Print Layout button** above the status bar).
- Click the **Zoom Control arrow** on the Standard toolbar and select **Two Pages**. Release the mouse to view the completed document in Figure 2.9g.
- You may want to add additional blank lines at the top of the title page to move the title further down on the page.
- Save the document. Be sure that the document fits on two pages (the title page and text), then click the **Print button** on the Standard toolbar to print the document for your instructor.
- Exit Word if you do not want to continue with the next exercise at this time.

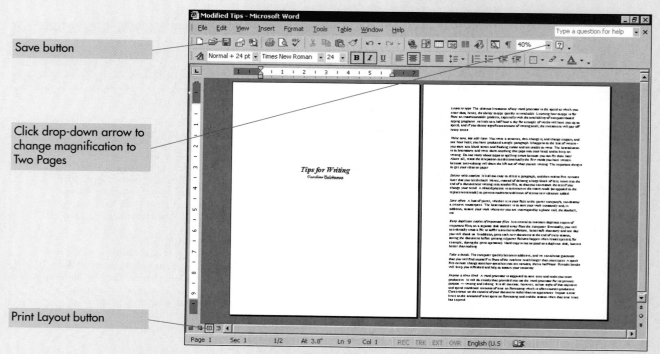

(g) The Completed Document (step 7)

FIGURE 2.9 *Hands-on Exercise 2 (continued)*

THE PAGE SETUP COMMAND

The Page Setup command controls the margins of a document, and by extension, it controls the amount of text that fits on a page. Pull down the File menu and click the Page Setup command to display the Page Setup dialog box, click the Margins tab, then adjust the left and right (or top and bottom) margins to fit additional text on a page. Click the down arrow in the Apply to area to select the whole document. Click OK to accept the settings and close the dialog box.

PARAGRAPH FORMATTING

A change in typography is only one way to alter the appearance of a document. You can also change the alignment, indentation, tab stops, or line spacing for any paragraph(s) within the document. You can control the pagination and prevent the occurrence of awkward page breaks by specifying that an entire paragraph has to appear on the same page, or that a one-line paragraph (e.g., a heading) should appear on the same page as the next paragraph. You can include borders or shading for added emphasis around selected paragraphs.

All of these features are implemented at the paragraph level and affect all selected paragraphs. If no paragraphs are selected, the commands affect the entire current paragraph (the paragraph containing the insertion point), regardless of the position of the insertion point when the command is executed.

Alignment

Text can be aligned in four different ways as shown in Figure 2.10. It may be justified (flush left/flush right), left aligned (flush left with a ragged right margin), right aligned (flush right with a ragged left margin), or centered within the margins (ragged left and right).

Left aligned text is perhaps the easiest to read. The first letters of each line align with each other, helping the eye to find the beginning of each line. The lines themselves are of irregular length. There is uniform spacing between words, and the ragged margin on the right adds white space to the text, giving it a lighter and more informal look.

Justified text produces lines of equal length, with the spacing between words adjusted to align at the margins. It may be more difficult to read than text that is left aligned because of the uneven (sometimes excessive) word spacing and/or the greater number of hyphenated words needed to justify the lines.

Type that is centered or right aligned is restricted to limited amounts of text where the effect is more important than the ease of reading. Centered text, for example, appears frequently on wedding invitations, poems, or formal announcements. Right aligned text is used with figure captions and short headlines.

Indents

Individual paragraphs can be indented so that they appear to have different margins from the rest of a document. Indentation is established at the paragraph level; thus different indentation can be in effect for different paragraphs. One paragraph may be indented from the left margin only, another from the right margin only, and a third from both the left and right margins. The first line of any paragraph may be indented differently from the rest of the paragraph. And finally, a paragraph may be set with no indentation at all, so that it aligns on the left and right margins.

The indentation of a paragraph is determined by three settings: the *left indent*, the *right indent*, and a *special indent* (if any). There are two types of special indentation, first line and hanging, as will be explained shortly. The left and right indents are set to zero by default, as is the special indent, and produce a paragraph with no indentation at all as shown in Figure 2.11a. Positive values for the left and right indents offset the paragraph from both margins as shown in Figure 2.11b.

The *first line indent* (Figure 2.11c) affects only the first line in the paragraph and is implemented by pressing the Tab key at the beginning of the paragraph. A *hanging indent* (Figure 2.11d) sets the first line of a paragraph at the left indent and indents the remaining lines according to the amount specified. Hanging indents are often used with bulleted or numbered lists.

We, the people of the United States, in order to form a more perfect Union, establish justice, insure domestic tranquillity, provide for the common defense, promote the general welfare, and secure the blessings of liberty to ourselves and our posterity, do ordain and establish this Constitution for the United States of America.

(a) Justified (flush left/flush right)

We, the people of the United States, in order to form a more perfect Union, establish justice, insure domestic tranquillity, provide for the common defense, promote the general welfare, and secure the blessings of liberty to ourselves and our posterity, do ordain and establish this Constitution for the United States of America.

(b) Left Aligned (flush left/ragged right)

We, the people of the United States, in order to form a more perfect Union, establish justice, insure domestic tranquillity, provide for the common defense, promote the general welfare, and secure the blessings of liberty to ourselves and our posterity, do ordain and establish this Constitution for the United States of America.

(c) Right Aligned (ragged left/flush right)

We, the people of the United States, in order to form a more perfect Union, establish justice, insure domestic tranquillity, provide for the common defense, promote the general welfare, and secure the blessings of liberty to ourselves and our posterity, do ordain and establish this Constitution for the United States of America.

(d) Centered (ragged left/ragged right)

FIGURE 2.10 *Alignment*

The left and right indents are defined as the distance between the text and the left and right margins, respectively. Both parameters are set to zero in this paragraph and so the text aligns on both margins. Different indentation can be applied to different paragraphs in the same document.

(a) No Indents

> Positive values for the left and right indents offset a paragraph from the rest of a document and are often used for long quotations. This paragraph has left and right indents of one-half inch each. Different indentation can be applied to different paragraphs in the same document.

(b) Left and Right Indents

    A first line indent affects only the first line in the paragraph and is implemented by pressing the Tab key at the beginning of the paragraph. The remainder of the paragraph is aligned at the left margin (or the left indent if it differs from the left margin) as can be seen from this example. Different indentation can be applied to different paragraphs in the same document.

(c) First Line Indent

A hanging indent sets the first line of a paragraph at the left indent and indents the remaining lines according to the amount specified. Hanging indents are often used with bulleted or numbered lists. Different indentation can be applied to different paragraphs in the same document.

(d) Hanging (Special) Indent

FIGURE 2.11 *Indents*

Tabs

Anyone who has used a typewriter is familiar with the function of the Tab key; that is, press Tab and the insertion point moves to the next **tab stop** (a measured position to align text at a specific place). The Tab key is much more powerful in Word as you can choose from four different types of tab stops (left, center, right, and decimal). You can also specify a **leader character**, typically dots or hyphens, to draw the reader's eye across the page. Tabs are often used to create columns of text within a document.

The default tab stops are set every ½ inch and are left aligned, but you can change the alignment and/or position with the Format Tabs command. Figure 2.12 illustrates a dot leader in combination with a right tab to produce a Table of Contents. The default tab stops have been cleared in Figure 2.12a, in favor of a single right tab at 5.5 inches. The option button for a dot leader has also been checked. The resulting document is shown in Figure 2.12b.

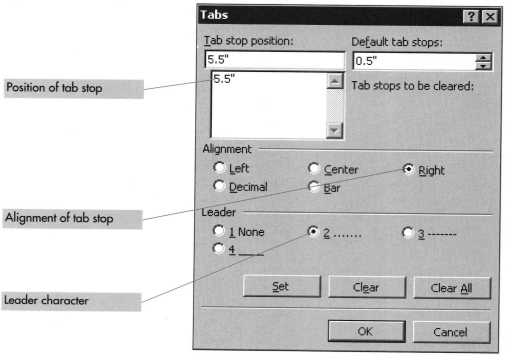

FIGURE 2.12 *Tabs*

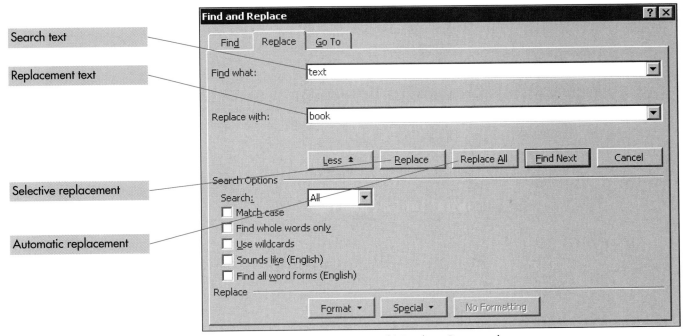

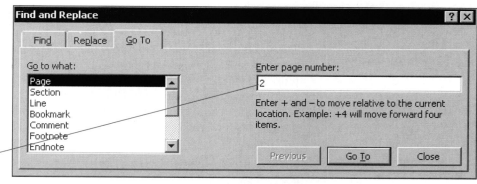

Hyphenation

Hyphenation gives a document a more professional look by eliminating excessive gaps of white space. It is especially useful in narrow columns and/or justified text. Hyphenation is implemented through the Language command in the Tools menu. You can choose to hyphenate a document automatically, in which case the hyphens are inserted as the document is created. (Microsoft Word will automatically rehyphenate the document to adjust for subsequent changes in editing.)

You can also hyphenate a document manually, to have Word prompt you prior to inserting each hyphen. Manual hyphenation does not, however, adjust for changes that affect the line breaks, and so it should be done only after the document is complete. And finally, you can fine-tune the use of hyphenation by preventing a hyphenated word from breaking if it falls at the end of a line. This is done by inserting a *nonbreaking hyphen* (press Ctrl+Shift+Hyphen) when the word is typed initially.

Line Spacing

Line spacing determines the space between the lines in a paragraph. Word provides complete flexibility and enables you to select any multiple of line spacing (single, double, line and a half, and so on). You can also specify line spacing in terms of points (there are 72 points per inch).

Line spacing is set at the paragraph level through the Format Paragraph command, which sets the spacing within a paragraph. The command also enables you to add extra spacing before the first line in a paragraph or after the last line. (Either technique is preferable to the common practice of single spacing the paragraphs within a document, then adding a blank line between paragraphs.)

FORMAT PARAGRAPH COMMAND

The *Format Paragraph command* is used to specify the alignment, indentation, line spacing, and pagination for the selected paragraph(s). As indicated, all of these features are implemented at the paragraph level and affect all selected paragraphs. If no paragraphs are selected, the command affects the entire current paragraph (the paragraph containing the insertion point).

The Format Paragraph command is illustrated in Figure 2.13. The Indents and Spacing tab in Figure 2.13a calls for a hanging indent, line spacing of 1.5 lines, and justified alignment. The preview area within the dialog box enables you to see how the paragraph will appear within the document.

The Line and Page Breaks tab in Figure 2.13b illustrates an entirely different set of parameters in which you control the pagination within a document. The check boxes in Figure 2.13b enable you to prevent the occurrence of awkward soft page breaks that detract from the appearance of a document.

You might, for example, want to prevent widows and orphans, terms used to describe isolated lines that seem out of place. A *widow* refers to the last line of a paragraph appearing by itself at the top of a page. An *orphan* is the first line of a paragraph appearing by itself at the bottom of a page.

You can also impose additional controls by clicking one or more check boxes. Use the Keep Lines Together option to prevent a soft page break from occurring within a paragraph and ensure that the entire paragraph appears on the same page. (The paragraph is moved to the top of the next page if it doesn't fit on the bottom of the current page.) Use the Keep with Next option to prevent a soft page break between the two paragraphs. This option is typically used to keep a heading (a one-line paragraph) with its associated text in the next paragraph.

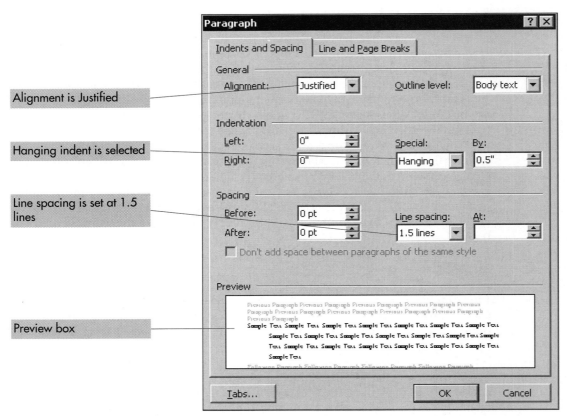

(a) Indents and Spacing

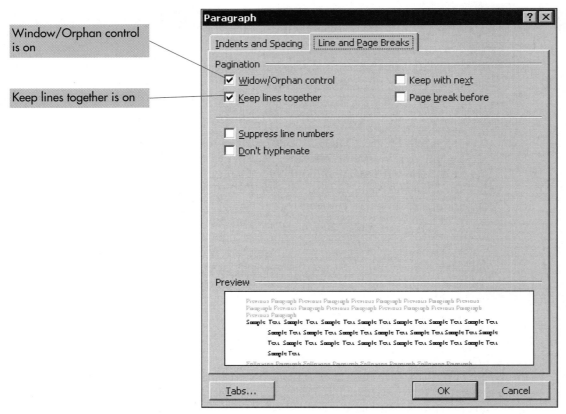

(b) Line and Page Breaks

FIGURE 2.13 *Format Paragraph Command*

Borders and Shading

The ***Borders and Shading command*** puts the finishing touches on a document and is illustrated in Figure 2.14. The command is applied to selected text within a paragraph, to the entire paragraph if no text is selected, or to the entire page if the Page Border tab is selected. Thus, you can create boxed and/or shaded text as well as place horizontal or vertical lines around different quantities of text.

You can choose from several different line styles in any color (assuming you have a color printer). You can place a uniform border around a paragraph (choose Box), or you can choose a shadow effect with thicker lines at the right and bottom. You can also apply lines to selected sides of a paragraph(s) by selecting a line style, then clicking the desired sides as appropriate.

The Page Border tab enables you to place a decorative border around one or more selected pages. As with a paragraph border, you can place the border around the entire page, or you can select one or more sides. The page border also provides an additional option to use preselected clip art instead of ordinary lines.

Shading is implemented independently of the border. Clear (no shading) is the default. Solid (100%) shading creates a solid box where the text is turned white so you can read it. Shading of 10 or 20 percent is generally most effective to add emphasis to the selected paragraph. The Borders and Shading command is implemented on the paragraph level and affects the entire paragraph (unless text has been selected within the paragraph)—either the current or selected paragraph(s).

The two command buttons at the bottom of the dialog box provide additional options. The Show Toolbar button displays the Tables and Borders toolbar that facilitates both borders and shading. The Horizontal Line button provides access to a variety of attractive designs.

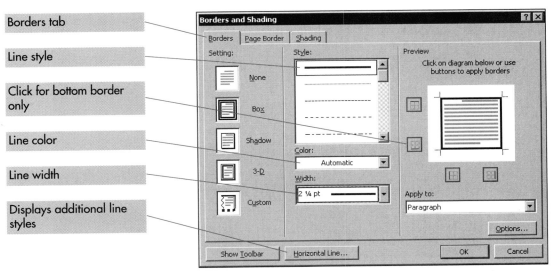

FIGURE 2.14 *Paragraph Borders and Shading*

FORMATTING AND THE PARAGRAPH MARK

The paragraph mark ¶ at the end of a paragraph does more than just indicate the presence of a hard return. It also stores all of the formatting in effect for the paragraph. Hence in order to preserve the formatting when you move or copy a paragraph, you must include the paragraph mark in the selected text. Click the Show/Hide ¶ button on the toolbar to display the paragraph mark and make sure it has been selected.

COLUMN FORMATTING

Columns add interest to a document and are implemented through the **Columns command** in the Format menu as shown in Figure 2.15. You specify the number of columns and, optionally, the space between columns. Microsoft Word does the rest, calculating the width of each column according to the left and right margins on the page and the specified (default) space between columns.

The dialog box in Figure 2.15 implements a design of three equal columns. The 2-inch width of each column is computed automatically based on left and right page margins of 1 inch each and the ¼-inch spacing between columns. The width of each column is determined by subtracting the sum of the margins and the space between the columns (a total of 2½ inches in this example) from the page width of 8½ inches. The result of the subtraction is 6 inches, which is divided by 3, resulting in a column width of 2 inches.

There is, however, one subtlety associated with column formatting, and that is the introduction of the **section**, which controls elements such as the orientation of a page (landscape or portrait), margins, page numbers, and/or the number of columns. All of the documents in the text thus far have consisted of a single section, and therefore section formatting was not an issue. It becomes important only when you want to vary an element that is formatted at the section level. You could, for example, use section formatting to create a document that has one column on its title page and two columns on the remaining pages. This requires you to divide the document into two sections through insertion of a **section break**. You then format each section independently and specify the number of columns in each section.

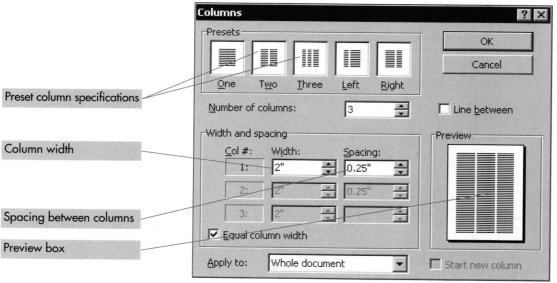

FIGURE 2.15 *The Format Columns Command*

THE SECTION VERSUS THE PARAGRAPH

Line spacing, alignment, tabs, and indents are implemented at the paragraph level. Change any of these parameters anywhere within the current (or selected) paragraph(s) and you change *only* those paragraph(s). Margins, page numbering, orientation, and columns are implemented at the section level. Change these parameters anywhere within a section and you change the characteristics of every page within that section.

HANDS-ON EXERCISE 3

PARAGRAPH FORMATTING

Objective To implement line spacing, alignment, and indents; to implement widow and orphan protection; to box and shade a selected paragraph. Use Figure 2.16 as a guide in the exercise.

Step 1: **Select-Then-Do**

➤ Open the **Modified Tips** document from the previous exercise. If necessary, change to the Print Layout view. Click the **Zoom drop-down arrow** and click **Two Pages** to match the view in Figure 2.16a.

➤ Select the entire second page as shown in the figure. Point to the selected text and click the **right mouse button** to produce the shortcut menu. Click **Paragraph**.

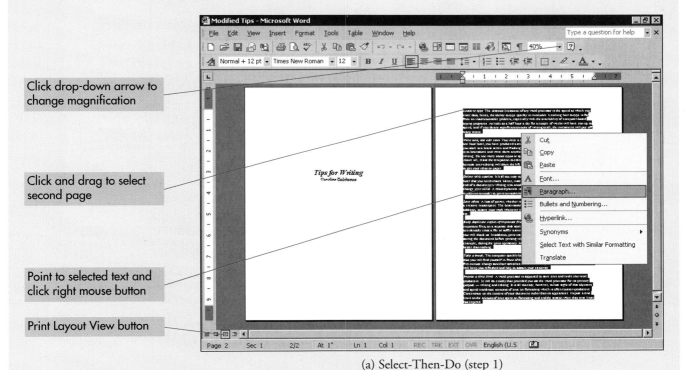

(a) Select-Then-Do (step 1)

FIGURE 2.16 *Hands-on Exercise 3*

SELECT TEXT WITH THE F8 (EXTEND) KEY

Move to the beginning of the text you want to select, then press the F8 (extend) key. The letters EXT will appear in the status bar. Use the arrow keys to extend the selection in the indicated direction; for example, press the down arrow key to select the line. You can also type any character—for example, a letter, space, or period—to extend the selection to the first occurrence of that character. Thus, typing a space or period is equivalent to selecting a word or sentence, respectively. Press Esc to cancel the selection mode.

Step 2: **Line Spacing, Justification, and Pagination**

➤ If necessary, click the **Indents and Spacing tab** to view the options in Figure 2.16b. Click the **down arrow** on the list box for Line Spacing and select **1.5 Lines**. Click the **down arrow** on the Alignment list box and select **Justified**.

➤ Click the tab for **Line and Page Breaks**. Check the box for **Keep Lines Together**. If necessary, check the box for **Widow/Orphan Control**. Click **OK** to accept all of the settings in the dialog box.

➤ Click anywhere in the document to deselect the text and see the effects of the formatting changes.

➤ Save the document.

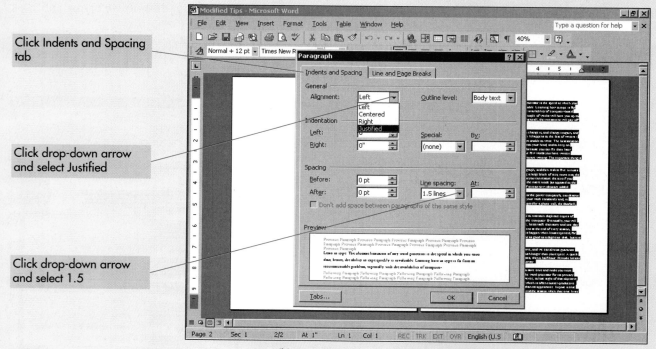

(b) Line Spacing, Justification, and Pagination (step 2)

FIGURE 2.16 *Hands-on Exercise 3 (continued)*

VIEW THE FORMATTING PROPERTIES

Open the task pane and click the down arrow in the title bar to select Formatting Properties to display complete information for the selected text in the document. The properties are displayed by Font, Paragraph, and Section, enabling you to click the plus or minus sign next to each item to view or hide the underlying details. The properties in each area are links to the associated dialog boxes. Click Alignment or Justification, for example, within the Paragraph area to open the associated dialog box, where you can change the indicated property.

Step 3: **Indents**

- Select the second paragraph as shown in Figure 2.16c. (The second paragraph will not yet be indented.)
- Pull down the **Format menu** and click **Paragraph** (or press the **right mouse button** to produce the shortcut menu and click **Paragraph**).
- If necessary, click the **Indents and Spacing tab** in the Paragraph dialog box. Click the **up arrow** on the Left Indentation text box to set the **Left Indent** to **.5** inch. Set the **Right indent** to **.5** inch. Click **OK**. Your document should match Figure 2.16c.
- Save the document.

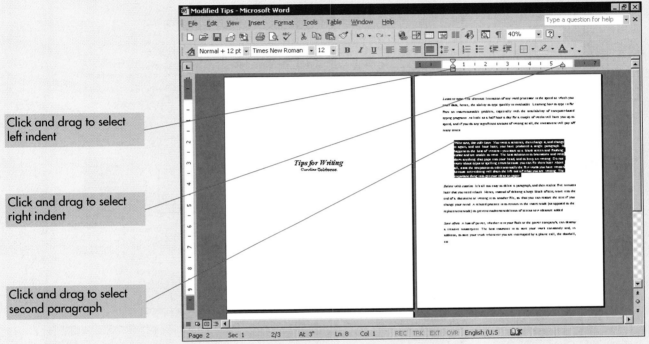

(c) Indents (step 3)

FIGURE 2.16 *Hands-on Exercise 3 (continued)*

INDENTS AND THE RULER

Use the ruler to change the special, left, and/or right indents. Select the paragraph (or paragraphs) in which you want to change indents, then drag the appropriate indent markers to the new location(s). If you get a hanging indent when you wanted to change the left indent, it means you dragged the bottom triangle instead of the box. Click the Undo button and try again. (You can always use the Format Paragraph command rather than the ruler if you continue to have difficulty.)

Step 4: **Borders and Shading**

➤ Pull down the **Format menu**. Click **Borders and Shading** to produce the dialog box in Figure 2.16d.
➤ If necessary, click the **Borders tab**. Select a style and width for the line around the box. Click the rectangle labeled **Box** under Setting. You can also experiment with a partial border by clicking in the Preview area to toggle a line on or off.
➤ Click the **Shading Tab**. Click the **down arrow** on the Style list box. Click **10%**.
➤ Click **OK** to accept the settings for both Borders and Shading. Click outside the paragraph.
➤ Save the document.

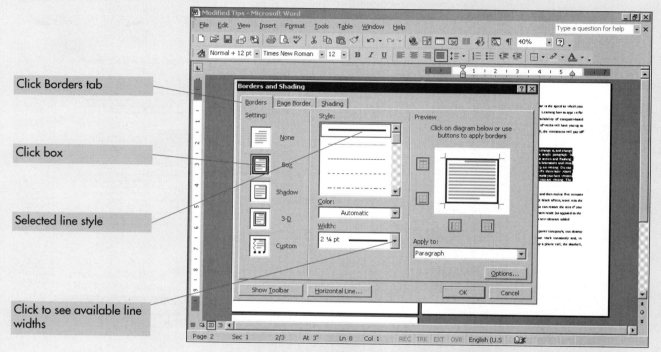

(d) Borders and Shading (step 4)

FIGURE 2.16 *Hands-on Exercise 3 (continued)*

SELECT NONCONTIGUOUS TEXT

Anyone who has used a previous version of Word will be happy to learn that you can select noncontiguous blocks of text, and then apply the same formatting to the selected text with a single command. Click and drag to select the first item, then press and hold the Ctrl key as you continue to drag the mouse over additional blocks of text. All of the selected text is highlighted within the document. Apply the desired formatting, then click anywhere in the document to deselect the text and continue working.

Step 5: **View Many Pages**

- Pull down the **View menu** and click **Zoom** to display the Zoom dialog box. Click the monitor icon in the Many Pages area, then click and drag to display three pages across. Release the mouse. Click **OK**.
- Your screen should match the one in Figure 2.16e, which displays all three pages of the document.
- The Print Layout view displays both a vertical and a horizontal ruler. The boxed and indented paragraph is clearly shown in the second page.
- The soft page break between pages two and three occurs between tips rather than within a tip; that is, the text of each tip is kept together on the same page.
- Save the document a final time. Print the document at this point in the exercise and submit it to your instructor.

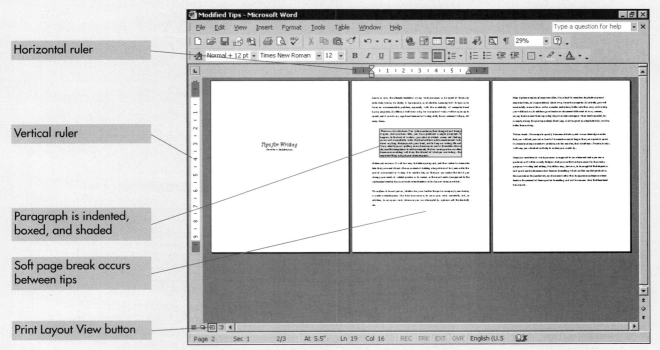

(e) View Many Pages (step 5)

FIGURE 2.16 *Hands-on Exercise 3 (continued)*

THE PAGE BORDER COMMAND

You can apply a border to the title page of your document, to every page except the title page, or to every page including the title page. Click anywhere on the page, pull down the Format menu, click Borders and Shading, and click the Page Borders tab. First design the border by selecting a style, color, width, and art (if any). Then choose the page(s) to which you want to apply the border by clicking the drop-down arrow in the Apply to list box. Close the Borders and Shading dialog box.

Step 6: **Change the Column Structure**

- Pull down the **File menu** and click the **Page Setup command** to display the Page Setup dialog box. Click the **Margins tab**, then change the Left and Right margins to 1" each. Click **OK** to accept the settings and close the dialog box.
- Click the **down arrow** on the Zoom list box and return to **Page Width**. Press the **PgUp** or **PgDn key** to scroll until the second page comes into view.
- Click anywhere in the paragraph, "Write Now but Edit Later". Pull down the **Format menu**, click the **Paragraph command**, click the **Indents and Spacing tab** if necessary, then change the left and right indents to 0.
- All paragraphs in the document should have the same indentation as shown in Figure 2.16f. Pull down the **Format menu** and click the **Columns command** to display the Columns dialog box.
- Click the icon for **three columns**. The default spacing between columns is .5", which leads to a column width of 1.83". Click in the Spacing list box and change the spacing to **.25"**, which automatically changes the column width to 2".
- Clear the box for the **Line Between** columns. Click **OK**.

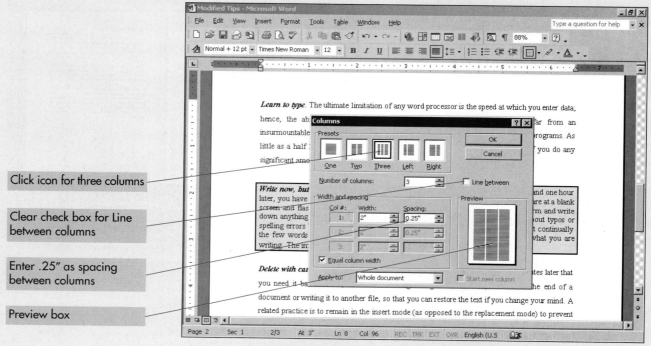

(f) Change the Column Structure (step 6)

FIGURE 2.16 *Hands-on Exercise 3 (continued)*

USE THE RULER TO CHANGE COLUMN WIDTH

Click anywhere within the column whose width you want to change, then point to the ruler and click and drag the right margin (the mouse pointer changes to a double arrow) to change the column width. Changing the width of one column in a document with equal-sized columns changes the width of all other columns so that they remain equal. Changing the width in a document with unequal columns changes only that column.

Step 7: **Insert a Section Break**

- Pull down the **View menu**, click the **Zoom command**, then click the **Many Pages option button**. Click and drag over 3 pages, then click **OK**. The document has switched to column formatting.
- Click at the beginning of the second page, immediately to the left of the first paragraph. Pull down the **Insert menu** and click **Break** to display the dialog box in Figure 2.16g.
- Click the **Continuous option button**, then click **OK** to accept the settings and close the dialog box.
- Click anywhere on the title page (before the section break you just inserted). Click the **Columns button**, then click the first column.
- The formatting for the first section of the document (the title page) should change to one column; the title of the document and your name are centered across the entire page.
- Print the document in this format for your instructor. Decide in which format you want to save the document—that is, as it exists now, or as it existed at the end of step 5. Exit Word.

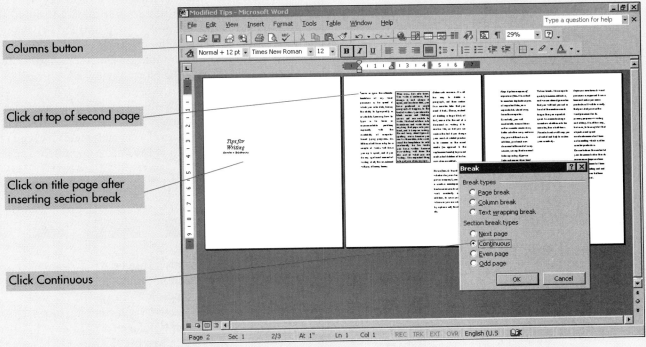

(g) Insert a Section Break (step 7)

FIGURE 2.16 *Hands-on Exercise 3 (continued)*

THE COLUMNS BUTTON

The Columns button on the Standard toolbar is the fastest way to create columns in a document. Click the button, drag the mouse to choose the number of columns, then release the mouse to create the columns. The toolbar lets you change the number of columns, but not the spacing between columns. The toolbar is also limited, in that you cannot create columns of different widths or select a line between the columns.

SUMMARY

Many operations in Word are done within the context of select-then-do; that is, select the text, then execute the necessary command. Text may be selected by dragging the mouse, by using the selection bar to the left of the document, or by using the keyboard. Text is deselected by clicking anywhere within the document.

The Find and Replace commands locate a designated character string and optionally replace one or more occurrences of that string with a different character string. The search may be case sensitive and/or restricted to whole words. The commands may also be applied to formatting and/or special characters.

Text is moved or copied through a combination of the Cut, Copy, and Paste commands and/or the drag-and-drop facility. The contents of the Windows clipboard are modified by any subsequent Cut or Copy command, but are unaffected by the Paste command; that is, the same text can be pasted into multiple locations. The Office clipboard retains up to 24 entries that were cut or copied.

The Undo command reverses the effect of previous commands. The Undo and Redo commands work in conjunction with one another; that is, every command that is undone can be redone at a later time.

Scrolling occurs when a document is too large to be seen in its entirety. Scrolling with the mouse changes what is displayed on the screen, but does not move the insertion point. Scrolling via the keyboard (for example, PgUp and PgDn) changes what is seen on the screen as well as the location of the insertion point.

The Print Layout view displays top and bottom margins, headers and footers, and other elements not seen in the Normal view. The Normal view is faster because Word spends less time formatting the display.

The Format Paragraph command determines the line spacing, alignment, indents, and text flow, all of which are set at the paragraph level. Borders and shading are also set at the character, paragraph, or page level. Margins, page size, and orientation are set in the Page Setup command and affect the entire document (or section).

KEY TERMS

Arial (p. 64)
Automatic replacement (p. 52)
Borders and Shading command (p. 83)
Case-insensitive search (p. 52)
Case-sensitive search (p. 52)
Columns command (p. 84)
Copy command (p. 50)
Courier New (p. 64)
Cut command (p. 50)
Find command (p. 51)
First line indent (p. 77)
Font (p. 64)
Format Font command (p. 65)
Format Painter (p. 73)
Format Paragraph command (p. 81)
Go To command (p. 51)
Hanging indent (p. 77)
Hard page break (p. 69)
Highlighting (p. 73)
Hyphenation (p. 81)
Landscape orientation (p. 69)
Leader character (p. 80)
Left indent (p. 77)
Line spacing (p. 81)
Margins tab (p. 69)
Monospaced typeface (p. 65)
Nonbreaking hyphen (p. 81)
Normal view (p. 55)
Office clipboard (p. 50)
Orphan (p. 81)
Page Setup command (p. 67)
Paste command (p. 50)
Paste Special command (p. 50)
Point (p. 65)
Portrait orientation (p. 69)
Print Layout view (p. 55)
Proportional typeface (p. 65)
Replace command (p. 51)
Right indent (p. 77)
Sans serif typeface (p. 64)
Scrolling (p. 53)
Section (p. 84)
Section break (p. 84)
Select-then-do (p. 50)
Selective replacement (p. 52)
Serif typeface (p. 64)
Soft page break (p. 69)
Special indent (p. 77)
Tab stop (p. 80)
Times New Roman (p. 64)
Typeface (p. 64)
Type size (p. 65)
Type style (p. 65)
Typography (p. 64)
Underlining (p. 72)
View menu (p. 55)
Whole words only (p. 52)
Widows (p. 81)
Wild card (p. 52)
Windows clipboard (p. 50)
Zoom command (p. 55)

MULTIPLE CHOICE

1. Which of the following commands does *not* place data onto the clipboard?
   (a) Cut
   (b) Copy
   (c) Paste
   (d) All of the above

2. What happens if you select a block of text, copy it, move to the beginning of the document, paste it, move to the end of the document, and paste the text again?
   (a) The selected text will appear in three places: at the original location, and at the beginning and end of the document
   (b) The selected text will appear in two places: at the beginning and end of the document
   (c) The selected text will appear in just the original location
   (d) The situation is not possible; that is, you cannot paste twice in a row without an intervening cut or copy operation

3. What happens if you select a block of text, cut it, move to the beginning of the document, paste it, move to the end of the document, and paste the text again?
   (a) The selected text will appear in three places: at the original location and at the beginning and end of the document
   (b) The selected text will appear in two places: at the beginning and end of the document
   (c) The selected text will appear in just the original location
   (d) The situation is not possible; that is, you cannot paste twice in a row without an intervening cut or copy operation

4. Which of the following are set at the paragraph level?
   (a) Alignment
   (b) Tabs and indents
   (c) Line spacing
   (d) All of the above

5. How do you change the font for *existing* text within a document?
   (a) Select the text, then choose the new font
   (b) Choose the new font, then select the text
   (c) Either (a) or (b)
   (d) Neither (a) nor (b)

6. The Page Setup command can be used to change:
   (a) The margins in a document
   (b) The orientation of a document
   (c) Both (a) and (b)
   (d) Neither (a) nor (b)

7. Which of the following is a true statement regarding indents?
   (a) Indents are measured from the edge of the page
   (b) The left, right, and first line indents must be set to the same value
   (c) The insertion point can be anywhere in the paragraph when indents are set
   (d) Indents must be set with the Format Paragraph command

8. The default tab stops are set to:
   (a) Left indents every ½ inch
   (b) Left indents every ¼ inch
   (c) Right indents every ½ inch
   (d) Right indents every ¼ inch

9. The spacing in an existing multipage document is changed from single spacing to double spacing throughout the document. What can you say about the number of hard and soft page breaks before and after the formatting change?
   (a) The number of soft page breaks is the same, but the number and/or position of the hard page breaks is different
   (b) The number of hard page breaks is the same, but the number and/or position of the soft page breaks is different
   (c) The number and position of both hard and soft page breaks is the same
   (d) The number and position of both hard and soft page breaks is different

10. Which of the following describes the Arial and Times New Roman fonts?
    (a) Arial is a sans serif font, Times New Roman is a serif font
    (b) Arial is a serif font, Times New Roman is a sans serif font
    (c) Both are serif fonts
    (d) Both are sans serif fonts

11. The find and replacement strings must be
    (a) The same length
    (b) The same case, either upper or lower
    (c) The same length and the same case
    (d) None of the above

12. You are in the middle of a multipage document. How do you scroll to the beginning of the document and simultaneously change the insertion point?
    (a) Press Ctrl+Home
    (b) Drag the scroll bar to the top of the scroll box
    (c) Both (a) and (b)
    (d) Neither (a) nor (b)

13. Which of the following substitutions can be accomplished by the Find and Replace command?
    (a) All occurrences of the words "Times New Roman" can be replaced with the word "Arial"
    (b) All text set in the Times New Roman font can be replaced by the Arial font
    (c) Both (a) and (b)
    (d) Neither (a) nor (b)

14. Which of the following deselects a selected block of text?
    (a) Clicking anywhere outside the selected text
    (b) Clicking any alignment button on the toolbar
    (c) Clicking the Bold, Italic, or Underline button
    (d) All of the above

15. Which view, and which magnification, lets you see the whole page, including top and bottom margins?
    (a) Print Layout view at 100% magnification
    (b) Print Layout view at Whole Page magnification
    (c) Normal view at 100% magnification
    (d) Normal view at Whole Page magnification

ANSWERS

1. c
2. a
3. b
4. d
5. a
6. c
7. c
8. a
9. b
10. a
11. d
12. a
13. c
14. a
15. b

PRACTICE WITH WORD

1. **Formatting 101:** The document in Figure 2.17 provides practice with basic formatting. Open the partially completed document in *Chapter 2 Practice 1* and then follow the instructions within the document itself to implement the formatting. You can implement the formatting in a variety of ways—by clicking the appropriate button on the Formatting toolbar, by pulling down the Format menu and executing the indicated command, or by using a keyboard shortcut. Add your name somewhere in the document, then print the completed document for your instructor. Do not be concerned if you do not have a color printer, but indicate this to your instructor in a note at the end of the document.

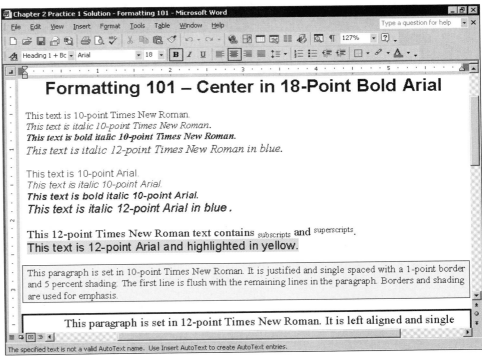

FIGURE 2.17 *Formatting 101 (Exercise 1)*

2. **Keyboard Shortcuts:** Keyboard shortcuts are especially useful if you type well because your hands can remain on the keyboard, as opposed to moving to the mouse. We never set out to memorize the shortcuts; we just learned them along the way as we continued to use Microsoft Office. It's much easier than you might think, because the same shortcuts apply to multiple applications. Ctrl+X, Ctrl+C, and Ctrl+V, for example, are the universal Windows shortcuts to cut, copy, and paste the selected text. The "X" is supposed to remind you of a pair of scissors, and the keys are located next to each other to link the commands to one another.

   Your assignment is to complete the document in *Chapter 2 Practice 2,* a portion of which can be seen in Figure 2.18. You can get the shortcut in one of two ways: by using the Help menu, or by displaying the shortcut in conjunction with the ScreenTip for the corresponding button on either the Standard or Formatting toolbar. (Pull down the Tools menu, click the Customize command to display the Customize dialog box, click the Options tab, then check the box to Show Shortcut Keys in ScreenTips.) Enter your name in the completed document, add the appropriate formatting, then submit the document to your instructor as proof you did this exercise.

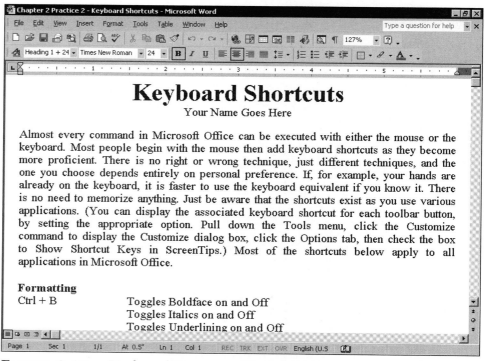

FIGURE 2.18 *Keyboard Shortcuts (Exercise 2)*

3. **Moving Text:** There are two basic ways to move text within a document. You can use a combination of the Cut and Paste commands, or you can simply click and drag text from one location to another. The latter technique tends to be easier if you are moving text a short distance, whereas cutting and pasting is preferable if the locations are far apart within a document. This exercise will give you a chance to practice both techniques.
   a. Open the partially completed document in *Chapter 2 Practice 3*, where you will find a list of the presidents of the United States together with the years that each man served.
   b. The list is out of order, and your assignment is to rearrange the names so that the presidents appear in chronological order. You don't have to be a historian to complete the exercise because you can use the years in office to determine the proper order.
   c. Use the Insert Hyperlink command (or click the corresponding button on the Standard toolbar) to insert a link to the White House Web site, where you can learn more about the presidents.
   d. Use the Format Columns command to display the presidents in two columns with a line down the middle as shown in Figure 2.19. You will have to implement a section break because the first two lines (the title and the hyperlink to the White House) are in one-column format, whereas the list of presidents is in two columns. You should create a second section break after the last president (George W. Bush) to balance the columns.
   e. Add your name to the completed document and submit it to your instructor as proof you completed this exercise.
   f. Start your Internet browser and connect to the White House. Select any president. Print the available biographical information for your instructor.

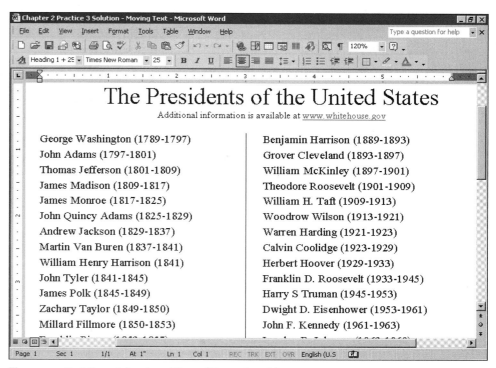

FIGURE 2.19 *Moving Text (Exercise 3)*

4. Tab Stops: Microsoft Word provides four different types of tabs that can be used to achieve a variety of formatting effects. Your assignment is to open the partially completed document in *Chapter 2 Practice 4,* then follow the instructions within the document to implement the formatting. The end result should be the document in Figure 2.20, which includes additional formatting in the opening paragraphs to boldface and italicize the key terms. Add your name somewhere in the document, then submit the completed exercise to your instructor.

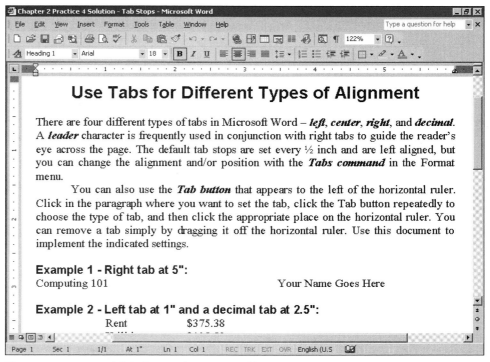

FIGURE 2.20 *Tab Stops (Exercise 4)*

5. **Inserting the Date and Time:** The document in Figure 2.21 describes the Insert Date and Time command and shows the various formats in which a date may appear. You need not duplicate our document exactly, but you are asked to insert the date multiple times, as both a fixed value and a field, in multiple formats. Divide your document into sections so that you can display the two sets of dates in adjacent columns. (Use the keyboard shortcut Ctrl+Shift+Enter to force a column break that will take you from the bottom of one column to the top of the next column.)

   Create your document on one day, then open it a day later to be sure that the dates that were entered as fields were updated correctly. Insert a section break after the last date to return to a single column format, then add a concluding paragraph that describes how to remove the shading from a date field. In essence any date that is entered as a field is shaded by default, as can be seen in our figure. To remove the shading, pull down the Tools menu, click the Options command, select the View tab, then click the drop-down arrow in the Field Shading list box to choose the desired option.)

   Add your name to the completed document and submit it to your instructor as proof you completed this exercise.

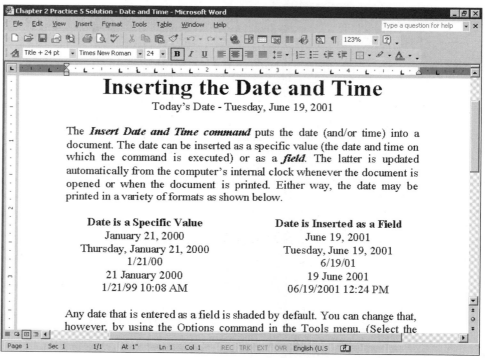

FIGURE 2.21 *Inserting the Date and Time (Exercise 5)*

6. **Tips for Internet Explorer:** A partially completed version of the document in Figure 2.22 can be found in the file *Chapter 2 Practice 6*. Your assignment is to open that document, then format the various tips for Internet Explorer in an attractive fashion. You need not follow our formatting exactly, but you are to apply uniform formatting throughout the document. Use one set of specifications for the heading of each tip (e.g., Arial 10-point bold) and a different format for the associated text. Use the Format Painter to copy formatting within the document. Insert a cover page that includes your name and the source of the information, then print the entire document for your instructor.

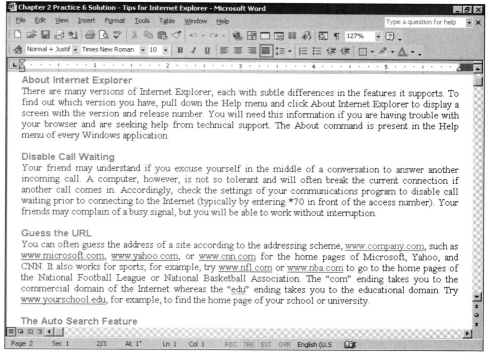

FIGURE 2.22 *Tips for Internet Explorer (Exercise 6)*

7. A Simple Newsletter: Create a simple newsletter such as the one in Figure 2.23. There is no requirement to write meaningful text, but your document will be more interesting if you choose a theme and follow it throughout. (The text of our document contains suggestions for creating the document.) The newsletter should have a meaningful name (e.g., "A Simple Newsletter"), and supporting headings for the various paragraphs ("Choose a Theme", "Create the Masthead", and so on.) The text within each paragraph can consist of the same sentences that are copied throughout the document. The design of the newsletter is up to you. The document in Figure 2.23 has a formal appearance, but you can modify that design in any way you like. Some suggestions:
   a. Change the default margins of your document before you begin. You are using columns, and left and right margins of a half or three-quarters of an inch are more appropriate than the default values of 1.25 inches. Reduce the top margins as well.
   b. Create a sample heading and associated text, and format both to be sure that you have all of the necessary specifications. You can use the Line and Page Breaks tab within the Format Paragraph command to force the heading to appear with the text, and further to force the text to appear together in one paragraph. That way you can avoid awkward column breaks.
   c. Use the Columns command to fine-tune the dimensions of the columns and/or to add a line between the columns. You can use columns of varying width to add interest to the document.
   d. Choose one or two important sentences and create a pull-quote within the newsletter. Not only does this break up the text, but it calls attention to an important point.
   e. You can create a reverse, light text on a dark background, as in the masthead of our newsletter, by specifying 100% shading within the Borders and Shading command. You can also use a right tab to force an entry (your name) to align with the right margin.
   f. Try to design your newsletter with pencil and paper before you get to the computer. Print the completed newsletter for your instructor.

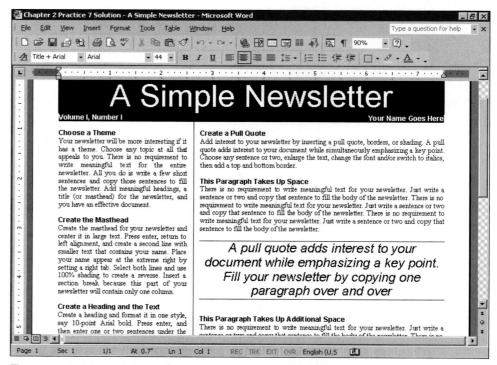

FIGURE 2.23   *A Simple Newsletter (Exercise 7)*

8. **Formatting at Three Levels:** You will find an unformatted version of the document in Figure 2.24 in the file *Chapter 2 Practice 8*. Open the document and match the formatting in Figure 2.24. (Use an appropriate point size for any formatting specification that is not visible within Figure 2.24.) Substitute your name as indicated, then print the completed document for your instructor.

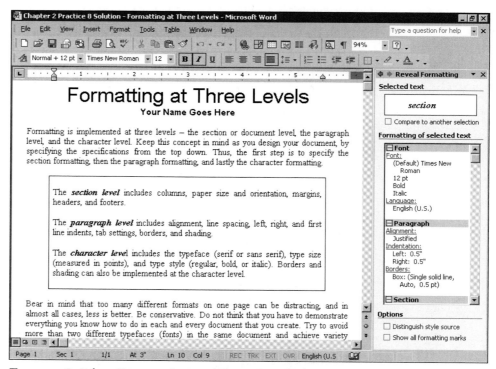

FIGURE 2.24   *Formatting at Three Levels (Exercise 8)*

9. **Viewing and Editing Comments:** The document in Figure 2.25 has been formatted for you. Look closely, however, and you will see that comments have been added to the first three tips. The comments appear in different colors to indicate that they were added by different people, in this case, Bob, Maryann, and a hypothetical student. Your assignment is to open the *Chapter 2 Practice 9* document and add additional comments of your own. Proceed as follows:
   a. Read the entire document, then choose at least five tips and insert your own comment. Click where you want the comment to go, then pull down the Insert menu, click Comment, and enter the text of your comment.
   b. Delete the comment that is associated with the Word Count toolbar.
   c. Add a title page that contains your name and today's date.
   d. Print the completed document to show the comments that you have added. Pull down the File menu, click the Print command, click the down arrow in the Print What area, and select Document Showing Markup.

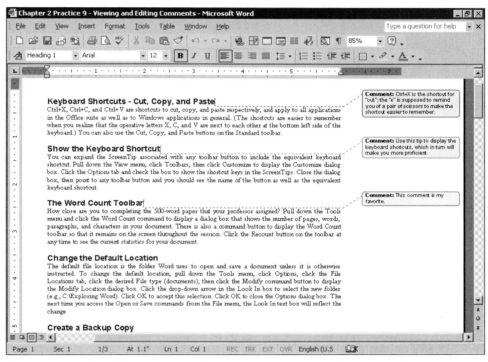

FIGURE 2.25 *Viewing and Editing Comments (Exercise 9)*

ON YOUR OWN

Your First Consultant's Job

Go to a real installation, such as a doctor's or an attorney's office, the company where you work, or the computer lab at school. Determine the backup procedures that are in effect, then write a one-page report indicating whether the policy is adequate and, if necessary, offering suggestions for improvement. Your report should be addressed to the individual in charge of the business, and it should cover all aspects of the backup strategy—that is, which files are backed up and how often, and what software is used for the backup operation. Use appropriate emphasis (for example, bold italics) to identify any potential problems. This is a professional document (it is your first consultant's job), and its appearance must be perfect in every way.

Computers Past and Present

The ENIAC was the scientific marvel of its day and the world's first operational electronic computer. It could perform 5,000 additions per second, weighed 30 tons, and took 1,500 square feet of floor space. The price was a modest $486,000 in 1946 dollars. The story of the ENIAC and other influential computers of the author's choosing is found in the file *History of Computers,* which we forgot to format, so we are asking you to do it for us.

Be sure to use appropriate emphasis for the names of the various computers. Create a title page in front of the document, then submit the completed assignment to your instructor. If you are ambitious, you can enhance this assignment by using your favorite search engine to look for computer museums on the Web. Visit one or two sites, and include this information on a separate page at the end of the document. One last task, and that is to update the description of Today's PC (the last computer in the document).

The Preamble to the Constitution

Use your favorite search engine to locate a Web site that contains the text of the United States constitution. Click and drag to select the text of the Preamble, use the Ctrl+C keyboard shortcut to copy the text to the Windows clipboard, start Word, and then paste the contents of the clipboard into the document. Format the Preamble in an attractive fashion, add a footnote that points to the Web page where you obtained the text, then add your name to the completed document.

To Hyphenate or Not to Hyphenate

The best way to learn about hyphenation is to experiment with an existing document. Open the *To Hyphenate or Not to Hyphenate* document that is on the data disk. The document is currently set in 12-point type with hyphenation in effect. Experiment with various formatting changes that will change the soft line breaks to see the effect on the hyphenation within the document. You can change the point size, the number of columns, and/or the right indent. You can also suppress hyphenation altogether, as described within the document. Summarize your findings in a short note to your instructor.

The Invitation

Choose an event and produce the perfect invitation. The possibilities are endless and limited only by your imagination. You can invite people to your wedding or to a fraternity party. Your laser printer and abundance of fancy fonts enable you to do anything a professional printer can do. Special paper will add the finishing touch. Go to it—this assignment is a lot of fun.

One Space after a Period

Touch typing classes typically teach the student to place two spaces after a period. The technique worked well in the days of the typewriter and monospaced fonts, but it creates an artificially large space when used with proportional fonts and a word processor. Select any document that is at least several paragraphs in length and print the document with the current spacing. Use the Find and Replace commands to change to the alternate spacing, then print the document a second time. Which spacing looks better to you? Submit both versions of the document to your instructor with a brief note summarizing your findings.

CHAPTER 3

Enhancing a Document: The Web and Other Resources

OBJECTIVES:

AFTER READING THIS CHAPTER YOU WILL BE ABLE TO:

1. Describe the resources in the Microsoft Media Gallery; insert clip art and/or a photograph into a document.
2. Use the Format Picture command to wrap text around a clip art image; describe various tools on the Picture toolbar.
3. Use WordArt to insert decorative text into a document; describe several tools on the WordArt toolbar.
4. Use the Drawing toolbar to create and modify lines and objects.
5. Download resources from the Internet for inclusion in a Word document; insert a footnote or endnote into a document to cite a reference.
6. Insert a hyperlink into a Word document; save a Word document as a Web page.
7. Use wizards and templates to create a document; list several wizards provided with Microsoft Word.
8. Define a mail merge; use the Mail Merge Wizard to create a set of form letters.

OVERVIEW

This chapter describes how to enhance a document using applications within Microsoft Office as well as resources from the Internet. We begin with a discussion of the Microsoft Media Gallery, a collection of clip art, sounds, photographs, and movies. We describe how Microsoft WordArt can be used to create special effects with text and how to create lines and objects through the Drawing toolbar.

These resources pale, however, in comparison to what is available via the Internet. Thus we also show you how to download a photograph from the Web and include it in an Office document. We describe how to add footnotes to give appropriate credit to your sources and how to further enhance a document through inclusion of hyperlinks. We also explain how to save a Word document as a Web page so that you can post the documents you create to a Web server or local area network. The chapter also describes some of the wizards and templates that are built into Word to help you create professionally formatted documents quickly and easily. We also introduce the concept of a mail merge.

ENHANCING A DOCUMENT

A Word document begins with text, but can be enhanced through the addition of objects. The document in Figure 3.1, for example, contains text, clip art, and WordArt (decorative text) and is the basis of a hands-on exercise that follows shortly. It also contains a scroll with additional text that was created through the Drawing toolbar, and the Windows logo that was added to the document through the Insert Symbol command. We describe each of these capabilities in the next few pages, then show you how to create the document.

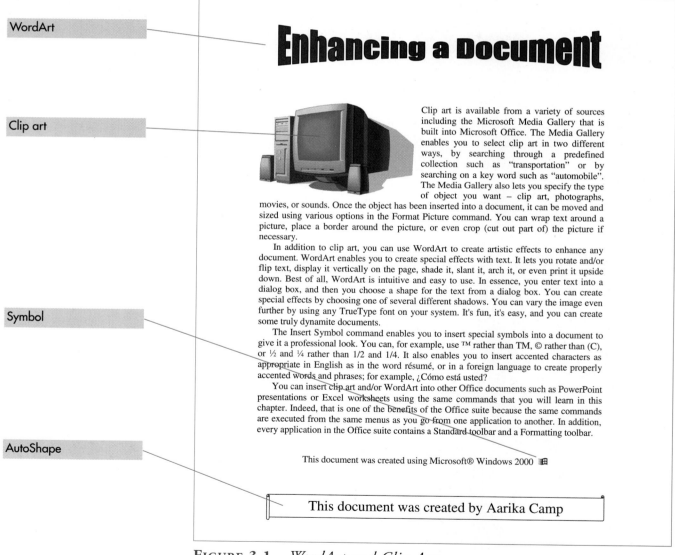

FIGURE 3.1 *WordArt and Clip Art*

The Media Gallery

The ***Media Gallery*** is accessible as a standalone application or from within multiple applications in Microsoft Office (Word, Excel, and PowerPoint, but not from Access). Either way, it is an excellent source for media objects such as clip art, sound files, photographs, and movies. There is an abundance of clip art from which to choose, so it is important to search through the available images efficiently, in order to choose the appropriate one.

The ***Insert Picture command*** displays a task pane in which you enter a key word (such as "basketball") that describes the picture you are looking for. The search returns a variety of potential clip art as shown in Figure 3.2a. Alternatively, you can click the down arrow in the task pane to select the Collection List as shown in Figure 3.2b. This method has you select (and then expand) various collections that may contain an appropriate image. We opened the Sports & Leisure collection where we found pictures of athletes in different sports, one of which was basketball. Look closely, and you will see that two of the pictures appear in both Figure 3.2a and 3.2b. Either way, you have ample images from which to choose.

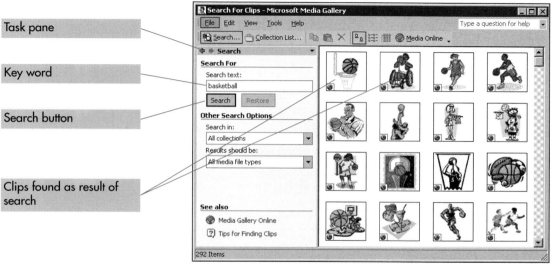

(a) Search by Key Word

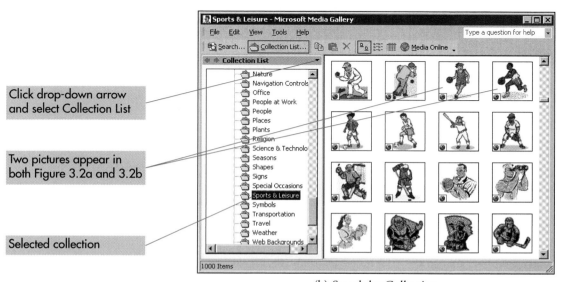

(b) Search by Collection

FIGURE 3.2 *The Media Gallery*

The Insert Symbol Command

The *Insert Symbol command* enables you to enter typographic symbols and/or foreign language characters into a document in place of ordinary typing—for example, ® rather than (R), © rather than (c), ½ and ¼ rather than 1/2 and 1/4, or é rather than e (as used in the word résumé). These special characters give a document a very professional look. You may have already discovered that some of this formatting can be done automatically through the *AutoCorrect* feature that is built into Word. If, for example, you type the letter "c" enclosed in parentheses, it will automatically be converted to the copyright symbol. Other symbols, such as accented letters like the é in résumé or those in a foreign language (e.g., ¿Cómo está usted?) have to be entered through the Insert Symbol command. (You could also create a macro, based on the Insert Symbol command, to simplify the process.)

Microsoft Office installs a variety of fonts onto your computer, each of which contains various symbols that can be inserted into a document. Selecting "normal text", however, as was done in Figure 3.3, provides access to the accented characters as well as other common symbols. Other fonts—especially the Wingdings, Webdings, and Symbols fonts—contain special symbols, including the Windows logo.

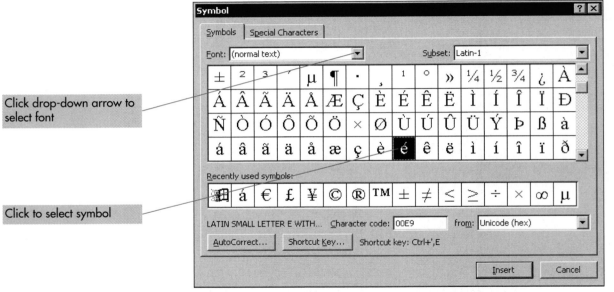

FIGURE 3.3 *Insert Symbol Command*

USE SYMBOLS AS CLIP ART

The Wingdings, Webdings, and Symbols fonts are among the best-kept secrets in Microsoft Office. Each font contains a variety of symbols that are actually pictures. You can insert any of these symbols into a document as text, select the character and enlarge the point size, change the color, then copy the modified character to create a truly original document. See practice exercise 2 at the end of the chapter.

Microsoft WordArt

Microsoft WordArt is an application within Microsoft Office that creates decorative text that can be used to add interest to a document. You can use ***WordArt*** in addition to clip art within a document, or in place of clip art if the right image is not available. You can rotate text in any direction, add three-dimensional effects, display the text vertically down the page, slant it, arch it, or even print it upside down. In short, you are limited only by your imagination.

WordArt is intuitively easy to use. In essence, you choose a style for the text from among the selections in Figure 3.4a. Then, you enter the specific text in a subsequent dialog box, after which the result is displayed as in Figure 3.4b. The finished WordArt is an object that can be moved and sized within a document, just like any other object. It's fun, it's easy, and you can create some truly unique documents.

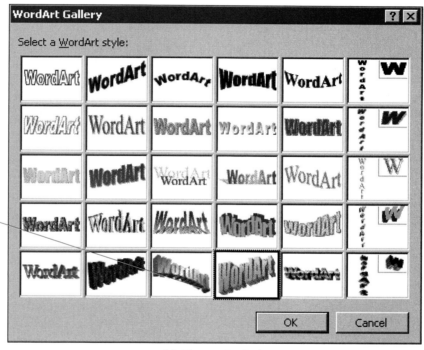

(a) Choose the Style

(b) Completed Entry

FIGURE 3.4 *Microsoft WordArt*

THE DRAWING TOOLBAR

All clip art is created from basic shapes, such as lines, and other basic tools that are found on the ***Drawing toolbar***. Select the Line tool, for example, then click and drag to create a line. Once the line has been created, you can select it and change its properties (such as thickness, style, or color) by using other tools on the Drawing toolbar. Draw a second line, or a curve, then depending on your ability, you have a piece of original clip art. You do not have to be an artist in order to use the basic tools to enhance any document.

The ***drawing canvas*** appears automatically whenever you select a tool from the Drawing toolbar and is indicated by a hashed line as shown in Figure 3.5. Each object within the canvas can be selected, at which point it displays its own ***sizing handles***. The blue rectangle is selected in Figure 3.5. You can size an object by clicking and dragging any one of the sizing handles. We don't expect you to create clip art comparable to the images within the Media Gallery, but you can use the tools on the Drawing toolbar to modify an existing image and/or create simple shapes of your own.

The Shift key has special significance when used in conjunction with the Line, Rectangle, and Oval tools. Press and hold the Shift key as you drag the line tool horizontally or vertically to create a perfectly straight line in either direction. Press and hold the Shift key as you drag the Rectangle and Oval tool to create a square or circle, respectively. The AutoShapes button contains a series of selected shapes, such as a callout or banner, and is very useful to create simple drawings. And, as with any other drawing object, you can change the thickness, color, or fill by selecting the object and choosing the appropriate tool. It is fun and it is easy. Just be flexible and willing to experiment. We think you will be pleased at what you will be able to do.

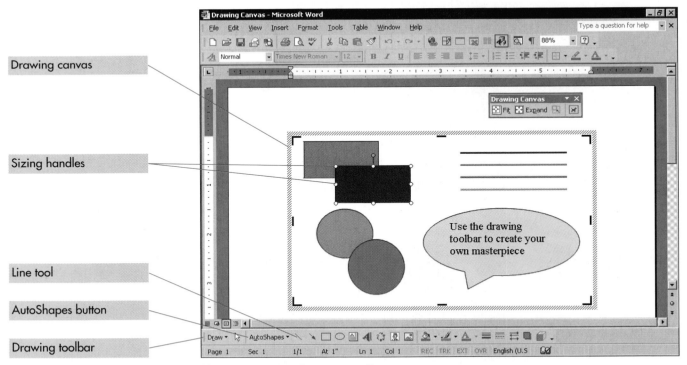

FIGURE 3.5 *Drawing Canvas*

HANDS-ON EXERCISE 1

CLIP ART AND WORDART

Objective To insert clip art and WordArt into a document; to use the Insert Symbol command to add typographical symbols. Use Figure 3.6 as a guide in completing the exercise.

Step 1: **Insert the Clip Art**

➤ Start Word. Open the **Clip Art** and **WordArt** document in the Exploring Word folder. Save the document as **Modified Clip Art and WordArt**.

➤ Check that the insertion point is at the beginning of the document. Pull down the **Insert menu**, click (or point to) **Picture**, then click **Clip Art**. The task pane opens (if it is not already open) and displays the Media Gallery Search pane as shown in Figure 3.6a.

➤ Click in the **Search text box**. Type **computer** to search for any clip art image that is indexed with this key word, then click the **Search button** or press **enter**.

➤ The images are displayed in the Results box. Point to an image to display a drop-down arrow to its right. Click the arrow to display a context menu.

➤ Click **Insert** to insert the image into the document. Do not be concerned about its size or position at this time. Close the task pane.

➤ Save the document.

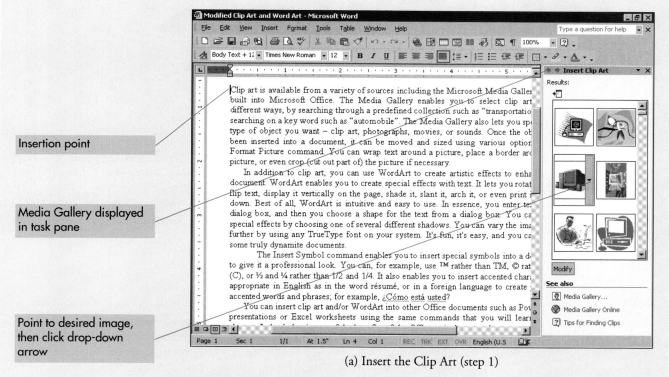

(a) Insert the Clip Art (step 1)

FIGURE 3.6 *Hands-on Exercise 1*

Step 2: **Move and Size the Picture**

- Point to the picture, click the **right mouse button** to display the context-sensitive menu, then click the **Format Picture command** to display the Format Picture dialog box in Figure 3.6b.
- You must change the layout in order to move and size the object. Click the **Layout tab**, choose the **Square layout**, then click the option button for **Left alignment**. Click **OK** to close the dialog box.
- To size the picture, click and drag a corner handle (the mouse pointer changes to a double arrow) to change the length and width simultaneously. This keeps the picture in proportion.
- To move the picture, point to any part of the image except a sizing handle (the mouse pointer changes to a four-sided arrow), then click and drag to move the image elsewhere in the document.
- Save the document.

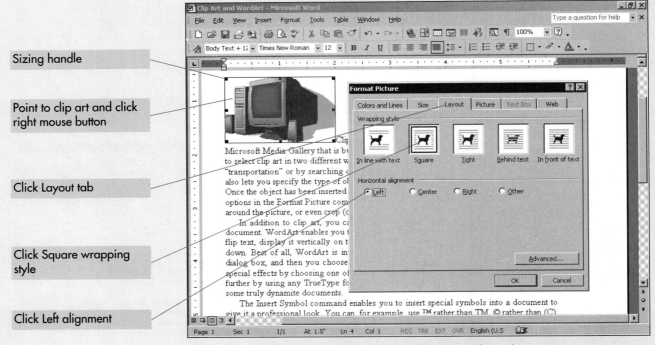

(b) Move and Size the Picture (step 2)

FIGURE 3.6 *Hands-on Exercise 1 (continued)*

SEARCH BY COLLECTION

The Media Gallery organizes its contents by collections and provides another way to select clip art. Pull down the Insert menu, click (or point to) the Picture command, then click Clip Art to open the task pane, where you can enter a key word to search for clip art. Instead of searching, however, click the link to Media Gallery at the bottom of the task pane to display the Media Gallery dialog box. Close the My Collections folder if it is open, then open the Office Collections folder, where you can explore the available images by collection.

Step 3: **WordArt**

- Press **Ctrl+End** to move to the end of the document. Pull down the **Insert menu**, click **Picture**, then click **WordArt** to display the WordArt Gallery dialog box in Figure 3.6c.
- Select the WordArt style you like (you can change it later). Click **OK**. You will see a second dialog box in which you enter the text. Enter **Enhancing a Document**. Click **OK**.
- The WordArt object appears in your document in the style you selected. Point to the WordArt object and click the **right mouse button** to display a shortcut menu. Click **Format WordArt** to display the Format WordArt dialog box.
- Click the **Layout tab**, then select **Square** as the Wrapping style. Click **OK**. It is important to select this wrapping option to facilitate placing the WordArt at the top of the document.
- Save the document.

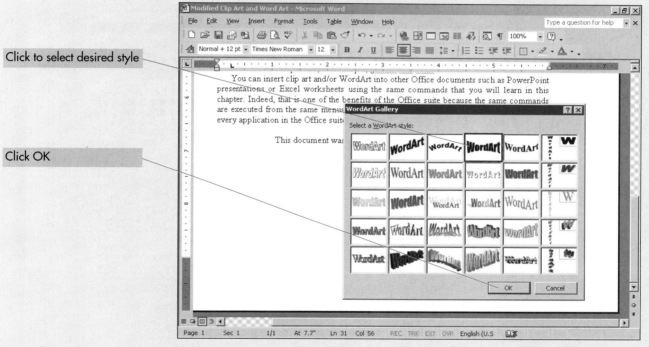

(c) WordArt (step 3)

FIGURE 3.6 *Hands-on Exercise 1 (continued)*

FORMATTING WORDART

The WordArt toolbar offers the easiest way to execute various commands associated with a WordArt object. It is displayed automatically when a WordArt object is selected and is suppressed otherwise. As with any other toolbar, you can point to a button to display a ScreenTip containing the name of the button, which is indicative of its function. The WordArt toolbar contains buttons to display the text vertically, change the style or shape, and/or edit the text.

Step 4: **WordArt Continued**

- Click and drag the WordArt object to move it the top of the document. (The Format WordArt dialog box is not yet visible.)
- Point to the WordArt object, click the **right mouse button** to display a shortcut menu, then click **Format WordArt** to display the Format WordArt dialog box as shown in Figure 3.6d.
- Click the **Colors and Lines tab**, then click the **Fill Color drop-down arrow** to display the available colors. Select a different color (e.g., blue).
- Move and/or size the WordArt as necessary. Click the **Undo button** if necessary to cancel the action and start again.
- Save the document.

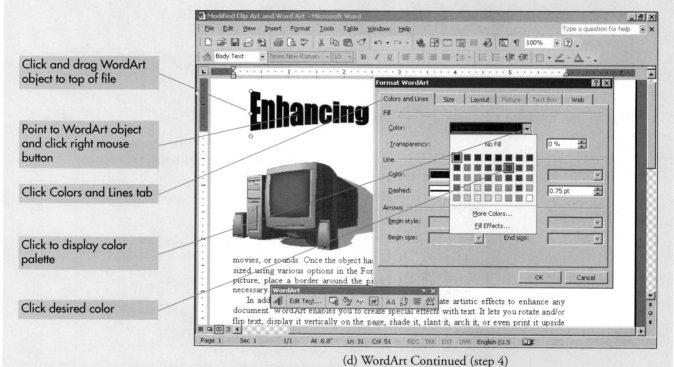

(d) WordArt Continued (step 4)

FIGURE 3.6 *Hands-on Exercise 1 (continued)*

THE THIRD DIMENSION

You can make your WordArt images even more dramatic by adding 3-D effects. You can tilt the text up or down, right or left, increase or decrease the depth, and change the shading. Pull down the View menu, click Toolbars, click Customize to display the complete list of available toolbars, then check the box to display the 3-D Settings toolbar. Select the WordArt object, then experiment with various tools and special effects. The results are even better if you have a color printer.

Step 5: **The Insert Symbol Command**

- Press **Ctrl+End** to move to the end of the document as shown in Figure 3.6e. If necessary, change the sentence to reflect the version of Windows that you are using, for example Windows 98, rather than Windows 2000.
- Click at the end of the sentence, pull down the **Insert menu**, click **Symbol**, then choose **Wingdings** from the font list box. Click the **Windows logo** (the last character in the last line), click **Insert**, then close the dialog box.
- Click and drag to select the newly inserted symbol. Change the font size to **24**. (One of our favorite shortcuts is to press and hold the Ctrl key as you press the square bracket to increase the font size; that is, Ctrl+] increases the font size, whereas Ctrl+[decreases the font size.)
- Click after the word Microsoft in the same sentence, type **(r)**, and try to watch the screen as you enter the text. The (r) will be converted to the ® registered trademark symbol by the AutoCorrect feature.
- Save the document.

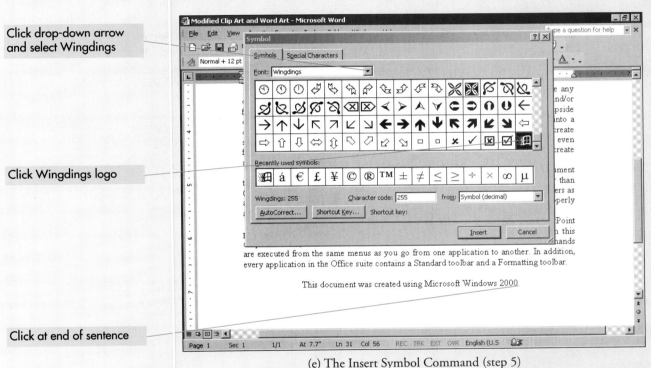

(e) The Insert Symbol Command (step 5)

FIGURE 3.6 *Hands-on Exercise 1 (continued)*

AUTOCORRECT AND AUTOFORMAT

The AutoCorrect feature not only corrects mistakes as you type by substituting one character string for another (e.g., "the" for "teh"), but it will also substitute symbols for typewritten equivalents such as © for (c), provided the entries are included in the table of substitutions. The AutoFormat feature is similar in concept and replaces common fractions such as 1/2 or 1/4 with ½ or ¼. It also converts ordinal numbers such as 1st or 2nd to 1st or 2nd.

Step 6: **Create the AutoShape**

- Pull down the **View menu**, click the **Toolbars command** to display the list of available toolbars, then click **Drawing toolbar**.
- Press the **End key** to move to the end of the line, then press the **enter key** once or twice to move below the last sentence in the document.
- Click the **down arrow** on the AutoShapes tool to display the menu. Click the **Stars and Banners submenu** and select the **Horizontal Scroll**.
- The mouse pointer changes to a tiny crosshair, and you will see a drawing canvas with an indication to create your drawing here. Press **Esc** to remove the drawing canvas (we find it easier to work without it), which in turn moves you to the bottom of the page.
- Click and drag to create a scroll, as shown in Figure 3.6f. Release the mouse. Right click in the scroll, click **Add Text**, change the font size to 18 point, then enter the text **This document was created by** (your name). Center the text.
- Click and drag the sizing handle (a circle) at the bottom of the scroll to make it narrow. Click and drag the yellow diamond at the left of the scroll to change the appearance of the scroll. The green dot at the top of the scroll allows you to rotate the scroll. Click off the scroll. Save the document.

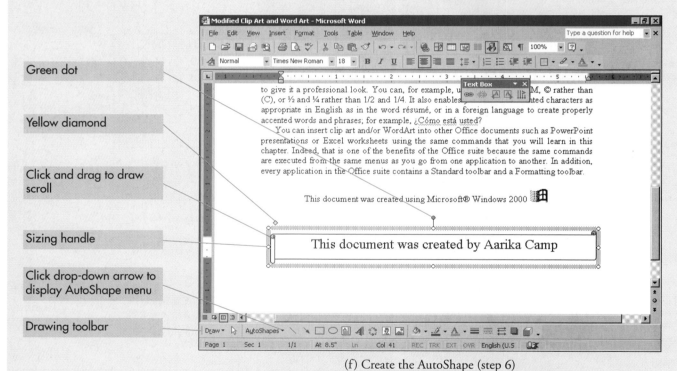

(f) Create the AutoShape (step 6)

FIGURE 3.6 *Hands-on Exercise 1 (continued)*

ORGANIZATION CHARTS AND OTHER DIAGRAMS

Microsoft Office includes a tool to create organization charts and other types of diagrams. Pull down the Insert menu and click the Diagram command to display the Diagram Gallery dialog box, where you choose the type of diagram. Click the Organization Chart, for example, and you are presented with a default chart that is the basis of a typical corporate organization chart. See practice exercise 11 at the end of the chapter.

Step 7: **The Completed Document**

- Pull down the **File menu** and click the **Page Setup command** to display the Page Setup dialog box. (You can also double click the ruler below the Formatting toolbar to display the dialog box.)
- Click the **Margins tab** and change the top margin to **1.5 inches** (to accommodate the WordArt at the top of the document). Click **OK**.
- Click the **drop-down arrow** on the Zoom box and select **Whole Page** to preview the completed document as shown in Figure 3.6g. You can change the size and position of the objects from within this view. For example:
  - Select (click) the clip art to select this object and display its sizing handles.
  - Select (click) the banner to select it (and deselect the previous selection).
  - Move and size either object as necessary.
- Print the document and submit it to your instructor as proof that you did the exercise. Save the document. Close the document.
- Exit Word if you do not want to continue with the next exercise at this time.

Click drop-down arrow and select Whole Page

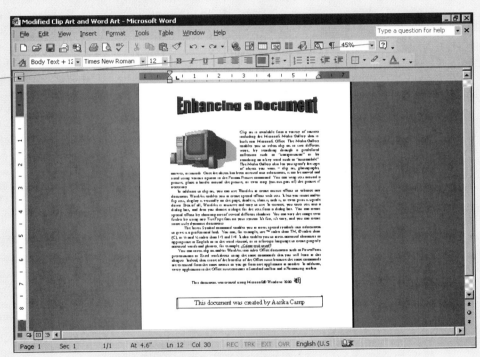

(g) The Completed Document (step 7)

FIGURE 3.6 *Hands-on Exercise 1 (continued)*

ANIMATE TEXT ON SCREEN

Select the desired text, pull down the Format menu, click the Font command to display the Font dialog box, then click the Text Effects tab. Select the desired effect such as "Blinking Background" or "Las Vegas Lights", then click OK to accept the settings and close the dialog box. The selected text should be displayed with the specified effect, which appears on the screen, but not when the document is printed. To cancel the effect, select the text, display the Font dialog box, click the Text Effects tab, select "None" as the effect then click OK.

MICROSOFT WORD AND THE INTERNET

The Internet and World Wide Web have totally changed society. Perhaps you are already familiar with the basic concepts that underlie the Internet but, if not, a brief review is in order. The ***Internet*** is a network of networks that connects computers anywhere in the world. The ***World Wide Web*** (WWW or simply, the Web) is a very large subset of the Internet, consisting of those computers that store a special type of document known as a ***Web page*** or ***HTML document***.

The interesting thing about a Web page is that it contains references called ***hyperlinks*** to other Web pages, which may in turn be stored on a different computer that is located anywhere in the world. And therein lies the fascination of the Web, in that you simply click on link after link to go effortlessly from one document to the next. You can start your journey on your professor's home page, then browse through any set of links you wish to follow.

Web pages are developed in a special language called ***HTML (HyperText Markup Language)***. Initially, the only way to create a Web page was to learn HTML. Microsoft Office simplifies the process because you can create the document in Word, then simply save it as a Web page. In other words, you start Word in the usual fashion, enter the text of the document with basic formatting, then use the ***Save As Web Page command*** to convert the document to HTML. Microsoft Word does the rest and generates the HTML statements for you. You can continue to enter text and/or change the formatting for existing text just as you can with an ordinary document.

Figure 3.7 contains the Web page you will create in the next hands-on exercise. The exercise begins by having you search the Web to locate a suitable photograph for inclusion into the document. You then download the picture to your PC and use the Insert Picture command to insert the photograph into your document. You add formatting, hyperlinks, and footnotes as appropriate, then you save the document as a Web page. The exercise is easy to do, and it will give you an appreciation for the various Web capabilities that are built into Office XP.

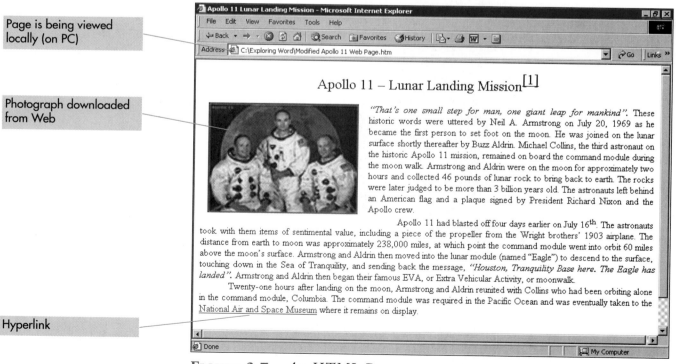

FIGURE 3.7   *An HTML Document*

Even if you do not place your page on the Web, you can still view it locally on your PC. This is the approach we follow in the next hands-on exercise, which shows you how to save a Word document as a Web page, then see the results of your effort in a Web browser. The Web page is stored on a local drive (e.g., on drive A or drive C) rather than on an Internet server, but it can still be viewed through Internet Explorer (or any other browser).

The ability to create links to local documents and to view those pages through a Web browser has created an entirely new way to disseminate information. Organizations of every size are taking advantage of this capability to develop an *intranet* in which Web pages are placed on a local area network for use within the organizations. The documents on an intranet are available only to individuals with access to the local area network on which the documents are stored.

THE WEB PAGE WIZARD

The Save As Web Page command converts a Word document to the equivalent HTML document for posting on a Web server. The Web Page Wizard extends the process to create a multipage Web site, complete with navigation and a professionally designed theme. The navigation options let you choose between horizontal and vertical frames so that the user can see the links and content at the same time. The design themes are quite varied and include every element on a Web page.

Copyright Protection

A *copyright* provides legal protection to a written or artistic work, giving the author exclusive rights to its use and reproduction, except as governed under the fair use exclusion as explained below. Anything on the Internet should be considered copyrighted unless the document specifically says it is in the *public domain*, in which case the author is giving everyone the right to freely reproduce and distribute the material.

Does this mean you cannot quote in your term papers statistics and other facts you find while browsing the Web? Does it mean you cannot download an image to include in your report? The answer to both questions depends on the amount of the material and on your intended use of the information. It is considered *fair use*, and thus not an infringement of copyright, to use a portion of the work for educational, nonprofit purposes, or for the purpose of critical review or commentary. In other words, you can use a quote, downloaded image, or other information from the Web, provided you cite the original work in your footnotes and/or bibliography. Facts themselves are not covered by copyright, so you can use statistical and other data without fear of infringement. You should, however, cite the original source in your document through appropriate footnotes or endnotes.

A *footnote* provides additional information about an item, such as its source, and appears at the bottom of the page where the reference occurs. An *endnote* is similar in concept but appears at the end of a document. A horizontal line separates the notes from the rest of the document. You can also convert footnotes to endnotes or vice versa.

The *Insert Reference command* inserts either a footnote or an endnote into a document, and automatically assigns the next sequential number to that note. The command adjusts for last-minute changes, either in your writing or in your professor's requirements. It will, for example, renumber all existing notes to accommodate the addition or deletion of a footnote or endnote. Existing notes are moved (or deleted) within a document by moving (deleting) the reference mark rather than the text of the footnote.

HANDS-ON EXERCISE 2

MICROSOFT WORD AND THE WEB

Objective To download a picture from the Internet for use in a Word document; to insert a hyperlink into a Word document; to save a Word document as a Web page. Use Figure 3.8 as a reference. The exercise requires an Internet connection.

Step 1: **Search the Web**

➤ Start Internet Explorer. It does not matter which page you see initially, as long as you are able to connect to the Internet and start Internet Explorer. Click the **Maximize button** so that Internet Explorer takes the entire screen.
➤ Click the **Search button** on the Standard Buttons toolbar to display the Search pane in the Explorer bar at the left of the Internet Explorer window. The option button to find a Web page is selected by default.
➤ Enter **Apollo 11** in the Find a Web page text box, then click the **Search button**. The results of the search are displayed in the left pane as shown in Figure 3.8a. The results you obtain will be different from ours.
➤ Check to see which search engine you used, and if necessary click the **down arrow** to the right of the Next button to select **Lycos** (the engine we used). The search will be repeated with this engine. Your results can still be different from ours, because new pages are continually added to the Web.
➤ Select (click) the link to Apollo 11 Home. (Enter the URL www.nasm.edu/apollo/AS11 manually if your search engine does not display this link.)
➤ Click the **Close button** to close the Search pane, so that your selected document takes the entire screen.

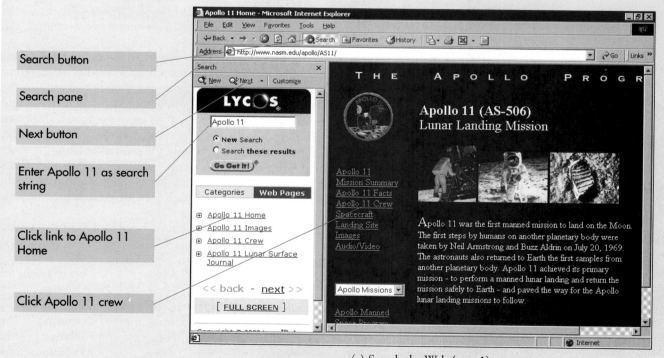

(a) Search the Web (step 1)

FIGURE 3.8 *Hands-on Exercise 2*

CHAPTER 3: ENHANCING A DOCUMENT

Step 2: **Save the Picture**

➤ Click the link to **Apollo 11 Crew** from the previous page to display the page in Figure 3.8b. Point to the picture of the astronauts, click the **right mouse button** to display a shortcut menu, then click the **Save Picture As command** to display the Save As dialog box.
- Click the **drop-down arrow** in the Save in list box to specify the drive and folder in which you want to save the graphic (e.g., in the **Exploring Word folder**).
- Internet Explorer supplies the file name and file type for you. You may change the name, but you cannot change the file type.
- Click the **Save button** to download the image. Remember the file name and location, as you will need to access the file in the next step.

➤ The Save As dialog box will close automatically after the picture has been downloaded. Click the **Minimize button** in the Internet Explorer window, since you are temporarily finished using the browser.

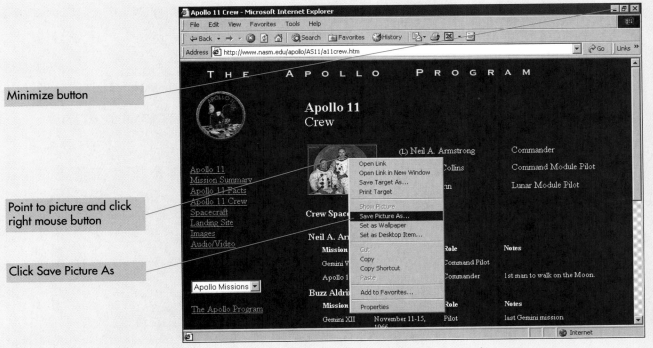

(b) Save the Picture (step 2)

FIGURE 3.8   *Hands-on Exercise 2 (continued)*

THE AUTOSEARCH FEATURE

The fastest way to initiate a search is to click in the Address box, enter the key word "go" followed by the topic you are searching for (e.g., go University of Miami), then press the enter key. Internet Explorer automatically invokes the MSN search engine and returns the relevant documents. You can also guess a Web address by typing, www.company.com, where you supply the name of the company.

Step 3: **Insert the Picture**

➤ Start Word and open the **Apollo 11 document** in the **Exploring Word folder**. Save the document as **Modified Apollo** so that you can return to the original document if necessary.

➤ Pull down the **View menu** to be sure that you are in the **Print Layout view** (or else you will not see the picture after it is inserted into the document). Pull down the **Insert menu**, point to (or click) **Picture command**, then click **From File** to display the Insert Picture dialog box shown in Figure 3.8c.

➤ Click the **down arrow** on the Look in text box to select the drive and folder where you previously saved the picture. Click the **down arrow** on the Files of type list box and specify **All files**.

➤ Select (click) **AS11_crew**, which is the file containing the picture that you downloaded earlier. Click the **drop-down arrow** on the **Views button**, then click **Preview** to display the picture prior to inserting. Click **Insert**.

➤ Save the document.

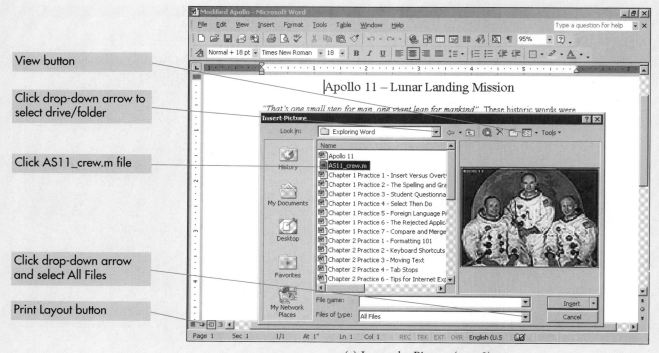

(c) Insert the Picture (step 3)

FIGURE 3.8 *Hands-on Exercise 2 (continued)*

CHANGE THE DEFAULT LOCATION

The default file location is the folder Word uses to open and save a document unless it is otherwise instructed. To change the default location, pull down the Tools menu, click Options, click the File Locations tab, click the desired File type (documents), then click the Modify command button to display the Modify Location dialog box. Click the drop-down arrow in the Look In box to select the new folder (e.g., C:\Exploring Word). Click OK to accept this selection. Click OK to close the Options dialog box. The next time you access the Open or Save commands from the File menu, the Look In text box will reflect the change.

Step 4: **Move and Size the Picture**

- Point to the picture after it is inserted into the document, click the **right mouse button** to display a shortcut menu, then click the **Format Picture command** to display the Format Picture dialog box.
- Click the **Layout tab** and choose **Square** in the Wrapping style area. Click **OK** to accept the settings and close the Format Picture dialog box.
- Move and/or size the picture so that it approximates the position in Figure 3.8d. Click the **Undo button** anytime that you are not satisfied with the result.
- Save the document.

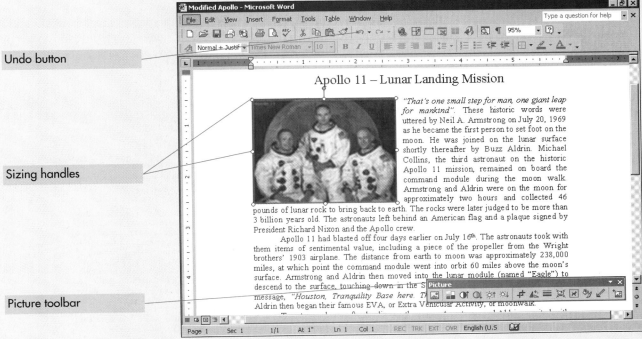

(d) Move and Size the Picture (step 4)

FIGURE 3.8 *Hands-on Exercise 2 (continued)*

CROPPING A PICTURE

The Crop tool is one of the most useful tools when dealing with a photograph as it lets you eliminate (crop) part of a picture. Select (click) the picture to display the Picture toolbar and sizing handles. (If you do not see the Picture toolbar, right click the picture to display a context-sensitive menu, then click the Show Picture Toolbar command. Click the Crop tool (the ScreenTip will display the name of the tool), then click and drag a sizing handle to crop the part of the picture you want to eliminate. Click the Crop button a second time to turn the feature off.

Step 5: **Insert a Footnote**

- Press **Ctrl+Home** to move to the beginning of the document, then click after Lunar Landing Mission in the title of the document. This is where you will insert the footnote.
- Pull down the **Insert menu**. Click **Reference**, then choose **Footnote** to display the Footnote and Endnote dialog as shown in Figure 3.8e. Check that the option buttons for **Footnotes** is selected and that the numbering starts at one. Click **Insert**.
- The insertion point moves to the bottom of the page, where you type the text of the footnote, which should include the Web site from where you downloaded the picture.
- Press **Ctrl+Home** to move to the beginning of the page, where you will see a reference for the footnote you just created. If necessary, you can move (or delete) a footnote by moving (deleting) the reference mark rather than the text of the footnote.
- Save the document.

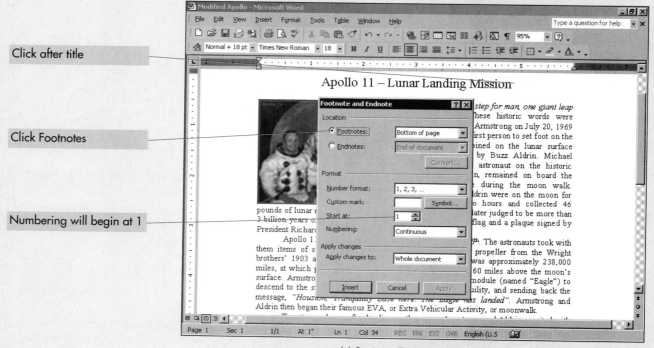

(e) Insert a Footnote (step 5)

FIGURE 3.8 *Hands-on Exercise 2 (continued)*

COPY THE WEB ADDRESS

Use the Copy command to enter a Web address from Internet Explorer into a Word document. Not only do you save time by not having to type the address yourself, but you also ensure that it is entered correctly. Click in the Address bar of Internet Explorer to select the URL, then pull down the Edit menu and click the Copy command (or use the Ctrl+C keyboard shortcut). Switch to the Word document, click at the place in the document where you want to insert the URL, pull down the Edit menu, and click the Paste command (or use the Ctrl+V keyboard shortcut).

Step 6: **Insert a Hyperlink**

➤ Scroll to the bottom of the document, then click and drag to select the text **National Air and Space Museum**.
➤ Pull down the **Insert menu** and click the **Hyperlink command** (or click the **Insert Hyperlink button** on the Standard toolbar) to display the Insert Hyperlink dialog box as shown in Figure 3.8f.
➤ National Air and Space Museum is entered automatically in the Text to display text box. Click in the Address text box to enter the address. Type **http://www.nasm.edu**.
➤ Click **OK** to accept the settings and close the dialog box. The hyperlink should appear as an underlined entry in the document.
➤ Save the document.

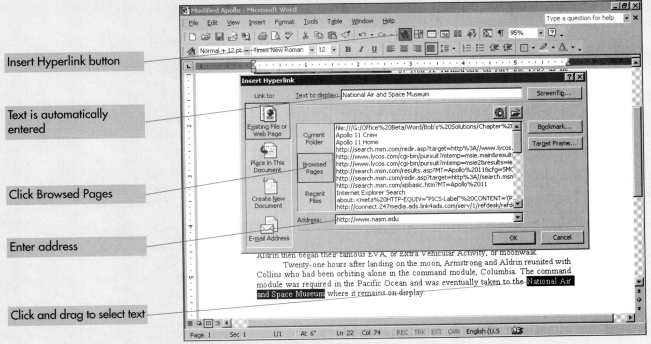

(f) Insert a hyperlink (step 6)

FIGURE 3.8 *Hands-on Exercise 2 (continued)*

CLICK TO EDIT, CTRL+CLICK TO FOLLOW

Point to a hyperlink within a Word document and you see a ToolTip that says to press and hold the Ctrl key (Ctrl+Click) to follow the link. This is different from what you usually do, because you normally just click a link to follow it. What if, however, you wanted to edit the link? Word modifies the convention so that clicking a link enables you to edit the link. Alternatively, you can right click the hyperlink to display a context-sensitive menu from where you can make the appropriate choice.

MICROSOFT WORD 2002

Step 7: **Create the Web Page**

➤ Pull down the **File menu** and click the **Save As Web Page command** to display the Save As dialog box as shown in Figure 3.8g. Click the **drop-down arrow** in the Save In list box to select the appropriate drive, then open the **Exploring Word folder**.

➤ Change the name of the Web page to **Modified Apollo 11 Web Page** (to differentiate it from the Word document of the same name). Click the **Change Title button** to display a dialog box in which you change the title of the Web page as it will appear in the title bar of the Web browser.

➤ Click the **Save button**. You will see a message indicating that the pictures will be left aligned. Click **Continue**.

➤ The title bar changes to reflect the name of the Web page. There are now two versions of this document in the Exploring Word folder, Modified Apollo 11, and Modified Apollo 11 Web Page. The latter has been saved as a Web page (in HTML format).

➤ Click the **Print button** on the Standard toolbar to print this page for your instructor from within Microsoft Word.

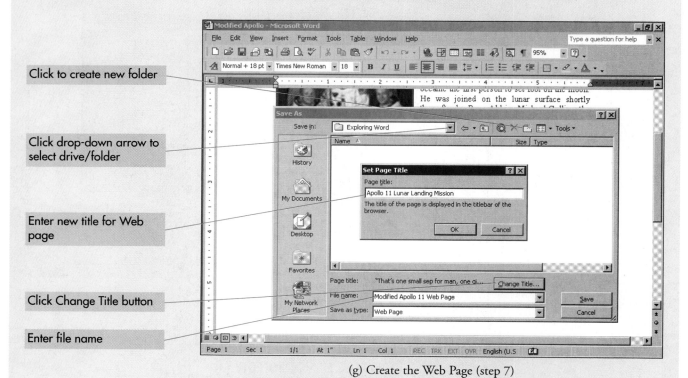

(g) Create the Web Page (step 7)

FIGURE 3.8 *Hands-on Exercise 2 (continued)*

CREATE A NEW FOLDER

Do you work with a large number of documents? If so, it may be useful to store those documents in different folders, perhaps one folder for each course you are taking. Pull down the File menu, click the Save As command to display the Save As dialog box, then click the Create New Folder button to display the associated dialog box. Enter the name of the folder, then click OK. Once the folder has been created, use the Look In box to change to that folder the next time you open that document.

Step 8: **View the Web Page**

- You can view the Web page you just created even though it has not been saved on a Web server. Click the button for Internet Explorer on the Windows taskbar to return to the browser.
- Pull down the **File menu** and click the **Open command** to display the Open dialog box. Click the **Browse button**, then select the folder (e.g., Exploring Word) where you saved the Web page. Select (click) the **Modified Apollo 11 Web Page** document, click **Open**, then click **OK** to open the document.
- You should see the Web page that was created earlier as shown in Figure 3.8h, except that you are viewing the page in Internet Explorer rather than Microsoft Word. The Address bar reflects the local address (in the Exploring Word folder) of the document.
- Click the **Print button** on the Internet Explorer toolbar to print this page for your instructor. Does this printed document differ from the version that was printed from within Microsoft Word at the end of the previous step?

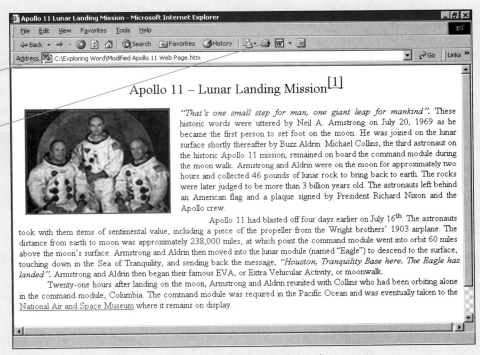

(h) View the Web Page (step 8)

FIGURE 3.8 *Hands-on Exercise 2 (continued)*

AN EXTRA FOLDER

Look carefully at the contents of the Exploring Word folder within the Open dialog box. You see the HTML document you just created, as well as a folder that was created automatically by the Save As Web Page command. The latter folder contains the objects that are referenced by the page, such as the crew's picture and a horizontal line above the footnotes. Be sure to copy the contents of this folder to the Web server in addition to your Web page if you decide to post the page.

Step 9: **Test the Web Page**

- This step requires an Internet connection because you will be verifying the addresses you entered earlier.
- Click the hyperlink to the **National Air and Space Museum** to display the Web page in Figure 3.8i. You can explore this site, or you can click the **Back button** to return to your Web page.
- If you are unable to connect to the Museum site, click in the Address bar and enter a different URL to see if you can connect to that site. If you connect to one site, but not the other, you should return to your original document to correct the URL.
- Click the **Word button** on the taskbar to return to the Web page, **right click** the hyperlink to display a context-sensitive menu, click **Edit Hyperlink**, and make the necessary correction. Save the corrected document.
- Click the **Browser button** on the Windows taskbar to return to the browser, click the **Refresh button** to load the corrected page, then try the hyperlink a second time.
- Close Internet Explorer. Exit Word if you do not want to continue with the next exercise at this time.

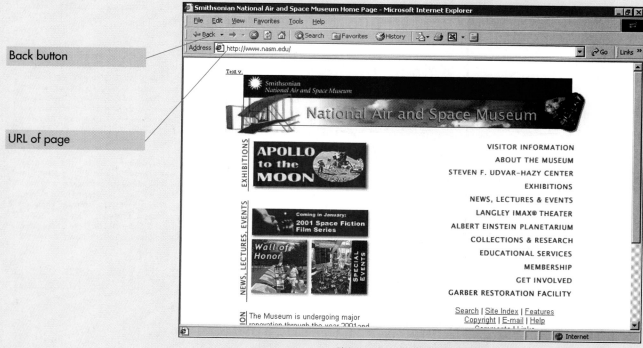

(i) Test the Web Page (step 9)

FIGURE 3.8   *Hands-on Exercise 2 (continued)*

VIEW THE HTML SOURCE CODE

Pull down the View menu in Internet Explorer and click the Source command to view the HTML statements that comprise the Web page. The statements are displayed in their own window, which is typically the Notepad accessory. Pull down the File menu in Notepad and click the Print command to print the HTML code. Do you see any relationship between the HTML statements and the Web page?

WIZARDS AND TEMPLATES

We have created some very interesting documents throughout the text, but in every instance we have formatted the document entirely on our own. It is time now to see what is available in terms of "jump starting" the process by borrowing professional designs from others. Accordingly, we discuss the wizards and templates that are built into Microsoft Word.

A *template* is a partially completed document that contains formatting, text, and/or graphics. It may be as simple as a memo or as complex as a résumé or newsletter. Microsoft Word provides a variety of templates for common documents, including a résumé, agenda, and fax cover sheet. You simply open the template, then modify the existing text as necessary, while retaining the formatting in the template. A *wizard* makes the process even easier by asking a series of questions, then creating a customized template based on your answers. A template or wizard creates the initial document for you. It's then up to you to complete the document by entering the appropriate information.

Figure 3.9 illustrates the use of wizards and templates in conjunction with a résumé. You can choose from one of three existing styles (contemporary, elegant, and professional) to which you add personal information. Alternatively, you can select the **Résumé Wizard** to create a customized template, as was done in Figure 3.9a.

After the Résumé Wizard is selected, it prompts you for the information it needs to create a basic résumé. You specify the style in Figure 3.9b, enter the requested information in Figure 3.9c, and choose the categories in Figure 3.9d. The wizard continues to ask additional questions (not shown in Figure 3.9), after which it displays the (partially) completed résumé based on your responses. You then complete the résumé by entering the specifics of your employment and/or additional information. As you edit the document, you can copy and paste information within the résumé, just as you would with a regular document. You can also change the formatting. It takes a little practice, but the end result is a professionally formatted résumé in a minimum of time.

Microsoft Word contains templates and wizards for a variety of other documents. (Look carefully at the tabs within the dialog box of Figure 3.9a and you can infer that Word will help you to create letters, faxes, memos, reports, legal pleadings, publications, and even Web pages. The Office Web site, www.microsoft.com/office, contains additional templates.) Consider, too, Figure 3.10, which displays four attractive documents that were created using the respective wizards. Realize, however, that while wizards and templates will help you to create professionally designed documents, they are only a beginning. *The content is still up to you.*

THIRTY SECONDS IS ALL YOU HAVE

Thirty seconds is the average amount of time a personnel manager spends skimming your résumé and deciding whether or not to call you for an interview. It doesn't matter how much training you have had or how good you are if your résumé and cover letter fail to project a professional image. Know your audience and use the vocabulary of your targeted field. Be positive and describe your experience from an accomplishment point of view. Maintain a separate list of references and have it available on request. Be sure all information is accurate. Be conscientious about the design of your résumé and proofread the final documents very carefully.

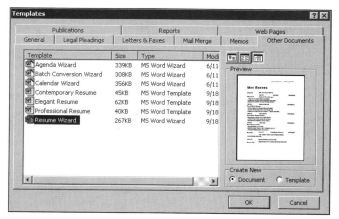
(a) Résumé Wizard

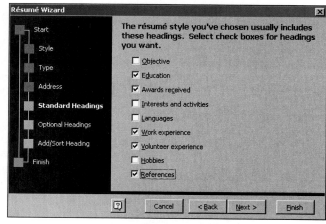
(d) Choose the Headings

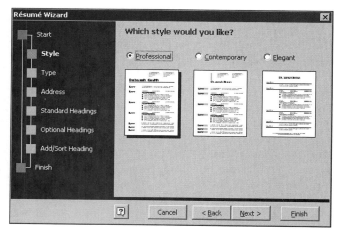
(b) Choose the Style

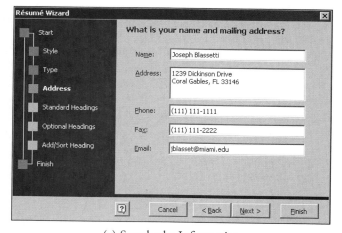
(c) Supply the Information

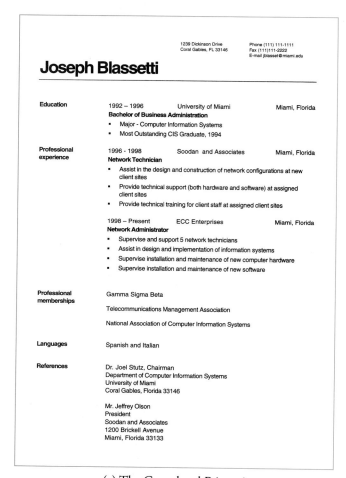
(e) The Completed Résumé

FIGURE 3.9 *Creating a Résumé*

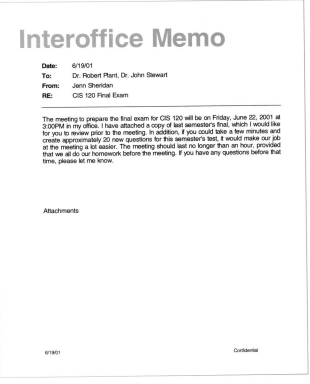

FIGURE 3.10   *What You Can Do with Wizards*

MAIL MERGE

A ***mail merge*** can create any type of standardized document, but it is used most frequently to create a set of ***form letters***. In essence, it creates the same letter many times, changing the name, address, and other information as appropriate from letter to letter. You might use a mail merge to look for a job upon graduation, when you send essentially the same cover letter to many different companies. The concept is illustrated in Figure 3.11, in which John Smith has written a letter describing his qualifications, then merges that letter with a set of names and addresses, to produce the individual letters.

The mail merge process uses two files as input, a main document and a data source. A set of form letters is created as output. The ***main document*** (e.g., the cover letter in Figure 3.11a) contains standardized text, together with one or more ***merge fields*** that serve as place holders for the variable data that will be inserted in the individual letters. The data source (the set of names and addresses in Figure 3.11b) contains the information that varies from letter to letter.

The first row in the data source is called the header row and identifies the fields in the remaining rows. Each additional row contains the data to create one letter and is called a data record. Every data record contains the same fields in the same order—for example, Title, FirstName, LastName, and so on. (The fields can also be specified collectively, but for purposes of illustration, we will show the fields individually.)

The main document and the data source work in conjunction with one another, with the merge fields in the main document referencing the corresponding fields in the data source. The first line in the address of Figure 3.11a, for example, contains three entries in angle brackets, <<Title>> <<FirstName>> <<LastName>>. (These entries are not typed explicitly but are entered through special commands, as described in the hands-on exercise that follows shortly.) The merge process examines each record in the data source and substitutes the appropriate field values for the corresponding merge fields as it creates the individual form letters. For example, the first three fields in the first record will produce Mr. Eric Simon. The same fields in the second record will produce Dr. Lauren Howard, and so on.

In similar fashion, the second line in the address of the main document contains the <<Company>> field. The third line contains the <<JobTitle>> field. The fourth line references the <<Address1>> field, and the last line contains the <<City>>, <<State>, and <<PostalCode>> fields. The salutation repeats the <<Title>> and <<LastName>> fields. The first sentence in the letter uses the <<Company>> field a second time. The mail merge prepares the letters one at a time, with one letter created for every record in the data source until the file of names and addresses is exhausted. The individual form letters are shown in Figure 3.11c. Each letter begins automatically on a new page.

The implementation of a mail merge is accomplished through the ***Mail Merge Wizard***, which will open in the task pane and guide you through the various steps in the mail merge process. In essence there are three things you must do:

1. Create and save the main document
2. Create and save the data source
3. Merge the main document and data source to create the individual letters

The same data source can be used to create multiple sets of form letters. You could, for example, create a marketing campaign in which you send an initial letter to the entire list, and then send follow-up letters at periodic intervals to the same mailing list. Alternatively, you could filter the original mailing list to include only a subset of names, such as the individuals who responded to the initial letter. You could also use the wizard to create a different set of documents, such as envelopes and/or e-mail messages. Note, too, that you can sort the addresses to print the documents in a specified sequence, such as zip code to take advantage of bulk mail.

John Doe Computing

1239 Dickinson Drive • Coral Gables, FL 33146 • (305) 666-5555

June 22, 2001

«Title» «FirstName» «LastName»
«JobTitle»
«Company»
«Address1»
«City», «State» «PostalCode»

Dear «Title» «LastName»:

I would like to inquire about a position with «Company» as an entry-level programmer. I have graduated from the University of Miami with a Bachelor's Degree in Computer Information Systems (May 2001) and I am very interested in working for you. I am proficient in all applications in Microsoft Office and also have experience with Visual Basic, C++, and Java. I have had the opportunity to design and implement a few Web applications, both as a part of my educational program, and during my internship with Personalized Computer Designs, Inc.

I am eager to put my skills to work and would like to talk with you at your earliest convenience. I have enclosed a copy of my résumé and will be happy to furnish the names and addresses of my references. You may reach me at the above address and phone number. I look forward to hearing from you.

Sincerely,

John Doe
President

(a) The Form Letter

Title	FirstName	LastName	Company	JobTitle	Address1	City	State	PostalCode
Mr.	Eric	Simon	Arnold and Joyce Computing	President	10000 Sample Road	Coral Springs	FL	33071
Dr.	Lauren	Howard	Unique Systems	President	475 LeJeune Road	Coral Springs	FL	33071
Mr.	Peter	Gryn	Gryn Computing	Director of Human Resources	1000 Federal Highway	Miami	FL	33133
Ms.	Julie	Overby	The Overby Company	President	100 Savona Avenue	Coral Gables	FL	33146

(b) The Data Source

FIGURE 3.11 *A Mail Merge*

(c) The Printed Letters

FIGURE 3.11   *A Mail Merge (continued)*

Hands-on Exercise 3

Mail Merge

Objective: To create a main document and associated data source; to implement a mail merge and produce a set of form letters. Use Figure 3.12 as a guide.

Step 1: **Open the Form Letter**

> Open the **Form Letter** document in the Exploring Word folder. If necessary, change to the **Print Layout view** and zoom to **Page Width** as shown in Figure 3.12a.
>
> Modify the letterhead to reflect your name and address. Select **"Your Name Goes Here"**, then type a new entry to replace the selected text. Add your address information to the second line.
>
> Click immediately to the left of the first paragraph, then press the **enter key** twice to insert two lines. Press the **up arrow** two times to return to the first line you inserted.
>
> Pull down the **Insert menu** and click the **Date and Time command** to display the dialog box in Figure 3.12a. Select (click) the date format you prefer and, if necessary, check the box to **Update automatically**. Click **OK** to close the dialog box.
>
> Save the document as **Modified Form Letter** so that you can return to the original document if necessary.

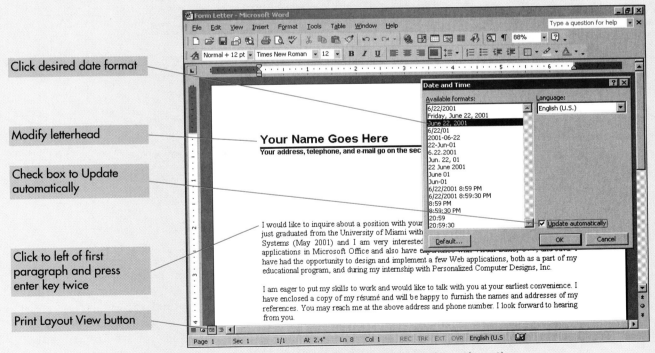

(a) Open the Form Letter (step 1)

Figure 3.12 *Hands-on Exercise 3*

MICROSOFT WORD 2002

Step 2: **The Mail Merge Wizard**

➤ Pull down the **Tools menu**, click **Letters and Mailings**, then click **Mail Merge Wizard** to open the task pane.
➤ The option button for **Letters** is selected by default as shown in Figure 3.12b. Click **Next: Starting Document** to begin creating the document.
➤ The option button to **Use the current document** is selected. (We began the exercise by providing you with the text of the document, as opposed to having you create the entire form letter.) Click **Next: Select Recipients** to enter the list of names and addresses.
➤ Click the option button to **Type a New List**, then click the link to **Create** that appears within the task pane. This brings you to a new screen, where you enter the data for the recipients of your form letter.

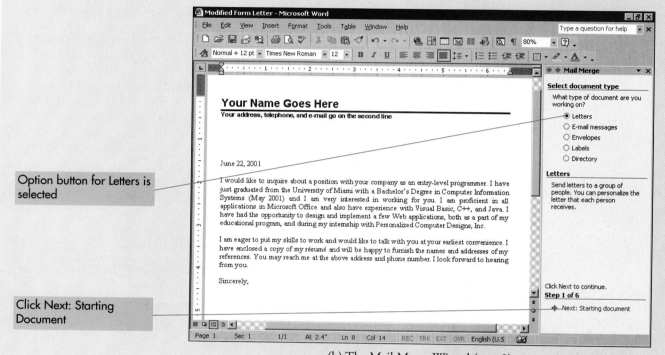

(b) The Mail Merge Wizard (step 2)

FIGURE 3.12   *Hands-on Exercise 3 (continued)*

THE MAIL MERGE WIZARD

The Mail Merge Wizard simplifies the process of creating form letters and other types of merge documents through step-by-step directions that appear automatically in the task pane. The options for the current step appear in the top portion of the task pane and are self-explanatory. Click the link to the next step at the bottom of the pane to move forward in the process, or click the link to the previous step to correct any mistakes you might have made.

Step 3: **Select the Recipients**

➤ Enter data for the first record, using your name and address as shown in Figure 3.12c. Type **Mr.** or **Ms.** in the Title field, then press **Tab** to move to the next (FirstName) field and enter your first name. Complete the first record.
➤ Click the **New Entry button** to enter the data for the next person. Enter your instructor's name and a hypothetical address. Enter data for one additional person, real or fictitious as you see fit. Click **Close** when you have completed the data entry.
➤ You will see the Save Address List dialog box, where you will be prompted to save the list of names and addresses. Save the file as **Names and Addresses** in the **Exploring Word folder**. The file type is specified as a Microsoft Office Address list.
➤ You will see a dialog box showing all of the records you have just entered. Click **OK** to close the dialog box. Click **Next: Write Your Letter** to continue.

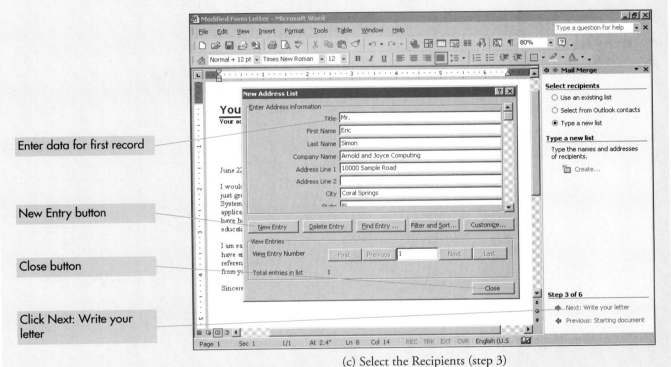

(c) Select the Recipients (step 3)

FIGURE 3.12 *Hands-on Exercise 3 (continued)*

THREE DIFFERENT FILES

A mail merge works with three different files. The main document and data source are input to the mail merge, which creates a set of merged letters as output. You can use the same data source (e.g., a set of names and addresses) with different main documents (a form letter and an envelope) and/or use the same main document with multiple data sources. You typically save, but do not print, the main document(s) and the data source(s). Conversely, you print the set of merged letters, but typically do not save them.

Step 4: **Write (Complete) the Letter**

- The text of the form letter is already written, but it is still necessary to insert the fields within the form letter.
- Click below the date and press the **enter key** once or twice. Click the link to the **Address block** to select a single entry that is composed of multiple fields (Street, City, ZipCode, and so on). Click **OK**. The AddressBlock field is inserted into the document as shown in Figure 3.12d.
- Press the **enter key** twice to leave a blank line after the address block. Click the link to the **Greeting line** to display the Greeting Line dialog box in Figure 3.12d.
- Choose the type of greeting you want. Change the comma that appears after the greeting to a colon since this is a business letter. Click **OK**. The GreetingLine field is inserted into the document and enclosed in angled brackets.
- Press **enter** to enter a blank line. Save the document. Click **Next: Preview Your Letters** to continue.

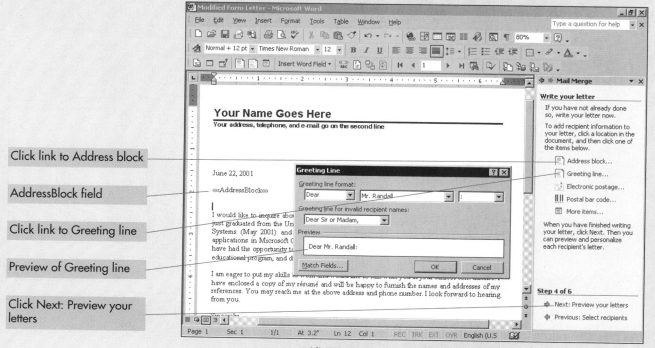

(d) Write (Complete) the Letter (step 4)

FIGURE 3.12 *Hands-on Exercise 3 (continued)*

BLOCKS VERSUS INDIVIDUAL FIELDS

The Mail Merge Wizard simplifies the process of entering field names into a form letter by supplying two predefined entries, AddressBlock and GreetingLine, which contain multiple fields that are typical of the ways in which an address and salutation appear in a conventional letter. You can still insert individual fields, by clicking in the document where you want the field to go, then clicking the Insert Merge Fields button on the Mail Merge toolbar. The blocks are easier.

Step 5: **View and Print the Letters**

- You should see the first form letter as shown. You can click the >> or << button in the task pane (not shown in Figure 3.12e) to move to the next or previous letter, respectively. You can also use the navigation buttons that appear on the Mail Merge toolbar.
- View the records individually to be sure that the form letter is correct and that the data has been entered correctly. Make corrections if necessary.
- Click **Next: Complete the Merge**, then click **Print** to display the dialog box in Figure 3.12e. Click **OK**, then **OK** again, to print the form letters.
- Click the **<<abc>>** button to display the field codes. Pull down the **File menu** and click the **Print command** to display the Print dialog box. Check the option to print the current page. Click **OK**. Submit this page to your instructor as well.
- Pull down the **File menu** and click the **Close command** to close the Modified Form Letter and the associated set of names and addresses. Save the documents if you are prompted to do so.

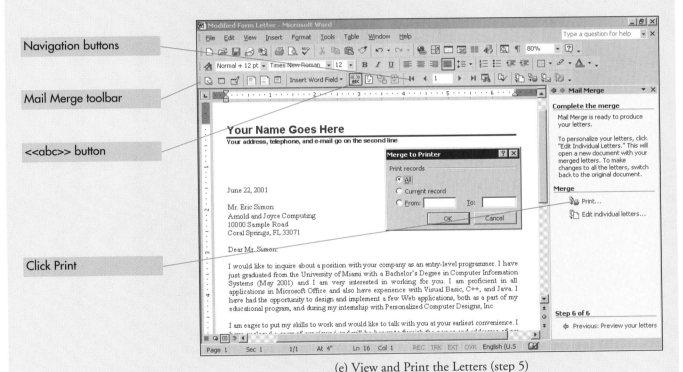

(e) View and Print the Letters (step 5)

FIGURE 3.12 *Hands-on Exercise 3 (continued)*

THE MAIL MERGE TOOLBAR

The Mail Merge toolbar appears throughout the mail merge process and contains various buttons that apply to different steps within the process. Click the <<abc>> button to display field values rather than field codes. Click the button a second time and you switch back to field codes from field values. Click the <<abc>> button to display the field values, then use the navigation buttons to view the different letters. Click the ▶ button, for example, and you move to the next letter. Click the ▶| button to display the form letter for the last record.

MICROSOFT WORD 2002   **137**

Step 6: **Open the Contemporary Merge Letter**

➤ Pull down the **File menu** and click the **New command** (or click the **New button** on the Standard toolbar). If necessary, pull down the **View menu** and open the **task pane**.

➤ Click the link to **General Templates** in the task pane to display the Templates dialog box, then click the **Mail Merge tab** to display the Templates dialog box in Figure 3.12f.

➤ Select (click) the **Contemporary Merge Letter**. Click the **Preview button** (if necessary) to see a thumbnail view of this document.

➤ Be sure that the **Document option button** is selected. Click **OK** to select this document and begin the merge process. You will see a form letter with the AddressBlock and GreetingLine fields already entered.

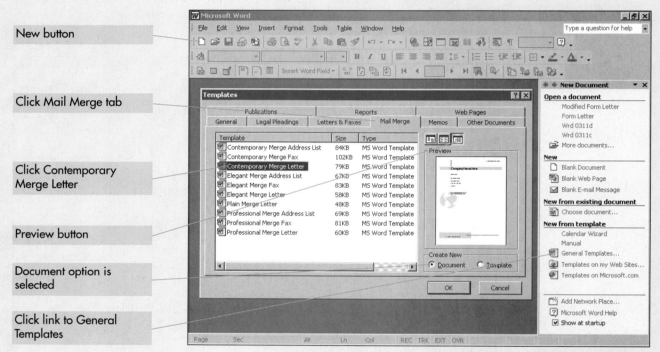

(f) Open the Contemporary Merge Letter (step 6)

FIGURE 3.12 *Hands-on Exercise 3 (continued)*

PAPER MAKES A DIFFERENCE

Most of us take paper for granted, but the right paper can make a significant difference in the effectiveness of the document, especially when you are trying to be noticed. Reports and formal correspondence are usually printed on white paper, but you would be surprised how many different shades of white there are. Other types of documents lend themselves to a specialty paper for additional impact. Consider the use of a specialty paper the next time you have an important project.

Step 7: **Select the Recipients**

- The option button to **Use an existing list** is selected. Click the link to **Browse** for the existing list to display the Select Data Source dialog box.
- We will use the same data source that you created earlier. (You could also use the list of Outlook contacts or an Access database as the source of your data.)
- Click the **down arrow** on the Look in box to select the Exploring Word folder. Click the **down arrow** on the File type list box to select **Microsoft Office Address Lists**. Select the **Names and Addresses** file from step 3. Click **Open**.
- You should see the Mail Merge Recipients dialog box in Figure 3.12g that contains the records you entered earlier. You can use this dialog box to modify existing data, to change the order in which the form letters will appear, and/or to choose which recipients are to receive the form letter.
- Click **OK** to close the dialog box. Click **Next: Write Your Letter** to continue.

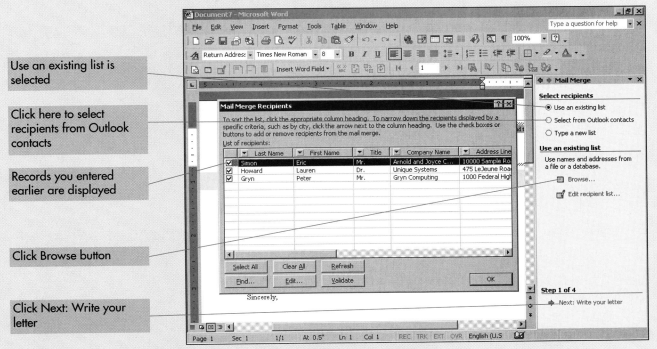

(g) Select the Recipients (step 7)

FIGURE 3.12 *Hands-on Exercise 3 (continued)*

USE OUTLOOK AS THE DATA SOURCE

Think of Microsoft Outlook as a desktop manager, or personal assistant, that keeps track of all types of information for you. You can use Outlook to maintain an address book (contact list), schedule appointments, create a task list, or send and receive e-mail. And since Outlook is an integral part of Office XP, its components are easily linked to other Office applications. You could, for example, use the Outlook address book as the data source for a mail merge in Word. Just start the mail merge in the usual way, then click the option button to select recipients from the Outlook contacts.

MICROSOFT WORD 2002   139

Step 8: **Write the Form Letter**

➤ The form letter has been created as a template as shown in Figure 3.12h. The AddressBlock and GreetingLine fields have been entered for you.

➤ Click and drag in **Type Your Letter Here** that appears in the form letter and enter two or three sentences of your own choosing. Our letter indicates that we are seeking to acquire one or more of the consulting firms on the mailing list.

➤ Continue to personalize the form letter by replacing the text in the original template. Click at the upper right of the letter and enter your return address. Use your name for the company name.

➤ Replace the signature lines with your name and title. Select the line at the bottom of the page that reads [Click here and type a slogan] and enter a slogan of your own.

➤ Save the document as **Contemporary Merge Letter** in the Exploring Word folder. Click the link to **Next: Preview your letter**.

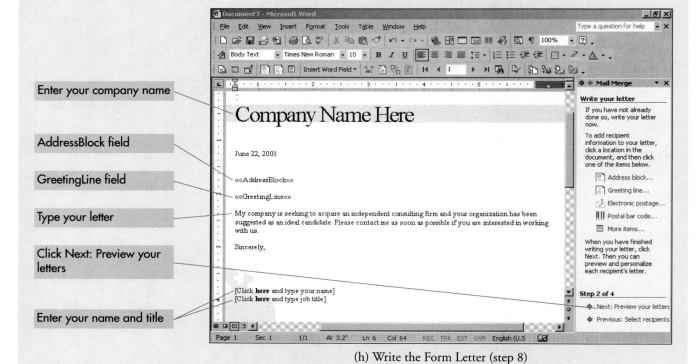

(h) Write the Form Letter (step 8)

FIGURE 3.12 *Hands-on Exercise 3 (continued)*

ENVELOPES AND MAILING LABELS

The set of form letters is only the first step in a true mail merge because you also have to create the envelopes and/or mailing labels to physically mail the letters. Start the Mail Merge wizard as you normally do, but this time specify labels (or envelopes) instead of a form letter. Follow the instructions provided by the wizard using the same data source as for the form letters. See practice exercise 8 at the end of the chapter.

Step 9: **Complete the Merge**

- You should see the first form letter. The name and address of this recipient are the same as in the set of form letters created earlier. Click **Next: Complete the Merge** to display the screen in Figure 3.12i.
- Use the navigation buttons on the Mail Merge toolbar to view the three form letters. Click the link to **Print . . .** (or click the **Merge to Printer button** on the Mail Merge toolbar).
- Click the option button to print all the letters, then click **OK** to display the Print dialog box. Click **OK** to print the individual form letters.
- Exit Word. Save the form letter and/or the names and addresses document if you are asked to do so.

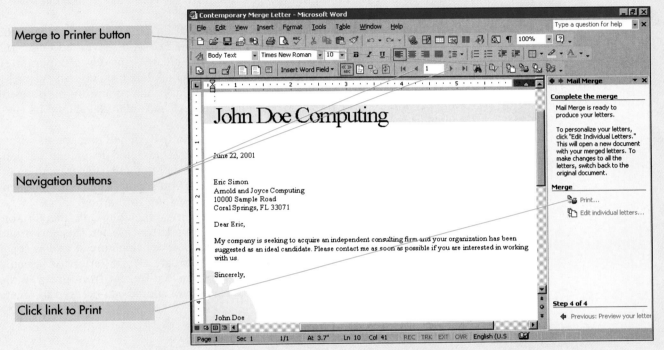

(i) Complete the Merge (step 9)

FIGURE 3.12 *Hands-on Exercise 3 (continued)*

EDIT THE INDIVIDUAL LETTERS

Click the Merge to File button (or click the link to Edit individual letters in the task pane) to create a third document (called Letters1 by default) consisting of the individual form letters. There are as many pages in this document as there are records in the address file. You can view and/or edit the individual letters from this document, then print the entire set of merged letters. You need not save this document, however, unless you actually made changes to the individual letters.

SUMMARY

The Microsoft Media Gallery contains clip art, sound files, photographs, and movies and it is accessible from any application in Microsoft Office. The Insert Picture command is used to insert clip art into a document through the task pane. Microsoft WordArt is an application within Microsoft Office that creates decorative text, which can be used to add interest to a document.

The Insert Symbol command provides access to special characters, making it easy to place typographic characters into a document. The symbols can be taken from any font and can be displayed in any point size.

Resources (such as clip art or photographs) can be downloaded from the Web for inclusion in a Word document. Web pages are written in a language called HTML (HyperText Markup Language). The Save As Web Page command saves a Word document as a Web page.

A copyright provides legal protection to a written or artistic work, giving the author exclusive rights to its use and reproduction, except as governed under the fair use exclusion. Anything on the Internet should be considered copyrighted unless the document specifically says it is in the public domain. The fair use exclusion enables you to use a portion of the work for educational, nonprofit purposes, or for the purpose of critical review or commentary. All such material should be cited through an appropriate footnote or endnote.

Wizards and templates help create professionally designed documents with a minimum of time and effort. A template is a partially completed document that contains formatting and other information. A wizard is an interactive program that creates a customized template based on the answers you supply. The resulting documentation can be modified with respect to content and/or formatting.

A mail merge creates the same letter many times, changing only the variable data, such as the addressee's name and address, from letter to letter. It is performed in conjunction with a main document and a data source, which are stored as separate documents. The mail merge can be used to create a form letter for selected records, and/or print the form letters in a sequence different from the way the records are stored in the data source. The same data source can be used to create multiple sets of form letters.

KEY TERMS

AutoCorrect (p. 106)
AutoFormat (p. 113)
Clip art (p. 109)
Copyright (p. 117)
Drawing canvas (p. 108)
Drawing toolbar (p. 108)
Endnote (p. 117)
Fair use (p. 117)
Footnote (p. 117)
Form letter (p. 130)
HTML document (p. 116)
Hyperlink (p. 116)

HyperText Markup Language (HTML) (p. 116)
Insert Picture command (p. 105)
Insert Reference command (p. 117)
Insert Symbol command (p. 106)
Internet (p. 116)
Intranet (p. 117)
Mail merge (p. 130)
Mail Merge Wizard (p. 130)
Main document (p. 130)
Media Gallery (p. 105)
Merge fields (p. 130)

Microsoft WordArt (p. 107)
Public domain (p. 117)
Résumé Wizard (p. 127)
Save As Web Page command (p. 116)
Sizing handle (p. 108)
Template (p. 127)
Web page (p. 116)
Wizard (p. 127)
WordArt (p. 107)
World Wide Web (p. 116)

MULTIPLE CHOICE

1. How do you change the size of a selected object so that the height and width change in proportion to one another?
   (a) Click and drag any of the four corner handles in the direction you want to go
   (b) Click and drag the sizing handle on the top border, then click and drag the sizing handle on the left side
   (c) Click and drag the sizing handle on the bottom border, then click and drag the sizing handle on the right side
   (d) All of the above

2. The Microsoft Media Gallery:
   (a) Is accessed through the Insert Picture command
   (b) Is available in Microsoft Word, Excel, and PowerPoint
   (c) Enables you to search for a specific piece of clip art by specifying a key word in the description of the clip art
   (d) All of the above

3. How do you search for clip art using the Microsoft Media Gallery?
   (a) By entering a key word that describes the image you want
   (b) By browsing through various collections
   (c) Both (a) and (b)
   (d) Neither (a) nor (b)

4. Which of the following objects can be inserted into a document from the Microsoft Media Gallery?
   (a) Clip art
   (b) Sound
   (c) Photographs
   (d) All of the above

5. Which of the following is true about a mail merge?
   (a) The same form letter can be used with different data sources
   (b) The same data source can be used with different form letters
   (c) Both (a) and (b)
   (d) Neither (a) nor (b)

6. Which of the following best describes the documents that are associated with a mail merge?
   (a) The main document is typically saved, but not necessarily printed
   (b) The names and addresses are typically saved, but not necessarily printed
   (c) The individual form letters are printed, but not necessarily saved
   (d) All of the above

7. Which of the following is true about footnotes or endnotes?
   (a) The addition of a footnote or endnote automatically renumbers the notes that follow
   (b) The deletion of a footnote or endnote automatically renumbers the notes that follow
   (c) Both (a) and (b)
   (d) Neither (a) nor (b)

8. Which of the following is true about the Insert Symbol command?
   (a) It can insert a symbol in different type sizes
   (b) It can access any font installed on the system
   (c) Both (a) and (b)
   (d) Neither (a) nor (b)

9. Which of the following is true regarding objects and the associated toolbars?
   (a) Clicking on a WordArt object displays the WordArt toolbar
   (b) Clicking on a Picture displays the Picture toolbar
   (c) Both (a) and (b)
   (d) Neither (a) nor (b)

10. Which of the following objects can be downloaded from the Web for inclusion in a Word document?
    (a) Clip art
    (b) Photographs
    (c) Sound and video files
    (d) All of the above

11. What is the significance of the Shift key in conjunction with various tools on the Drawing toolbar?
    (a) It will draw a circle rather than an oval using the Oval tool
    (b) It will draw a square rather than a rectangle using the Rectangle tool
    (c) It will draw a horizontal or vertical line with the Line tool
    (d) All of the above

12. What happens if you enter the text www.intel.com into a document?
    (a) The entry is converted to a hyperlink, and further, the text will be underlined and displayed in a different color
    (b) The associated page will be opened provided your computer has access to the Internet
    (c) Both (a) and (b)
    (d) Neither (a) nor (b)

13. Which of the following is a true statement about wizards?
    (a) They are accessed from the General Templates link on the task pane
    (b) They always produce a finished document with no further modification necessary
    (c) Both (a) and (b)
    (d) Neither (a) nor (b)

14. Which of the following is true about an HTML document that was created from within Microsoft Word?
    (a) It can be viewed locally
    (b) It can be viewed via the Web provided it is uploaded onto a Web server
    (c) Both (a) and (b)
    (d) Neither (a) nor (b)

15. Which of the following are created as a result of the Save As Web Page command, given that the document is called "My Home Page"?
    (a) An HTML document called "My Home Page"
    (b) A "My Home Page" folder that contains the objects that appear on the associated page
    (c) Both (a) and (b)
    (d) Neither (a) nor (b)

ANSWERS

1. a	6. d	11. d
2. d	7. c	12. a
3. c	8. c	13. a
4. d	9. c	14. c
5. c	10. d	15. c

PRACTICE WITH WORD

1. Travel World: You have been hired as an intern for the Travel World agency and asked to create a flyer to distribute on campus. The only requirement is to include the travel agent's name and e-mail address. (Our information appears at the bottom of the page and is not visible in Figure 3.13.) Use any combination of clip art or photographs to make the flyer as attractive as possible. Be sure to spell check the completed flyer, then print the document for your instructor.

FIGURE 3.13 *Travel World (Exercise 1)*

2. Symbols as Clip Art: The installation of Microsoft Windows and/or Microsoft Office provides multiple fonts that are accessible from any application. Two of the fonts, Symbols and Wingdings, contain a variety of special characters that can be inserted to create some unusual documents as shown in Figure 3.14. The "art" in these documents is not clip art per se, but symbols that are added to a document through the Insert Symbol command. Use your imagination, coupled with the fact that a font is scalable to any size, to recreate the documents in Figure 3.14. Better yet, create two documents of your own design that utilize these special fonts.

3. Create a Home Page: It's easy to create a home page. Start a new document and enter its text just as you would any other document. Use any and all formatting commands to create a document similar to the one in Figure 3.15. We suggest you use clip art, as opposed to a real picture. Pull down the File menu and use the Save As Web Page command to convert the Word document to an HTML document. Complete the document as described below:
   a. Use the Insert Hyperlink command to create a list of 3 to 5 hyperlinks. Be sure to enter accurate Web addresses for each of your sites.
   b. Select all of the links after they have been entered, then click the Bullets button on the Formatting toolbar to create a bulleted list.

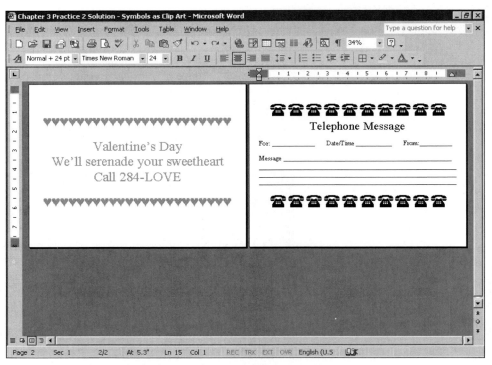

FIGURE 3.14 *Symbols as Clip Art (Exercise 2)*

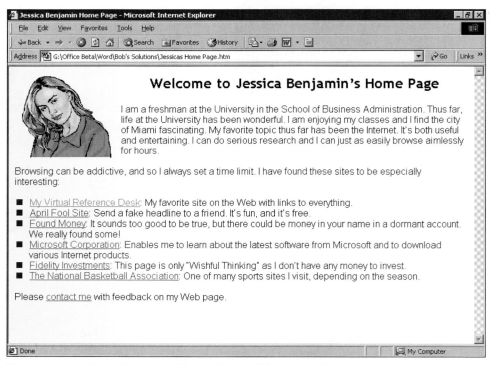

FIGURE 3.15 *Create a Home Page (Exercise 3)*

c. Pull down the Format menu, click the Themes command, then select a professionally chosen design for your Web page.
d. Save the document a final time, then exit Word. Start Windows Explorer, then go to the folder containing your home page, and double click the file you just created. Internet Explorer will start automatically (because your document was saved as a Web page). You should see your document within Internet Explorer as shown in Figure 3.15. Look carefully at the Address bar and note the local address on drive C, as opposed to a Web address. Print the document for your instructor as proof you completed the assignment.
e. Creating the home page and viewing it locally is easy. Placing the page on the Web where it can be seen by anyone with an Internet connection is not as straightforward. You will need additional information from your instructor about how to obtain an account on a Web server (if that is available at your school), and further how to upload the Web page from your PC to the server.

4. **A Commercial Web Page:** Create a home page for a real or hypothetical business as shown in Figure 3.16, using the same general procedure as in the previous exercise. Start Word and enter the text of a new document that describes your business. Use clip art, bullets, hyperlinks, and other formatting as appropriate. Save the completed document as a Web page. Start Windows Explorer and locate the newly created document. Double click the document to open the default browser and display the page as shown in Figure 3.16, then print the page from within the browser.

There is no requirement to upload the page to the Web, but it is worth doing if you have the capability. You will need additional information from your instructor about how to obtain an account on a Web server (if that is available at your school), and further how to upload the Web page from your PC to the server.

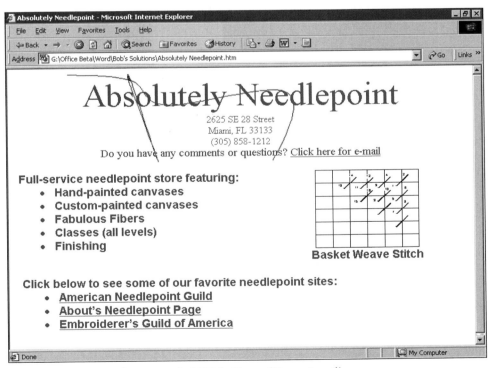

FIGURE 3.16   *Commercial Web Page (Exercise 4)*

MICROSOFT WORD 2002   **147**

5. **Presidential Anecdotes:** Figure 3.17 displays the finished version of a document containing ten presidential anecdotes. The anecdotes were taken from the book *Presidential Anecdotes,* by Paul F. Boller, Jr., published by Penguin Books (New York, 1981). Open the *Chapter 3 Practice 5* document that is found in the Exploring Word folder, then make the following changes:
   a. Add a footnote after Mr. Boller's name, which appears at the end of the second sentence, citing the information about the book. This, in turn, renumbers all existing footnotes in the document.
   b. Switch the order of the anecdotes for Lincoln and Jefferson so that the presidents appear in order. The footnotes for these references are changed automatically.
   c. Convert all of the footnotes to endnotes, as shown in the figure.
   d. Go to the White House Web site at www.whitehouse.gov and download a picture of any of the ten presidents, then incorporate that picture into a cover page. Remember to cite the reference with an appropriate footnote.
   e. Submit the completed document to your instructor.

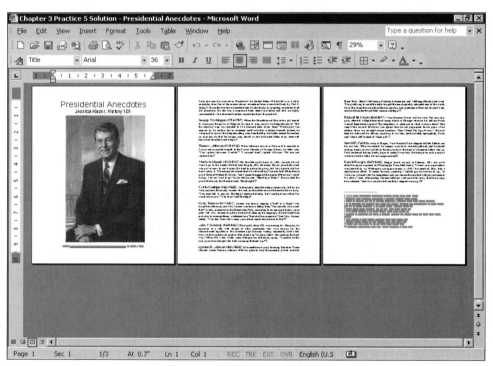

FIGURE 3.17 *Presidential Anecdotes (Exercise 5)*

6. **The iCOMP® Index:** The iCOMP® index was created by Intel to compare the speeds of various microprocessors to one another. This assignment asks you to search the Web to find a chart of the current index (3.0 or later), download the chart to your PC, then insert the picture into the document in Figure 3.18. You will find the text of that document in the file, *Chapter 3 Practice 6*, in the Exploring Word folder. Be sure to format the document completely, including the registration mark. Add your name to the completed document and submit it to your instructor.

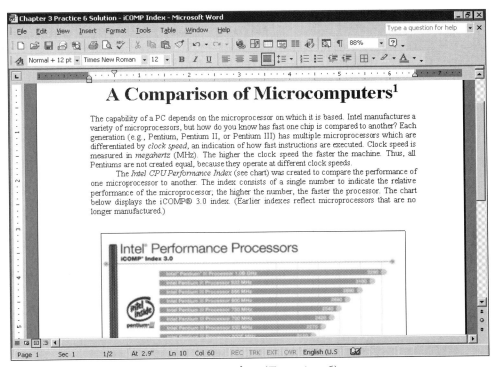

FIGURE 3.18   *The iCOMP® Index (Exercise 6)*

7. **Create an Envelope:** The Résumé Wizard will step you through the process of creating a résumé, but you need an envelope in which to mail it. Pull down the Tools menu, click Letters and Mailings, click the Envelopes and Labels command, click the Envelopes tab, then enter the indicated information. You can print the envelope and/or include it permanently in the document as shown in Figure 3.19. Do not, however, attempt to print the envelope in a computer lab at school unless envelopes are available for the printer.

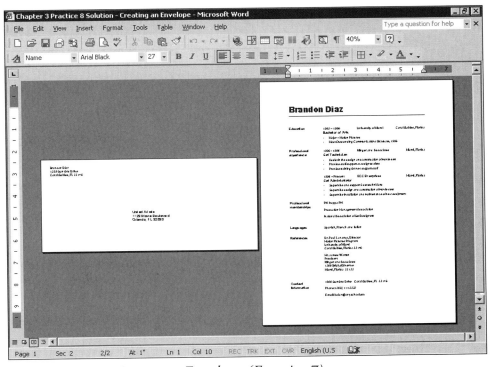

FIGURE 3.19   *Create an Envelope (Exercise 7)*

8. **Mailing Labels:** A mail merge creates the form letters for a mailing. That is only the first step, however, because you also have to create envelopes and/or mailing labels to physically mail the letters. Start the Mail Merge Wizard as you did in the third hands-on exercise, but this time, specify labels instead of a form letter. Follow the instructions provided by the wizard to create a set of mailing labels using the same data source as you did in the hands-on exercise. Do *not* attempt to print the labels, however, unless you actually have mailing labels for the printer.

    You can, however, capture the screen in Figure 3.20 to prove to your instructor that you created the labels. It's easy. Use the mail merge to create the labels as shown in Figure 3.20. Press the Print Screen key to capture the screen to the Windows clipboard. Start a new Word document, then click the Paste button to paste the screen into the document. Add a sentence or two that describes the assignment. Include a cover page with your name, then print the completed document for your instructor.

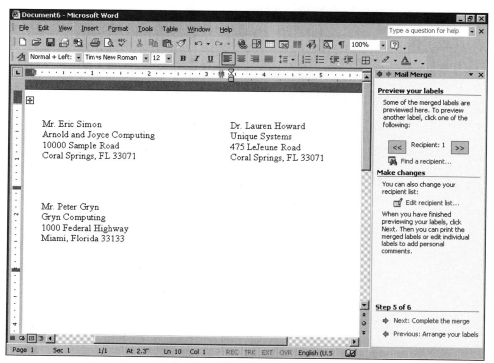

FIGURE 3.20 *Mailing Labels (exercise 8)*

9. **Organization Charts:** The document in Figure 3.21 shows how you can create organization charts within Microsoft Office. Pull down the Insert menu and click the Diagrams command to display the Diagram Gallery dialog box from where you can select the Organization chart. The default chart consists of four boxes that are displayed on two levels. The lower level has three subordinates reporting to the single box on the top level. You can modify the chart by adding (removing) boxes at various levels using the Insert Shape button on the Organization Chart toolbar. You can also click in any box to add the appropriate descriptive text. The organization chart is a single object that can be moved and sized within the document, just like any Windows object.

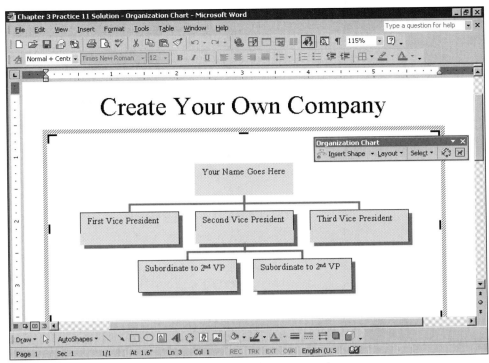

FIGURE 3.21   *Organization Chart (Exercise 9)*

10. **The Calendar Wizard:** The Calendar Wizard is one of several wizards that are built into Microsoft Office. Pull down the File menu, click the New command, then click General Templates in the task pane to display the associated dialog box. Click the Other documents tab to access the Calendar Wizard and create a calendar for the current month as shown in Figure 3.22.

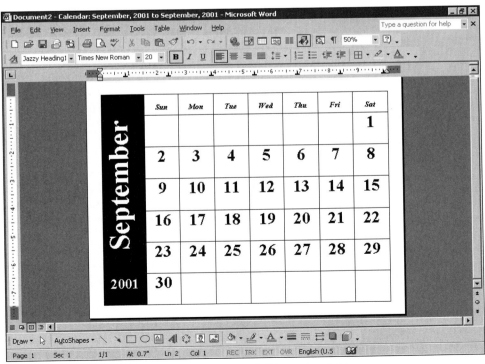

FIGURE 3.22   *The Calendar Wizard (Exercise 10)*

ON YOUR OWN

The Cover Page

Use WordArt and/or the Media Gallery to create a truly original cover page that you can use with all of your assignments. The cover page should include the title of the assignment, your name, course information, and date. (Use the Insert Date and Time command to insert the date as a field so that it will be updated automatically every time you retrieve the document.) The formatting is up to you. Print the completed cover page and submit it to your instructor, then use the cover page for all future assignments.

My Favorite Recording Group

The Web is a source of infinite variety, including music from your favorite rock group. You can also find music, which you can download and play, provided you have the necessary hardware. Use any search engine to find one or more sites about your favorite rock group. Try to find biographical information as well as a picture. Incorporate the results of your research into a short paper for your instructor.

The Résumé

Use your imagination to create a résumé for Benjamin Franklin or Leonardo Da Vinci, two acknowledged geniuses. The résumé is limited to one page and will be judged for content (yes, you have to do a little research on the Web) as well as appearance. You can intersperse fact and fiction as appropriate; for example, you may want to leave space for a telephone and/or a fax number, but could indicate that these devices have not yet been implemented. You can choose a format for the résumé using the Résumé Wizard, or better yet, design your own.

Macros

The Insert Symbol command can be used to insert foreign characters into a document, but this technique is too slow if you use these characters with any frequency. It is much more efficient to develop a series of macros (keyboard shortcuts) that will insert the characters for you. You could, for example, create a macro to insert an accented e, then invoke that macro through the Ctrl+e keyboard shortcut. Parallel macros could be developed for the other vowels or special characters that you use frequently. Use the Help menu to learn about macros, then summarize your findings in a short note to your instructor.

The Letterhead

A well-designed letterhead adds impact to your correspondence. Collect samples of professional stationery, then design your own letterhead, which includes your name, address, phone, and any other information you deem relevant. Include a fax number and/or e-mail address as appropriate. Use your imagination and design the letterhead for your planned career. Try different fonts and/or the Format Border command to add horizontal line(s) under the text. Consider a graphic logo, but keep it simple. You might also want to decrease the top margin so that the letterhead prints closer to the top of the page.

CHAPTER 4

Advanced Features: Outlines, Tables, Styles, and Sections

OBJECTIVES

AFTER READING THIS CHAPTER YOU WILL BE ABLE TO:

1. Create a bulleted or numbered list; create an outline using a multilevel list.
2. Describe the Outline view; explain how this view facilitates moving text within a document.
3. Describe the tables feature; create a table and insert it into a document.
4. Explain how styles automate the formatting process and provide a consistent appearance to common elements in a document.
5. Use the AutoFormat command to apply styles to an existing document; create, modify, and apply a style to selected elements of a document.
6. Define a section; explain how section formatting differs from character and paragraph formatting.
7. Create a header and/or a footer; establish different headers or footers for the first, odd, or even pages in the same document.
8. Insert page numbers into a document; use the Edit menu's Go To command to move directly to a specific page in a document.
9. Create an index and a table of contents.

OVERVIEW

This chapter presents a series of advanced features that will be especially useful the next time you have to write a term paper with specific formatting requirements. We show you how to create a bulleted or numbered list to emphasize important items within a term paper, and how to create an outline for that paper. We also introduce the tables feature, which is one of the most powerful features in Microsoft Word as it provides an easy way to arrange text, numbers, and/or graphics.

The second half of the chapter develops the use of styles, or sets of formatting instructions that provide a consistent appearance to similar elements in a document. We describe the AutoFormat command that assigns styles to an existing document and greatly simplifies the formatting process. We show you how to create a new style, how to modify an existing style, and how to apply those styles to text within a document. We introduce the Outline view, which is used in conjunction with styles to provide a condensed view of a document. We also discuss several items associated with longer documents, such as page numbers, headers and footers, a table of contents, and an index.

BULLETS AND LISTS

A list helps you organize information by highlighting important topics. A ***bulleted list*** emphasizes (and separates) the items. A ***numbered list*** sequences (and prioritizes) the items and is automatically updated to accommodate additions or deletions. An ***outline*** (or ***outline numbered list***) extends a numbered list to several levels, and it too is updated automatically when topics are added or deleted. Each of these lists is created through the ***Bullets and Numbering command*** in the Format menu, which displays the Bullets and Numbering dialog box in Figure 4.1.

The tabs within the Bullets and Numbering dialog box are used to choose the type of list and customize its appearance. The Bulleted tab selected in Figure 4.1a enables you to specify one of several predefined symbols for the bullet. Typically, that is all you do, although you can use the Customize button to change the default spacing (of ¼ inch) of the text from the bullet and/or to choose a different symbol for the bullet.

The Numbered tab in Figure 4.1b lets you choose Arabic or Roman numerals, or upper- or lowercase letters, for a Numbered list. As with a bulleted list, the Customize button lets you change the default spacing, the numbering style, and/or the punctuation before or after the number or letter. Note, too, the option buttons to restart or continue numbering, which become important if a list appears in multiple places within a document. In other words, each occurrence of a list can start numbering anew, or it can continue from where the previous list left off.

The Outline Numbered tab in Figure 4.1c enables you to create an outline to organize your thoughts. As with the other types of lists, you can choose one of several default styles, and/or modify a style through the Customize command button. You can also specify whether each outline within a document is to restart its numbering, or whether it is to continue numbering from the previous outline.

The List Styles tab (not shown in Figure 4.1) lets you change the style (formatting specifications) associated with a list. You can change the font size, use a picture or symbol for a bullet, add color, and so on. Styles are discussed later in the chapter.

CREATING AN OUTLINE

Our next exercise explores the Bullets and Numbering command in conjunction with creating an outline for a hypothetical paper on the United States Constitution. The exercise begins by having you create a bulleted list, then asking you to convert it to a numbered list, and finally to an outline. The end result is the type of outline your professor may ask you to create prior to writing a term paper.

As you do the exercise, remember that a conventional outline is created as an outline numbered list within the Bullets and Numbering command. Text for the outline is entered in the Print Layout or Normal view, *not* the Outline view. The latter provides a completely different capability—a condensed view of a document that is used in conjunction with styles and is discussed later in the chapter. We mention this to avoid confusion should you stumble into the Outline view.

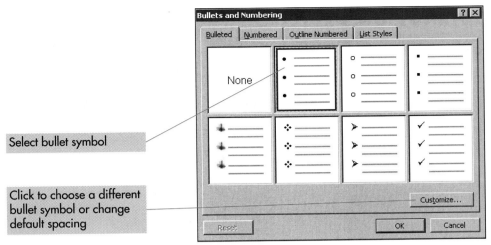

(a) Bulleted List

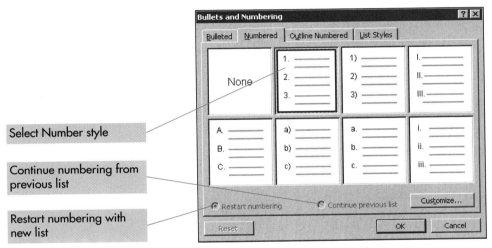

(b) Numbered List

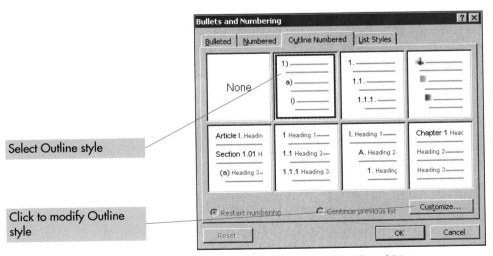

(c) Outline Numbered List

FIGURE 4.1   *Bullets and Numbering*

MICROSOFT WORD 2002

HANDS-ON EXERCISE 1

BULLETS, LISTS, AND OUTLINES

Objective To use the Bullets and Numbering command to create a bulleted list, a numbered list, and an outline. Use Figure 4.2 as a guide.

Step 1: **Create a Bulleted List**

➤ Start Word and begin a new document. Type **Preamble**, the first topic in our list, and press **enter**.

➤ Type the three remaining topics, **Article I—Legislative Branch**, **Article II—Executive Branch**, and **Article III—Judicial Branch**. Do not press enter after the last item.

➤ Click and drag to select all four topics as shown in Figure 4.2a. Pull down the **Format menu** and click the **Bullets and Numbering command** to display the Bullets and Numbering dialog box.

➤ If necessary, click the **Bulleted tab**, select the type of bullet you want, then click **OK** to accept this setting and close the dialog box. Bullets have been added to the list.

➤ Click after the words **Judicial Branch** to deselect the list and also to position the insertion point at the end of the list. Press **enter** to begin a new line. A bullet appears automatically.

➤ Type **Amendments**. Press **enter** to end this line and begin the next, which already has a bullet.

➤ Press **enter** a second time to terminate the bulleted list.

➤ Save the document as **US Constitution** in the **Exploring Word folder**.

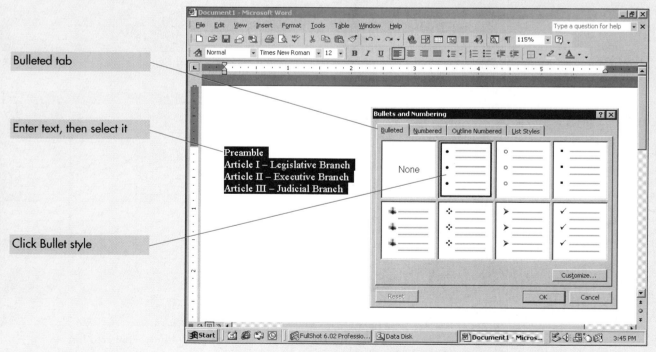

(a) Create a Bulleted List (step 1)

FIGURE 4.2 *Hands-on Exercise 1*

Step 2: **Modify a Numbered List**

> Click and drag to select the five items in the bulleted list, then click the **Numbering button** on the Standard toolbar.
> The bulleted list has been converted to a numbered list as shown in Figure 4.2b. (The last two items have not yet been added to the list.)
> Click immediately after the last item in the list and press **enter** to begin a new line. Word automatically adds the next sequential number to the list.
> Type **History** and press **enter**. Type **The Constitution Today** as the seventh (and last) item.
> Click in the selection area to the left of the sixth item, **History** (only the text is selected). Now drag the selected text to the beginning of the list, in front of *Preamble*. Release the mouse.
> The list is automatically renumbered. *History* is now the first item, *Preamble* is the second item, and so on.
> Save the document.

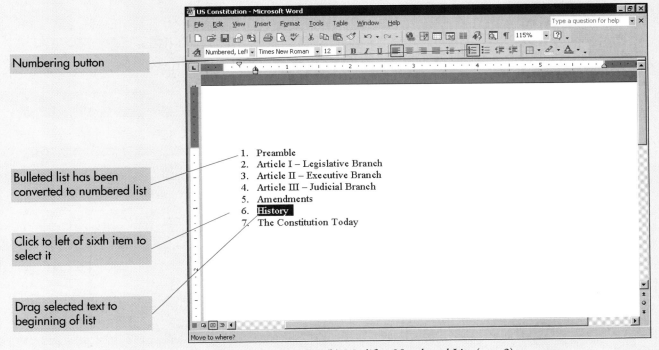

(b) Modify a Numbered List (step 2)

FIGURE 4.2 *Hands-on Exercise 1 (continued)*

THE BULLETS AND NUMBERING BUTTONS

Select the items for which you want to create a list, then click the Numbering or Bullets button on the Formatting toolbar to create a numbered or bulleted list, respectively. The buttons function as toggle switches; that is, click the button once (when the items are selected) and the list formatting is in effect. Click the button a second time and the bullets or numbers disappear. The buttons also enable you to switch from one type of list to another; that is, selecting a bulleted list and clicking the Numbering button changes the list to a numbered list, and vice versa.

Step 3: **Convert to an Outline**

- Click and drag to select the entire list, then click the **right mouse button** to display a context-sensitive menu.
- Click the **Bullets and Numbering command** to display the Bullets and Numbering dialog box in Figure 4.2c.
- Click the **Outline Numbered tab**, then select the type of outline you want. (Do not be concerned if the selected formatting does not display Roman numerals as we customize the outline later in the exercise.)
- Click **OK** to accept the formatting and close the dialog box. The numbered list has been converted to an outline, although that is difficult to see at this point.
- Click at the end of the third item, **Article I—Legislative Branch**. Press **enter**. The number 4 is generated automatically for the next item in the list.
- Press the **Tab key** to indent this item and automatically move to the next level of numbering (a lowercase *a*). Type **House of Representatives**.
- Press **enter**. The next sequential number (a lowercase *b*) is generated automatically. Type **Senate**.
- Save the document.

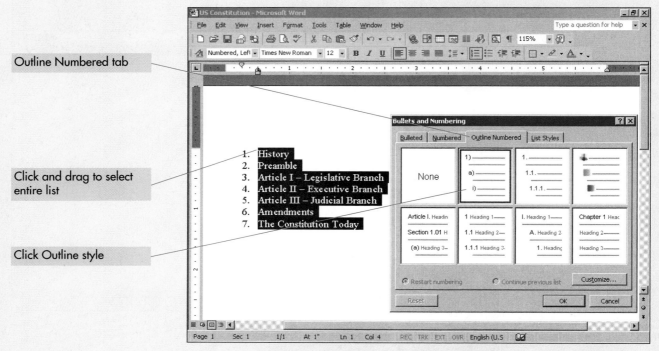

(c) Convert to an Outline (step 3)

FIGURE 4.2 *Hands-on Exercise 1 (continued)*

THE TAB AND SHIFT+TAB KEYS

The easiest way to enter text into an outline is to type continually from one line to the next, using the Tab and Shift+Tab keys as necessary. Press the enter key after completing an item to move to the next item, which is automatically created at the same level, then continue typing if the item is to remain at this level. To change the level, press the Tab key to demote the item (move it to the next lower level), or the Shift+Tab combination to promote the item (move it to the next higher level).

Step 4: **Enter Text into the Outline**

- Your outline should be similar in appearance to Figure 4.2d, except that you have not yet entered most of the text. Click at the end of the line containing *House of Representatives*.
- Press **enter** to start a new item (which begins with a lowercase *b*). Press **Tab** to indent one level, changing the letter to a lowercase *i*. Type **Length of term**. Press **enter**. Type **Requirements for office**.
- Click at the end of the line containing the word *Senate*. Press **enter** to start a new line (which begins with the letter *c*). Press **Tab** to indent one level, changing the letter to an *i*. Type **Length of term**, press **enter**, type **Requirements for office**, and press **enter**.
- Press **Shift+Tab** to move up one level. Enter the remaining text as shown in Figure 4.2.d, using the **Tab** and **Shift+Tab** keys to demote and promote the items.
- Save the document.

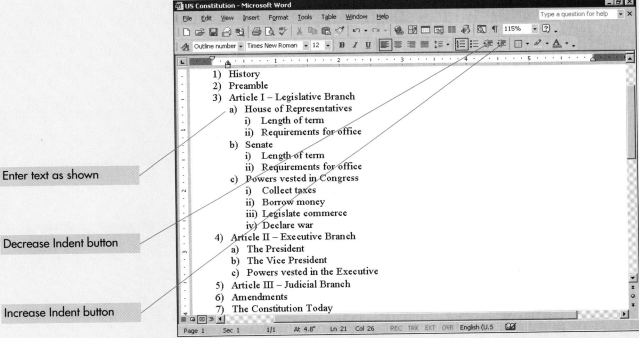

(d) Enter Text into the Outline (step 4)

FIGURE 4.2 *Hands-on Exercise 1 (continued)*

THE INCREASE AND DECREASE INDENT BUTTONS

The Increase and Decrease Indent buttons on the Standard toolbar are another way to change the level within an outline. Click anywhere within an item, then click the appropriate button to change the level within the outline. Indentation is implemented at the paragraph level, and hence you can click the button without selecting the entire item. You can also click and drag to select multiple item(s), then click the desired button.

Step 5: **Customize the Outline**

➤ Select the entire outline, pull down the **Format menu**, then click **Bullets and Numbering** to display the Bullets and Numbering dialog box.
➤ If necessary, click the **Outline Numbered tab** and click **Customize** to display the Customize dialog box as shown in Figure 4.2e. Level **1** should be selected in the Level list box.
  • Click the **drop-down arrow** in the Number style list box and select **I, II, III** as the style.
  • Click in the Number format text box, which now contains the Roman numeral I followed by a right parenthesis. Click and drag to select the parenthesis and replace it with a period.
  • Click the **drop-down arrow** in the Number position list box. Click **right** to right-align the Roman numerals that will appear in your outline.
➤ Click the number **2** in the Level list box and select **A, B, C** as the Number style. Click in the Number format text box and replace the right parenthesis with a period.
➤ Click the number **3** in the Level list box and select **1, 2, 3** as the Number style. Click in the Number format text box and replace the right parenthesis with a period.
➤ Click **OK** to accept these settings and close the dialog box. The formatting of your outline has changed to match the customization in this step.
➤ Save the document.

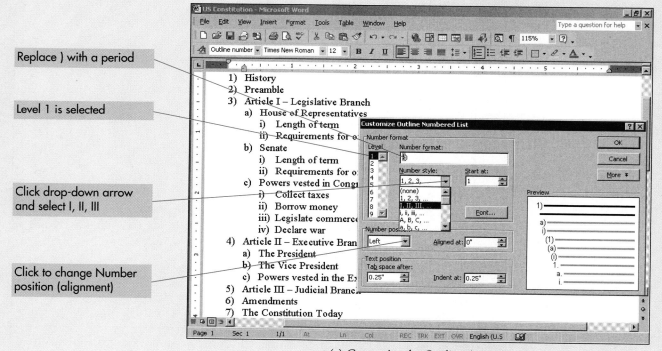

(e) Customize the Outline (step 5)

FIGURE 4.2 *Hands-on Exercise 1 (continued)*

Step 6: **The Completed Outline**

➤ Press **Ctrl+Home** to move to the beginning of the outline. The insertion point is after Roman numeral I, in front of the word *History*. Type **The United States Constitution**. Press **enter**.

➤ The new text appears as Roman numeral I and all existing entries have been renumbered appropriately. The insertion point is immediately before the word *History*. Press **enter** to create a blank line (for your name).

➤ The blank line is now Roman numeral II and *History* has been moved to Roman numeral III. Move the insertion point to the blank line.

➤ Press the **Tab key** so that the blank line (which will contain your name) is item A. This also renumbers *History* as Roman numeral II.

➤ Enter your name as shown in Figure 4.2f. Save the document, then print the outline and submit it to your instructor as proof you did this exercise.

➤ Close the document. Exit Word if you do not want to continue with the next exercise at this time.

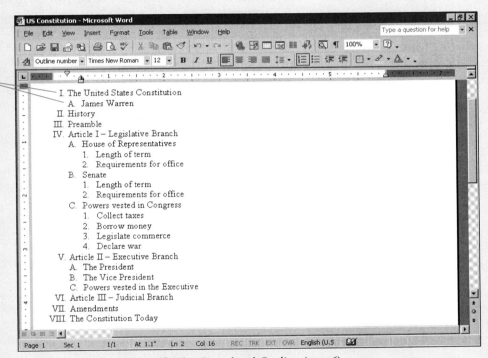

(f) The Completed Outline (step 6)

FIGURE 4.2 *Hands-on Exercise 1 (continued)*

AUTOMATIC CREATION OF A NUMBERED LIST

Word automatically creates a numbered list when you begin a paragraph with a number or letter, followed by a period, tab, or right parenthesis. Press the enter key at the end of the line and you see the next item in the sequence. To end the list, press the backspace key once, or press the enter key twice. You can also turn off the automatic numbering feature by clicking the AutoCorrect Options button that appears when you create the second item in the list and selecting Undo Automatic Numbering.

TABLES

The ***tables feature*** is one of the most powerful in Word and is the basis for an almost limitless variety of documents. The study schedule in Figure 4.3a, for example, is actually a 12 × 8 (12 rows and 8 columns) table as can be seen from the underlying structure in Figure 4.3b. The completed table looks quite impressive, but it is very easy to create once you understand how a table works. You can use the tables feature to create almost any type of document. (See the practice exercises at the end of the chapter for other examples.)

The rows and columns in a table intersect to form ***cells***. Each cell is formatted independently of every other cell and may contain text, numbers and/or graphics. Commands operate on one or more cells. Individual cells can be joined together to form a larger cell as was done in the first and last rows of Figure 4.3a. Conversely, a single cell can be split into multiple cells. The rows within a table can be different heights, just as each column can be a different width. You can specify the height or width explicitly, or you can let Word determine it for you.

A cell can contain anything, even clip art as in the bottom right corner of Figure 4.3a. Just click in the cell where you want the clip art to go, then use the Insert Picture command as you have throughout the text. Use the sizing handles once the clip art has been inserted to move and/or position it within the cell.

A table is created through the ***Insert Table command*** in the ***Table menu***. The command produces a dialog box in which you enter the number of rows and columns. Once the table has been defined, you enter text in individual cells. Text wraps as it is entered within a cell, so that you can add or delete text in a cell without affecting the entries in other cells. You can format the contents of an individual cell the same way you format an ordinary paragraph; that is, you can change the font, use boldface or italics, change the alignment, or apply any other formatting command. You can select multiple cells and apply the formatting to all selected cells at once.

You can also modify the structure of a table after it has been created. The ***Insert*** and ***Delete commands*** in the Table menu enable you to add new rows or columns, or delete existing rows or columns. You can invoke other commands to shade and/or border selected cells or the entire table.

You can work with a table using commands in the Table menu, or you can use the various tools on the Tables and Borders toolbar. (Just point to a button to display a ScreenTip indicative of its function.) Some of the buttons are simply shortcuts for commands within the Table menu. Other buttons offer new and intriguing possibilities, such as the button to Change Text Direction. Note, for example, how we drew an "X" to reserve Sunday morning (for sleeping).

It's easy, and as you might have guessed, it's time for another hands-on exercise in which you create the table in Figure 4.3.

LEFT	**CENTER**	**RIGHT**
Many documents call for left, centered, and/or right aligned text on the same line, an effect that is achieved through setting tabs, or more easily through a table. To achieve the effect shown in the heading of this box, create a 1 × 3 table (one row and three columns), type the text in the three cells as needed, then use the buttons on the Formatting toolbar to left align, center, and right align the respective cells. Select the table, pull down the Format menu, click Borders and Shading, then specify None as the Border setting.		

Weekly Class and Study Schedule

	Monday	Tuesday	Wednesday	Thursday	Friday	Saturday	Sunday
8:00AM							X
9:00AM							
10:00AM							
11:00AM							
12:00PM							
1:00PM							
2:00PM							
3:00PM							
4:00PM							
Notes							

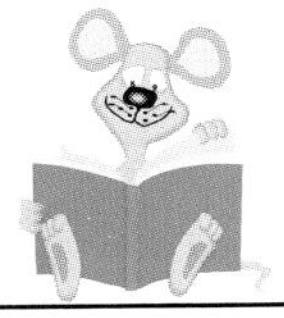

(a) Completed Table

(b) Underlying Structure

FIGURE 4.3 *The Tables Feature*

MICROSOFT WORD 2002

Hands-on Exercise 2

Tables

Objective To create a table; to change row heights and column widths; to merge cells; to apply borders and shading to selected cells. Use Figure 4.4 as a guide for the exercise.

Step 1: **The Page Setup Command**

- Start Word. Click the **Tables and Borders button** on the Standard toolbar to display the Tables and Borders toolbar as shown in Figure 4.4a.
- The button functions as a toggle switch—click it once and the toolbar is displayed. Click the button a second time and the toolbar is suppressed. Click and drag the title bar at the left of the toolbar to anchor it under the Formatting toolbar.
- Pull down the **File menu** and click the **Page Setup command** to display the dialog box in Figure 4.4a.
- Click the **Margins tab** and click the **Landscape icon**. Change the top and bottom margins to **.75** inch.
- Change the left and right margins to **.5** inch each. Click **OK** to accept the settings and close the dialog box.
- Save the document as **My Study Schedule** in the **Exploring Word folder** that you have used throughout the text.
- Change to the **Print Layout view**. Zoom to **Page Width**. You are now ready to create the table.

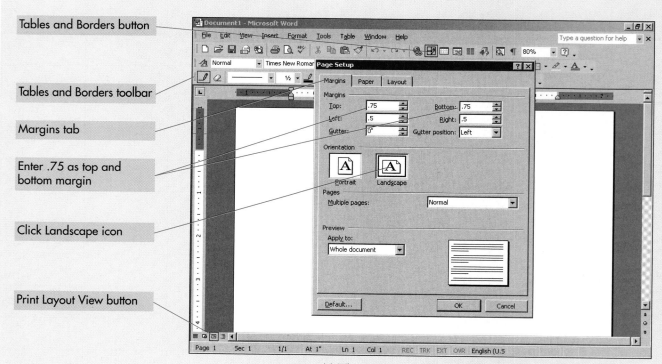

(a) The Page Setup Command (step 1)

FIGURE 4.4 *Hands-on Exercise 2*

Step 2: **Create the Table**

➤ Pull down the **Table menu**, click **Insert**, and click **Table** to display the dialog box in Figure 4.4b. Enter **8** and **12** as the number of columns and rows, respectively. Click **OK** and the table will be inserted into the document.
➤ Practice selecting various elements from the table, something that you will have to do in subsequent steps:
  • To select a single cell, click inside the left grid line (the pointer changes to an arrow when you are in the proper position).
  • To select a row, click outside the table to the left of the first cell in that row.
  • To select a column, click just above the top of the column (the pointer changes to a small black arrow).
  • To select adjacent cells, drag the mouse over the cells.
  • To select the entire table, drag the mouse over the table or click the box that appears at the upper left corner of the table.
➤ Click outside the table. Save the table.

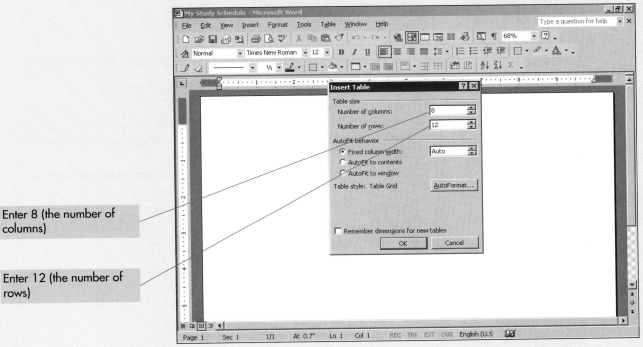

Enter 8 (the number of columns)

Enter 12 (the number of rows)

(b) Create the Table (step 2)

FIGURE 4.4 *Hands-on Exercise 2 (continued)*

TABS AND TABLES

The Tab key functions differently in a table than in a regular document. Press the Tab key to move to the next cell in the current row (or to the first cell in the next row if you are at the end of a row). Press Tab when you are in the last cell of a table to add a new blank row to the bottom of the table. Press Shift+Tab to move to the previous cell in the current row (or to the last cell in the previous row). You must press Ctrl+Tab to insert a regular tab character within a cell.

Step 3: **Merge the Cells**

- ➤ This step merges the cells in the first and last rows of the table. Click outside the table to the left of the first cell in the first row to select the entire first row as shown in Figure 4.4c.
- ➤ Pull down the **Table menu** and click **Merge Cells** (or click the **Merge Cells button** on the Tables and Borders toolbar). Click in the second row to deselect the first row, which now consists of a single cell.
- ➤ Click in the merged cell. Type **Weekly Class and Study Schedule** and format the text in **24 point Arial bold**. Click the **Center button** on the Formatting toolbar to center the title of the table.
- ➤ Click outside the table to the left of the first cell in the last row to select the entire row as shown in Figure 4.4c. Click the **Merge Cells button** on the Tables and Borders toolbar to merge these cells.
- ➤ Click outside the cell to deselect it, then click in the cell and type **Notes**. Press the **enter key** five times. The height of the cell increases to accommodate the blank lines. Click and drag to select the text, then format the text in **12 point Arial bold**.
- ➤ Save the table.

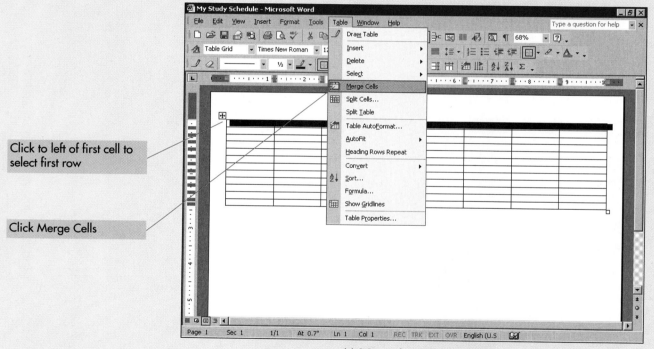

Click to left of first cell to select first row

Click Merge Cells

(c) Merge the Cells (step 3)

FIGURE 4.4 *Hands-on Exercise 2 (continued)*

SPLITTING A CELL

Splitting cells is the opposite of merging them. Click in any cell that you want to split, pull down the Table menu, and click the Split Cells command (or click the Split Cells button on the Tables and Borders toolbar) to display the associated dialog box. Enter the number of rows and columns that should appear after the split. Click OK to accept the settings and close the dialog box.

Step 4: **Enter the Days and Hours**

- Click the second cell in the second row. Type **Monday**. Press the **Tab** (or **right arrow**) **key** to move to the next cell. Type **Tuesday**. Continue until the days of the week have been entered.
- Select the entire row. Use the various tools on the Formatting toolbar to change the text to **10 point Arial Bold**. Click the **Center button** on the Formatting toolbar to center each day within the cell.
- Click the first cell in the third row. Type **8:00AM**. Press the **down arrow key** to move to the first cell in the fourth row. Type **9:00AM**. Continue in this fashion until you have entered the hourly periods up to 4:00PM. Format as appropriate. (We right aligned the time periods and changed the font to Arial bold.)
- Select the cells containing the hours of the day. Pull down the **Table menu**. Click **Table Properties**, then click the **Row tab** to display the Table Properties dialog box in Figure 4.4d.
- Click the **Specify height** check box. Click the **up arrow** until the height is **.5″**. Click the **drop-down arrow** on the Row height list box and select **Exactly**.
- Click the **Cell tab** in the Tables Properties dialog box, then click the **Center button**. Click **OK** to accept the settings and close the dialog box. Save the table.

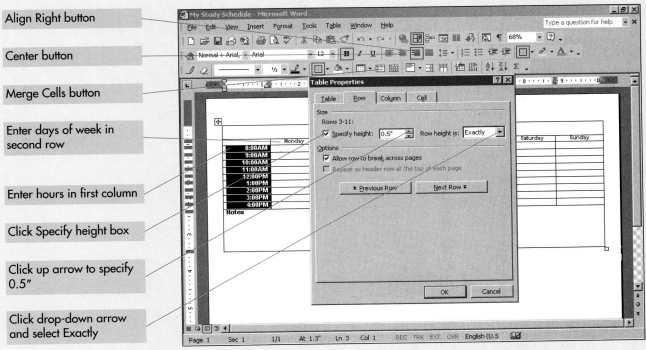

(d) Enter the Days and Hours (step 4)

FIGURE 4.4 *Hands-on Exercise 2 (continued)*

THE AUTOTEXT FEATURE

Type the first few letters of any day in the week and you will see a ScreenTip telling you to press enter to insert the completed day into your document. The days of the week are examples of AutoText (shorthand) entries that are built into Word. Pull down the Insert menu and click the AutoText command to explore the complete set of entries. You can also add your own entries to create a personal shorthand.

Step 5: **Borders and Shading**

- Select (click) the cell containing the title of your table. Click the **Shading Color button** on the Table and Borders toolbar to display a color palette, then choose a background color. We selected red.
- Click and drag to select the text within the cell. Click the **down arrow** on the **Font Color button** to display its palette, then choose **white** (that is, we want white letters on a dark background).
- Click and drag to select the first four cells under "Sunday", then click the **Merge Cells button** to merge these cells.
- Click the **down arrow** on the Line Weight tool and select **3** pt. Click the **down arrow** on the Border Color tool and select the same color you used to shade the first row.
- Click in the upper-left corner of the merged cell, then click and drag to draw a diagonal line as shown in Figure 4.4e. Click and drag to draw a second line to complete the cell. Save the table.

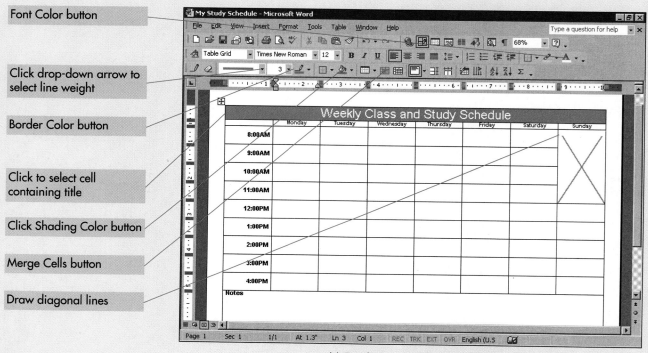

(e) Borders and Shading (step 5)

FIGURE 4.4 *Hands-on Exercise 2 (continued)*

THE AUTOFORMAT COMMAND

The AutoFormat command does not do anything that could not be done through individual formatting commands, but it does provide inspiration by suggesting attractive designs. Click anywhere in the table, pull down the Table menu, and click the Table AutoFormat command to display the associated dialog box. Choose (click) any style, click the Modify button if you want to change any aspect of the formatting, then click the Apply button to format your table in the selected style. See practice exercise 3 at the end of the chapter.

Step 6: **Insert the Clip Art**

- Click anywhere in the merged cell in the last row of the table. Pull down the **Insert menu**, click (or point to) **Picture**, then click **Clip Art**. The task pane opens and displays the Media Gallery Search pane as shown in Figure 4.4f.
- Click in the **Search** text box. Type **books** to search for any clip art image that is indexed with this key word, then click the **Search button** or press **enter**.
- The images are displayed in the Results box. Point to an image to display a drop-down arrow to its right. Click the arrow to display a context menu. Click **Insert** to insert the image into the document.
- Do not be concerned about the size or position of the image at this time.
- Close the task pane. Save the document.

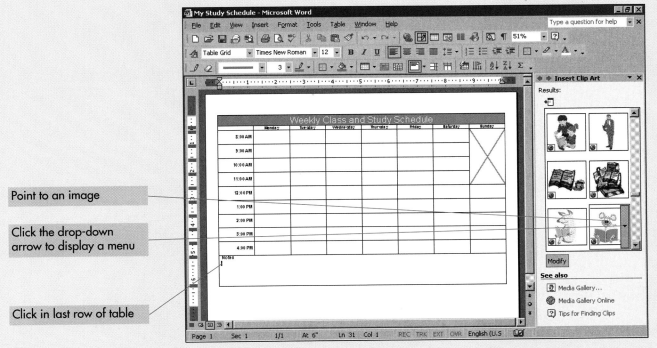

Point to an image

Click the drop-down arrow to display a menu

Click in last row of table

(f) Insert the Clip Art (step 6)

FIGURE 4.4 *Hands-on Exercise 2 (continued)*

SEARCH BY COLLECTION

The Media Gallery organizes its contents by collections and thus provides another way to select clip art other than by a key word. Pull down the Insert menu, click (or point to) the Picture command, then click Clip Art to open the task pane, where you can enter a key word to search for clip art. Instead of searching, however, click the link to Media Gallery at the bottom of the task pane to display the Media Gallery dialog box. Collapse the My Collections folder if it is open, then expand the Office Collections folder, where you can explore the available images by collection.

Step 7: **The Finishing Touches**

➤ Select the newly inserted clip art to display the Picture toolbar, then click the **Format Picture button** to display the Format Picture dialog box. Click the **Layout tab** and choose the **Square layout**. Click **OK** to close the dialog box.
➤ Select (click) the clip art to display its sizing handles as shown in Figure 4.4g. Move and size the image as necessary within its cell.
➤ Click anywhere in the first row of the table. Pull down the **Table menu** and click the **Table Properties command** to display the associated dialog box. Change the row height to exactly **.5 inch**.
➤ Click the **down arrow** next to the **Align button** on the Tables and Borders toolbar and select **center alignment** to center the text vertically.
➤ Use the **Table Properties command** to change the row height of the second row to **.25 inch**. Center these entries vertically as well.
➤ Save the table, then print it for your instructor. Exit Word if you do not want to continue with the next exercise at this time.

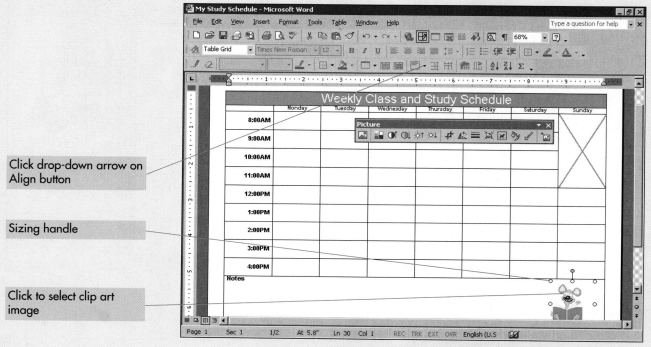

(g) The Finishing Touches (step 7)

FIGURE 4.4 *Hands-on Exercise 2 (continued)*

INSERTING OR DELETING ROWS AND COLUMNS

You can insert or delete rows and columns after a table has been created. To insert a row, click in any cell above or below where the new row should go, pull down the Table menu, click the Insert command, then choose rows above or below as appropriate. Follow a similar procedure to insert a column, choosing whether you want the new column to go to the left or right of the selected cell.

One characteristic of a professional document is the uniform formatting that is applied to similar elements throughout the document. Different elements have different formatting. Headings may be set in one font, color, style, and size, and the text under those headings may be set in a completely different design. The headings may be left aligned, while the text is fully justified. Lists and footnotes can be set in entirely different styles.

One way to achieve uniformity throughout the document is to use the Format Painter to copy the formatting from one occurrence of each element to the next, but this is tedious and inefficient. And if you were to change your mind after copying the formatting throughout a document, you would have to repeat the entire process all over again. A much easier way to achieve uniformity is to store the formatting information as a *style*, then apply that style to multiple occurrences of the same element within the document. Change the style and you automatically change all text defined by that style.

Styles are created on the character or paragraph level. A ***character style*** stores character formatting (font, size, and style) and affects only the selected text. A ***paragraph style*** stores paragraph formatting (such as alignment, line spacing, indents, tabs, text flow, and borders and shading, as well as the font, size, and style of the text in the paragraph). A paragraph style affects the current paragraph or multiple paragraphs if several paragraphs are selected. Styles are created and applied through the **Styles and Formatting command** in the Format menu as shown in Figure 4.5.

The document in Figure 4.5a consists of multiple tips for Microsoft Word. Each tip begins with a one-line heading, followed by the associated text. The task pane in the figure displays all of the styles that are in use in the document. The ***Normal style*** contains the default paragraph settings (left aligned, single spacing, and a default font) and is automatically assigned to every paragraph unless a different style is specified. The Clear Formatting style removes all formatting from selected text. It is the ***Heading 1*** and ***Body Text styles***, however, that are of interest to us, as these styles have been applied throughout the document to the associated elements. (The style assignments are done automatically through the AutoFormat command as will be explained shortly.)

The specifications for the Heading 1 and Body Text styles are shown in Figures 4.5b and 4.5c, respectively. The current settings within the Heading 1 style call for 16 point Arial bold type in blue. The text is left justified, and the heading will always appear on the same page as the next paragraph. The Body Text style is in 10 point Times New Roman and is fully justified. The preview box in both figures shows how paragraphs formatted in the style will appear. You can change the specifications of either style using any combination of buttons or associated menu commands. (Clicking the Format button in either dialog box provides access to the various commands in the Format menu.) And as indicated earlier, any changes to the style are automatically reflected in all elements that are defined by that style.

Styles automate the formatting process and provide a consistent appearance to a document. Any type of character or paragraph formatting can be stored within a style, and once a style has been defined, it can be applied to multiple occurrences of the same element within a document to produce identical formatting.

STYLES AND PARAGRAPHS

A paragraph style affects the entire paragraph; that is, you cannot apply a paragraph style to only part of a paragraph. To apply a style to an existing paragraph, place the insertion point anywhere within the paragraph, pull down the Style list box on the Formatting toolbar, then click the name of the style you want.

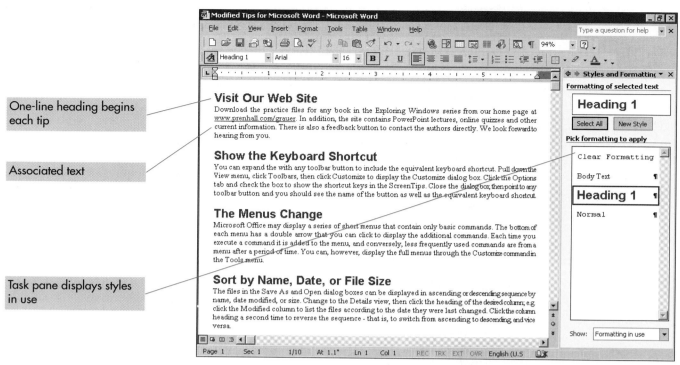

(a) The Document

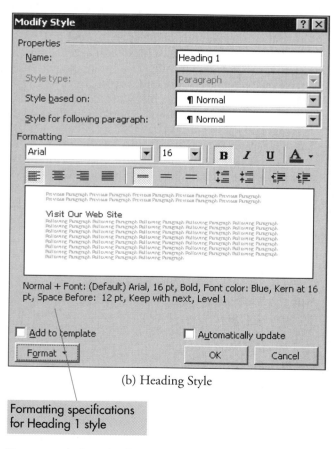

(b) Heading Style

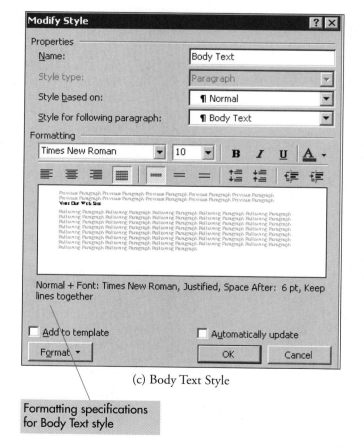

(c) Body Text Style

FIGURE 4.5 *Styles*

172   CHAPTER 4: ADVANCED FEATURES

THE OUTLINE VIEW

One additional advantage of styles is that they enable you to view a document in the ***Outline view***. The Outline view does not display a conventional outline (such as the multilevel list created earlier in the chapter), but rather a structural view of a document that can be collapsed or expanded as necessary. Consider, for example, Figure 4.6, which displays the Outline view of a document that will be the basis of the next hands-on exercise. The document consists of a series of tips for Microsoft Word 2002. The heading for each tip is formatted according to the Heading 1 style. The text of each tip is formatted according to the Body Text style.

The advantage of the Outline view is that you can collapse or expand portions of a document to provide varying amounts of detail. We have, for example, collapsed almost the entire document in Figure 4.6, displaying the headings while suppressing the body text. We also expanded the text for two tips (Visit Our Web Site and Moving Within a Document) for purposes of illustration.

Now assume that you want to move the latter tip from its present position to immediately below the first tip. Without the Outline view, the text would stretch over two pages, making it difficult to see the text of both tips at the same time. Using the Outline view, however, you can collapse what you don't need to see, then simply click and drag the headings to rearrange the text within the document.

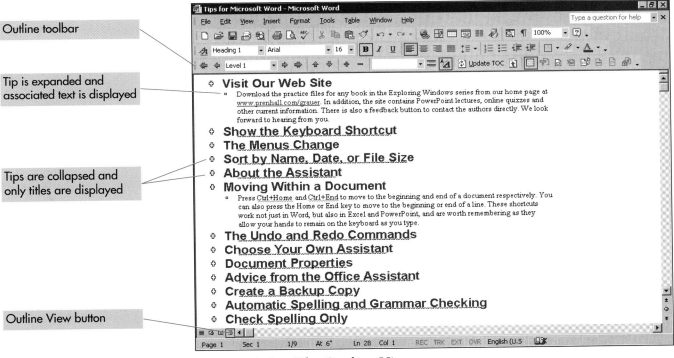

FIGURE 4.6 *The Outline View*

THE OUTLINE VERSUS THE OUTLINE VIEW

A conventional outline is created as a multilevel list within the Bullets and Numbering command. Text for the outline is entered in the Print Layout or Normal view, *not* the Outline view. The latter provides a condensed view of a document that is used in conjunction with styles.

The AutoFormat Command

Styles are extremely powerful. They enable you to impose uniform formatting within a document and they let you take advantage of the Outline view. What if, however, you have an existing and/or lengthy document that does not contain any styles (other than the default Normal style, which is applied to every paragraph)? Do you have to manually go through every paragraph in order to apply the appropriate style? The AutoFormat command provides a quick solution.

The ***AutoFormat command*** enables you to format lengthy documents quickly, easily, and in a consistent fashion. In essence, the command analyzes a document and formats it for you. Its most important capability is the application of styles to individual paragraphs; that is, the command goes through an entire document, determines how each paragraph is used, then applies an appropriate style to each paragraph. The formatting process assumes that one-line paragraphs are headings and applies the predefined Heading 1 style to those paragraphs. It applies the Body Text style to ordinary paragraphs and can also detect lists and apply a numbered or bullet style to those lists.

The AutoFormat command will also add special touches to a document if you request those options. It can replace "ordinary quotation marks" with "smart quotation marks" that curl and face each other. It will replace ordinal numbers (1st, 2nd, or 3rd) with the corresponding superscripts (1st, 2nd, or 3rd), or common fractions (1/2 or 1/4) with typographical symbols (½ or ¼).

The AutoFormat command will also replace Internet references (Web addresses and e-mail addresses) with hyperlinks. It will recognize, for example, any entry beginning with http: or www. as a hyperlink and display the entry as underlined blue text (www.microsoft.com). This is not merely a change in formatting, but an actual hyperlink to a document on the Web or corporate Intranet. It also converts entries containing an @ sign, such as rgrauer@umiami.miami.edu to a hyperlink as well. (All Word documents are Web enabled. Unlike a Web document, however, you need to press and hold the Ctrl key to follow the link and display the associated page. This is different from what you usually do, because you normally just click a link to follow it. What if, however, you wanted to edit the link? Accordingly, Word modifies the convention so that clicking a link enables you to edit the link.)

The various options for the AutoFormat command are controlled through the AutoCorrect command in the Tools menu. Once the options have been set, all formatting is done automatically by selecting the AutoFormat command from the Format menu. The changes are not final, however, as the command gives you the opportunity to review each formatting change individually, then accept the change or reject it as appropriate. (You can also format text automatically as it is entered according to the options specified under the AutoFormat As You Type tab.)

AUTOMATIC BORDERS AND LISTS

The AutoFormat As You Type option applies sophisticated formatting as text is entered. It automatically creates a numbered list any time a number is followed by a period, tab, or right parenthesis (press enter twice in a row to turn off the feature). It will also add a border to a paragraph any time you type three or more hyphens, equal signs, or underscores followed by the enter key. Pull down the Tools menu, click the AutoCorrect command, then click the AutoFormat As You Type tab and select the desired features.

Hands-on Exercise 3

Styles

Objective To use the AutoFormat command to apply styles to an existing document; to modify existing styles; to create a new style. Use Figure 4.7 as a guide for the exercise.

Step 1: **The AutoFormat Command**

> Start Word. Open the document **Tips for Microsoft Word** in the **Exploring Word folder**. Save the document as **Modified Tips for Microsoft Word** so that you can return to the original if necessary.
> Press **Ctrl+Home** to move to the beginning of the document. Pull down the **Format menu**. Click **AutoFormat** to display the dialog box in Figure 4.7a.
> Click the **Options command button**. Be sure that every check box is selected to implement the maximum amount of automatic formatting. Click the **OK button** in the AutoCorrect dialog box to close the dialog box.
> If necessary, check the option to **AutoFormat now**, then click the **OK command button** to format the document.
> The status bar indicates the progress of the formatting operation, after which you will see a newly formatted document.
> Save the document.

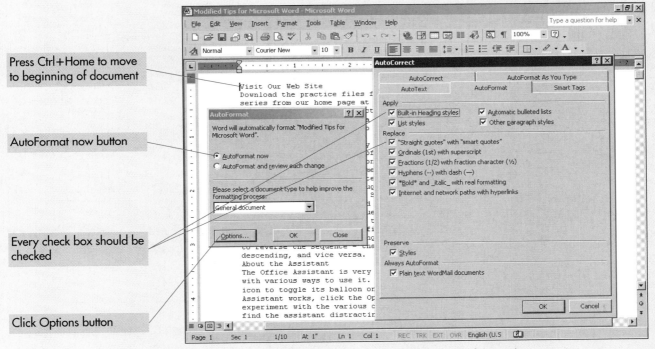

(a) The AutoFormat Command (step 1)

Figure 4.7 *Hands-on Exercise 3*

Step 2: **Formatting Properties**

- Pull down the **Format menu** and click the **Reveal Formatting command** to open the task pane as shown in Figure 4.7b.
- Press **Ctrl+Home** to move to the beginning of the document.
- The task pane displays the formatting properties for the first heading in your document. Heading 1 is specified as the paragraph style within the task pane. The name of the style for the selected text (Heading 1) also appears in the Style list box at the left of the Formatting toolbar.
- Click in the text of the first tip to view the associated formatting properties. This time Body Text is specified as the paragraph style in the task pane. Click the title of any tip and you will see the Heading 1 style in the Style box. Click the text of any tip and you will see the Body Text style in the Style box.
- Click the **down arrow** to the left of the Close button in the task pane and click **Styles and Formatting** to show the styles in your document. If necessary, click the **down arrow** in the Show list box to show just the formatting in use. You will see Heading 1 and Body Text styles and an option to clear formatting.

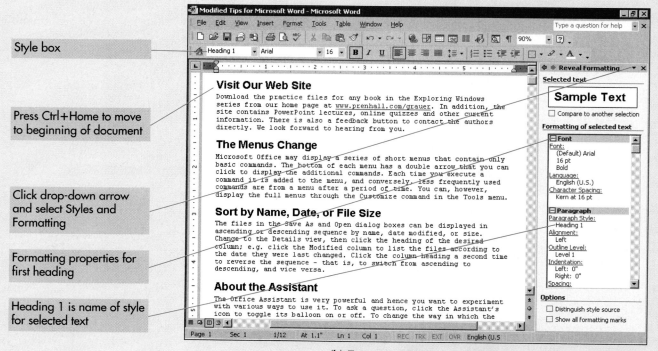

(b) Formatting Properties (step 2)

FIGURE 4.7 *Hands-on Exercise 3 (continued)*

STYLES AND THE AUTOFORMAT COMMAND

The AutoFormat command applies the Heading 1 and Body Text styles to single- and multiple-line paragraphs, respectively. Thus, all you have to do to change the appearance of the headings or paragraphs throughout the document is change the associated style. Change the Heading 1 style, for example, and you automatically change every heading throughout the document. Change the Body Text style and you change every paragraph.

Step 3: **Modify the Body Text Style**

- Point to the **Body Text style** in the task pane, click the **down arrow** that appears to display a context-sensitive menu, and click **Modify** to display the Modify Style dialog box.
- Click the **Justify button** to change the alignment of every similar paragraph in the document. Change the font to **Times New Roman**.
- Click the **down arrow** next to the **Format button**, then click **Paragraph** to display the Paragraph dialog box in Figure 4.7c. If necessary, click the **Line and Page Breaks tab**.
- The box for Widow/Orphan control is checked by default. This ensures that any paragraph defined by the Body Text style will not be split to leave a single line at the bottom or top of a page.
- Check the box to **Keep Lines Together**. This is a more stringent requirement and ensures that the entire paragraph is not split. Click **OK** to close the Paragraph dialog box. Click **OK** to close the Modify Style dialog box.
- All of the paragraphs in the document change automatically to reflect the new definition of the Body Text style, which includes justification and ensures that the paragraph is not split across pages. Save the document.

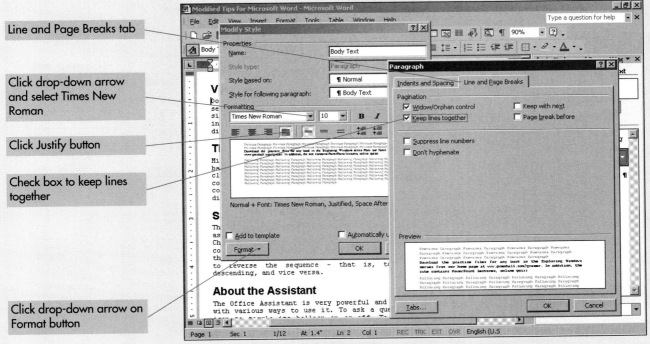

(c) Modify the Body Text Style (step 3)

FIGURE 4.7 *Hands-on Exercise 3 (continued)*

BE CAREFUL WHERE YOU CLICK

If you click the style name instead of the down arrow, you will apply the style to the selected text instead of modifying it. We know because we made this mistake. Click the Undo button to cancel the command. Click the down arrow next to the style name to display the associated menu, and click the Modify command to display the Modify Style dialog box.

MICROSOFT WORD 2002

Step 4: **Modify the Heading 1 Style**

- Point to the **Heading 1 style** in the task pane, click the **down arrow** that appears, then click **Modify** to display the Modify Style dialog box.
- Click the **Font Color button** to display the palette in Figure 4.7d. Click **Blue** to change the color of all of the headings in the document. The change will not take effect until you click the OK button to accept the settings and close the dialog box.
- Click the **Format button** toward the bottom of the dialog box, then click **Paragraph** to display the Paragraph dialog box. Click the **Indents and Spacing tab**. Change the **Spacing After** to 0. Click **OK** to accept the settings and close the Paragraph dialog box.
- Click **OK** to close the Modify Style dialog box. The formatting in your document has changed to reflect the changes in the Heading 1 style.
- Save the document.

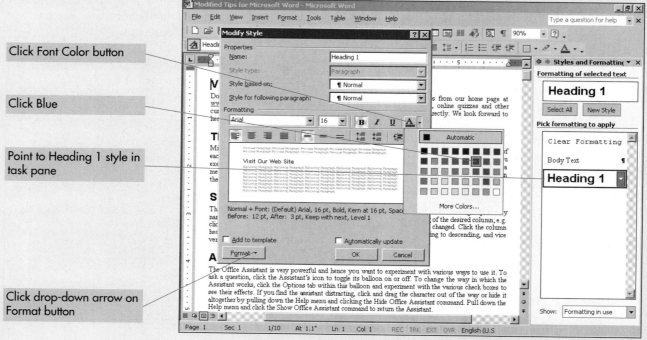

(d) Modify the Heading 1 Style (step 4)

FIGURE 4.7 *Hands-on Exercise 3 (continued)*

SPACE BEFORE AND AFTER

It's common practice to press the enter key twice at the end of a paragraph (once to end the paragraph, and a second time to insert a blank line before the next paragraph). The same effect can be achieved by setting the spacing before or after the paragraph using the Spacing Before or After list boxes in the Format Paragraph command. The latter technique gives you greater flexibility in that you can specify any amount of spacing (e.g., 6 points) to leave only half a line before or after a paragraph. It also enables you to change the spacing between paragraphs more easily because the spacing information can be stored within the paragraph style.

Step 5: **The Outline View**

- Close the task pane. Pull down the **View menu** and click **Outline** (or click the **Outline View button** above the status bar) to display the document in Outline view.
- Pull down the **Edit menu** and click **Select All** (or press **Ctrl+A**) to select the entire document. Click the **Collapse button** on the Outlining toolbar to collapse the entire document so that only the headings are visible.
- If necessary, scroll down in the document until you can click in the heading of the tip entitled "Show the Keyboard Shortcut" as shown in Figure 4.7e. Click the **Expand button** on the Outlining toolbar to see the subordinate items under this heading.
- Click and drag to select the tip **Show the Keyboard Shortcut**. Point to the **plus sign** next to the selected tip (the mouse pointer changes to a double arrow), then click and drag to move the tip toward the top of the document, immediately below the first tip. Release the mouse.
- Save the document.

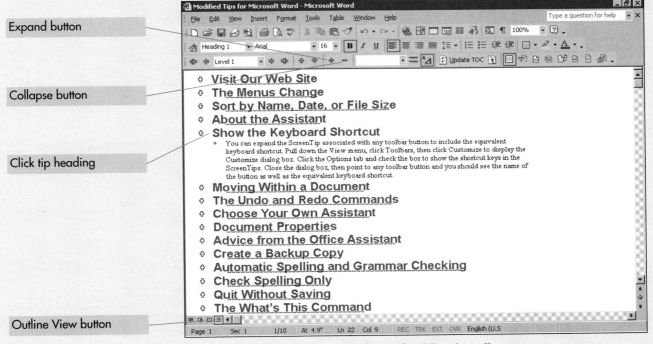

(e) The Outline View (step 5)

FIGURE 4.7   *Hands-on Exercise 3 (continued)*

THE DOCUMENT MAP

The Document Map helps you to navigate within a large document. Click the Document Map button on the Standard toolbar to divide the screen into two panes. The headings in a document are displayed in the left pane, and the text of the document is visible in the right pane. To go to a specific point in a document, click its heading in the left pane, and the insertion point is moved automatically to that point in the document, which is visible in the right pane. Click the Map button a second time to turn the feature off.

Step 6: **Create a Paragraph Style**

- Pull down the **View menu** and change to the **Normal view**. Pull down the **Format menu** and click **Styles and Formatting** to open the task pane as shown in Figure 4.7f.
- Press **Ctrl+Home** to move the insertion point to the beginning of the document, then press **Ctrl+Enter** to create a page break for a title page.
- Press the **up arrow** to move the insertion point to the left of the page break. Press the **enter key** twice and press the **up arrow** to move above the page break. Select the two blank lines and click **Clear Formatting** in the task pane. Press the **up arrow**.
- Enter the title of the document, **Tips for Microsoft Word** in **24 Points**. Change the text to **Arial Bold** in **blue**. Click the **Center button** on the Formatting toolbar. Press **enter**.
- The task pane displays the specifications for the text you just entered. You have created a new style, but the style is as yet unnamed. Point to the specification for the title (Arial, 24 pt, Centered) to display a down arrow, then click the arrow as shown in Figure 4.7f.
- Click the **Modify Style command** to display the Modify Style dialog box. Click in the **Name** text box in the Properties area and enter **Report Title** (the name of the new style). Click **OK**.
- Enter your name below the report title. Add a second line that references the authors of the textbook, Robert Grauer and Maryann Barber. Use a smaller point size, change the font color to blue, and center the lines. Name the associated style **Report Author**.
- Save the document.

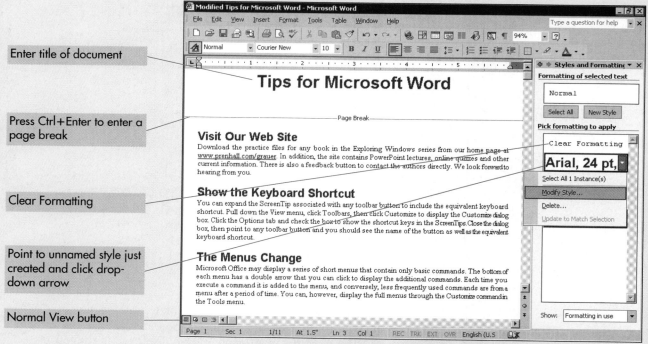

(f) Create a Paragraph Style (step 6)

FIGURE 4.7 *Hands-on Exercise 3 (continued)*

Step 7: **Create a Character Style**

- Click and drag to select the words **Screen Tip** (that appear within the second tip). Click the **Bold** and **Italic buttons** on the Formatting toolbar so that the selected text appears in bold and italics.
- Once again, you have created a style as can be seen in the task pane. Point to the right of the formatting specification in the task pane, click the **down arrow**, then click the **Modify Style command** to display the Modify Style dialog box in Figure 4.7g.
- Click in the **Name** text box in the Properties area and enter **Emphasize** as the name of the style. Click the **down arrow** in the Style type list box and select **Character**. Click **OK**.
- Click and drag to select the words **practice files** that appear in the first tip, click the **down arrow** in the Style List box on the Formatting toolbar, and apply the newly created Emphasize character style to the selected text.
- Save the document. Close the task pane.

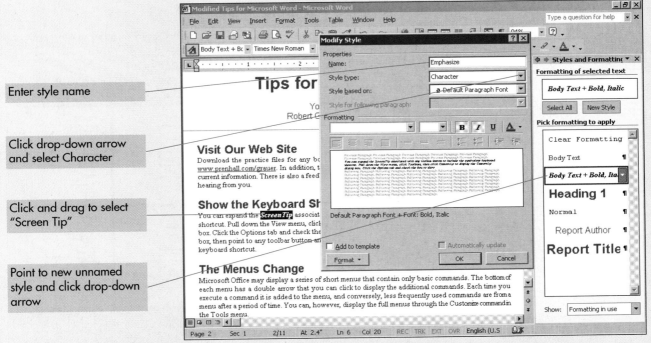

(g) Create a Character Style (step 7)

FIGURE 4.7 *Hands-on Exercise 3 (continued)*

SHOW THE KEYBOARD SHORTCUT

You can expand the ScreenTip associated with any toolbar button to include the equivalent keyboard shortcut. Pull down the View menu, click Toolbars, then click Customize to display the Customize dialog box. Click the Options tab and check the box to show the shortcut keys in the ScreenTips. Close the dialog box, then point to any toolbar button, and you should see the name of the button as well as the equivalent keyboard shortcut. There is no need to memorize the shortcuts, but they do save time.

Step 8: **The Completed Document**

- Change to the **Print Layout view**. Pull down the **View menu** and click the **Zoom command** to display the Zoom dialog box. Click the option button next to **Many Pages**, then click and drag the computer icon to display multiple pages. Click **OK**.
- You should see a multipage display similar to Figure 4.7h. The text on the individual pages is too small to read, but you can see the page breaks and overall document flow.
- The various tips should all be justified. Moreover each tip should fit completely on one page without spilling over to the next page according to the specifications in the Body Text style.
- Click above the title on the first page and press the **enter key** (if necessary) to position the title further down the page. Conversely, you could press the **Del key** to remove individual lines and move the title up the page.
- Save the document. Print the document only if you do not intend to do the next hands-on exercise. Exit Word if you do not want to continue with the next exercise at this time.

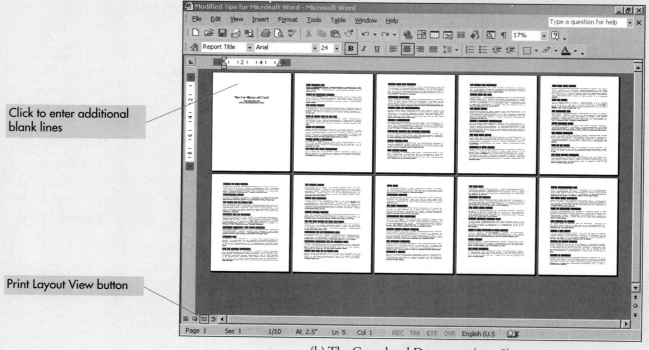

(h) The Completed Document (step 8)

FIGURE 4.7   *Hands-on Exercise 3 (continued)*

PRINT SELECTED PAGES

Why print an entire document if you want only a few pages? Pull down the File menu and click Print as you usually do, to initiate the printing process. Click the Pages option button, then enter the page numbers and/or page ranges you want; for example, 3, 6-8 will print page three and pages six through eight. You can also print multiple copies by entering the appropriate number in the Number of copies list box.

WORKING IN LONG DOCUMENTS

Long documents, such as term papers or reports, require additional formatting for better organization. These documents typically contain page numbers, headers and/or footers, a table of contents, and an index. Each of these elements is discussed in turn and will be illustrated in a hands-on exercise.

Page Numbers

The *Insert Page Numbers command* is the easiest way to place *page numbers* into a document and is illustrated in Figure 4.8. The page numbers can appear at the top or bottom of a page, and can be left, centered, or right aligned. Word provides additional flexibility in that you can use Roman rather than Arabic numerals, and you need not start at page number one.

The Insert Page Number command is limited, however, in that it does not provide for additional text next to the page number. You can overcome this restriction by creating a header or footer that contains the page number.

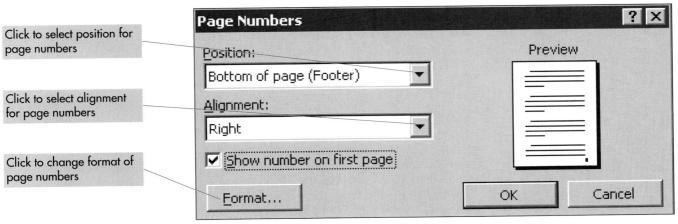

FIGURE 4.8 *Page Numbers*

Headers and Footers

Headers and footers give a professional appearance to a document. A *header* consists of one or more lines that are printed at the top of every page. A *footer* is printed at the bottom of the page. A document may contain headers but not footers, footers but not headers, or both headers and footers.

Headers and footers are created from the View menu. (A simple header or footer is also created automatically by the Insert Page Number command, depending on whether the page number is at the top or bottom of a page.) Headers and footers are formatted like any other paragraph and can be centered, left or right aligned. They can be formatted in any typeface or point size and can include special codes to automatically insert the page number, date, and/or time a document is printed.

The advantage of using a header or footer (over typing the text yourself at the top or bottom of every page) is that you type the text only once, after which it appears automatically according to your specifications. In addition, the placement of the headers and footers is adjusted for changes in page breaks caused by the insertion or deletion of text in the body of the document.

Headers and footers can change continually throughout a document. The Page Setup dialog box (in the File menu) enables you to specify a different header or footer for the first page, and/or different headers and footers for the odd and even pages. If, however, you wanted to change the header (or footer) midway through a document, you would need to insert a section break at the point where the new header (or footer) is to begin.

Sections

Formatting in Word occurs on three levels. You are already familiar with formatting at the character and paragraph levels that have been used throughout the text. Formatting at the section level controls headers and footers, page numbering, page size and orientation, margins, and columns. All of the documents in the text so far have consisted of a single *section*, and thus any section formatting applied to the entire document. You can, however, divide a document into sections and format each section independently.

Formatting at the section level gives you the ability to create more sophisticated documents. You can use section formatting to:

- Change the margins within a multipage letter, where the first page (the letterhead) requires a larger top margin than the other pages in the letter.
- Change the orientation from portrait to landscape to accommodate a wide table at the end of the document.
- Change the page numbering to use Roman numerals at the beginning of the document for a table of contents and Arabic numerals thereafter.
- Change the number of columns in a newsletter, which may contain a single column at the top of a page for the masthead, then two or three columns in the body of the newsletter.

In all instances, you determine where one section ends and another begins by using the **Insert menu** to create a *section break*. You also have the option of deciding how the section break will be implemented on the printed page; that is, you can specify that the new section continue on the same page, that it begin on a new page, or that it begin on the next odd or even page even if a blank page has to be inserted.

Word stores the formatting characteristics of each section in the section break at the end of a section. Thus, deleting a section break also deletes the section formatting, causing the text above the break to assume the formatting characteristics of the next section.

Figure 4.9 displays a multipage view of a ten-page document. The document has been divided into two sections, and the insertion point is currently on the fourth page of the document (page four of ten), which is also the first page of the second section. Note the corresponding indications on the status bar and the position of the headers and footers throughout the document.

Figure 4.9 also displays the Header and Footer toolbar, which contains various icons associated with these elements. As indicated, a header or footer may contain text and/or special codes—for example, the word "page" followed by a code for the page number. The latter is inserted into the header by clicking the appropriate button on the Header and Footer toolbar. Remember, headers and footers are implemented at the section level. Thus, changing a header or footer within a document requires the insertion of a section break.

Table of Contents

A *table of contents* lists headings in the order they appear in a document and the page numbers where the entries begin. Word will create the table of contents automatically, provided you have identified each heading in the document with a built-in heading style (Heading 1 through Heading 9). Word will also update the table automatically to accommodate the addition or deletion of headings and/or changes in page numbers brought about through changes in the document.

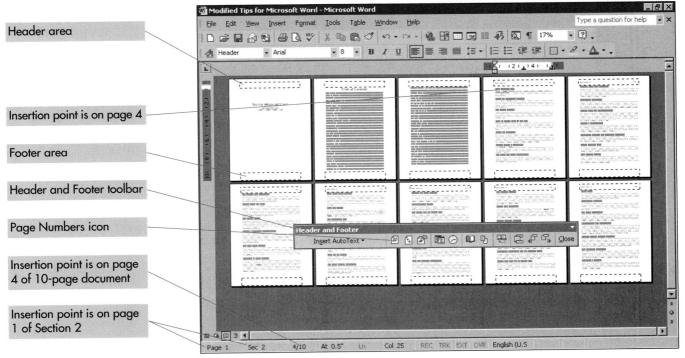

FIGURE 4.9   *Headers and Footers*

The table of contents is created through the ***Index and Tables command*** from the Insert menu as shown in Figure 4.10a. You have your choice of several predefined formats and the number of levels within each format; the latter correspond to the heading styles used within the document. You can also choose the ***leader character*** and whether or not to right align the page numbers.

Creating an Index

An ***index*** is the finishing touch in a long document. Word will create an index automatically provided that the entries for the index have been previously marked. This, in turn, requires you to go through a document, and one by one, select the terms to be included in the index and mark them accordingly. It's not as tedious as it sounds. You can, for example, select a single occurrence of an entry and tell Word to mark all occurrences of that entry for the index. You can also create cross-references, such as "see also Internet."

After the entries have been specified, you create the index by choosing the appropriate settings in the Index and Tables command as shown in Figure 4.10b. You can choose a variety of styles for the index just as you can for the table of contents. Word will put the index entries in alphabetical order and will enter the appropriate page references. You can also create additional index entries and/or move text within a document, then update the index with the click of a mouse.

The Go To Command

The ***Go To command*** moves the insertion point to the top of a designated page. The command is accessed from the Edit menu by pressing the F5 function key, or by double clicking the Page number on the status bar. After the command has been executed, you are presented with a dialog box in which you enter the desired page number. You can also specify a relative page number—for example, P +2 to move forward two pages, or P −1 to move back one page.

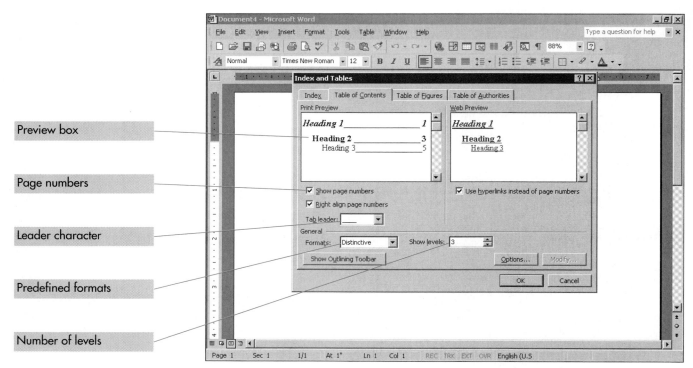

(a) Table of Contents

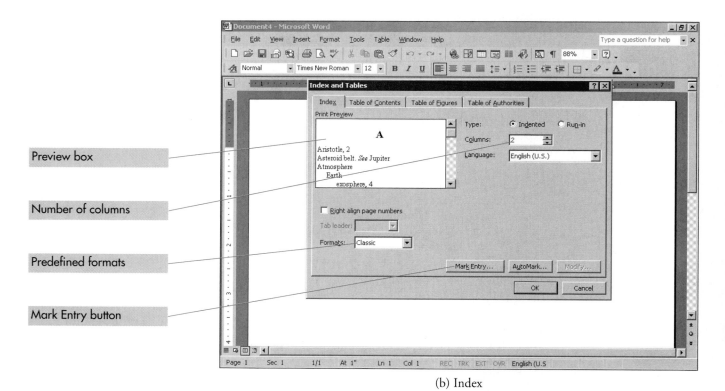

(b) Index

FIGURE 4.10 *Index and Tables Command*

HANDS-ON EXERCISE 4

WORKING IN LONG DOCUMENTS

Objective To create a header (footer) that includes page numbers; to insert and update a table of contents; to add an index entry; to insert a section break and demonstrate the Go To command; to view multiple pages of a document. Use Figure 4.11 as a guide for the exercise.

Step 1: **Applying a Style**

> ➤ Open the **Modified Tips for Word document** from the previous exercise. Zoom to **Page Width**. Scroll to the top of the second page.
> ➤ Click to the left of the first tip title. (If necessary, click the **Show/Hide ¶ button** on the Standard toolbar to hide the paragraph marks.)
> ➤ Type **Table of Contents**. Press the **enter key** two times.
> ➤ Click anywhere within the phrase "Table of Contents". Click the **down arrow** on the **Styles** list box to pull down the styles for this document as shown in Figure 4.11a.
> ➤ Click **Report Title** (the style you created at the end of the previous exercise). "Table of Contents" is centered in 24 point blue Arial bold according to the definition of Report Title.

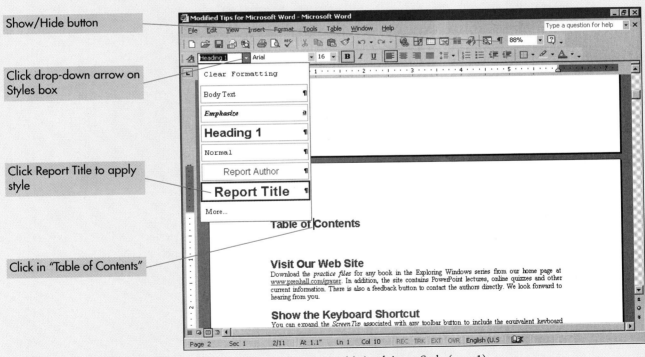

(a) Applying a Style (step 1)

FIGURE 4.11 *Hands-on Exercise 4*

Step 2: **Table of Contents**

➤ If necessary, change to the **Print Layout view**. Click the line immediately under the title for the table of contents. Pull down the **View menu**. Click **Zoom** to display the associated dialog box.
➤ Click the **monitor icon**. Click and drag the **page icons** to display two pages down by five pages across as shown in the figure. Release the mouse.
➤ Click **OK**. The display changes to show all eleven pages in the document.
➤ Pull down the **Insert menu**. Click **Reference**, then click **Index and Tables**. If necessary, click the **Table of Contents tab** to display the dialog box in Figure 4.11b.
➤ Check the boxes to **Show Page Numbers** and to **Right Align Page Numbers**.
➤ Click the **down arrow** on the Formats list box, then click **Distinctive**. Click the **arrow** in the **Tab Leader list box**. Choose a dot leader. Click **OK**. Word takes a moment to create the table of contents, which extends to two pages.
➤ Save the document.

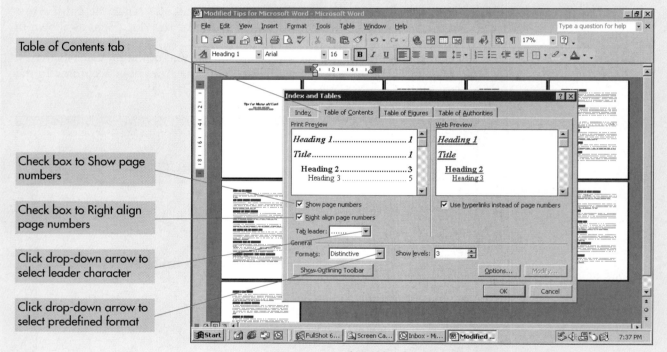

(b) Table of Contents (step 2)

FIGURE 4.11 *Hands-on Exercise 4 (continued)*

AUTOFORMAT AND THE TABLE OF CONTENTS

Word will create a table of contents automatically, provided you use the built-in heading styles to define the items for inclusion. If you have not applied the styles to the document, the AutoFormat command will do it for you. Once the heading styles are in the document, pull down the Insert command, click Reference, then click Index and Tables, then click the Table of Contents command.

Step 3: **Field Codes and Field Text**

➤ Click the **arrow** on the **Zoom Control box** on the Standard toolbar. Click **Page Width** in order to read the table of contents as in Figure 4.11c.
➤ Use the **up arrow key** to scroll to the beginning of the table of contents. Press **Alt+F9**. The table of contents is replaced by an entry similar to {TOC \o "1-3"} to indicate a field code. The exact code depends on the selections you made in step 2.
➤ Press **Alt+F9** a second time. The field code for the table of contents is replaced by text.
➤ Pull down the **Edit menu**. Click **Go To** to display the dialog box in Figure 4.11c.
➤ Type **3** and press the **enter key** to go to page 3, which contains the second page of the table of contents. Click **Close**.

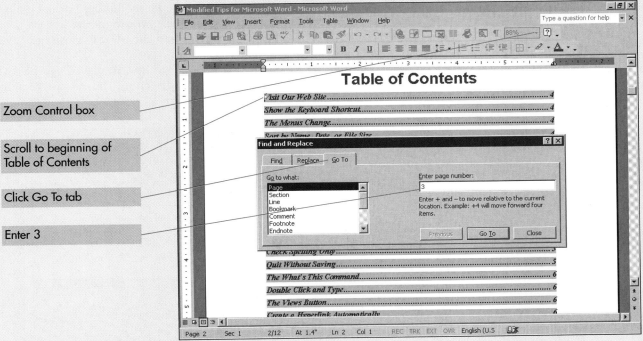

(c) Field Codes and Field Text (step 3)

FIGURE 4.11 *Hands-on Exercise 4 (continued)*

THE GO TO AND GO BACK COMMANDS

The F5 key is the shortcut equivalent of the Go To command and displays a dialog box to move to a specific location (a page or section) within a document. The Shift+F5 combination executes the Go Back command and returns to a previous location of the insertion point; press Shift+F5 repeatedly to cycle through the last three locations of the insertion point.

Step 4: **Insert a Section Break**

➤ Scroll down page 3 until you are at the end of the table of contents. Click to the left of the first tip heading as shown in Figure 4.11d.

➤ Pull down the **Insert menu**. Click **Break** to display the Break dialog box. Click the **Next Page button** under Section Break types. Click **OK** to create a section break, simultaneously forcing the first tip to begin on a new page.

➤ The first tip, Visit Our Web Site, moves to the top of the next page (page 4 in the document). If the status bar already displays Page 1 Section 2, a previous user has changed the default numbering to begin each section on its own page and you can go to step 6. If not, you need to change the page numbering.

➤ Pull down the **Insert menu** and click **Page Numbers** to display the Page Numbers dialog box. Click the **drop-down arrow** in the Position list box to position the page number at the top of page (in the header).

➤ Click the **Format command button** to display the Page Number Format dialog box. Click the option button to **Start at** page 1 (i.e., you want the first page in the second section to be numbered as page 1), and click **OK** to close the Page Number Format box.

➤ Close the Page Numbers dialog box. The status bar now displays Page 1 Sec 2 to indicate that you are on page 1 in the second section. The entry 4/12 indicates that you are physically on the fourth page of a 12-page document.

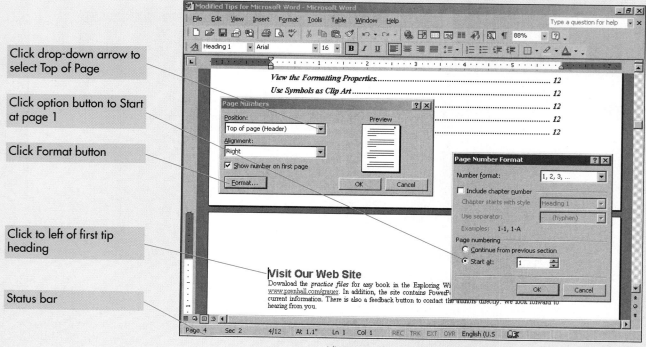

(d) Insert a Section Break (step 4)

FIGURE 4.11   *Hands-on Exercise 4 (continued)*

Step 5: **The Page Setup Command**

- Pull down the **File menu** and click the **Page Setup command** (or double click the **ruler**) to display the Page Setup dialog box.
- Click the **Layout tab** to display the dialog box in Figure 4.11e.
- If necessary, clear the box for Different Odd and Even Pages and for Different First Page, as all pages in this section (Section 2) are to have the same header. Click **OK**.
- Save the document.

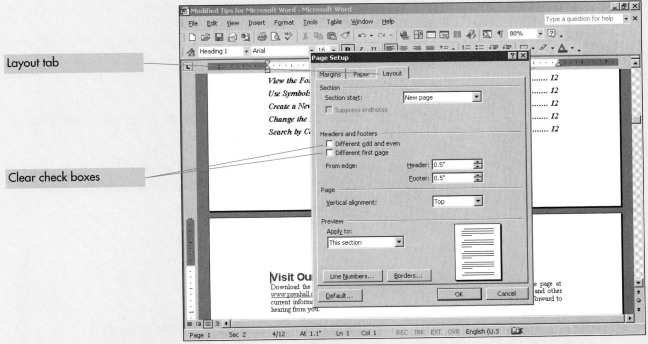

(e) The Page Setup Command (step 5)

FIGURE 4.11 *Hands-on Exercise 4 (continued)*

MOVING WITHIN LONG DOCUMENTS

Double click the page indicator on the status bar to display the dialog box for the Go To command from where you can go directly to any page within the document. You can also Ctrl+Click an entry in the table of contents to go directly to the text of that entry. And finally, you can use the Ctrl+Home and Ctrl+End keyboard shortcuts to move to the beginning or end of the document, respectively. The latter are universal shortcuts and apply to other Office documents as well.

Step 6: **Create the Header**

> ➤ Pull down the **View menu**. Click **Header and Footer** to produce the screen in Figure 4.11f. The text in the document is faded to indicate that you are editing the header, as opposed to the document.
> ➤ The "Same as Previous" indicator is on since Word automatically uses the header from the previous section.
> ➤ Click the **Same as Previous button** on the Header and Footer toolbar to toggle the indicator off and to create a different header for this section.
> ➤ If necessary, click in the header. Click the **arrow** on the Font list box on the Formatting toolbar. Click **Arial**. Click the **arrow** on the Font size box. Click **8**. Type **Tips for Microsoft Word**.
> ➤ Press the **Tab key** twice. Type **PAGE**. Press the **space bar**. Click the **Insert Page Number button** on the Header and Footer toolbar.
> ➤ Click the **Close button** on the Header and Footer toolbar. The header is faded, and the document text is available for editing.

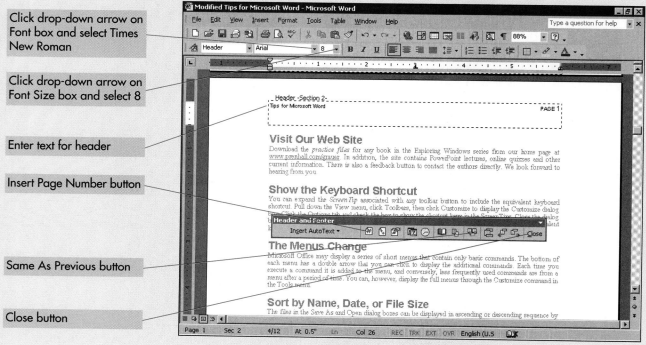

(f) Create the Header (step 6)

FIGURE 4.11 *Hands-on Exercise 4 (continued)*

HEADERS AND FOOTERS

If you do not see a header or footer, it is most likely because you are in the wrong view. Headers and footers are displayed in the Print Layout view but not in the Normal view. (Click the Print Layout button on the status bar to change the view.)

HANDS-ON EXERCISE 4

WORKING IN LONG DOCUMENTS

Objective   To create a header (footer) that includes page numbers; to insert and update a table of contents; to add an index entry; to insert a section break and demonstrate the Go To command; to view multiple pages of a document. Use Figure 4.11 as a guide for the exercise.

Step 1:   **Applying a Style**

> ➤ Open the **Modified Tips for Word document** from the previous exercise. Zoom to **Page Width**. Scroll to the top of the second page.
> ➤ Click to the left of the first tip title. (If necessary, click the **Show/Hide ¶ button** on the Standard toolbar to hide the paragraph marks.)
> ➤ Type **Table of Contents**. Press the **enter key** two times.
> ➤ Click anywhere within the phrase "Table of Contents". Click the **down arrow** on the **Styles** list box to pull down the styles for this document as shown in Figure 4.11a.
> ➤ Click **Report Title** (the style you created at the end of the previous exercise). "Table of Contents" is centered in 24 point blue Arial bold according to the definition of Report Title.

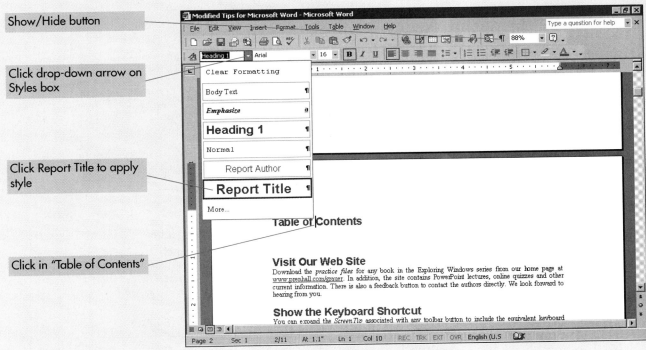

(a) Applying a Style (step 1)

FIGURE 4.11   *Hands-on Exercise 4*

Step 2: **Table of Contents**

- If necessary, change to the **Print Layout view**. Click the line immediately under the title for the table of contents. Pull down the **View menu**. Click **Zoom** to display the associated dialog box.
- Click the **monitor icon**. Click and drag the **page icons** to display two pages down by five pages across as shown in the figure. Release the mouse.
- Click **OK**. The display changes to show all eleven pages in the document.
- Pull down the **Insert menu**. Click **Reference**, then click **Index and Tables**. If necessary, click the **Table of Contents tab** to display the dialog box in Figure 4.11b.
- Check the boxes to **Show Page Numbers** and to **Right Align Page Numbers**.
- Click the **down arrow** on the Formats list box, then click **Distinctive**. Click the **arrow** in the **Tab Leader list box**. Choose a dot leader. Click **OK**. Word takes a moment to create the table of contents, which extends to two pages.
- Save the document.

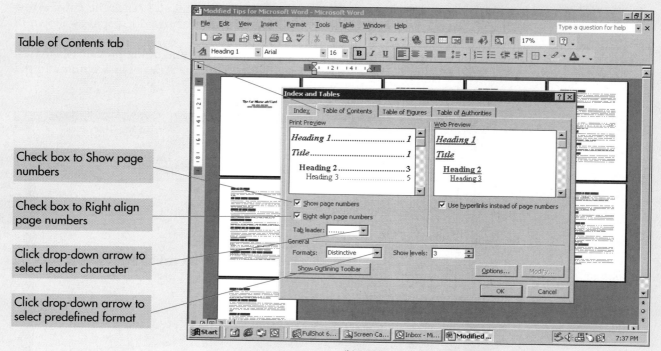

(b) Table of Contents (step 2)

FIGURE 4.11 *Hands-on Exercise 4 (continued)*

AUTOFORMAT AND THE TABLE OF CONTENTS

Word will create a table of contents automatically, provided you use the built-in heading styles to define the items for inclusion. If you have not applied the styles to the document, the AutoFormat command will do it for you. Once the heading styles are in the document, pull down the Insert command, click Reference, then click Index and Tables, then click the Table of Contents command.

Step 3: **Field Codes and Field Text**

- Click the **arrow** on the **Zoom Control box** on the Standard toolbar. Click **Page Width** in order to read the table of contents as in Figure 4.11c.
- Use the **up arrow key** to scroll to the beginning of the table of contents. Press **Alt+F9**. The table of contents is replaced by an entry similar to {TOC \o "1-3"} to indicate a field code. The exact code depends on the selections you made in step 2.
- Press **Alt+F9** a second time. The field code for the table of contents is replaced by text.
- Pull down the **Edit menu**. Click **Go To** to display the dialog box in Figure 4.11c.
- Type **3** and press the **enter key** to go to page 3, which contains the second page of the table of contents. Click **Close**.

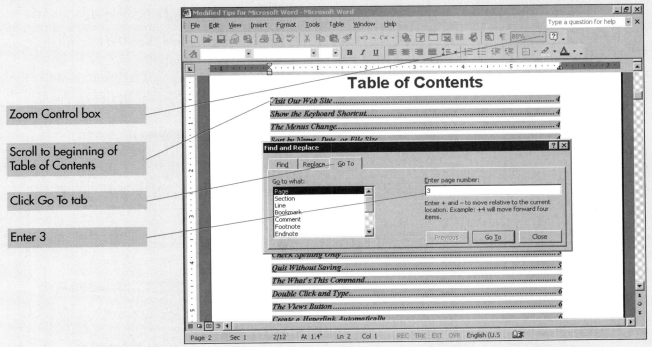

(c) Field Codes and Field Text (step 3)

FIGURE 4.11 *Hands-on Exercise 4 (continued)*

THE GO TO AND GO BACK COMMANDS

The F5 key is the shortcut equivalent of the Go To command and displays a dialog box to move to a specific location (a page or section) within a document. The Shift+F5 combination executes the Go Back command and returns to a previous location of the insertion point; press Shift+F5 repeatedly to cycle through the last three locations of the insertion point.

Step 4: **Insert a Section Break**

➤ Scroll down page 3 until you are at the end of the table of contents. Click to the left of the first tip heading as shown in Figure 4.11d.

➤ Pull down the **Insert menu**. Click **Break** to display the Break dialog box. Click the **Next Page button** under Section Break types. Click **OK** to create a section break, simultaneously forcing the first tip to begin on a new page.

➤ The first tip, Visit Our Web Site, moves to the top of the next page (page 4 in the document). If the status bar already displays Page 1 Section 2, a previous user has changed the default numbering to begin each section on its own page and you can go to step 6. If not, you need to change the page numbering.

➤ Pull down the **Insert menu** and click **Page Numbers** to display the Page Numbers dialog box. Click the **drop-down arrow** in the Position list box to position the page number at the top of page (in the header).

➤ Click the **Format command button** to display the Page Number Format dialog box. Click the option button to **Start at** page 1 (i.e., you want the first page in the second section to be numbered as page 1), and click **OK** to close the Page Number Format box.

➤ Close the Page Numbers dialog box. The status bar now displays Page 1 Sec 2 to indicate that you are on page 1 in the second section. The entry 4/12 indicates that you are physically on the fourth page of a 12-page document.

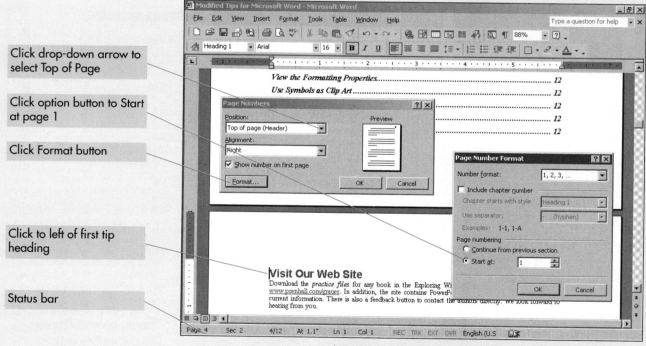

Click drop-down arrow to select Top of Page

Click option button to Start at page 1

Click Format button

Click to left of first tip heading

Status bar

(d) Insert a Section Break (step 4)

FIGURE 4.11   *Hands-on Exercise 4 (continued)*

Step 5: **The Page Setup Command**

- Pull down the **File menu** and click the **Page Setup command** (or double click the **ruler**) to display the Page Setup dialog box.
- Click the **Layout tab** to display the dialog box in Figure 4.11e.
- If necessary, clear the box for Different Odd and Even Pages and for Different First Page, as all pages in this section (Section 2) are to have the same header. Click **OK**.
- Save the document.

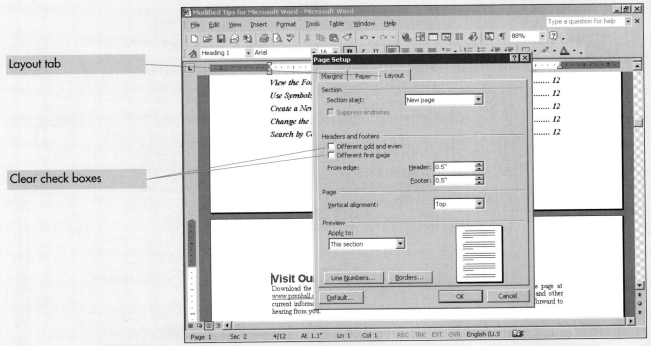

(e) The Page Setup Command (step 5)

FIGURE 4.11 *Hands-on Exercise 4 (continued)*

MOVING WITHIN LONG DOCUMENTS

Double click the page indicator on the status bar to display the dialog box for the Go To command from where you can go directly to any page within the document. You can also Ctrl+Click an entry in the table of contents to go directly to the text of that entry. And finally, you can use the Ctrl+Home and Ctrl+End keyboard shortcuts to move to the beginning or end of the document, respectively. The latter are universal shortcuts and apply to other Office documents as well.

Step 6: **Create the Header**

> ➤ Pull down the **View menu**. Click **Header and Footer** to produce the screen in Figure 4.11f. The text in the document is faded to indicate that you are editing the header, as opposed to the document.
>
> ➤ The "Same as Previous" indicator is on since Word automatically uses the header from the previous section.
>
> ➤ Click the **Same as Previous button** on the Header and Footer toolbar to toggle the indicator off and to create a different header for this section.
>
> ➤ If necessary, click in the header. Click the **arrow** on the Font list box on the Formatting toolbar. Click **Arial**. Click the **arrow** on the Font size box. Click **8**. Type **Tips for Microsoft Word**.
>
> ➤ Press the **Tab key** twice. Type **PAGE**. Press the **space bar**. Click the **Insert Page Number button** on the Header and Footer toolbar.
>
> ➤ Click the **Close button** on the Header and Footer toolbar. The header is faded, and the document text is available for editing.

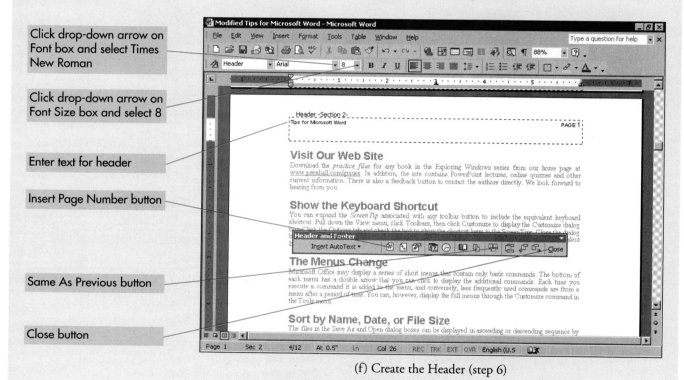

(f) Create the Header (step 6)

FIGURE 4.11 *Hands-on Exercise 4 (continued)*

HEADERS AND FOOTERS

If you do not see a header or footer, it is most likely because you are in the wrong view. Headers and footers are displayed in the Print Layout view but not in the Normal view. (Click the Print Layout button on the status bar to change the view.)

CHAPTER 4: ADVANCED FEATURES

Step 7: **Update the Table of Contents**

- Press **Ctrl+Home** to move to the beginning of the document. The status bar indicates Page 1, Sec 1.
- Click the **Select Browse Object button** on the Vertical scroll bar, then click the **Browse by Page** icon.
- If necessary, click the **Next Page button** or **Previous Page button** on the vertical scroll bar (or press **Ctrl+PgDn**) to move to the page containing the table of contents.
- Click to the left of the first entry in the Table of Contents. Press the **F9 key** to update the table of contents. If necessary, click the **Update Entire Table button** as shown in Figure 4.11g, then click **OK**.
- The pages are renumbered to reflect the actual page numbers in the second section.

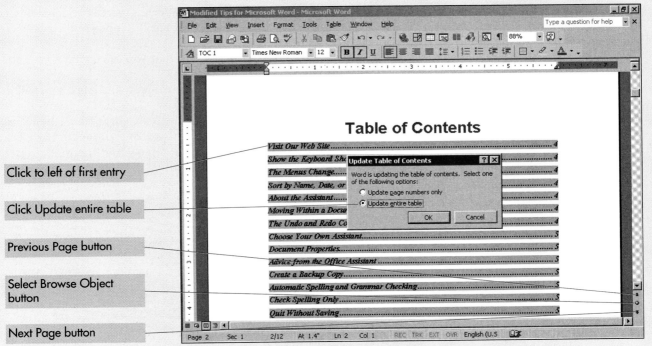

(g) Update the Table of Contents (step 7)

FIGURE 4.11 *Hands-on Exercise 4 (continued)*

SELECT BROWSE OBJECT

Click the Select Browse Object button toward the bottom of the vertical scroll bar to display a menu in which you specify how to browse through a document. Typically you browse from one page to the next, but you can browse by footnote, section, graphic, table, or any of the other objects listed. Once you select the object, click the Next or Previous buttons on the vertical scroll bar (or press Ctrl+PgDn or Ctrl+PgUp) to move to the next or previous occurrence of the selected object.

Step 8: **Create an Index Entry**

➤ Press **Ctrl+Home** to move to the beginning of the document. Pull down the **Edit menu** and click the **Find command**. Search for the first occurrence of the text "Ctrl+Home" within the document, as shown in Figure 4.11h. Close the Find and Replace dialog box.

➤ Click the **Show/Hide ¶ button** so you can see the nonprinting characters in the document, which include the index entries that have been previously created by the authors. (The index entries appear in curly brackets and begin with the letters XE.)

➤ Check that the text "Ctrl+Home" is selected within the document, then press **Alt+Shift+X** to display the Mark Index Entry dialog box. (Should you forget the shortcut, pull down the **Insert menu**, click Reference, click the **Index and Tables command**, click the **Index tab**, then click the **Mark Entry command button**.)

➤ Click the **Mark command button** to create the index entry, after which you see the field code, {XE "Ctrl+Home"}, to indicate that the index entry has been created.

➤ The Mark Index Entry dialog box stays open so that you can create additional entries by selecting additional text.

➤ Click the option button to create a **cross-reference**. Type **keyboard shortcut** in the associated text box. Click **Mark**.

➤ Click in the document, click and drag to select the text "Ctrl+End," then click in the dialog box, and the Main entry changes to Ctrl+End automatically. Click the **Mark command button** to create the index entry. Close the Mark Index Entry dialog box.

➤ Save the document.

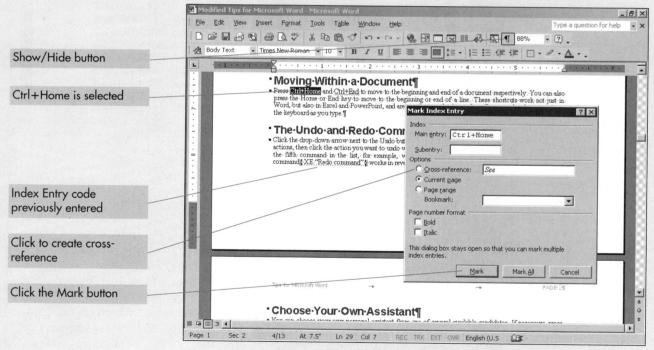

(h) Create an Index Entry (step 8)

FIGURE 4.11 *Hands-on Exercise 4 (continued)*

Step 9: **Create the Index**

➤ Press **Ctrl+End** to move to the end of the document, where you will insert the index.
➤ Press **enter** to begin a new line.
➤ Pull down the **Insert menu**, click **Reference**, then click the **Index and Tables command** to display the Index and Tables dialog box in Figure 4.11i. Click the **Index tab** if necessary.
➤ Choose the type of index you want. We selected a **classic format** over **two columns**. Click **OK** to create the index. Click the **Undo button** if you are not satisfied with the appearance of the index, then repeat the process to create an index with a different style.
➤ Save the document.

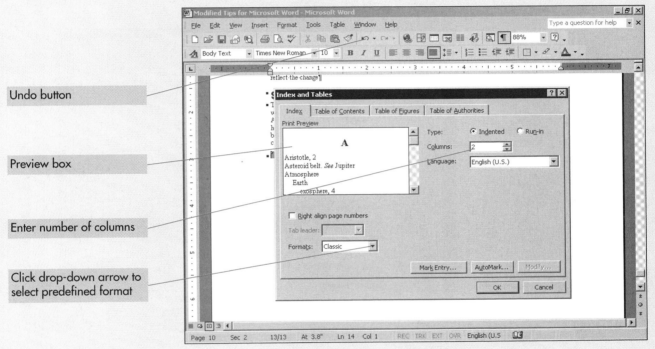

(i) Create the Index (step 9)

FIGURE 4.11 *Hands-on Exercise 4 (continued)*

AUTOMARK INDEX ENTRIES

The AutoMark command will, as the name implies, automatically mark all occurrences of all entries for inclusion in an index. To use the feature, you have to create a separate document that lists the terms you want to reference, then you execute the AutoMark command from the Index and Tables dialog box. The advantage is that it is fast. The disadvantage is that every occurrence of an entry is marked in the index so that a commonly used term may have too many page references. You can, however, delete superfluous entries by manually deleting the field codes. Click the Show/Hide button if you do not see the entries in the document.

Step 10: **Complete the Index**

➤ Scroll to the beginning of the index and click to the left of the letter "A." Pull down the **File menu** and click the **Page Setup command** to display the Page Setup dialog box and click the **Layout tab**.
➤ Click the **down arrow** in the Section start list box and specify **New page**. Click the **down arrow** in the Apply to list box and specify **This section**. Click **OK**. The index moves to the top of a new page.
➤ Click anywhere in the index, which is contained in its own section since it is displayed over two columns. The status bar displays Page 1, Section 3, 13/13 as shown in Figure 4.11j.
➤ Save the document.

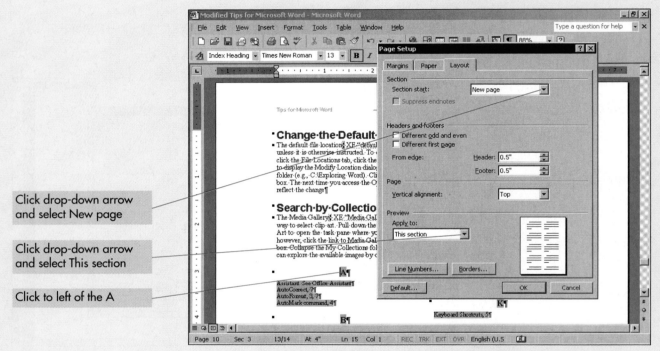

(j) Complete the Index (step 10)

FIGURE 4.11 *Hands-on Exercise 4 (continued)*

SECTION FORMATTING

Page numbering and orientation, headers, footers, and columns are implemented at the section level. Thus the index is automatically placed in its own section because it contains a different number of columns from the rest of the document. The notation on the status bar, Page 1, Section 3, 13/13 indicates that the insertion point is on the first page of section three, corresponding to the 13th page of a 13-page document.

Step 11: **The Completed Document**

- Pull down the **View menu**. Click **Zoom**. Click **Many Pages**. Click the **monitor icon**. Click and drag the page icon within the monitor to display two pages down by five pages. Release the mouse. Click **OK**.
- The completed document is shown in Figure 4.11k. The index appears by itself on the last (13th) page of the document.
- Save the document, then print the completed document to prove to your instructor that you have completed the exercise.
- Congratulations on a job well done. You have created a document with page numbers, a table of contents, and an index. Exit Word.

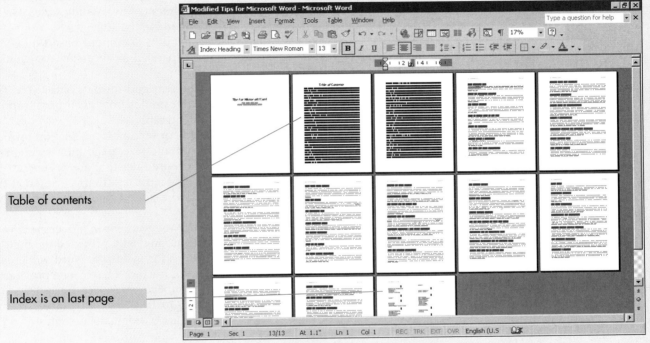

(k) The Completed Document (step 11)

FIGURE 4.11 *Hands-on Exercise 4 (continued)*

UPDATING THE TABLE OF CONTENTS

Use a shortcut menu to update the table of contents. Point to any entry in the table of contents, then press the right mouse button to display a shortcut menu. Click Update Field, click the Update Entire Table command button, and click OK. The table of contents will be adjusted automatically to reflect page number changes as well as the addition or deletion of any items defined by any built-in heading style.

SUMMARY

A list helps to organize information by emphasizing important topics. A bulleted or numbered list can be created by clicking the appropriate button on the Formatting toolbar or by executing the Bullets and Numbering command in the Format menu. An outline extends a numbered list to several levels.

Tables represent a very powerful capability within Word and are created through the Insert Table command in the Table menu or by using the Insert Table button on the Standard toolbar. Each cell in a table is formatted independently and may contain text, numbers, and/or graphics.

A style is a set of formatting instructions that has been saved under a distinct name. Styles are created at the character or paragraph level and provide a consistent appearance to similar elements throughout a document. Any existing styles can be modified to change the formatting of all text defined by that style.

The Outline view displays a condensed view of a document based on styles within the document. Text may be collapsed or expanded as necessary to facilitate moving text within long documents.

The AutoFormat command analyzes a document and formats it for you. The command goes through an entire document, determines how each paragraph is used, then applies an appropriate style to each paragraph.

Formatting occurs at the character, paragraph, or section level. Section formatting controls margins, columns, page orientation and size, page numbering, and headers and footers. A header consists of one or more lines that are printed at the top of every (designated) page in a document. A footer is text that is printed at the bottom of designated pages. Page numbers may be added to either a header or footer.

A table of contents lists headings in the order they appear in a document with their respective page numbers. It can be created automatically, provided the built-in heading styles were previously applied to the items for inclusion. Word will create an index automatically, provided that the entries for the index have been previously marked. This, in turn, requires you to go through a document, select the appropriate text, and mark the entries accordingly. The Edit Go To command enables you to move directly to a specific page, section, or bookmark within a document.

KEY TERMS

AutoFormat command (p. 174)
AutoMark (p. 195)
Body Text style (p. 171)
Bulleted list (p. 154)
Bullets and Numbering command (p. 154)
Cell (p. 162)
Character style (p. 171)
Delete command (p. 162)
Footer (p. 183)
Go To command (p. 185)
Header (p. 183)

Heading 1 style (p. 171)
Index (p. 185)
Index and Tables command (p. 185)
Insert menu (p. 184)
Insert Page Numbers command (p. 183)
Insert Table command (p. 162)
Leader character (p. 185)
Mark Index entry (p. 194)
Outline numbered list (p. 154)
Normal style (p. 171)
Numbered list (p. 154)

Outline (p. 154)
Outline view (p. 173)
Page numbers (p. 183)
Paragraph style (p. 171)
Section (p. 184)
Section break (p. 184)
Style (p. 171)
Styles and Formatting command (p. 171)
Table menu (p. 162)
Table of contents (p. 184)
Tables feature (p. 162)

MULTIPLE CHOICE

1. Which of the following can be stored within a paragraph style?
   (a) Tabs and indents
   (b) Line spacing and alignment
   (c) Shading and borders
   (d) All of the above

2. What is the easiest way to change the alignment of five paragraphs scattered throughout a document, each of which has been formatted with the same style?
   (a) Select the paragraphs individually, then click the appropriate alignment button on the Formatting toolbar
   (b) Select the paragraphs at the same time, then click the appropriate alignment button on the Formatting toolbar
   (c) Change the format of the existing style, which changes the paragraphs
   (d) Retype the paragraphs according to the new specifications

3. The AutoFormat command will do all of the following except:
   (a) Apply styles to individual paragraphs
   (b) Apply boldface italics to terms that require additional emphasis
   (c) Replace ordinary quotes with smart quotes
   (d) Substitute typographic symbols for ordinary letters—such as © for (C)

4. Which of the following is used to create a conventional outline?
   (a) The Bullets and Numbering command
   (b) The Outline view
   (c) Both (a) and (b)
   (d) Neither (a) nor (b)

5. In which view do you see headers and/or footers?
   (a) Print Layout view
   (b) Normal view
   (c) Both (a) and (b)
   (d) Neither (a) nor (b)

6. Which of the following numbering schemes can be used with page numbers?
   (a) Roman numerals (I, II, III . . . or i, ii, iii)
   (b) Regular numbers (1, 2, 3, . . .)
   (c) Letters (A, B, C . . . or a, b, c)
   (d) All of the above

7. Which of the following is true regarding headers and footers?
   (a) Every document must have at least one header
   (b) Every document must have at least one footer
   (c) Both (a) and (b)
   (d) Neither (a) nor (b)

8. Which of the following is a *false* statement regarding lists?
   (a) A bulleted list can be changed to a numbered list and vice versa
   (b) The symbol for the bulleted list can be changed to a different character
   (c) The numbers in a numbered list can be changed to letters or roman numerals
   (d) The bullets or numbers cannot be removed

9. Page numbers can be specified in:
   (a) A header but not a footer
   (b) A footer but not a header
   (c) A header or a footer
   (d) Neither a header nor a footer

10. Which of the following is true regarding the formatting within a document?
    (a) Line spacing and alignment are implemented at the section level
    (b) Margins, headers, and footers are implemented at the paragraph level
    (c) Both (a) and (b)
    (d) Neither (a) nor (b)

11. What happens when you press the Tab key from within a table?
    (a) A Tab character is inserted just as it would be for ordinary text
    (b) The insertion point moves to the next column in the same row or the first column in the next row if you are at the end of the row
    (c) Both (a) and (b)
    (d) Neither (a) nor (b)

12. Which of the following is true, given that the status bar displays Page 1, Section 3, followed by 7/9?
    (a) The document has a maximum of three sections
    (b) The third section begins on page 7
    (c) The insertion point is on the very first page of the document
    (d) All of the above

13. The Edit Go To command enables you to move the insertion point to:
    (a) A specific page
    (b) A relative page forward or backward from the current page
    (c) A specific section
    (d) Any of the above

14. Once a table of contents has been created and inserted into a document:
    (a) Any subsequent page changes arising from the insertion or deletion of text to existing paragraphs must be entered manually
    (b) Any additions to the entries in the table arising due to the insertion of new paragraphs defined by a heading style must be entered manually
    (c) Both (a) and (b)
    (d) Neither (a) nor (b)

15. Which of the following is *false* about the Outline view?
    (a) It can be collapsed to display only headings
    (b) It can be expanded to show the entire document
    (c) It requires the application of styles
    (d) It is used to create a conventional outline

ANSWERS

1. d	**6.** d	**11.** b
2. c	**7.** d	**12.** b
3. b	**8.** d	**13.** d
4. a	**9.** c	**14.** d
5. a	**10.** d	**15.** d

PRACTICE WITH MICROSOFT WORD

1. **The Résumé:** Microsoft Word includes a Résumé Wizard, but you can achieve an equally good result through the tables feature. Start a new document and create a two-column table with approximately ten rows. Merge the two cells in the first row to enter your name in a distinctive font as shown in Figure 4.12. Complete the résumé by entering the various categories in the left cell of each row and the associated information in the right cell of the corresponding row. Our résumé, for example, uses right alignment for the category, but left aligns the detailed information. Select the entire table and remove the borders surrounding the individual cells. (Figure 4.12 displays gridlines, which—unlike borders—do not appear in the printed document.) Print the completed résumé for your instructor.

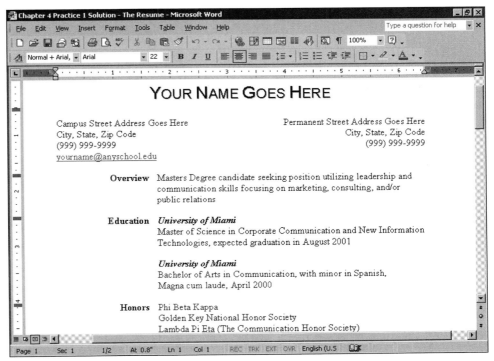

FIGURE 4.12 *The Résumé (Exercise 1)*

2. **The Employment Application:** A table can be the basis of almost any type of document. Use the tables feature to create a real or hypothetical employment application similar to the document in Figure 4.13a. You can follow our design or you can create your own, but try to develop an effective and attractive document. (We created the check box next to the highest degree by choosing the symbol from the Wingdings font within the Insert Symbol command.) Use appropriate spacing throughout the table, so that the completed application fills the entire page. Print the finished document for your instructor.

    Use this exercise to practice your file management skills by saving the solution in a new folder. Complete the employment application, then pull down the File menu and click the Save As command to display the Save As dialog box. Change to the Exploring Word folder, click the New folder button on the toolbar, then create a new folder as shown in Figure 4.13b. Click OK to create the folder, then click the Save button to save the document in the newly created folder. Additional folders become quite useful as you work with large numbers of documents.

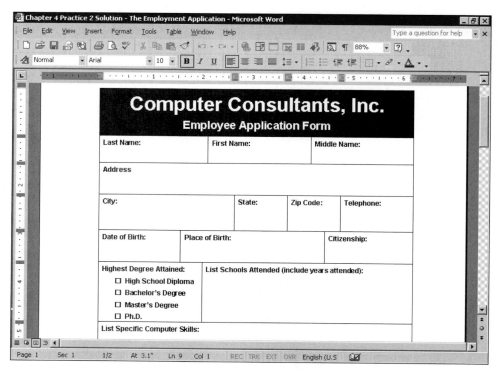

(a) The Employment Application

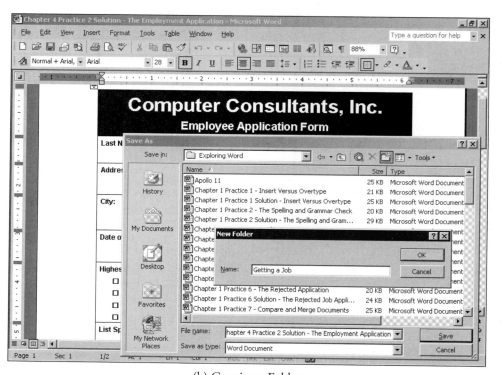

(b) Creating a Folder

FIGURE 4.13   *The Employment Application (Exercise 2)*

3. **Buying a PC:** The PC today is a commodity that allows the consumer to select each component. Thus, it is important to compare competing systems with respect to their features and cost. To that end, we suggest that you create a table similar to the one in Figure 4.14. You need not follow our design exactly, but you are required to leave space for at least two competing systems. Use the Table AutoFormat command to apply a format to the table. (We used the Colorful 2 design and modified the design to include a grid within the table.) You might also want to insert clip art to add interest to your table. If so, you will need to use the Format Picture command to change the layout and/or the order of the objects (the computer is to appear in front of the text.)

    The table is to appear within an existing document, *Chapter 4 Practice 3*, that contains a set of tips to consider when purchasing a PC. Open the document in the Exploring Word folder and insert a page break at the beginning of the existing document. The new page is to contain the table you see in Figure 4.14. The second page will contain the tips that we provide, but it is up to you to complete the formatting. Print the document for your instructor.

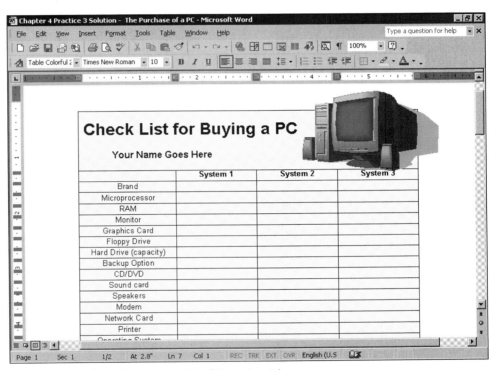

FIGURE 4.14 *Buying a PC (Exercise 3)*

BUILDS ON

HANDS-ON
EXERCISE 2
PAGES 164–170

4. **Section Formatting:** Formatting in Microsoft Word takes place at the character, paragraph, and/or section level, with the latter controlling margins and page orientation. This assignment asks you to create the study schedule that is described in the second hands-on exercise in the chapter, after which you are to insert a title page in front of the table as shown in Figure 4.13. The title page uses portrait orientation, whereas the table uses landscape. This in turn requires you to insert a section break after the title page in order to print each section with the appropriate orientation. Print the entire document for your instructor.

5. **Tips for Healthy Living:** Figure 4.16 displays the first several tips in a document that contains several tips for healthier living. The unformatted version of this document can be found in the *Chapter 4 Practice 5* document in the Exploring Word folder. Open that document, then use the AutoFormat command to apply the Heading 1 and Body Text styles throughout the document. Complete the document as you see fit.

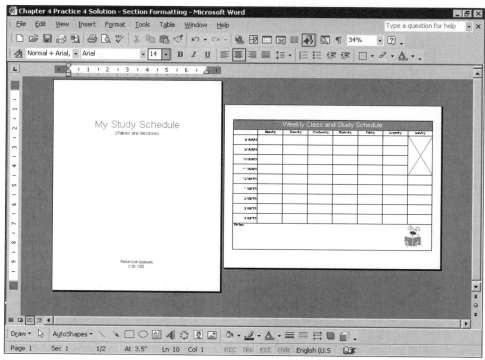

FIGURE 4.15 *Section Formatting (Exercise 4)*

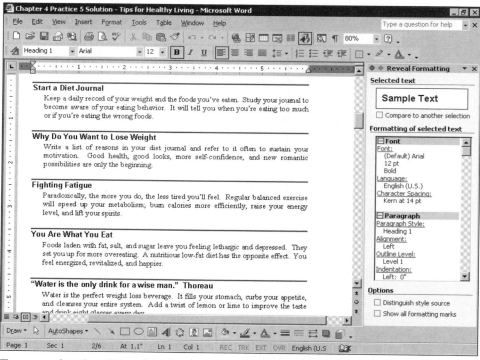

FIGURE 4.16 *Tips for Healthy Living (Exercise 5)*

CHAPTER 4: ADVANCED FEATURES

6. **Exporting an Outline:** An outline is the basis of a PowerPoint presentation, regardless of whether it (the outline) is created in Word or PowerPoint. Open the *Chapter 4 Practice 6* document in the Exploring Word folder to display an outline for a presentation that describes e-mail. Proceed as follows:
   a. Pull down the File menu, click the Send to command, then click Microsoft PowerPoint as shown in Figure 4.17. This in turn will start the PowerPoint application and create a presentation based on the Word outline.
   b. Remain in PowerPoint. Pull down the Format menu and click the Slide Design command to open the task pane and view the available templates. Point to any design that is appealing to you, then click the arrow that appears after you select the design and click the Apply to All Slides command. Your presentation will be reformatted according to the selected design.
   c. Select (click) the slide miniature of the first slide at the extreme left of the PowerPoint window. Pull down the Format menu and click the Slide Layout command to change the display in the task pane to the various layouts. Point to the layout at the top left of the Text Layout section (a ScreenTip will say "Title Slide"), click the arrow that appears, and click the Apply to Selected Slides command. Click in the slide where it says to "Click to add subtitle" and enter your name.
   d. Pull down the File menu and click the Print command to display the Print dialog box. Click the down arrow in the Print What area and choose handouts, then specify 6 slides per page. Check the box to Frame slides, then click OK to print the handouts for your instructor.
   e. Congratulations, you have just created your first PowerPoint presentation.

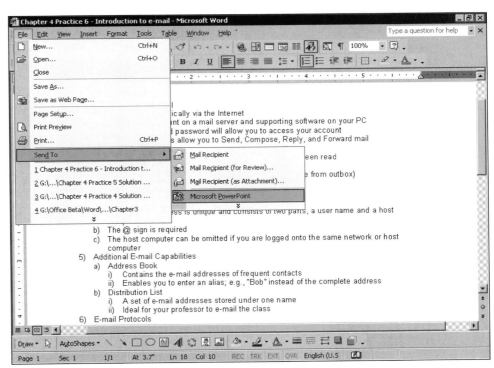

FIGURE 4.17 *Exporting an Outline (Exercise 6)*

7. **Introduction to the Internet:** The presentation in Figure 4.18 was created from the *Chapter 4 Practice 7* document in the Exploring Excel folder using the same instructions as in the previous exercise. This time, however, we have created a short presentation on basic Internet concepts. Add your name to the title slide and print the audience handouts for your instructor.

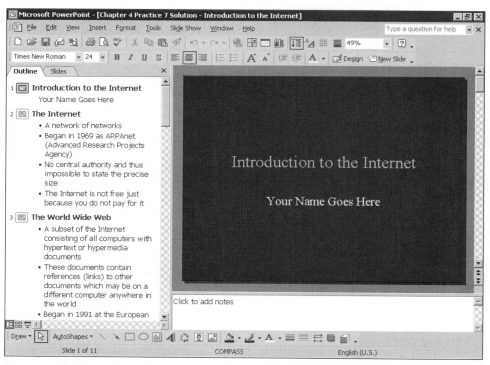

FIGURE 4.18 *Introduction to the Internet (Exercise 7)*

8. **The Constitution:** Use your favorite search engine to locate the text of the United States Constitution. There are many such sites on the Internet, one of which is shown in Figure 4.19. (We erased the URL, or else the assignment would be too easy.) Once you locate the document, expand the outline created in the first hands-on exercise to include information about the other provisions in the Constitution (Articles IV through VII, the Bill of Rights, and the other amendments). Submit the completed outline to your professor.

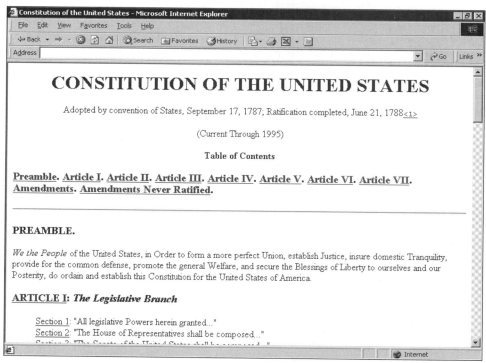

FIGURE 4.19 *The Constitution (Exercise 8)*

Tips for Windows 2000

Open the *Tips for Windows 2000* document that can be found in the Exploring Word folder. The tips are not formatted, so we would like you to use the AutoFormat command to create an attractive document. There are lots of tips, so a table of contents is also appropriate. Add a cover page with your name and date, then submit the completed document to your instructor.

Milestones in Communications

We take for granted immediate news of everything that is going on in the world, but it was not always that way. Did you know, for example, that it took five months for Queen Isabella to hear of Columbus' discovery, or that it took two weeks for Europe to learn of Lincoln's assassination? We've done some research on milestones in communications and left the file for you (*Milestones in Communications*). It runs for two, three, or four pages, depending on the formatting, which we leave to you. We would like you to include a header, and we think you should box the quotations that appear at the end of the document (it's your call as to whether to separate the quotations or group them together). Please be sure to number the completed document and don't forget a title page.

The Term Paper

Go to your most demanding professor and obtain the formatting requirements for the submission of a term paper. Be as precise as possible; for example, ask about margins, type size, and so on. What are the requirements for a title page? Is there a table of contents? Are there footnotes or endnotes, headers or footers? What is the format for the bibliography? Summarize the requirements, then indicate the precise means of implementation within Microsoft Word.

Forms, Forms, and More Forms

Every business uses a multitude of forms. Job applicants submit an employment application, sales personnel process order forms, and customers receive invoices. Even telephone messages have a form of their own. The office manager needs forms for everything, and she has come to you for help. You remember reading something about a tables feature and suggest that as a starting point. She needs more guidance, so you sit down with her and quickly design two forms that meet with her approval. Bring the two forms to class and compare your work with that of your classmates.

Writing Style

Use your favorite search engine to locate documents that describe suggested writing style for research papers. You will find different guidelines for traditional documents versus those that are published on the Web. Can you create a sample document that implements the suggested specifications? Summarize your findings in a brief note to your professor.

CHAPTER 5

Desktop Publishing: Creating a Newsletter and Other Documents

OBJECTIVES

AFTER READING THIS CHAPTER YOU WILL BE ABLE TO:

1. Design and implement a multicolumn newsletter; explain how sections are used to vary the number of columns in a document.
2. Define a pull quote and a reverse; explain how to implement these features using Microsoft Word.
3. Define typography; explain how styles can be used to implement changes in typography throughout a document.
4. Use the Insert Picture command to insert clip art into a document; explain how the Format Picture command is used to move and size a graphic.
5. Discuss the importance of a grid in the design of a document; describe the use of white space as a design element.
6. Use the Drawing toolbar to add objects to a Word document; describe the function of at least four different drawing tools.
7. Use object linking to create a Word document that contains an Excel worksheet and an Excel chart.

OVERVIEW

Desktop publishing evolved through a combination of technologies, including faster computers, laser printers, and sophisticated page composition software to manipulate text and graphics. Desktop publishing was initially considered a separate application, but today's generation of word processors has matured to such a degree that it is difficult to tell where word processing ends and desktop publishing begins. Microsoft Word is, for all practical purposes, a desktop publishing program that can be used to create all types of documents.

The essence of ***desktop publishing*** is the merger of text with graphics to produce a professional-looking document without reliance on external services. Desktop publishing will save you time and money because you are doing the work yourself rather than sending it out as you did in traditional publishing. That is the good news. The bad news is that desktop publishing is not as easy as it sounds, precisely because you are doing work that was done previously by skilled professionals. Nevertheless, with a little practice, and a basic knowledge of graphic design, you will be able to create effective and attractive documents.

Our chapter begins with the development of a simple newsletter in which we create a multicolumn document, import clip art and other objects, and position those objects within a document. The newsletter also reviews material from earlier chapters on bullets and lists, borders and shading, and section formatting. The second half of the chapter presents additional tools that you can use to enhance your documents. We describe the Drawing toolbar and explain how it is used to add objects to a Word document. We also describe how to create a Word document that contains an Excel worksheet and an Excel chart. The document is created in such a way that any changes to the underlying Excel workbook are automatically reflected in the Word document.

THE NEWSLETTER

The newsletter in Figure 5.1 demonstrates the basics of desktop publishing and provides an overview of the chapter. The material is presented conceptually, after which you implement the design in two hands-on exercises. We provide the text and you do the formatting. The first exercise creates a simple newsletter from copy that we provide. The second exercise uses more sophisticated formatting as described by the various techniques mentioned within the newsletter. Many of the terms are new, and we define them briefly in the next few paragraphs.

A ***reverse*** (light text on a dark background) is a favorite technique of desktop publishers to emphasize a specific element. It is used in the ***masthead*** (the identifying information) at the top of the newsletter and provides a distinctive look to the publication. The number of the newsletter and the date of publication also appear in the masthead in smaller letters.

A ***pull quote*** is a phrase or sentence taken from an article to emphasize a key point. It is typically set in larger type, often in a different typeface and/or italics, and may be offset with parallel lines at the top and bottom.

A ***dropped-capital letter*** is a large capital letter at the beginning of a paragraph. It, too, catches the reader's eye and calls attention to the associated text.

Clip art, used in moderation, will catch the reader's eye and enhance almost any newsletter. It is available from a variety of sources including the Microsoft Media Gallery, which is included in Office XP. Clip art can also be downloaded from the Web, but be sure you are allowed to reprint the image. The banner at the bottom of the newsletter is not a clip art image per se, but was created using various tools on the ***Drawing toolbar***.

Borders and shading are effective individually, or in combination with one another, to emphasize important stories within the newsletter. Simple vertical and/or horizontal lines are also effective. The techniques are especially useful in the absence of clip art or other graphics and are a favorite of desktop publishers.

Lists, whether bulleted or numbered, help to organize information by emphasizing important topics. A ***bulleted list*** emphasizes (and separates) the items. A ***numbered list*** sequences (and prioritizes) the items and is automatically updated to accommodate additions or deletions.

All of these techniques can be implemented with commands you already know, as you will see in the hands-on exercise, which follows shortly.

Creating a Newsletter

Volume I, Number 2   Spring 2001

Desktop publishing is easy, but there are several points to remember. This chapter will take you through the steps in creating a newsletter. The first hands-on exercise creates a simple newsletter with a masthead and three-column design. The second exercise creates a more attractive document by exploring different ways to emphasize the text.

Clip Art and Other Objects
Clip art is available from a variety of sources. You can also use other types of objects such as maps, charts, or organization charts, which are created by other applications, then brought into a document through the Insert Object command. A single dominant graphic is usually more appealing than multiple smaller graphics.

Techniques to Consider
Our finished newsletter contains one or more examples of each of the following desktop publishing techniques. Can you find where each technique is used, and further, explain, how to implement that technique in Microsoft Word?
1. Pull Quotes
2. Reverse
3. Drop Caps
4. Tables
5. Styles
6. Bullets and Numbering
7. Borders and Shading
8. The Drawing Toolbar

Newspaper-Style Columns
The essence of a newsletter is the implementation of columns in which text flows continuously from the bottom of one column to the top of the next. You specify the number of columns, and optionally, the space between columns. Microsoft Word does the rest. It will compute the width of each column based on the number of columns and the margins.

Beginners often specify margins that are too large and implement too much space between the columns. Another way to achieve a more sophisticated look is to avoid the standard two-column design. You can implement columns of varying width and/or insert vertical lines between the columns.

The number of columns will vary in different parts of a document. The masthead is typically a single column, but the body of the newsletter will have two or three. Remember, too, that columns are implemented at the section level and hence, section breaks are required throughout a document.

Typography
Typography is the process of selecting typefaces, type styles, and type sizes, and is a critical element in the success of any document. Type should reinforce the message and should be consistent with the information you want to convey. More is not better, especially in the case of too many typefaces and styles, which produce cluttered documents that impress no one. Try to limit yourself to a maximum of two typefaces per document, but choose multiple sizes and/or styles within those typefaces. Use boldface or italics for emphasis, but do so in moderation, because if you use too many different elements, the effect is lost.

A pull quote adds interest to a document while simultaneously emphasizing a key point. It is implemented by increasing the point size, changing to italics, centering the text, and displaying a top and bottom border on the paragraph.

Use Styles as Appropriate
Styles were covered in the previous chapter, but that does not mean you cannot use them in conjunction with a newsletter. A style stores character and/or paragraph formatting and can be applied to multiple occurrences of the same element within a document. Change the style and you automatically change all text defined by that style. You can also use styles from one edition of your newsletter to the next to insure consistency.

Borders and Shading
Borders and shading are effective individually or in combination with one another. Use a thin rule (one point or less) and light shading (five or ten percent) for best results. The techniques are especially useful in the absence of clip art or other graphics and are a favorite of desktop publishers.

All the News That Fits

FIGURE 5.1   *The Newsletter*

Typography

Typography is the process of selecting typefaces, type styles, and type sizes. It is a critical, often subtle, element in the success of a document, and its importance cannot be overstated. You would not, for example, use the same design to announce a year-end bonus and a plant closing. Indeed, good typography goes almost unnoticed, whereas poor typography calls attention to itself and detracts from a document. Our discussion reviews basic concepts and terminology.

A ***typeface*** (or ***font***) is a complete set of characters (upper- and lowercase letters, numbers, punctuation marks, and special symbols). Typefaces are divided into two general categories, serif and sans serif. A ***serif typeface*** has tiny cross lines at the ends of the characters to help the eye connect one letter with the next. A ***sans serif typeface*** (sans from the French for *without*) does not have these lines. A commonly accepted practice is to use serif typefaces with large amounts of text and sans serif typefaces for smaller amounts. The newsletter in Figure 5.1, for example, uses **Times New Roman** (a serif typeface) for the text and ***Arial*** (a sans serif typeface) for the headings.

A second characteristic of a typeface is whether it is monospaced or proportional. A ***monospaced typeface*** (e.g., Courier New) uses the same amount of space for every character regardless of its width. A ***proportional typeface*** (e.g., Times New Roman or Arial) allocates space according to the width of the character. Monospaced fonts are used in tables and financial projections where items must be precisely lined up, one beneath the other. Proportional typefaces create a more professional appearance and are appropriate for most documents.

Any typeface can be set in different styles (such as bold or italic) to create *Times New Roman Italic*, **Arial bold**, or `Courier New Bold Italic`. Other effects are also possible, such as small caps, shadow, and outline, but these should be used with moderation.

Type size is a vertical measurement and is specified in points. One ***point*** is equal to $1/72$ of an inch. The text in most documents is set in 10- or 12-point type. (The book you are reading is set in 10 point.) Different elements in the same document are often set in different type sizes to provide suitable emphasis. A variation of at least two points, however, is necessary for the difference to be noticeable. The headings in the newsletter, for example, were set in 12-point type, whereas the text of the articles is in 10-point type. The introduction of columns into a document poses another concern in that the type size should be consistent with the width of a column. Nine-point type, for example, is appropriate in columns that are two inches wide, but much too small in a single-column term paper. In other words, longer lines or wider columns require larger type sizes.

There are no hard and fast rules for the selection of type, only guidelines and common sense. Your objective should be to create a document that is easy to read and visually appealing. You will find that the design that worked so well in one document may not work at all in a different document. Good typography is often the result of trial and error, and we encourage you to experiment freely.

USE MODERATION AND RESTRAINT

More is not better, especially in the case of too many typefaces and styles, which produce cluttered documents that impress no one. Try to limit yourself to a maximum of two typefaces per document, but choose multiple sizes and/or styles within those typefaces. Use boldface or italics for emphasis, but do so in moderation, because if you emphasize too many elements, the effect is lost.

The Columns Command

The columnar formatting in a newsletter is implemented through the **Columns command** as shown in Figure 5.2. Start by selecting one of the preset designs, and Microsoft Word takes care of everything else. It calculates the width of each column based on the number of columns, the left and right margins on the page, and the specified (default) space between columns.

Consider, for example, the dialog box in Figure 5.2, in which a design of three equal columns is selected with a spacing of ¼ inch between each column. The 2-inch width of each column is computed automatically based on left and right margins of 1 inch each and the ¼-inch spacing between columns. The width of each column is computed by subtracting the sum of the margins and the space between the columns (a total of 2½ inches in this example) from the page width of 8½ inches. The result of the subtraction is 6 inches, which is divided by 3, resulting in a column width of 2 inches.

You can change any of the settings in the Columns dialog box, and Word will automatically make the necessary adjustments. The newsletter in Figure 5.1, for example, uses a two-column layout with wide and narrow columns. We prefer this design to columns of uniform width, as we think it adds interest to our document. Note, too, that once columns have been defined, text will flow continuously from the bottom of one column to the top of the next

Return for a minute to the newsletter in Figure 5.1, and notice that the number of columns varies from one part of the newsletter to another. The masthead is displayed over a single column at the top of the page, whereas the remainder of the newsletter is formatted in two columns of different widths. The number of columns is specified at the section level, and thus a *section break* is required whenever the column specification changes. A section break is also required at the end of the last column to balance the text within the columns.

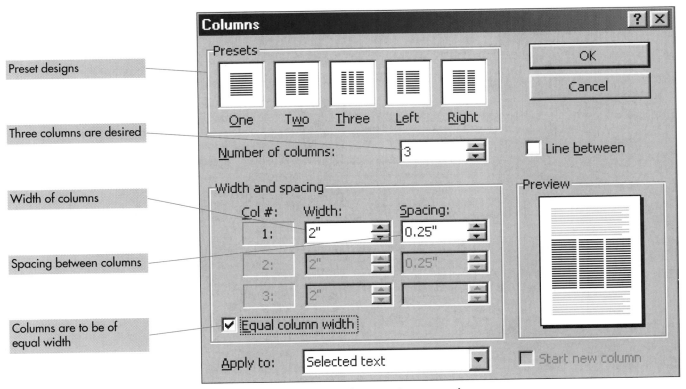

FIGURE 5.2 *The Columns Command*

Hands-on Exercise 1

Newspaper Columns

Objective To create a basic newsletter through the Format Columns command; to use section breaks to change the number of columns. Use Figure 5.3.

Step 1: **The Page Setup Command**

➤ Start Word. Open the **Text for Newsletter document** in the Exploring Word folder. Save the document as **Modified Newsletter**.
➤ Pull down the **File menu**. Click **Page Setup** to display the Page Setup dialog box in Figure 5.3a. Change the top, bottom, left, and right margins to .75.
➤ Click **OK** to accept these settings and close the Page Setup dialog box. If necessary, click the **Print Layout View button** above the status bar. Set the magnification (zoom) to **Page Width**.

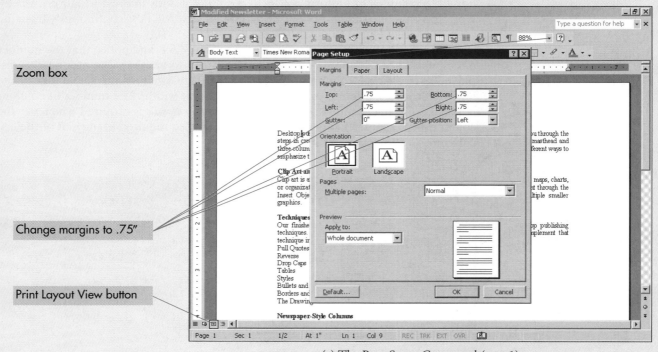

(a) The Page Setup Command (step 1)

FIGURE 5.3 *Hands-on Exercise 1*

CHANGE THE MARGINS

The default margins of 1 inch at the top and bottom of a page, and 1¼ inches on the sides, are fine for a typical document. A multicolumn newsletter, however, looks better with smaller margins, which in turn enables you to create wider columns. Margins are defined at the section level, and hence it's easiest to change the margins at the very beginning, when a document consists of only a single section.

Step 2: **Check the Document**

- Pull down the **Tools menu**, click **Options**, and click the **Spelling and Grammar tab**. Click the **drop-down arrow** on the Writing style list box and select **Grammar & Style**. Click **OK** to close the Options dialog box.
- Click the **Spelling and Grammar button** on the Standard toolbar to check the document for errors.
- The first error detected by the spelling and grammar check is the omitted hyphen between the words *three* and *column* as shown in Figure 5.3b. (This is a subtle mistake and emphasizes the need to check a document using the tools provided by Word.) Click **Change** to accept the indicated suggestion.
- Continue checking the document, accepting (or rejecting) the suggested corrections as you see fit.
- Save the document.

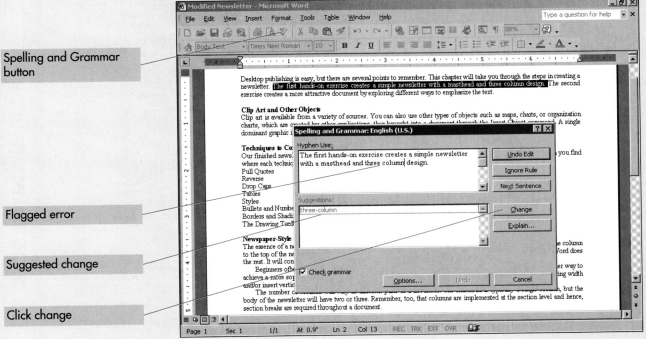

(b) Check the Document (step 2)

FIGURE 5.3 *Hands-on Exercise 1 (continued)*

USE THE SPELLING AND GRAMMAR CHECK

Our eyes are less discriminating than we would like to believe, allowing misspellings and simple typos to go unnoticed. To prove the point, count the number of times the letter f appears in this sentence, *"Finished files are the result of years of scientific study combined with the experience of years."* The correct answer is six, but most people find only four or five. Checking your document takes only a few minutes. Do it!

Step 3: **Implement Column Formatting**

➤ Pull down the **Format menu**. Click **Columns** to display the dialog box in Figure 5.3c. Click the **Presets icon** for **Two**. The column width for each column and the spacing between columns will be determined automatically from the existing margins.

➤ If necessary, clear the **Line between box**. Click **OK** to accept the settings and close the Columns dialog box.

➤ The text of the newsletter should be displayed in two columns. If you do not see the columns, it is probably because you are in the wrong view. Click the **Print Layout View button** above the status bar to change to this view.

➤ Save the document.

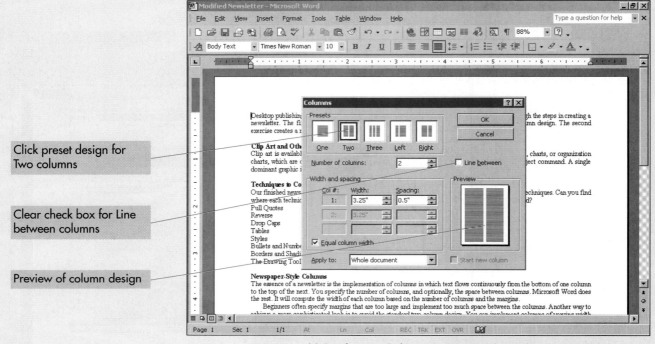

(c) Implement Column Formatting (step 3)

FIGURE 5.3 *Hands-on Exercise 1 (continued)*

PAGE BREAKS, COLUMN BREAKS, AND LINE BREAKS

Force Word to begin the next entry on a new page or column by inserting the proper type of break. Pull down the Insert menu, click the Break command to display the Break dialog box, then choose the option button for a page break or column break, respectively. It's easier, however, to use the appropriate shortcut, Ctrl+Enter or Shift+Ctrl+Enter, for a page or column break, respectively. You can also use Shift+Enter to force a line break, where the next word begins on a new line within the same paragraph. Click the Show/Hide button to display the hidden codes to see how the breaks are implemented.

Step 4: **Balance the Columns**

- Use the **Zoom box** on the Standard toolbar to zoom to **Whole Page** to see the entire newsletter as shown in Figure 5.3d. Do not be concerned if the columns are of different lengths.
- Press **Ctrl+End** to move the insertion point to the end of the document. Pull down the **Insert menu**. Click **Break** to display the Break dialog box in Figure 5.3d. Select the **Continuous option button** under Section breaks.
- Click **OK** to accept the settings and close the dialog box. The columns should be balanced, although one column may be one line longer than the other.
- Save the document.

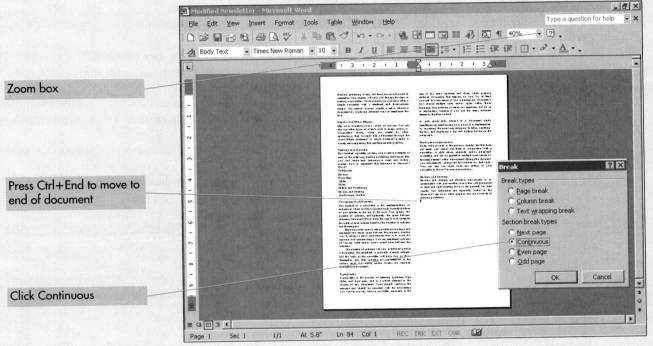

(d) Balance the Columns (step 4)

FIGURE 5.3 *Hands-on Exercise 1 (continued)*

USE THE RULER TO CHANGE COLUMN WIDTH

Click anywhere within the column whose width you want to change, then point to the ruler and click and drag the right column margin (the mouse pointer changes to a double arrow) to change the column width. Changing the width of one column in a document with equal-sized columns changes the width of all other columns so that they remain equal. Changing the width in a document with unequal columns changes only that column. You can also double click the ruler to display the Page Setup dialog box, then click the Margins tab to change the left and right margins, which in turn will change the column width.

Step 5: **Create the Masthead**

- Use the **Zoom box** on the Standard toolbar to change to **Page Width**. Click the **Show/Hide ¶ button** to display the paragraph and section marks.
- Press **Ctrl+Home** to move the insertion point to the beginning of the document. Pull down the **Insert menu**, click **Break**, select the **Continuous option button**, and click **OK**. You should see a double dotted line indicating a section break as shown in Figure 5.3e.
- Click immediately to the left of the dotted line, which will place the insertion point to the left of the line. Check the status bar to be sure you are in section one.
- Change the format for this section to a single column by clicking the **Columns button** on the Standard toolbar and selecting one column. (Alternatively, you can pull down the **Format menu**, click **Columns**, and choose **One** from the Presets column formats.)
- Type **Creating a Newsletter** and press the **enter key** twice. Select the newly entered text, click the **Center button** on the Formatting toolbar. Change the font to **48-point Arial Bold**.
- Click underneath the masthead (to the left of the section break). Pull down the **Table menu**, click **Insert** to display a submenu, then click **Table**. Insert a table with one row and two columns as shown in Figure 5.3e.
- Click in the left cell of the table. Type **Volume I, Number 1**. Click in the right cell (or press the **Tab key** to move to this cell and type the current semester (for example, **Spring 2001**). Click the **Align Right button**.
- Save the document.

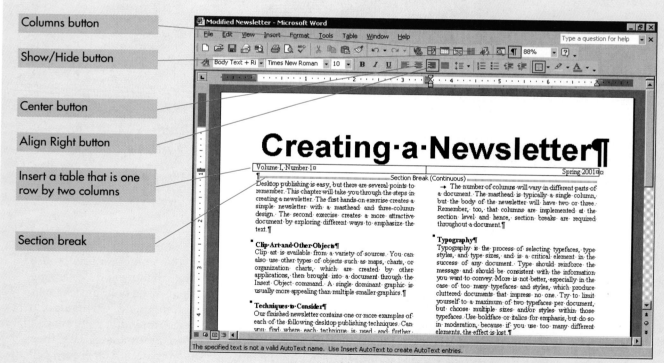

(e) Create the Masthead (step 5)

FIGURE 5.3 *Hands-on Exercise 1 (continued)*

Step 6: **Create a Reverse**

- Press **Ctrl+Home** to move the insertion point to the beginning of the newsletter. Click anywhere within the title of the newsletter.
- Pull down the **Format menu**, click **Borders and Shading** to display the Borders and Shading dialog box, then click the **Shading tab** in Figure 5.3f.
- Click the **drop-down arrow** in the Style list box (in the Patterns area) and select **Solid (100%)** shading. Click **OK** to accept the setting and close the dialog box. Click elsewhere in the document to see the results.
- The final step is to remove the default border that appears around the table. Click in the selection area to the left of the table to select the entire table.
- Pull down the **Format menu**, click **Borders and Shading**, and if necessary click the **Borders tab**. Click the **None icon** in the Presets area. Click **OK**. Click elsewhere in the document to see the result.

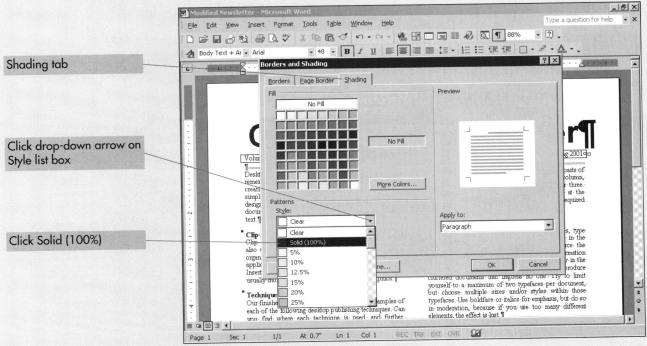

(f) Create a Reverse (step 6)

FIGURE 5.3 *Hands-on Exercise 1 (continued)*

LEFT ALIGNED **CENTERED** **RIGHT ALIGNED**

Many documents call for left-, centered, and/or right-aligned text on the same line, an effect that is achieved through setting tabs, or more easily through a table. To achieve the effect shown at the top of this box, create a 1 × 3 table (one row and three columns), type the text in the cells, then use the buttons on the Formatting toolbar to left-align, center, and right-align the cells. Select the table, pull down the Format menu, click Borders and Shading, then specify None as the Border setting.

Step 7: **Modify the Heading Style**

- Two styles have been implemented for you in the newsletter. Click in any text paragraph, and you see the Body Text style name displayed in the Style box on the Formatting toolbar. Click in any heading, and you see the Heading 1 style.
- Pull down the **View menu** and click the **Task Pane command** to open the task pane. Click the **down arrow** within the task pane and choose **Style and Formatting**.
- Point to the **Heading 1** style to display a down arrow, then click the **Modify command** to display the **Modify Style** dialog box shown in Figure 5.3g.
- Change the font to **Arial** and the font size to **12**. Click **OK** to accept the settings and close the dialog box. All of the headings in the document are changed automatically to reflect the changes in the Heading 1 style.
- Experiment with other styles as you see fit. (You can remove the formatting of existing text by clicking within a paragraph, then clicking **Clear Formatting** within the task pane.)
- Save the newsletter. Close the task pane.

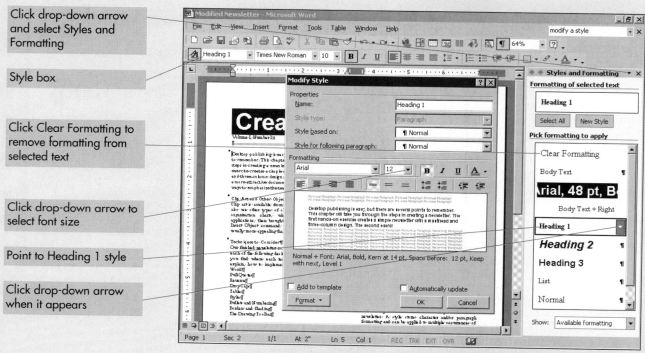

(g) Modify the Heading Style (step 7)

FIGURE 5.3 *Hands-on Exercise 1 (continued)*

USE STYLES AS APPROPRIATE

Styles were covered in the previous chapter, but that does not mean you cannot use them in conjunction with a newsletter. A style stores character and/or paragraph formatting and can be applied to multiple occurrences of the same element within a document. Change the style and you automatically change all text defined by that style. Use the same styles from one edition of your newsletter to the next to ensure consistency. Use styles for any document to promote uniformity and increase flexibility.

Step 8: **The Print Preview Command**

- Pull down the **File menu** and click **Print Preview** (or click the **Print Preview button** on the Standard toolbar) to view the newsletter as in Figure 5.3h. This is a basic two-column newsletter with the masthead appearing as a reverse and stretching over a single column.
- Click the **Print button** to print the newsletter at this stage so that you can compare this version with the finished newsletter at the end of the next exercise.
- Click the **Close button** on the Print Preview toolbar to close the Preview view and return to the Page Layout view.
- Exit Word if you do not want to continue with the next exercise at this time.

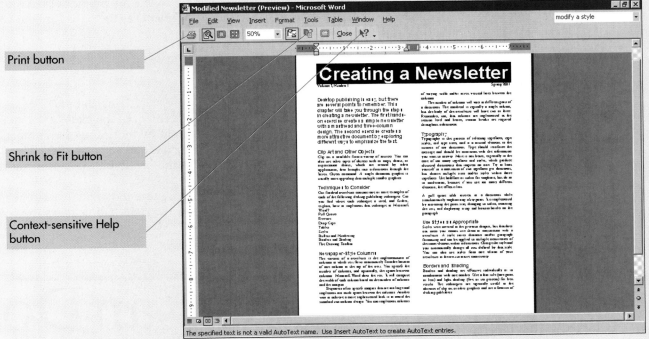

(h) The Print Preview Command (step 8)

FIGURE 5.3 *Hands-on Exercise 1 (continued)*

THE PRINT PREVIEW TOOLBAR

The Print Preview toolbar appears automatically when you switch to this view, and it contains several tools that are helpful prior to printing a document. The Shrink to Fit button is especially useful if a small portion of a document spills over to a second page—click this button, and it uniformly reduces the fonts throughout a document to eliminate the extra page. The context-sensitive Help button, on the extreme right of the toolbar, explains the function of the other buttons. Click the button (the mouse pointer changes to an arrow and a question mark), then click any other button for an explanation of its function. We suggest that you avoid the Full Screen button and that you close the full screen immediately if you wind up in this view.

ELEMENTS OF GRAPHIC DESIGN

We trust you have completed the first hands-on exercise without difficulty and that you were able to duplicate the initial version of the newsletter. That, however, is the easy part of desktop publishing. The more difficult aspect is to develop the design in the first place because the mere availability of a desktop publishing program does not guarantee an effective document, any more than a word processor will turn its author into another Shakespeare. Other skills are necessary, and so we continue with a brief introduction to graphic design.

Much of what we say is subjective, and what works in one situation will not necessarily work in another. Your eye is the best judge of all, and you should follow your own instincts. Experiment freely and realize that successful design is the result of trial and error. Seek inspiration from others by collecting samples of real documents that you find attractive, then use those documents as the basis for your own designs.

The Grid

The design of a document is developed on a ***grid***, an underlying, but *invisible,* set of horizontal and vertical lines that determine the placement of the major elements. A grid establishes the overall structure of a document by indicating the number of columns, the space between columns, the size of the margins, the placement of headlines, art, and so on. The grid does *not* appear in the printed document or on the screen.

Figure 5.4 shows the "same" document in three different designs. The left half of each design displays the underlying grid, whereas the right half displays the completed document.

(a) Three-column Grid

FIGURE 5.4 *The Grid System of Design*

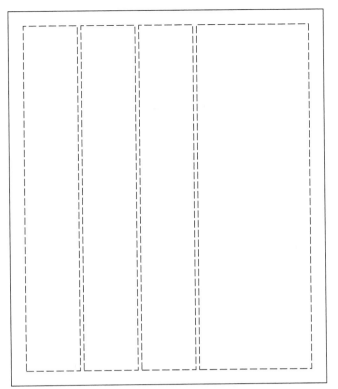

(b) Four-column Grid

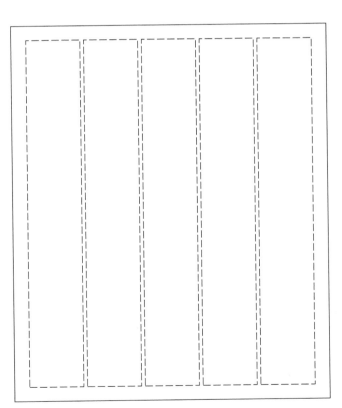

(c) Five-column Grid

FIGURE 5.4   *The Grid System of Design (continued)*

A grid may be simple or complex, but it is always distinguished by the number of columns it contains. The three-column grid of Figure 5.4a is one of the most common and utilitarian designs. Figure 5.4b shows a four-column design for the same document, with unequal column widths to provide interest. Figure 5.4c illustrates a five-column grid that is often used with large amounts of text. Many other designs are possible as well. A one-column grid is used for term papers and letters. A two-column, wide-and-narrow format is appropriate for textbooks and manuals. Two- and three-column formats are used for newsletters and magazines.

The simple concept of a grid should make the underlying design of any document obvious, which in turn gives you an immediate understanding of page composition. Moreover, the conscious use of a grid will help you organize your material and result in a more polished and professional-looking publication. It will also help you to achieve consistency from page to page within a document (or from issue to issue of a newsletter). Indeed, much of what goes wrong in desktop publishing stems from failing to follow or use the underlying grid.

Emphasis

Good design makes it easy for the reader to determine what is important. As indicated earlier, *emphasis* can be achieved in several ways, the easiest being variations in type size and/or type style. Headings should be set in type sizes (at least two points) larger than body copy. The use of **boldface** is effective as are *italics,* but both should be done in moderation. (UPPERCASE LETTERS and underlining are alternative techniques that we believe are less effective.)

Boxes and/or shading call attention to selected articles. Horizontal lines are effective to separate one topic from another or to call attention to a pull quote. A reverse can be striking for a small amount of text. Clip art, used in moderation, will catch the reader's eye and enhance almost any newsletter.

Clip Art

Clip art is available from a variety of sources including the Microsoft Media Gallery and Microsoft Web site. The Media Gallery can be accessed in a variety of ways, most easily through the *Insert Picture command*. Once clip art has been inserted into a document, it can be moved and sized just like any other Windows object, as will be illustrated in our next hands-on exercise.

The *Format Picture command* provides additional flexibility in the placement of clip art. The Text Wrapping tab, in the Advanced Layout dialog box, determines the way text is positioned around a picture. The Top and Bottom option (no wrapping) is selected in Figure 5.5a, and the resulting document is shown in Figure 5.5b. The sizing handles around the clip art indicate that it is currently selected, enabling you to move and/or resize the clip art using the mouse. (You can also use the Size and Position tabs in the Format Picture dialog box for more precision with either setting.) Changing the size or position of the object, however, does not affect the way in which text wraps around the clip art.

The document in Figure 5.5c illustrates a different wrapping selection in which text is wrapped on both sides. Figure 5.5c also uses an option on the Colors and Lines tab to draw a blue border around the clip art. The document in Figure 5.5d eliminates the border and chooses the tight wrapping style so that the text is positioned as closely as possible to the figure in a free-form design. Choosing among the various documents in Figure 5.5 is one of personal preference. Our point is simply that Word provides multiple options, and it is up to you, the desktop publisher, to choose the design that best suits your requirements.

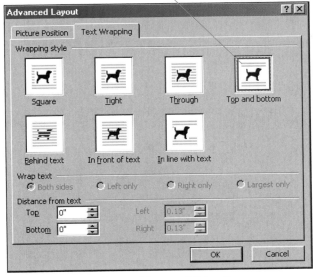

(a) Advanced Layout Dialog Box

(b) Top and Bottom Wrapping

(c) Square Wrapping (both sides)

(d) Tight Wrapping (both sides)

FIGURE 5.5 *The Format Picture Command*

THE DRAWING TOOLBAR

Did you ever stop to think how the images in the Media Gallery were developed? Undoubtedly they were drawn by someone with artistic ability who used basic shapes, such as lines and curves in various combinations, to create the images. The Drawing toolbar in Figure 5.6a contains all of the tools necessary to create original clip art. Select the Line tool for example, then click and drag to create the line. Once the line has been created, you can select it, then change its properties (such as thickness, style, or color) by using other tools on the Drawing toolbar. Draw a second line, or a curve—then, depending on your ability, you have a piece of original clip art.

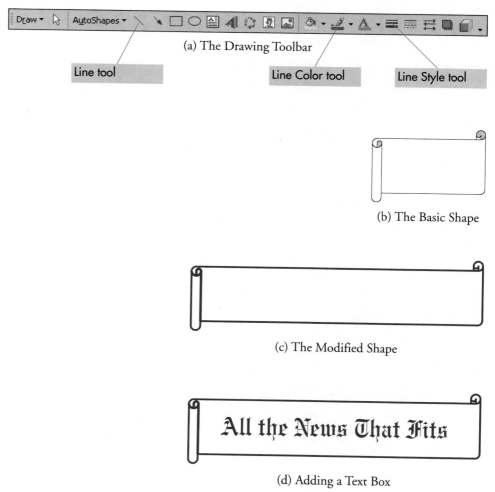

FIGURE 5.6 *The Drawing Toolbar*

We don't expect you to create clip art comparable to the images within the Media Gallery. You can, however, use the tools on the Drawing toolbar to modify an existing image and/or create simple shapes of your own that can enhance any document. One tool that is especially useful is the AutoShapes button that displays a series of predesigned shapes. Choose a shape (the banner in Figure 5.6b), change its size and color (Figure 5.6c), then use the Textbox tool to add an appropriate message.

The Drawing toolbar is displayed through the Toolbars command in the View menu. The following exercise has you use the toolbar to create the banner and text in Figure 5.6d. It's fun, it's easy; just be flexible and willing to experiment. We think you will be pleased at what you will be able to do.

HANDS-ON EXERCISE 2

COMPLETE THE NEWSLETTER

Objective   To insert clip art into a newsletter; to format a newsletter using styles, borders and shading, pull quotes, and lists. Use Figure 5.7a as a guide in the exercise.

Step 1:   **Change the Column Layout**

> ➤ Open the **Modified Newsletter** from the previous exercise. Click in the masthead and change the number of this edition from 1 to **2**.
> ➤ Click anywhere in the body of the newsletter. The status bar should indicate that you are in the second section. Pull down the **Format menu**. Click **Columns** to display the dialog box in Figure 5.7a. Click the **Left Preset icon**.
> ➤ Change the width of the first column to **2.25** and the space between columns to **.25**. Check (click) the **Line Between box**. Click **OK**.
> ➤ Save the newsletter.

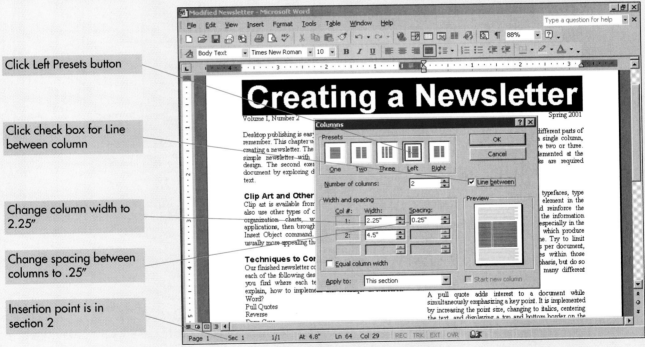

(a) Change the Column Layout (step 1)

FIGURE 5.7   *Hands-on Exercise 2*

EXPERIMENT WITH THE DESIGN

The number, width, and spacing of the columns in a newsletter is the single most important element in its design. Experiment freely. Good design is often the result of trial and error. Use the Undo command as necessary to restore the document.

Step 2: **Bullets and Numbering**

- Scroll in the document until you come to the list within the **Techniques to Consider** paragraph. Select the entire list as shown in Figure 5.7b.
- Pull down the **Format menu** and click **Bullets and Numbering** to display the Bullets and Numbering dialog box.
- If necessary, click the **Numbered tab** and choose the numbering style with Arabic numbers followed by periods. Click **OK** to accept these settings and close the Bullets and Numbering dialog box.
- Click anywhere in the newsletter to deselect the text.
- Save the newsletter.

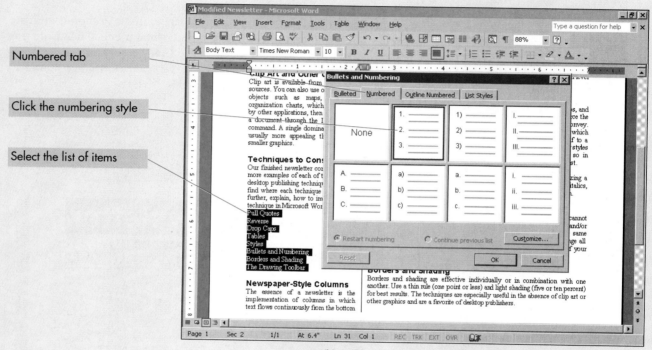

(b) Bullets and Numbering (step 2)

FIGURE 5.7   *Hands-on Exercise 2 (continued)*

LISTS AND THE FORMATTING TOOLBAR

The Formatting toolbar contains four buttons for use with bulleted and numbered lists. The Increase Indent and Decrease Indent buttons move the selected items one tab stop to the right and left, respectively. The Bullets button creates a bulleted list from unnumbered items or converts a numbered list to a bulleted list. The Numbering button creates a numbered list or converts a bulleted list to numbers. The Bullets and Numbering buttons also function as toggle switches; for example, clicking the Bullets button when a bulleted list is already in effect will remove the bullets.

Step 3: **Insert the Clip Art**

- Click immediately to the left of the article beginning **Clip Art and Other Objects**. Pull down the **Insert menu**, click **Picture**, then click **Clip Art** to display the Insert Clip Art task pane.
- Click in the **Search text box** and type **goals** to search for all pictures that have been catalogued to describe this attribute. Click the **Search button**. The search begins and the various pictures appear individually within the task pane.
- Point to the image you want in your newsletter, click the **down arrow** that appears, then click **Insert** to insert the clip art.
- The picture should appear in the document, where it can be moved and sized as described in the next several steps. Click the **Close button** on the task pane.
- Save the document.

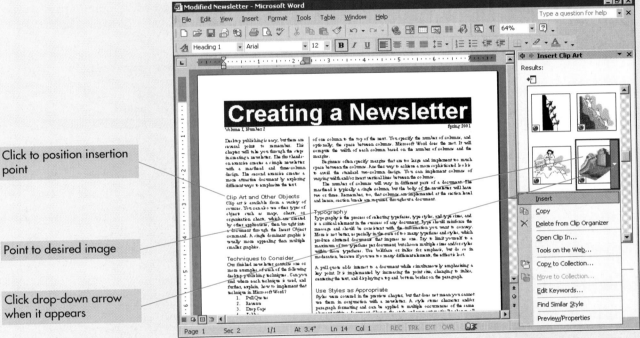

(c) Insert the Clip Art (step 3)

FIGURE 5.7 *Hands-on Exercise 2 (continued)*

CLIPS ONLINE

Why settle for the same old clip art when you can get new images from the Microsoft Web site? Pull down the Insert menu, click the Picture command, then choose Clip Art to open the task pane. Click the Clips Online button to connect to the Microsoft site, where you have your choice of clip art, photographs, sounds, and motion clips in a variety of categories. Click any image to see a preview, then click the preview box to download the image to your PC. Click the image after it has been downloaded to display a shortcut menu from where you insert the image into your document.

Step 4: **Move and Size the Clip Art**

- Click the **drop-down arrow** on the Zoom list box and select **Whole Page**.
- Point to the picture, click the **right mouse button** to display a context-sensitive menu, then click the **Format Picture command** to display the Format Picture dialog box as shown in Figure 5.7d.
- Click the **Layout tab**, choose the **Square layout**, then click the option button for left or right alignment. Click **OK** to close the dialog box. You can now move and size the clip art just like any other Windows object.
- To size the clip art, click anywhere within the clip art to select it and display the sizing handles. Drag a corner handle (the mouse pointer changes to a double arrow) to change the length and width of the picture simultaneously and keep the object in proportion.
- To move the clip art, click the object to select it and display the sizing handles. Point to any part of the object except a sizing handle (the mouse pointer changes to a four-sided arrow), then click and drag to move the clip art.
- Save the document.

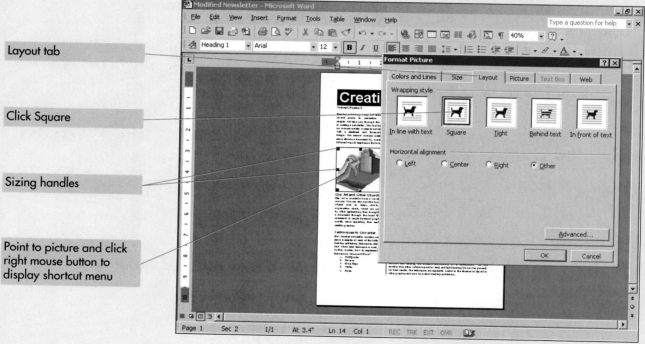

(d) Move and Size the Clip Art (step 4)

FIGURE 5.7 *Hands-on Exercise 2 (continued)*

CROPPING A PICTURE

Select a picture, and Word automatically displays the Picture toolbar, which enables you to modify the picture in subtle ways. The Crop tool enables you to eliminate (crop) part of a picture. Select the picture to display the Picture toolbar and display the sizing handles. Click the Crop tool (the ScreenTip will display the name of the tool), then click and drag a sizing handle to crop the part of the picture you want to eliminate.

Step 5: **Borders and Shading**

- Change to **Page Width** and click the **Show/Hide ¶ button** to display the paragraph marks. Press **Ctrl+End** to move to the end of the document, then select the heading and associated paragraph for Borders and Shading. (Do not select the ending paragraph mark.)
- Pull down the **Format menu**. Click **Borders and Shading**. If necessary, click the **Borders tab** to display the dialog box in Figure 5.4e. Click the **Box icon** in the Setting area. Click the **drop-down arrow** in the Width list box and select the **1 pt** line style.
- Click the **Shading tab**. Click the **drop-down arrow** in the Style list box (in the Patterns area) and select 5% shading. Click **OK** to accept the setting.
- Click elsewhere in the document to see the results. The heading and paragraph should be enclosed in a border with light shading.
- Save the document.

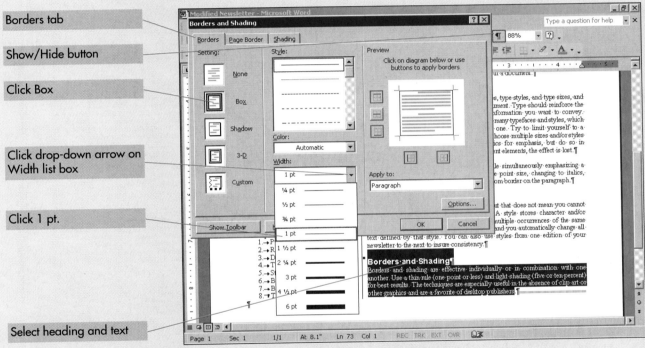

(e) Borders and Shading (step 5)

FIGURE 5.7 *Hands-on Exercise 2 (continued)*

USE THE TOOLBAR

The Outside Border button on the Formatting toolbar changes the style of the border for the selected text. That tool is also accessible from the Tables and Borders toolbar, which contains additional tools to insert or merge cells and/or to change the line style, thickness, or shading within a table. If the toolbar is not visible, point to any visible toolbar, click the right mouse button to show the list of toolbars, then click the Tables and Borders toolbar to display it on your screen.

Step 6: **Create a Pull Quote**

- Scroll to the bottom of the document until you find the paragraph describing a pull quote. Select the entire paragraph and change the text to **14-point Arial italic**.
- Click in the paragraph to deselect the text, then click the **Center button** to center the paragraph.
- Click the **drop-down arrow** on the **Border button** to display the different border styles as shown in Figure 5.7f.
- Click the **Top Border button** to add a top border to the paragraph.
- Click the **Bottom border button** to create a bottom border and complete the pull quote.
- Save the document.

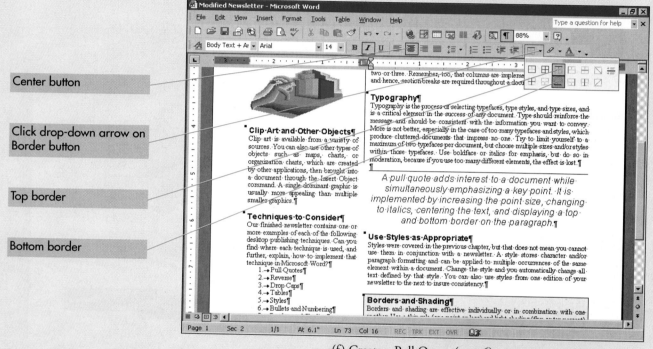

(f) Create a Pull Quote (step 6)

FIGURE 5.7 *Hands-on Exercise 2 (continued)*

EMPHASIZE WHAT'S IMPORTANT

Good design makes it easy for the reader to determine what is important. A pull quote (a phrase or sentence taken from an article) adds interest to a document while simultaneously emphasizing a key point. Boxes and shading are also effective in catching the reader's attention. A simple change in typography, such as increasing the point size, changing the typeface, and/or the use of boldface or italics, calls attention to a heading and visually separates it from the associated text.

Step 7: **Create a Drop Cap**

- Scroll to the beginning of the newsletter. Click immediately before the D in *Desktop publishing*.
- Pull down the **Format menu**. Click the **Drop Cap command** to display the dialog box in Figure 5.7g.
- Click the **Position icon** for **Dropped** as shown in the figure. We used the default settings, but you can change the font, size (lines to drop), or distance from the text by clicking the arrow on the appropriate list box.
- Click **OK** to create the Drop Cap dialog box. Click outside the frame around the drop cap.
- Save the newsletter.

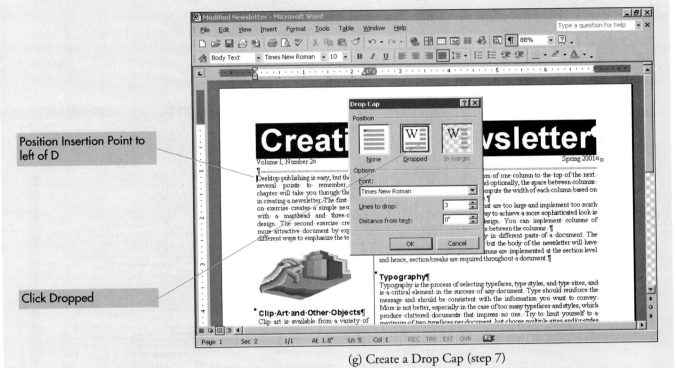

(g) Create a Drop Cap (step 7)

FIGURE 5.7 *Hands-on Exercise 2 (continued)*

MODIFYING A DROP CAP

Select (click) a dropped-capital letter to display a thatched border known as a frame, then click the border or frame to display its sizing handles. You can move and size a frame just as you can any other Windows object; for example, click and drag a corner sizing handle to change the size of the frame (and the drop cap it contains). Experiment with different fonts to increase the effectiveness of the dropped-capital letter, regardless of its size. To delete the frame (and remove the drop cap), press the delete key.

Step 8: **Create the AutoShape**

- Click the **Show/Hide button** to hide the nonprinting characters. Pull down the **View menu**, click (or point to) the **Toolbars command** to display the list of available toolbars, then click the **Drawing toolbar** to display this toolbar.
- Press **Ctrl+End** to move to the end of the document. Click the **down arrow** on the AutoShapes button to display the AutoShapes menu. Click the **Stars and Banners submenu** and select (click) the **Horizontal scroll**.
- Press **Esc** to remove the drawing canvas. The mouse pointer changes to a tiny crosshair. Click and drag the mouse at the bottom of the newsletter to create the scroll as shown in Figure 5.7h.
- Release the mouse. The scroll is still selected as can be seen by the sizing handles. (You can click and drag the yellow diamond to change the thickness of the scroll.)
- Click the **Line Style tool** to display this menu as shown in Figure 5.7h. Select a thicker line (we chose **3 points**). Click the **down arrow** on the **Line color tool** to display the list of colors (if you have access to a color printer. We selected **blue**).

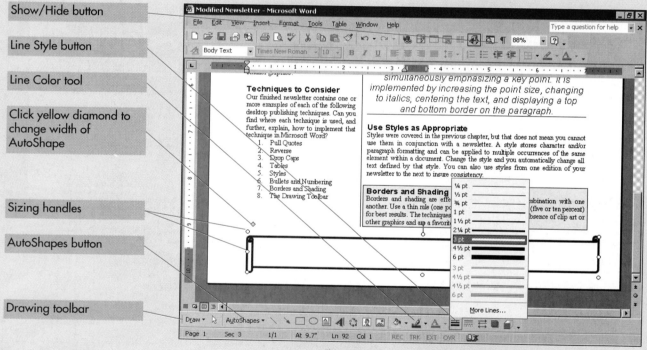

(h) Create the AutoShape (step 8)

FIGURE 5.7 *Hands-on Exercise 2 (continued)*

DISPLAY THE AUTOSHAPE TOOLBARS

Click the down arrow on the AutoShapes button on the Drawing toolbar to display a cascaded menu listing the various types of AutoShapes, then click and drag the menu's title bar to display the menu as a floating toolbar. Click any tool on the AutoShapes toolbar (such as Stars and Banners), then click and drag its title bar to display the various stars and banners in their own floating toolbar.

Step 9: **Create the Text Box**

> ➤ Click the **Text Box tool**, then click and drag within the banner to create a text box as shown in Figure 5.7i. Type **All the News that Fits** as the text of the banner. Click the **Center button** on the Formatting toolbar.
> ➤ Click and drag to select the text, click the **down arrow** on the **Font Size list box**, and select a larger point size (22 or 24 points). If necessary, click and drag the bottom border of the text box, and/or the bottom border of the AutoShape, in order to see all of the text. Click the **down arrow** on the **Font list box** and choose a different font.
> ➤ Right click the text box to display a context-sensitive menu, then click the **Format Text Box command** to display the **Format Text Box dialog** box as shown in Figure 5.7i. Click the **Colors and Lines tab** (if necessary), click the **down arrow** next to Color in the Line section, click **No Line**, then click **OK** to accept the settings and close the dialog box.
> ➤ Click anywhere in the document to deselect the text box. Save the document.

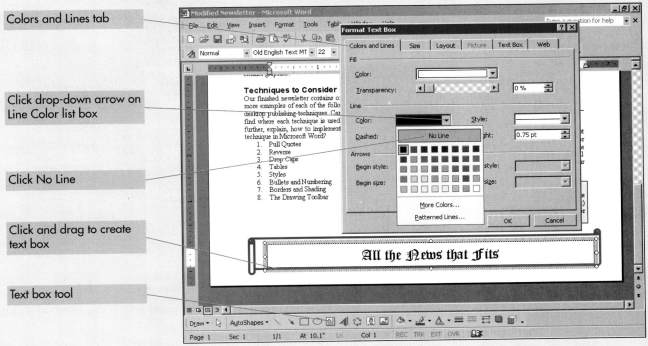

(i) Create the Text Box (step 9)

FIGURE 5.7 *Hands-on Exercise 2 (continued)*

DON'T FORGET WORDART

Microsoft WordArt is another way to create decorative text to add interest to a document. Pull down the Insert menu, click Picture, click WordArt, choose the WordArt style, and click OK. Enter the desired text, then click OK to create the WordArt object. You can click and drag the sizing handles to change the size or proportion of the text. Use any tool on the WordArt toolbar to further change the appearance of the object.

Step 10: **The Completed Newsletter**

➤ Zoom to **Whole Page** to view the completed newsletter as shown in Figure 5.7j. The newsletter should fit on a single page, but if not, there are several techniques that you can use:
- Pull down the **File menu**, click the **Page Setup command**, click the **Margins tab**, then reduce the top and/or bottom margins to .5 inch. Be sure to apply this change to the **Whole document** within the Page Setup dialog box.
- Change the **Heading 1 style** to reduce the point size to **10 points** and/or the space before the heading to **6 points**.
- Click the **Print Preview button** on the Standard toolbar, then click the **Shrink to Fit button** on the Print Preview toolbar.
- Save the document a final time. Print the completed newsletter and submit it to your instructor as proof you did this exercise. Congratulations on a job well done.

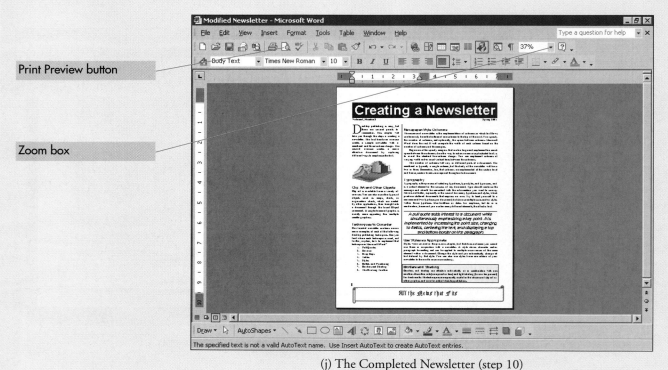

(j) The Completed Newsletter (step 10)

FIGURE 5.7 *Hands-on Exercise 2 (continued)*

A FINAL WORD OF ADVICE

Desktop publishing is not a carefree operation. It is time-consuming to implement, and you will be amazed at the effort required for even a simple document. Computers are supposed to save time, not waste it, and while desktop publishing is clearly justified for some documents, the extensive formatting is not necessary for most documents. And finally, remember that the content of a document is its most important element.

OBJECT LINKING AND EMBEDDING

Microsoft Office enables you to create documents that contain data (objects) from multiple applications. The document in Figure 5.8a, for example, was created in Microsoft Word, but it contains objects (a worksheet and a chart) that were developed in ***Microsoft Excel***. ***Object Linking and Embedding*** (***OLE***, pronounced "oh-lay") is the means by which you create the document.

Every Excel chart is based on numerical data that is stored in a worksheet. Figures 5.8b and 5.8c enlarge the worksheet and chart that appear in the document of Figure 5.8a. The worksheet shows the quarterly sales for each of three regions, East, West, and North. There are 12 ***data points*** (four quarterly values for each of three regions). The data points are grouped into ***data series*** that appear as rows or columns in the worksheet. (The chart was created through the Chart Wizard that prompts you for information about the source data and the type of chart you want. Any chart can be subsequently modified by choosing appropriate commands from the Chart menu.)

The data in the chart is plotted by rows or by columns, depending on the message you want to convey. Our data is plotted by rows to emphasize the amount of sales in each quarter, as opposed to the sales in each region. Note that when the data is plotted by rows, the first row in the worksheet will appear on the X axis of the chart, and the first column will appear as the legend. Conversely, if you plot the data by columns, the first column appears on the X axis, and the first row appears as a legend.

Look closely at Figures 5.8b and 5.8c to see the correspondence between the worksheet and the chart. The data is plotted by rows. Thus there are three rows of data (three data series), corresponding to the values in the Eastern, Western, and Northern regions, respectively. The entries in the first row appear on the X axis. The entries in the first column appear as a legend to identify the value of each column in the chart. The chart is a ***side-by-side column chart*** that shows the value of each data point separately. You could also create a ***stacked column chart*** for each quarter that would put the columns one on top of another. And, as with the stacked-column chart, you have your choice of plotting the data in rows or columns.

After the chart has been created, it is brought into the Word document through Object Linking and Embedding. The essential difference between linking and embedding depends on where the object is stored. An embedded object is physically within the Word document. A ***linked object***, however, is stored in its own file, which may in turn be tied to many documents. The same Excel chart, for example, can be linked to a Word document and a PowerPoint presentation or to multiple Word documents and/or to multiple presentations. Any changes to a linked object (the Excel chart) are automatically reflected in all of the documents to which it is linked. An ***embedded object***, however, is stored within the Word document and it is no longer tied to its source. Thus, any changes made in the original object or in the embedded object are not reflected in one another.

EMPHASIZE YOUR MESSAGE

A graph exists to deliver a message, and you want that message to be as clear as possible. One way to help put your point across is to choose a title that leads the audience. A neutral title such as *Sales Data* does nothing and requires the audience to reach its own conclusion. A better title might be *Eastern Region Has Record 3rd Quarter* to emphasize the results in the individual sales offices. Conversely, *Western Region Has a Poor Year* conveys an entirely different message. This technique is so simple that we wonder why it isn't used more frequently.

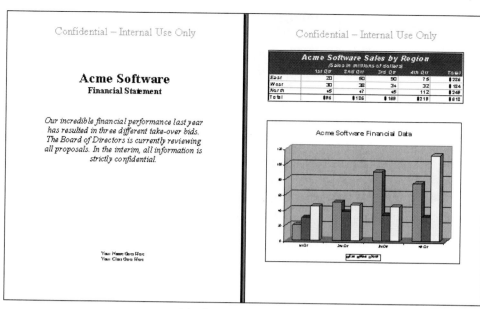

(a) The Word Document

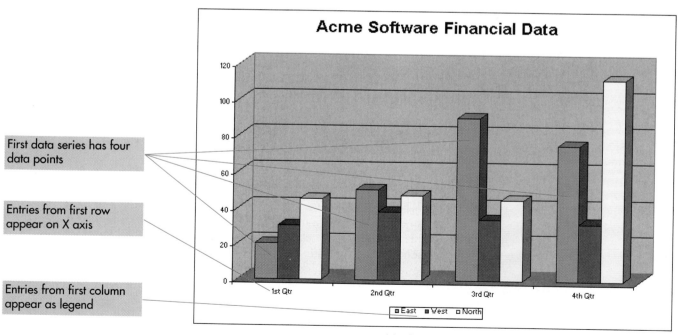

(b) The Excel Worksheet

(c) Alternate Chart

FIGURE 5.8 *Object Linking and Embedding*

HANDS-ON EXERCISE 3

Object Linking and Embedding

Objective Use object linking to create a Word document that contains an Excel worksheet and an Excel chart. Use Figure 5.9 as a guide in the exercise.

Step 1: **Create the Title Page**

- ➤ Start Word. Click the **New button** on the Standard toolbar to open a new document. Press the **enter key** several times, then enter the title of the document, **Acme Software Financial Statement**, **your name**, and the **course number** with appropriate formatting. Save the document as **Confidential Memo**.
- ➤ Click the **Print Layout View button** above the status bar, then click the **down arrow** on the Zoom list box and select **Two Pages**. Your document takes only a single page, however, as shown in Figure 5.9a.
- ➤ Pull down the **View menu** and click the **Header and Footer command** to display the Header and Footer toolbar. The text in the document (its title, your name, and class) is dim since you are working in the header and footer area of the document.
- ➤ Click the **down arrow** on the Font Size box and change to **28 points**. Click inside the header and enter **Confidential - Internal Use Only**. Center the text.
- ➤ Click the **Close button** on the Header and Footer toolbar to close the toolbar. The header you just created is visible, but dim.

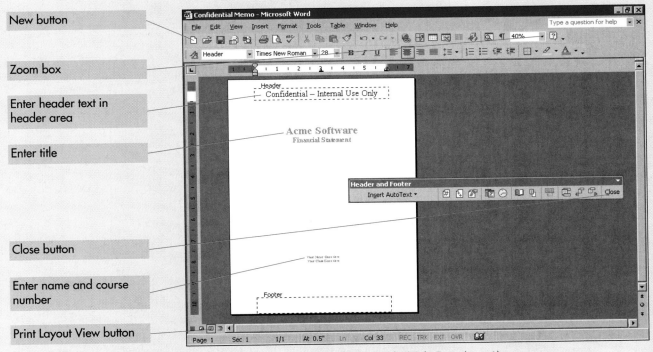

(a) Create the Title Page (step 1)

FIGURE 5.9 *Hands-on Exercise 3*

Step 2: **Copy the Worksheet**

➤ Click the **Start button**, click **Programs,** then click **Microsoft Excel** to start Excel. The taskbar now contains buttons for both Word and Excel. Click either button to move back and forth between the open applications. End in Excel.

➤ Pull down the **File menu** and click the **Open command** (or click the **Open button** on the Standard toolbar) to display the Open dialog box.

➤ Click the **down arrow** on the Look in list box to select the Exploring Word folder that you have used throughout the text. Open the **Acme Software workbook**.

➤ Click the **Sales Data** worksheet tab. Click and drag to select **cells A1 through F7** as shown in Figure 5.9b.

➤ Pull down the **Edit menu** and click the **Copy command** (or click the **Copy button** on the Standard toolbar). A moving border appears around the entire worksheet, indicating that it has been copied to the clipboard.

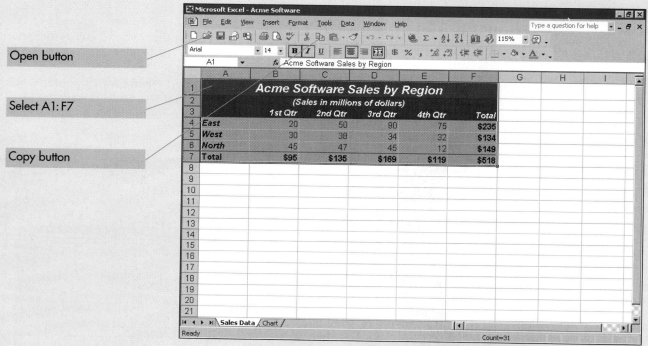

(b) Copy the Worksheet (step 2)

FIGURE 5.9   *Hands-on Exercise 3 (continued)*

THE COMMON USER INTERFACE

The common user interface provides a sense of familiarity from one Office application to the next. Even if you have never used Microsoft Excel, you will recognize many of the elements that are present in Word. The applications share a common menu structure with consistent ways to execute commands from those menus. The Standard and Formatting toolbars are present in both applications. Many keyboard shortcuts are also common; for example: Ctrl+X, Ctrl+C, and Ctrl+V to cut, copy, and paste, respectively.

Step 3: **Create the Link**

- Click the **Word button** on the taskbar to return to the document as shown in Figure 5.9c. Press **Ctrl+End** to move to the end of the document, which is where you will insert the Excel worksheet.
- Press **Ctrl+Enter** to create a page break, which adds a second page to the document. This page is blank except for the header, which appears automatically.
- Pull down the **Edit menu** and click **Paste Special** to display the dialog box in Figure 5.9c. Select **Microsoft Excel Worksheet Object**. Click the **Paste Link** option button. Click **OK** to insert the worksheet into the document.
- Do not be concerned about the size or position of the worksheet at this time. Press the **enter key** twice to create a blank line between the worksheet and the chart, which will be added later.
- Save the document.

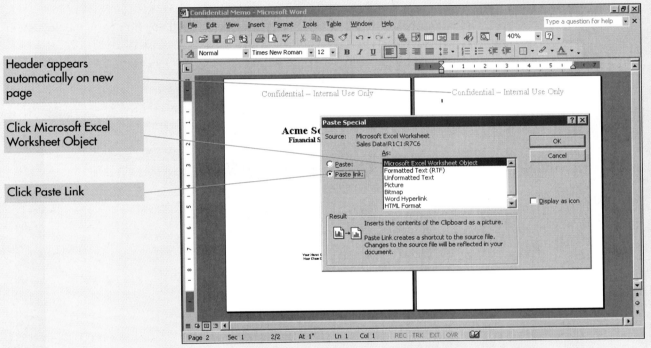

(c) Create the Link (step 3)

FIGURE 5.9 *Hands-on Exercise 3 (continued)*

THE WINDOWS TASKBAR

Multitasking, the ability to run multiple applications at the same time, is one of the primary advantages of the Windows environment. Each button on the taskbar appears automatically when its application or folder is opened and disappears upon closing. (The buttons are resized automatically according to the number of open windows.) You can customize the taskbar by right clicking an empty area to display a shortcut menu, then clicking the Properties command. You can resize the taskbar by pointing to the inside edge and then dragging when you see the double-headed arrow. You can also move the taskbar to the left or right edge, or to the top of the desktop, by dragging a blank area of the taskbar to the desired position.

Step 4: **Format the Object**

- Point to the newly inserted worksheet, click the **right mouse button** to display a context-sensitive menu, then click the **Format Object command** to display the dialog box in Figure 5.9d.
- Click the **Layout tab** and choose **Square**. Click the option button to **Center** the object. You can now move and size the object.
- Select (click on) the worksheet to display its sizing handles. Click and drag a corner sizing handle to enlarge the worksheet, keeping it in its original proportions.
- Click and drag any element except a sizing handle to move the worksheet within the document.
- Right click the worksheet a second time and click the **Format Object command** to display the associated dialog box. Click the **Colors and lines tab**, click the **drop-down arrow** next to color in the line area and choose **black**. Click the **Spin button** next to weight and choose **.25**.
- Click **OK** to accept these settings and close the dialog box. Save the document.

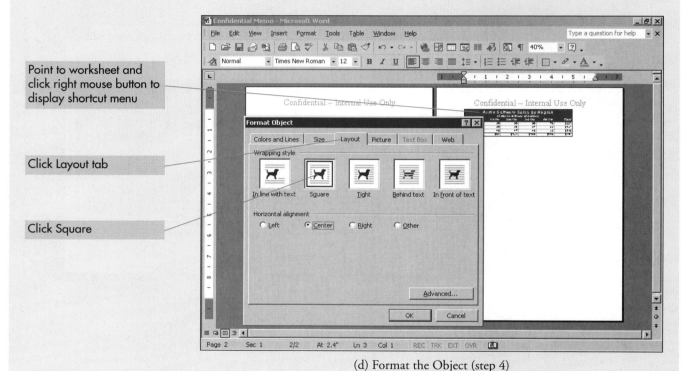

(d) Format the Object (step 4)

FIGURE 5.9 *Hands-on Exercise 3 (continued)*

TO CLICK OR DOUBLE CLICK

An Excel chart that is linked or embedded into a Word document retains its connection to Microsoft Excel for easy editing. Click the chart to select it within the Word document, then move and size the chart just as any other object. (You can also press the Del key to delete the graph from a document.) Click outside the chart to deselect it, then double click the chart to restart Microsoft Excel (the chart is bordered by a hashed line), at which point you can edit the chart using the tools of the original application.

Step 5: **Copy the Chart**

- Click the **Excel button** on the taskbar to return to the worksheet. Click outside the selected area (cells A1 through F7) to deselect the cells.
- Click the **Chart tab** to select the chart sheet. Point just inside the border of the chart, then click the left mouse button to select the chart. Be sure you have selected the entire chart as shown in Figure 5.9e.
- Pull down the **Edit menu** and click **Copy** (or click the **Copy button** on the Standard toolbar). Once again you see the moving border, indicating that the selected object (the chart in this example) has been copied to the clipboard.
- Click the **Word button** on the taskbar to return to the document.

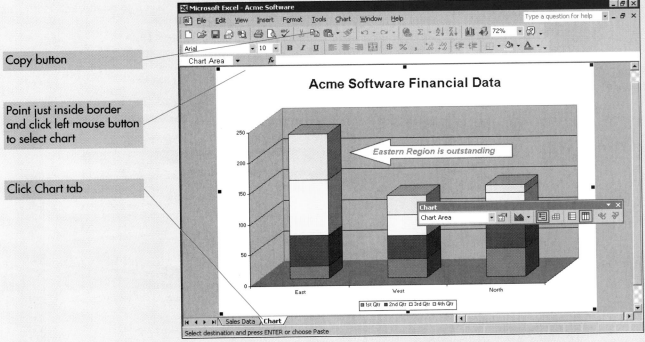

(e) Copy the Chart (step 5)

FIGURE 5.9 *Hands-on Exercise 3 (continued)*

KEEP IT SIMPLE

Microsoft Excel provides unlimited flexibility with respect to the charts it creates. You can, for example, right click any data series within a graph and click the Format Data Series command to change the color, fill pattern, or shape of a data series. There are other options, such as the 3-D View command that lets you fine-tune the graph by controlling the rotation, elevation, and other parameters. It's fun to experiment, but the best advice is to keep it simple and set a time limit, at which point the project is finished. Use the Undo command at any time to cancel your last action(s).

Step 6: **Complete the Word Document**

- You should be back in the Word document, where you may need to insert a few blank lines, so that the insertion point is beneath the spreadsheet. Press **Ctrl+End** to move to the end of the document, where you will insert the chart.
- Pull down the **Edit menu**, click the **Paste Special command**, and click the **Paste Link** option button. If necessary, click **Microsoft Excel Chart Object**.
- Click **OK** to insert the chart into the document. (Do not be concerned if you do not see the entire chart.)
- Select the chart, pull down the **Format menu**, and click the **Object command**.
- Select the **Layout tab** and change the layout to **Square**. **Center** the chart. Select the **Color and Lines tab**, and add a **.25" black line**.
- Move and size the chart as shown in Figure 5.9f. Save the document. Print this version of the document for your instructor.

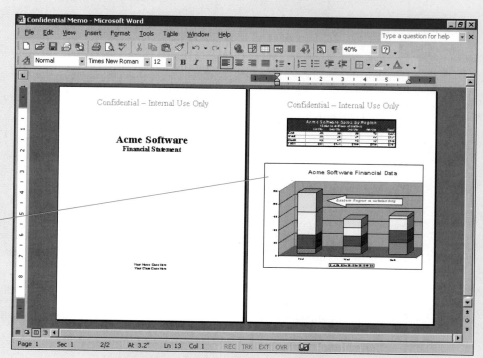

(f) Complete the Word Document (step 6)

FIGURE 5.9 *Hands-on Exercise 3 (continued)*

LINKING VERSUS EMBEDDING

The Paste Special command will link or embed an object, depending on whether the Paste Link or Paste Option button is checked. Linking stores a pointer to the file containing the object together with a reference to the server application, and changes to the object are automatically reflected in all documents that are linked to the object. Embedding stores a copy of the object with a reference to the server application, but changes to the object are not reflected in the document that contains the embedded (rather than linked) object.

Step 7: **Modify the Chart**

> ► Click the **Excel button** on the taskbar to return to Excel. Click the **Sales Data tab** to return to the worksheet.
> ► Click in **cell E6**, the cell containing the sales data for the Northern region in the fourth quarter. Type **112**, then press **enter**. The sales totals for the region and quarter change to 249 and 219, respectively.
> ► Click the tab for the chart sheet. The chart has changed automatically to reflect the change in the underlying data. The bars for the Eastern and Northern regions are approximately the same size.
> ► Pull down the **Chart menu** and click the **Chart Type command** to display the Chart Type dialog box. Click the **Standard Types tab**. Select the **Clustered Column Chart with 3D Visual Effect** (the first chart in the second row).
> ► Click **OK** to accept this chart type and close the dialog box. The chart type changes to side-by-side columns as shown in Figure 5.9g.
> ► Select the arrow on the chart. Press the **Del key** since the text is no longer applicable. Save the workbook.

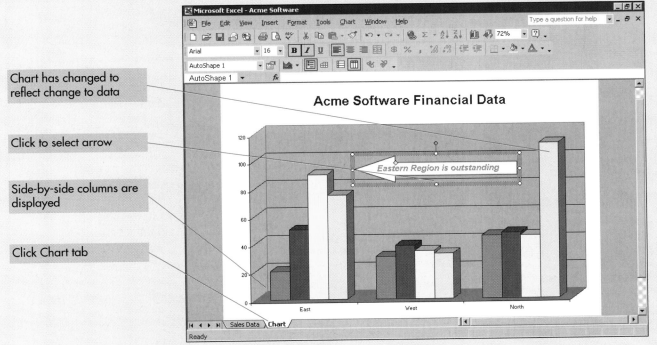

(g) Modify the Chart (step 7)

FIGURE 5.9 *Hands-on Exercise 3 (continued)*

THE DRAWING TOOLBAR

The Drawing toolbar is common to all applications in Microsoft Office. Click the down arrow next to the AutoShapes button to display the various shapes, then click Block Arrows to display the arrows that are available. Select an arrow. The mouse pointer changes to a tiny crosshair that you click and drag to create the arrow within the document. Right click the arrow, then click the Add Text command to enter text within the arrow. Use the other buttons to change the color or other properties.

Step 8: **The Modified Document**

- Click the **Word button** on the taskbar to return to the Word document, which should automatically reflect the new chart. (If this is not the case, right click the chart and click the **Update Link command**.)
- Move and/or resize the chart and spreadsheet as necessary. Save the document.
- Complete the document by adding text as appropriate as shown in Figure 5.9h. You can use the text in our document that describes a confidential takeover, or make up your own.
- Place a border around the chart. Click the chart to select it, pull down the **Format menu**, click the **Object command**, and click the **Colors and Lines tab**.
- Click the **down arrow** on the Color text box (in the Line section of the dialog box) and select a line color. Click **OK**. Click elsewhere in the document to deselect the graph.
- Save the document a final time. Click the **Print button** on the Standard toolbar to print the document for your instructor.

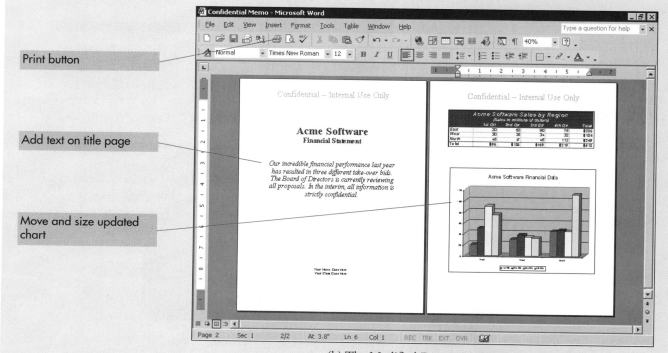

(h) The Modified Document (step 8)

FIGURE 5.9 *Hands-on Exercise 3 (continued)*

ALT+TAB STILL WORKS

Alt+Tab was a treasured shortcut in Windows 3.1 that enabled users to switch back and forth between open applications. The shortcut also works in all subsequent versions of Windows. Press and hold the Alt key while you press and release the Tab key repeatedly to cycle through the open applications, whose icons are displayed in a small rectangular window in the middle of the screen. Release the Alt key when you have selected the icon for the application you want.

SUMMARY

The essence of desktop publishing is the merger of text with graphics to produce a professional-looking document. Proficiency in desktop publishing requires knowledge of the associated commands in Microsoft Word, as well as familiarity with the basics of graphic design.

Typography is the process of selecting typefaces, type styles, and type sizes. A typeface (or font) is a complete set of characters (upper- and lowercase letters, numbers, punctuation marks, and special symbols). Type size is a vertical measurement and is specified in points. One point is equal to $\frac{1}{72}$ of an inch.

The design of a document is developed on a grid, an underlying but invisible set of horizontal and vertical lines that determine the placement of the major elements. A newsletter can be divided into any number of newspaper-style columns in which text flows from the bottom of one column to the top of the next. Columns are implemented by clicking the Columns button on the Standard toolbar or by selecting the Columns command from the Format menu. Sections are required if different column arrangements are present in the same document. The Page Layout view is required to see the columns displayed side by side.

Emphasis can be achieved in several ways, the easiest being variations in type size and/or type style. Boxes and/or shading call attention to selected articles in a document. Horizontal lines are effective in separating one topic from another or calling attention to a pull quote (a phrase or sentence taken from an article to emphasize a key point). A reverse (light text on a solid background) is striking for a small amount of text.

Clip art is available from a variety of sources, including the Microsoft Media Gallery, which is accessed most easily through the Insert Picture command. Once clip art has been inserted into a document, it can be moved and sized just like any other Windows object. The Format Picture command provides additional flexibility and precision in the placement of an object. The Drawing toolbar contains various tools that are used to insert and/or modify objects into a Word document.

Graphic design does not have hard and fast rules, only guidelines and common sense. Creating an effective document is an iterative process and reflects the result of trial and error. We encourage you to experiment freely with different designs.

Object linking and embedding enables the creation of a document containing data (objects) from multiple applications. The essential difference between linking and embedding is whether the object is stored within the document (embedding) or stored within its own file (linking). The advantage of linking is that any changes to the linked object are automatically reflected in every document that is linked to that object.

KEY TERMS

Arial (p. 212)
AutoShape (p. 234)
AutoShapes toolbar (p. 234)
Borders and Shading (p. 210)
Bulleted list (p. 210)
Clip art (p. 210)
Columns command (p. 213)
Common User Interface (p. 240)
Data points (p. 237)
Data series (p. 237)
Desktop publishing (p. 210)
Drawing toolbar (p. 210)
Dropped-capital letter (p. 210)
Embedded object (p. 237)
Emphasis (p. 224)
Font (p. 212)
Format Picture command (p. 224)
Grid (p. 222)
Insert Picture command (p. 224)
Linked object (p. 237)
Masthead (p. 210)
Microsoft Excel (p. 237)
Monospaced typeface (p. 212)
Numbered list (p. 210)
Object linking and embedding (OLE) (p. 237)
Paste Special command (p. 241)
Proportional typeface (p. 212)
Pull quote (p. 210)
Reverse (p. 210)
Sans serif typeface (p. 212)
Section break (p. 213)
Serif typeface (p. 212)
Side-by-side column chart (p. 237)
Stacked column chart (p. 237)
Text box (p. 235)
Times New Roman (p. 212)
Type size (p. 212)
Typeface (p. 212)
Typography (p. 212)

MULTIPLE CHOICE

1. Which of the following is a commonly accepted guideline in typography?
   (a) Use a serif typeface for headings and a sans serif typeface for text
   (b) Use a sans serif typeface for headings and a serif typeface for text
   (c) Use a sans serif typeface for both headings and text
   (d) Use a serif typeface for both headings and text

2. Which of the following best enables you to see a multicolumn document as it will appear on the printed page?
   (a) Normal view at 100% magnification
   (b) Normal view at whole page magnification
   (c) Print Layout view at 100% magnification
   (d) Print Layout view at whole page magnification

3. What is the width of each column in a document with two uniform columns, given 1¼-inch margins and ½-inch spacing between the columns?
   (a) 2½ inches
   (b) 2¾ inches
   (c) 3 inches
   (d) Impossible to determine

4. What is the minimum number of sections in a three-column newsletter whose masthead extends across all three columns, with text balanced in all three columns?
   (a) One
   (b) Two
   (c) Three
   (d) Four

5. Which of the following describes the Arial and Times New Roman fonts?
   (a) Arial is a sans serif font, Times New Roman is a serif font
   (b) Arial is a serif font, Times New Roman is a sans serif font
   (c) Both are serif fonts
   (d) Both are sans serif fonts

6. How do you balance the columns in a newsletter so that each column contains the same amount of text?
   (a) Check the Balance Columns box in the Format Columns command
   (b) Visually determine where the break should go, then insert a column break at the appropriate place
   (c) Insert a continuous section break at the end of the last column
   (d) All of the above

7. What is the effect of dragging one of the four corner handles on a selected object?
   (a) The length of the object is changed but the width remains constant
   (b) The width of the object is changed but the length remains constant
   (c) The length and width of the object are changed in proportion to one another
   (d) Neither the length nor width of the object is changed

8. Which type size is the most reasonable for columns of text, such as those appearing in the newsletter created in the chapter?
   (a) 6 point
   (b) 10 point
   (c) 14 point
   (d) 18 point

9. A grid is applicable to the design of:
    (a) Documents with one, two, or three columns and moderate clip art
    (b) Documents with four or more columns and no clip art
    (c) Both (a) and (b)
    (d) Neither (a) nor (b)

10. Which of the following can be used to add emphasis to a document?
    (a) Borders and shading
    (b) Pull quotes and reverses
    (c) Both (a) and (b)
    (d) Neither (a) nor (b)

11. Which of the following is a recommended guideline in the design of a typical newsletter?
    (a) Use at least three different clip art images in every newsletter
    (b) Use at least three different typefaces in a document to maintain interest
    (c) Use the same type size for the heading and text of an article
    (d) None of the above

12. Which of the following is implemented at the section level?
    (a) Columns
    (b) Margins
    (c) Both (a) and (b)
    (d) Neither (a) nor (b)

13. How do you size an object so that it maintains the original proportion between height and width?
    (a) Drag a sizing handle on the left or right side of the object to change its width, then drag a sizing handle on the top or bottom edge to change the height
    (b) Drag a sizing handle on any of the corners
    (c) Both (a) and (b)
    (d) Neither (a) nor (b)

14. A reverse is implemented:
    (a) By selecting 100% shading in the Borders and Shading command
    (b) By changing the Font color to black
    (c) Both (a) and (b)
    (d) Neither (a) nor (b)

15. The Format Picture command enables you to:
    (a) Change the way in which text is wrapped around a figure
    (b) Change the size of a figure
    (c) Place a border around a figure
    (d) All of the above

ANSWERS

1. b	**6.** c	**11.** d
2. d	**7.** c	**12.** c
3. b	**8.** b	**13.** b
4. c	**9.** c	**14.** a
5. a	**10.** c	**15.** d

PRACTICE WITH WORD

1. **Study Tips:** Create a simple newsletter similar to the document in Figure 5.10. There is no requirement to write meaningful text, but the headings in the newsletter should follow the theme of the graphic. The intent of this problem is to provide practice in graphic design.
   a. Develop an overall design away from the computer—that is, with pencil and paper. Use a grid to indicate the placement of the articles, headings, clip art, and masthead. You may be surprised to find that it is easier to master commands in Word than it is to design the newsletter; do not, however, underestimate the importance of graphic design in the ultimate success of every document you create.
   b. Use meaningful headings to give the document a sense of realism. The text under each heading can be a single sentence that repeats indefinitely to take up the allotted space. Your eye is the best judge of all, and you may need to decrease the default space between columns and/or change the type size to create an appealing document.
   c. Insert clip art to add interest to your document, then write one or two sentences in support of the clip art. Use the clip art within Microsoft Office or any other clip art you have available. You can also download pictures from the Web, but be sure to credit the source. The image you choose should be related to studying in some way.
   d. More is not better; that is, do not use too many fonts, styles, sizes, and clip art just because they are available. Don't crowd the page, and remember white space is a very effective design element. There are no substitutes for simplicity and good taste.
   e. Submit the completed newsletter to your instructor for inclusion in a class contest. Your instructor might want to select the five best designs as semifinalists and let the class vote on the overall winner.

FIGURE 5.10 *Study Tips (Exercise 1)*

250   CHAPTER 5: DESKTOP PUBLISHING

BUILDS ON
HANDS-ON
EXERCISE 2
PAGES 227–236

2. Alternate Design: The document in Figure 5.11 contains the same text as the newsletter that was developed in the chapter, but it is formatted differently. Start with the original text of the newsletter and implement a new design. You can match our formatting, or better yet, develop your own design. The completed newsletter must fit on a single page. Clip art is optional. Submit the completed newsletter to your instructor for inclusion in a class contest to pick the best design.

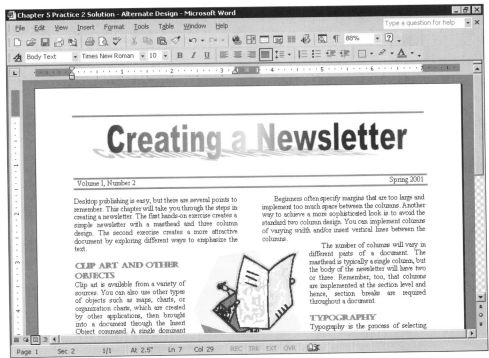

FIGURE 5.11   *Alternate Design (Exercise 2)*

3. A Guide for Smart Shopping: This problem is more challenging than the previous exercises in that you are asked to consider content as well as design. The objective is to develop a one- or two-page document with helpful tips to the novice on buying a computer as shown in Figure 5.12. We have, however, written the copy for you and put the file on the data disk.
   a. Open and print the *Chapter 5 Practice 3 document* on the data disk, which takes approximately a page and a half as presently formatted. Read our text and determine the tips you want to retain and those you want to delete. Add other tips as you see fit.
   b. Examine the available clip art through the Insert Picture command or through the Microsoft Media Gallery. There is no requirement, however, to include a graphic; that is, use clip art only if you think it will enhance the effectiveness of your document.
   c. Use an imaginary grid to develop a rough sketch of the completed document showing the masthead and the placement of the text and clip art. Do this away from the computer.
   d. Implement your design in Microsoft Word. Try to create a balanced publication, which completely fills the space allotted; that is, your document should take exactly one or two pages (rather than the page and a half in the original document on the data disk).
   e. Submit the completed newsletter to your instructor for inclusion in a class contest. Your instructor might want to select the five best designs as semifinalists and let the class vote on the overall winner.

FIGURE 5.12   *A Guide for Smart Shopping (Exercise 3)*

4. **The Equation Editor:** Microsoft Office includes several shared applications, each of which creates an object that can be inserted into a Word document. (WordArt and the Microsoft Media Gallery are both shared applications.) The newsletter in Figure 5.13 illustrates the Equation Editor, a shared application of particular use to math majors.
   a. Create a simple newsletter such as the two-column design in Figure 5.13. There is no requirement to write meaningful text, as the intent of this exercise is to illustrate the Equation Editor. Thus, all you need to do is write a sentence or two, then copy that sentence so that it fills the newsletter.
   b. To create the equation, pull down the Insert menu, click the Object command, click the Create New tab, then select Microsoft Equation to start the Equation Editor. This is a new application, and we do not provide instruction in its use. It does, however, follow the conventions of other Office applications, and with trial and error, and reference to the Help menu, you should be able to duplicate our equation.
   c. Once the equation (object) has been created, you can move and size it within the document. Clicking an object selects the object and displays the sizing handles to move, size, or delete the object. Double clicking an object loads the application that created it, and enables you to modify the object using the tools of the original application.

5. **The Flyer:** Any document can be enhanced through Microsoft WordArt, an application included in Microsoft Office to create decorative text. Pull down the Insert menu, click Picture, click WordArt, choose the desired style, then click OK. Enter the desired text, then click OK to insert the WordArt object into your document. Use the Format Object command to change to a square layout, then move and size the WordArt within the document.
   Create at least two flyers, consisting of text, clip art, and WordArt, similar to the flyers in Figure 5.14. You can duplicate the flyers in our figure, or better yet, create your own. The flyers can describe a real or hypothetical event. Submit your work to your instructor.

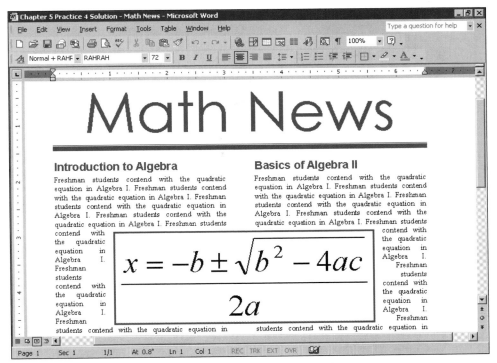

FIGURE 5.13 *The Equation Editor (Exercise 4)*

FIGURE 5.14 *The Flyer (Exercise 5)*

6. **My Favorite Car:** The document in Figure 5.15 consists of descriptive text, a photograph, and a worksheet to compute a car payment. We have created the spreadsheet, but you will have to obtain the other information via the Web.
   a. Choose any vehicle you like, then go to the Web to locate a picture of your vehicle together with descriptive material. *Be sure to credit your source in the completed document.* Start a new Word document. Enter a title for the document, then insert the photograph and descriptive information. Do not worry about the precise formatting at this time.
   b. Open the *Chapter 5 Practice 6 workbook* that is found in the Exploring Word folder. Enter the necessary information for your vehicle. The monthly payment will be computed automatically, based on the amount you are borrowing, the interest rate, and the term of your loan. Save the workbook.
   c. Click and drag to select the appropriate cells within the worksheet, return to Word, and use object linking and embedding to include this information in the document.
   d. Move and size the various objects to complete the document. Add your name and submit the completed document to your instructor.

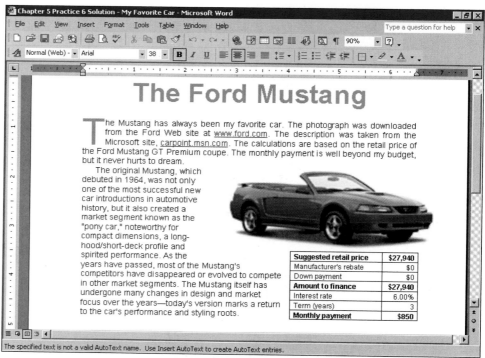

FIGURE 5.15 *My Favorite Car (Exercise 6)*

7. **PowerPoint Presentation:** The third hands-on exercise described how to create a Word document that contains an Excel worksheet and associated chart. Object linking and embedding can also be used to create a PowerPoint presentation similar to the examples in Figure 5.16. Moreover, the same object can be linked to both the PowerPoint presentation and the Word document, so that any changes to the worksheet are automatically reflected in both documents.
   a. Create a PowerPoint presentation similar to Figure 5.16a that is based on the worksheet from the third hands-on exercise. Add text and clip art as appropriate. You can duplicate our presentation or you can create your own.
   b. Create a second presentation similar to Figure 5.16b that is based on the Ford Mustang example in practice exercise 6.

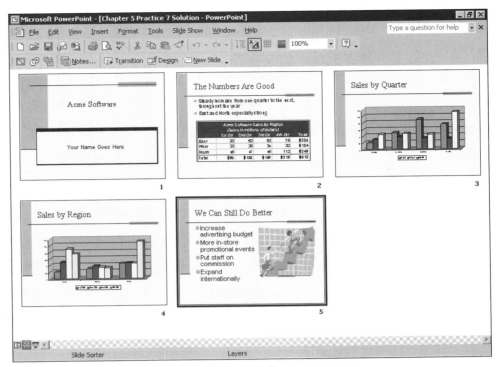

(a) Acme Software

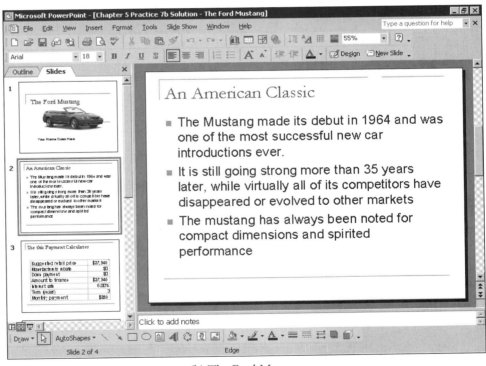

(b) The Ford Mustang

FIGURE 5.16 *PowerPoint Presentations (Exercise 7)*

ON YOUR OWN

Before and After

The best way to learn about the do's and don'ts of desktop publishing is to study the work of others. Choose a particular type of document—for example, a newsletter, résumé, or advertising flyer, then collect samples of that document. Choose one sample that is particularly bad and redesign the document. You need not enter the actual text, but you should keep all of the major headings so that the document retains its identity. Add or delete clip art as appropriate. Bring the before and after samples to class for your professor.

Clip Art

Clip art—you see it all the time, but where do you get it, and how much does it cost? Scan the computer magazines and find at least two sources for additional clip art. Better yet, use your favorite search engine to locate additional sources of clip art on the Web. Return to class with specific information on prices and types of the clip art.

Color Separations

It's difficult to tell where word processing stops and desktop publishing begins. One distinguishing characteristic of a desktop publishing program, however, is the ability to create color separations, which in turn enable you to print a document in full color. Use your favorite search engine to learn more about the process of color separations. Summarize the results of your research in a short paper to your instructor.

Photographs versus Clip Art

The right clip art can enhance any document, but there are times when clip art just won't do. It may be too juvenile or simply inappropriate. Photographs offer an alternative and are inserted into a presentation through the Insert Picture command. Once inserted into a presentation, photographs can be moved or sized just like any other Windows object. Use your favorite search engine to locate a photograph, then incorporate that photograph into the newsletter that was developed in this chapter.

Subscribe to a Newsletter

Literally thousands of regularly published newsletters are distributed in printed and/or electronic form. Some charge a subscription fee, but many are available just for the asking. Use your favorite search engine to locate a free newsletter in an area of interest to you. Download an issue, then summarize the results of your research in a brief note to your instructor.

CHAPTER 6

Introduction to HTML: Creating a Home Page and a Web Site

OBJECTIVES

AFTER READING THIS CHAPTER YOU WILL BE ABLE TO:

1. Define HTML and its role on the World Wide Web; describe HTML codes and explain how they control the appearance of a Web document.
2. Use the Insert Hyperlink command to include hyperlinks, bookmarks, and/or an e-mail address in a Word document.
3. Use the Save As Web page command to convert a Word document to HTML.
4. Use the Format Theme command to enhance the appearance of a Web document.
5. Use an FTP program to upload a document to a Web server; add a Web page to the catalog of a search engine.
6. Explain how to view HTML codes from within Internet Explorer; describe the use of the Telnet program that is built into Windows.
7. Use the Web Page Wizard to create a Web site with multiple pages.
8. Explain how the use of frames and/or bookmarks facilitates navigation between multiple documents.

OVERVIEW

Sooner or later anyone who cruises the World Wide Web wants to create a home page and/or a Web site of their own. That, in turn, requires an appreciation for *HyperText Markup Language (HTML)*, the language in which all Web pages are written. A Web page (HTML document) consists of text and graphics, together with a set of codes (or tags) that describe how the document is to appear when viewed in a Web browser such as Internet Explorer.

In the early days of the Web, anyone creating a Web document (home page) had to learn each of these codes and enter it explicitly. Today, however, it's much easier as you can create a Web document within any application in Microsoft Office XP. In essence, you enter the text of a document, apply basic formatting such as boldface or italics, then simply save the file as a Web document. Office XP also provides an FTP (File Transfer Protocol) capability that lets you upload your documents directly on to a Web server.

There are, of course, other commands that you will need to learn, but all commands are executed from within Word, through pull-down menus, toolbars, or keyboard shortcuts. You can create a single document (called a home page) or you can create multiple documents to build a simple Web site. Either way, the document(s) can be viewed locally within a Web browser such as Internet Explorer, and/or they can be placed on a Web server where they can be accessed by anyone with an Internet connection.

As always, the hands-on exercises are essential to our learn-by-doing philosophy since they enable you to apply the conceptual material at the computer. The exercises are structured in such a way that you can view the Web pages you create, even if you do not have access to the Internet. The last exercise introduces the Web Page Wizard to create a Web site that contains multiple Web pages.

LEARN MORE ABOUT HTML

Use your favorite Web search engine to search for additional information about HTML. One excellent place to begin is the resource page on HTML that is maintained by the Library of Congress at http://lcweb.loc.gov/global/html.html. This site contains links to several HTML tutorials and also provides you with information about the latest HTML standard.

INTRODUCTION TO HTML

Figure 6.1 displays a simple Web page that is similar to the one you will create in the hands-on exercise that follows shortly. Our page has the look and feel of Web pages you see when you access the World Wide Web. It includes different types of formatting, a bulleted list, underlined links, and a heading displayed in a larger font. All of these elements are associated with specific HTML codes that identify the appearance and characteristics of the item. Figure 6.1a displays the document as it would appear when viewed in Internet Explorer. Figure 6.1b shows the underlying HTML codes (*tags*) that are necessary to format the page.

Fortunately, however, it is not necessary to memorize the HTML tags since you can usually determine their meaning from the codes themselves. Nor is it even necessary for you to enter the tags, as Word will create the HTML tags for you based on the formatting in the document. Nevertheless, we think it worthwhile for you to gain an appreciation for HTML by comparing the two views of the document.

HTML codes become less intimidating when you realize that they are enclosed in angle brackets and are used consistently from document to document. Most tags occur in pairs, at the beginning and end of the text to be formatted, with the ending code preceded by a slash, such as <p and </p> to indicate the beginning and end of a paragraph. Links to other pages (which are known as hyperlinks) are enclosed within a pair of anchor tags <A and </A> in which you specify the URL address of the document through the HREF parameter.

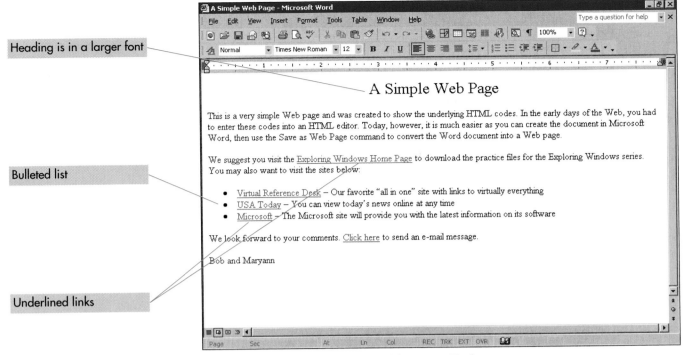

(a) Internet Explorer

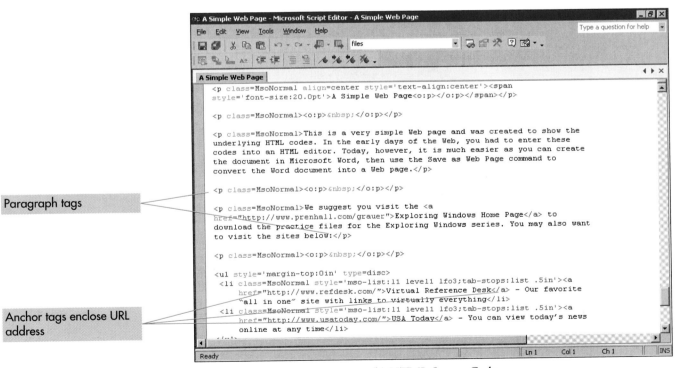

(b) HTML Source Code

FIGURE 6.1 *Introduction to HTML*

Microsoft Word

As indicated, there are different ways to create an HTML document. The original (and more difficult) method was to enter the codes explicitly in a text editor such as the Notepad accessory that is built into Windows. An easier way (and the only method you need to consider) is to use Microsoft Word to create the document for you, without having to enter or reference the HTML codes at all.

Figure 6.2 displays David Guest's ***home page*** in Microsoft Word. You can create a similar page by entering the text and formatting just as you would enter the text of an ordinary document. The only difference is that instead of saving the document in the default format (as a Word document), you use the ***Save As Web Page command*** to specify the HTML format. Microsoft Word does the rest, generating the HTML codes needed to create the document.

Hyperlinks are added through the Insert Hyperlink button on the Standard toolbar or through the corresponding ***Insert Hyperlink command*** in the Insert menu. You can format the elements of the document (the heading, bullets, text, and so on) individually, or you can select a ***theme*** from those provided by Microsoft Word. A theme (or template) is a set of unified design elements and color schemes that will save you time, while making your document more attractive.

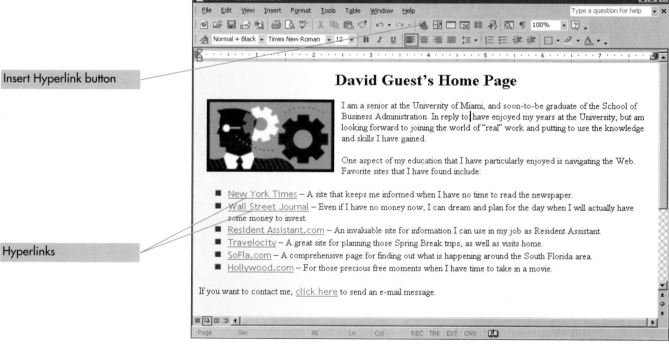

FIGURE 6.2 *A Student's Home Page*

ROUND-TRIP HTML

Each application in Microsoft Office XP lets you open an HTML document in both Internet Explorer and the application that created the Web page initially. In other words, you can start with a Word document and use the Save As Web Page command to convert the document to a Web page, then view that page in a Web browser. You can then reopen the Web page in Word (the original Office application) with full access to all Word commands, should you want to modify the document.

Hands-on Exercise 1

Introduction to HTML

Objective To use Microsoft Word to create a simple home page with clip art and multiple hyperlinks; to format a Web page by selecting a theme. Use Figure 6.3 as a guide in the exercise.

Step 1: **Enter the Text**

- Start Microsoft Word. Pull down the **View menu** and click the **Web Layout command**. Enter the text of your home page as shown in Figure 6.3a. Use any text you like and choose an appropriate font and type size. Center and enlarge the title for your page.
- Enter the text for our links (e.g., *New York Times* and the *Wall Street Journal* sites), or choose your own. You do not enter the URL addresses at this time.
- Click and drag to select all of your links, then click the **Bullets button** on the Formatting toolbar to precede each link with a bullet.
- The Bullets button functions as a toggle switch; that is, click it a second time and the bullets disappear. Click anywhere to deselect the text.
- Click the **Spelling and Grammar button** to check the document for spelling. (There is absolutely no excuse for not checking the spelling and grammar in a document, be it a printed document or a Web page.)

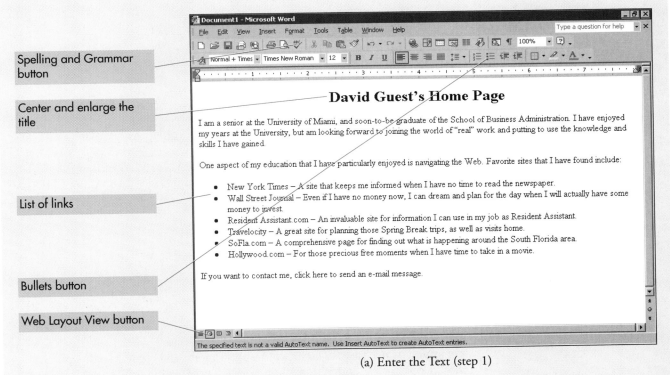

(a) Enter the Text (step 1)

FIGURE 6.3 *Hands-on Exercise 1*

Step 2: **Save the Document**

- Pull down the **File menu** and click the **Save As Web Page** command to display the Save As dialog box in Figure 6.3b.
- Click the **drop-down arrow** in the Save In list box to select the appropriate drive—drive C or drive A. Click to open the **Exploring Word folder** that contains the documents you have used throughout the text.
- Change the name of the Web page to **index**. (Use index as the name of the document to be consistent with the convention used by a Web browser—that is, to automatically display the index document, if it exists.)
- Click the **Change Title button** if you want to change the title of the Web page as it will appear in the Title bar of the Web browser. (The default title is the opening text in your document.)
- Click the **Save button**. The title bar reflects the name of the Web page (index), but the screen does not change in any other way.

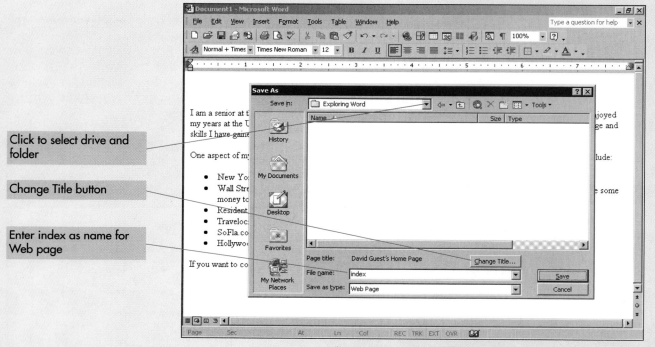

(b) Save the Document (step 2)

FIGURE 6.3 *Hands-on Exercise 1 (continued)*

THE FILE TYPES ARE DIFFERENT

Click the Start button, click (or point to) the Programs command, then start Windows Explorer. Select the drive and folder where you saved the index document. If necessary, pull down the View menu and change to the Details view. Look for the index document you just created, and note that it is displayed with the icon of a Web browser (Internet Explorer or Netscape Navigator) to indicate that it is an HTML document, rather than a Word document.

Step 3: **Insert the Clip Art**

- Pull down the **Insert menu**, click (or point to) **Picture**, then click **Clip Art** to display the Insert Clip Art task pane in Figure 6.3c.
- Click in the Search text box and type **man** to search for all pictures that have been catalogued to describe this attribute. Click the **Search button**. The search begins and the various pictures appear individually within the task pane.
- Point to the image you want in your newsletter, click the **down arrow** that appears, then click **Insert** to insert the clip art.
- The picture should appear in the document. Close the task pane.
- Point to the picture and click the **right mouse button** to display the context-sensitive menu. Click the **Format Picture command** to display the Format Picture dialog box.
- Click the **Layout tab**, choose the **Square layout**, then click the option button for Left or Right alignment. Click **OK** to close the dialog box. Click and drag the sizing handles on the picture as appropriate. Save the document.

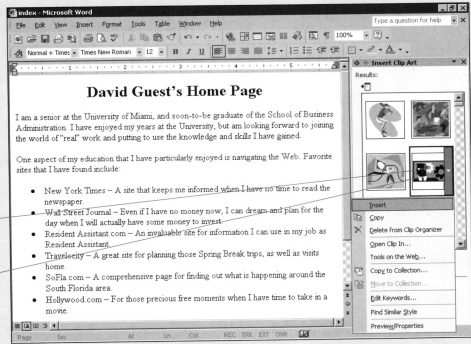

(c) Insert the Clip Art (step 3)

FIGURE 6.3 *Hands-on Exercise 1 (continued)*

SEARCH BY COLLECTION

The Media Gallery organizes its contents by collections and provides another way to select clip art. Pull down the Insert menu, click (or point to) the Picture command, then click Clip Art to open the task pane, where you can enter a key word to search for clip art. Instead of searching, however, click the link to Media Gallery at the bottom of the task pane to display the Media Gallery dialog box. Close the My Collections folder if it is open, then open the Office Collections folder, where you can explore the available images by collection.

Step 4: **Add the Hyperlinks**

- Select **New York Times** (the text for the first hyperlink). Pull down the **Insert menu** and click **Hyperlink** (or click the **Insert Hyperlink button**) to display the Insert Hyperlink dialog box in Figure 6.3d.
- The text to display (New York Times) is already entered because the text was selected prior to executing the Insert Hyperlink command. If necessary, click the icon for **Existing File or Web Page**, then click **Browsed Pages**.
- Click in the second text box and enter the address **www.nytimes.com** (the http is assumed). Click **OK**.
- Add the additional links in similar fashion. The addresses we used in our document are: **www.wsj.com**, **www.residentassistant.com**, **www.travelocity.com**, **www.sofla.com**, and **www.hollywood.com**.
- Click and drag to select the words, **click here**, then click the **Insert Hyperlink button** to display the Insert Hyperlink dialog box.
- Click the **E-mail Address icon**, then click in the E-mail Address text box and enter your e-mail address. Click **OK**.

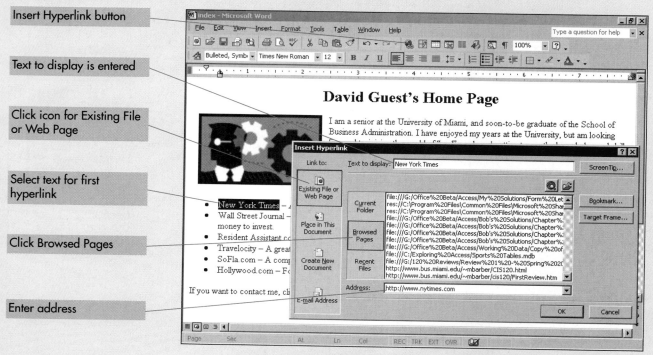

(d) Add the Hyperlinks (step 4)

FIGURE 6.3 *Hands-on Exercise 1 (continued)*

CLICK TO EDIT, CTRL+CLICK TO FOLLOW

Point to a hyperlink within a Word document, and you see a ToolTip that says to press and hold the Ctrl key (Ctrl+click) to follow the link. This is different from what you usually do, because you normally just click a link to follow it. What if, however, you wanted to edit the link? Word modifies the convention so that clicking a link enables you to edit the link. Alternatively, you can right click the hyperlink to display a context-sensitive menu from where you can make the appropriate choice.

Step 5: **Apply a Theme**

- You should see underlined hyperlinks in your document. Pull down the **Format menu** and click the **Theme command** to display the Theme dialog box in Figure 6.3e.
- Select (click) a theme from the list box on the left, and a sample of the design appears in the right. Only a limited number of the listed themes are installed by default, however, and thus you may be prompted for the Microsoft Office XP CD depending on your selection. Click **OK**.
- You can go from one theme to the next by clicking the new theme. There are approximately 65 themes to choose from, and they are all visually appealing. Every theme offers a professionally designed set of formatting specifications for the various headings, horizontal lines, bullets, and links.
- Make your decision as to which theme you will use. Save the document.

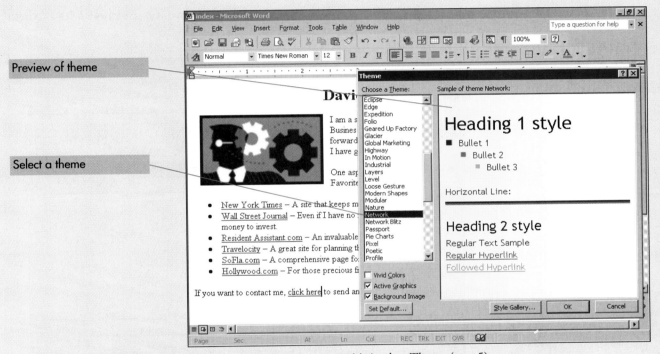

(e) Apply a Theme (step 5)

FIGURE 6.3 *Hands-on Exercise 1 (continued)*

KEEP IT SIMPLE

Too many would-be designers clutter a page unnecessarily by importing a complex background, which tends to obscure the text. The best design is a simple design—either no background or a very simple pattern. We also prefer light backgrounds with dark text (e.g., black or dark blue text on a white background), as opposed to the other way around. Design, however, is subjective, and there is no consensus as to what makes an attractive page. Variety is indeed the spice of life.

Step 6: **View the Web Page**

- Start your Web browser. Pull down the **File menu** and click the **Open command** to display the Open dialog box in Figure 6.3f. Click the **Browse button**, then select the drive folder (e.g., Exploring Word on drive C) where you saved the Web page.
- Select (click) the **index document**, click **Open**, then click **OK** to open the document. You should see the Web page that was just created except that you are viewing it in your browser rather than Microsoft Word.
- The Address bar shows the local address (C:\Exploring Word\index.htm) of the document. (You can also open the document from the Address bar, by clicking in the Address bar, then typing the address of the document—for example, c:\Exploring word\index.htm.)
- Click the **Print button** on the Internet Explorer toolbar to print this page for your instructor.
- Exit Word and Internet Explorer if you do not want to continue with the next exercise at this time.

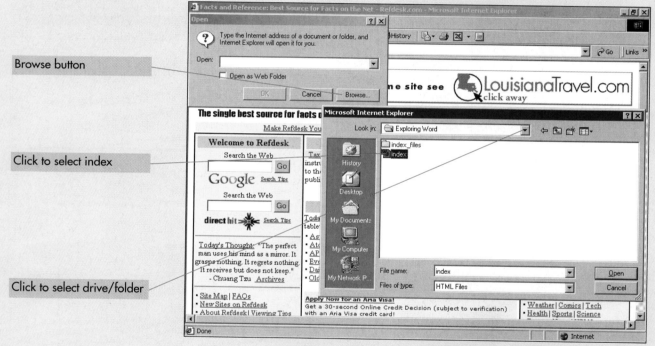

(f) View the Web Page (step 6)

FIGURE 6.3 *Hands-on Exercise 1 (continued)*

AN EXTRA FOLDER

Look carefully at the contents of the Exploring Word folder within the Open dialog box. You see the HTML document you just created as well as a folder that was created automatically by the Save As Web Page command. The latter folder contains the various objects that are referenced by the Web page. Be sure to copy the contents of this folder to the Web server in addition to your Web page if you decide to post the page.

PUBLISH YOUR HOME PAGE

The previous exercise described how to create a Web document and view it through a browser such as *Internet Explorer*. The document was stored locally and thus you are the only one who can view it. It's more fun, however, to place the document on a Web *server*, where it becomes part of the World Wide Web and is therefore accessible by anyone with an Internet connection. The next several pages tell you how.

To place your page on the Web you will need Internet access, and further, you will need permission to store your document on a computer connected to the Internet. Thus, you will need the address of the server, a username, and a password. You will also need the name of the folder where you are to store your document (typically the public_html folder) as well as the path to that folder. Your instructor will provide this information for you if, in fact, your school is able to offer you this service. If not, you can still do the exercise if you have a computer at home and your Internet Service Provider enables you to store documents on its server. (Most ISPs provide several megabytes of online storage in conjunction with their service.)

The procedure to place your document on the Web is straightforward. First you create the document as you did in the previous exercise. However, in addition to (and/or instead of) storing the document locally, you have to upload it to your Web server. This is accomplished through *FTP* (*File Transfer Protocol*), which transmits files between a PC and a server. (There is an FTP capability built into Office XP, but we find it easier to use an independent program.) There are many sites on the Web where you can download a shareware version. Once the file has been saved on the Web server, it may be viewed by anyone with Internet access. (FTP can also be used to upload duplicate copies of important files to a remote site to provide backup for these files.)

The following exercise takes you through the procedure in detail as we upload David Guest's home page to the Web server at the University of Miami. The username (or User ID) is specific to David, and the address of the server is that of the university. The combination of the two, homer.bus.miami.edu/~fze79cav, is the address of David's home page. (The university uses the ~ symbol in front of the username to indicate that the complete path to the folder need not be specified.) You will have to obtain similar information from your instructor or local Internet Service Provider.

You can upload and maintain a Web page entirely through a combination of Microsoft Word and FTP. Occasionally, however, it is advantageous to connect directly to the Web server and execute commands as though you were attached directly to that computer. This is known as a *terminal session* and it is established through a program called *Telnet*. (A terminal, unlike a PC, is a device without memory or disk storage that communicates with a computer via a keyboard and display.) Telnet is illustrated in the last step of the hands-on exercise.

THE INTRANET

The ability to create links to local documents and to view those pages through a Web browser has created an entirely new way to disseminate information. Indeed, many organizations are taking advantage of this capability to develop an *Intranet*, in which Web pages are placed on a local area network for use within the organization. The documents on an Intranet are available only to individuals with access to the LAN on which the documents are stored. This is in contrast to loading the pages onto a Web server where they can be viewed by anyone with access to the Web.

Hands-on Exercise 2

Publishing Your Home Page

Objective To use FTP to upload an HTML document onto a Web server; to demonstrate Telnet. Use Figure 6.4 as a guide in the exercise.

Step 1: **Start FTP**

- ➤ This exercise requires that you have an account on a Web server in order to store your Web page, and further that you have access to an FTP program to upload the files from your PC to the server. (You can use your favorite search engine if you are working at home to locate a shareware site where you can download the software.)
- ➤ Click the **Start button** to start the FTP program. The left side of the FTP window displays the contents of a folder on your computer. The right side of the window shows the content of the FTP site, which is currently empty.
- ➤ Click the **Connect button** to display the Session Properties dialog box in Figure 6.4a. The entries in the dialog box are specific to one student at the University of Miami. In any event, you will need to enter:
  - The address of the Web server (homer.bus.miami.edu in our example).
  - The User ID (or username, f2e79cav in our example).
  - Your Password (not visible in Figure 6.4a).
- ➤ Click the **OK button** to connect to the Web site. The Session Properties dialog box will close, and you should be connected to your server.
- ➤ You are ready to upload files from your PC to the Web server.

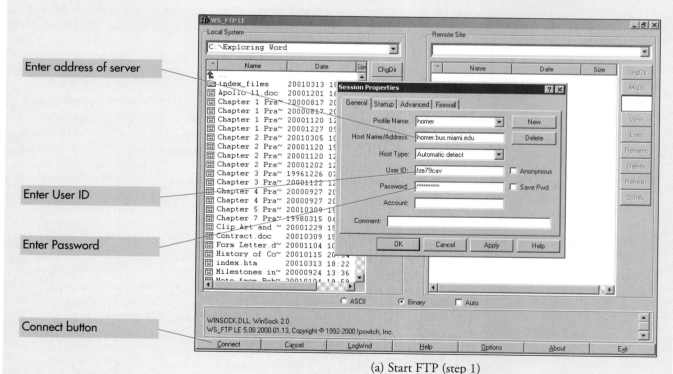

(a) Start FTP (step 1)

Figure 6.4 *Hands-on Exercise 2*

Step 2: **Upload the Files**

- Change to the folder on your computer that contains the home page you created earlier. We are using the Exploring Word folder on drive C, but your files may be in a different location.
- Change to the **public_html** (or similar) **folder** on the Web server. Typically, all you have to do is double click the public_html folder that appears when you first log into the site.
- You *must* be in the public_html folder or its equivalent on your system. Check the address bar in the right pane to see that it ends with the name of the folder as shown in Figure 6.4b.
- Select (click) the **index_files folder** that appears in the left pane. Press and hold the **Ctrl key** to select the **index.htm file** as well. Both files must be selected!
- Click the → button to upload the files. Click **Yes** if asked whether you want to transfer the selected folders and their contents.
- The file transfer should take a few seconds, after which the index.htm file and the associated file should appear in the right pane, indicating that the files have been transferred to the server. Close the FTP window.

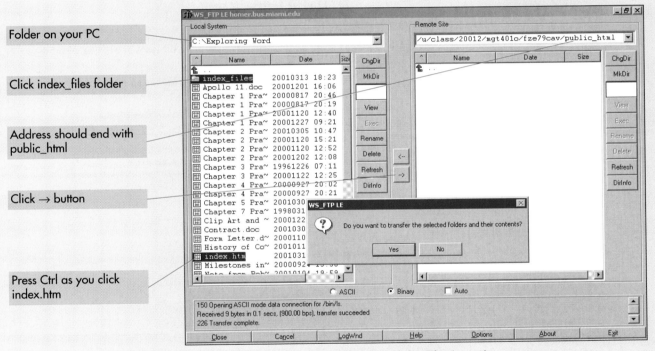

(b) Upload the Files (step 2)

FIGURE 6.4 *Hands-on Exercise 2 (continued)*

IT'S EASY TO MAKE A MISTAKE

It's easy to create a Web page. The hard part, if there is a hard part, is uploading the page and its associated folder to a Web server. Be sure that you select both the index.htm (home page) as well as the index_files folder (the supporting files) for the file transfer. One common mistake is to forget the folder, in which case you will not see the graphic elements when you view the page. In addition, be sure that you change to the public_html folder on the server prior to uploading the files.

Step 3: **View Your Home Page**

- Start your Internet browser. Click in the **Address bar** and enter the name of your server followed by your username (preceded by the ~ character); for example, homer.bus.miami.edu/~pze79cav. Your server may follow a different convention.
- You should see your home page as shown in Figure 6.4c. You do not have to specify the document name because the browser will automatically display a document called index.html if that document is present in the public_html folder.
- Click the hyperlinks on your page to be sure that they work as intended. (You should have tested the hyperlinks in the first exercise, but we suggest that you try one or two at this time anyway.)

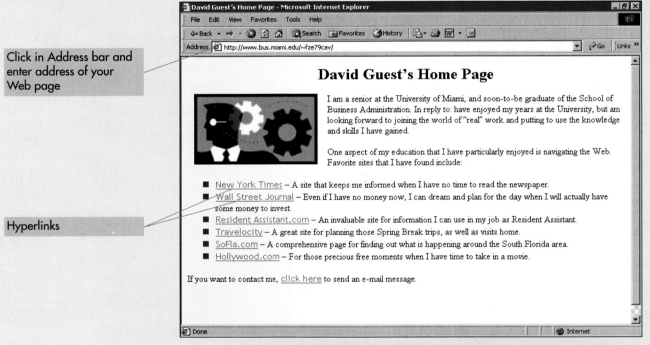

(c) View Your Home Page (step 3)

FIGURE 6.4   *Hands-on Exercise 2 (continued)*

HYPERLINKS BEFORE AND AFTER (INTERNET EXPLORER)

Hyperlinks are displayed in different colors, depending on whether (or not) the associated page has been displayed. You can change the default colors, however, to suit your personal preference. Pull down the Tools menu, click the Internet Options command to display the Internet Options dialog box, and click the General tab. Click the Colors button, then click the color box next to the visited or unvisited links to display a color palette. Select (click) the desired color, click OK to close the palette, click OK to close the Colors dialog box, then click OK to close the Internet Options dialog box.

Step 4: **Add Your Home Page to a Search Engine Database**

➤ Click in the **address bar**, enter **www.lycos.com/addasite.html**, then press the **enter key**. Scroll down the page until you see the section to add your site as shown in Figure 6.4d.

➤ Enter your URL (the address of your home page) and your e-mail address. Click the button to **Add Site to Lycos**, then read the additional information that is displayed on the screen.

➤ Wait a week or two, then see if your Web page has been entered into the Lycos database. Start Internet Explorer, then type **www.lycos.com** to access the Lycos search engine. (You can also access the Lycos search engine by clicking the **Search button** on the Internet Explorer toolbar.)

➤ Enter your name in the Search for text box at the top of the page, then click the **Search button**, to see if your page is recognized by the Lycos search engine. Try a different search engine and compare the results.

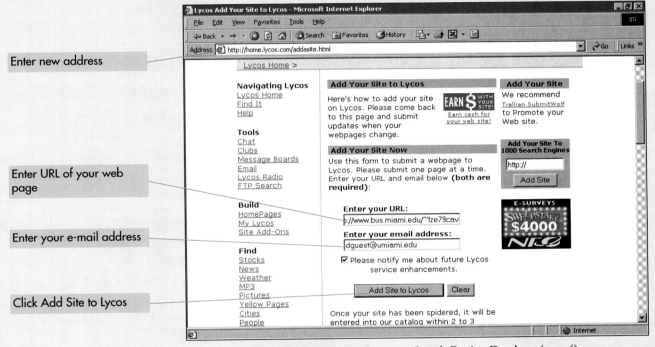

(d) Add Your Home Page to a Search Engine Database (step 4)

FIGURE 6.4 *Hands-on Exercise 2 (continued)*

ADD YOUR E-MAIL ADDRESS TO A DIRECTORY

Add your personal information to various directories on the Internet. Click the Address bar, type www.whowhere.com, press the enter key to display the form for one such directory, then enter your name and other information as appropriate. Click the Search button to see whether you are listed. If not, scroll down the page until you can click Adding Your Listing to add your e-mail address to the directory. As with all directories, there are advantages/disadvantages to an unlisted address.

Step 5: **Print the Web Document**

> ➤ Right click the **Back button**, then click the address of your home page. Pull down the **File menu**, click the **File menu**, and click the **Print command** to display the dialog box in Figure 6.4e.
> ➤ Click the **Options tab**. Check the box to **Print table of links**, then click **OK** to print your home page.
> ➤ Look closely at the printed document. You will see the text of your home page (as you did in the previous exercise). You will also see a list of the hyperlinks in your document together with the associated URLs (Web addresses).
> ➤ Look at the header and/or footer of the printed page. You should see the address of your Web page as well as today's date. If not, you can modify this information through the Page Setup command.
> ➤ Submit the printed document to your instructor.

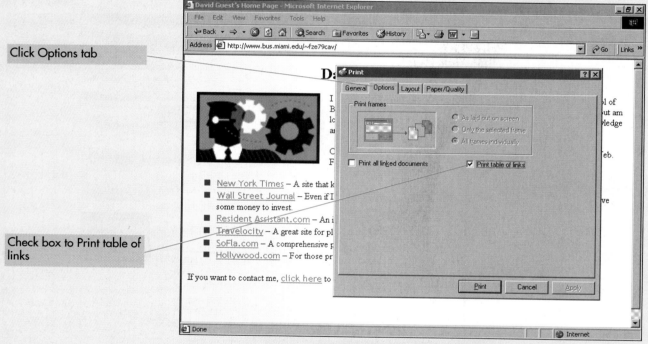

Click Options tab

Check box to Print table of links

(e) Print the Web Document (step 5)

FIGURE 6.4   *Hands-on Exercise 2 (continued)*

THE PAGE SETUP COMMAND

Pull down the File menu and click the Page Setup command to display the Page Setup dialog box that controls the appearance of the printed page. The contents of the header and footer text boxes are especially interesting as they contain information such as the URL of the page and/or the date the page was printed. The precise information that is printed is a function of the code that is entered in the text box—for example, &u and &d for the URL and date, respectively. Click the Help button (the question mark at the right of the title bar), then point to the Header or Footer text boxes to see the meanings of the various codes, then modify the codes as necessary.

Step 6: **Telnet to Your Account**

- Click the **Start button** on the Windows taskbar, click the **Run command** to display the Run dialog box. Type **Telnet**, then click **OK**.
- Pull down the **Connect menu**, click the **Remote Server command**, then enter the address of your Web server (homer.bus.miami.edu in our example) in the Host Name text box. Click **Connect**.
- Enter your username and password (the same entries you supplied earlier). The system will then display an opening message (e.g., University of Miami on our system), followed by a prompt (homer> on our system). Our Web server uses the Unix operating system as shown in Figure 6.4f.
- Type the command **ls −l** to display the files and folders that are stored in your account. You should see the public_html folder (the folder you specified when you uploaded your Web page).
- Type the command **cd public_html** to change to this directory, then type the command **pwd** to print the name of the working directory. You should see an address corresponding to the path you specified when you saved your Web page.
- Type the command **ls −l** to see the contents of the public_html directory. You should see index.htm (the name of your home page) and index files (the folder that contains the graphic elements on that page).
- Type **exit** to log out; you will then see a dialog box indicating that the connection to the host is lost. Click **OK**.
- Close the Telnet window. Close Internet Explorer. Close Microsoft Word if you do not want to do the next exercise at this time.

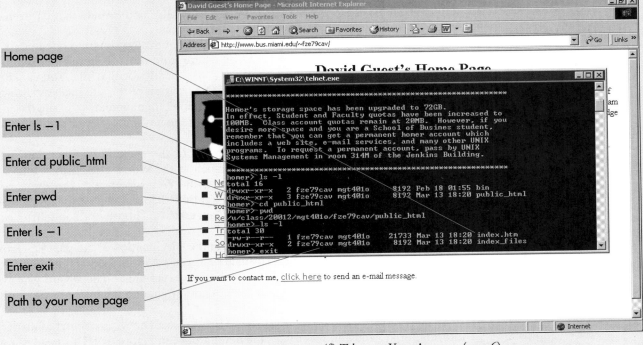

(f) Telnet to Your Account (step 6)

FIGURE 6.4 *Hands-on Exercise 2 (continued)*

MICROSOFT WORD 2002

CREATING A WEB SITE

Thus far, you have created a home page and have placed it on a Web server. The next logical task is to extend your home page to reference additional pages that are stored on the server. In other words, you want to create a **Web site**, as opposed to a single home page.

Figure 6.5 displays a Web site for a hypothetical travel agency. The Address bar in both Figures 6.5a and 6.5b shows that the document (named default.htm) is stored in the World Wide Travel folder on drive C. (The document would be subsequently renamed to index.htm prior to uploading it onto the Web.) The reference to drive C indicates that the site is being developed on a local machine and has not yet been uploaded to a Web server.

The default.htm document is divided into two vertical *frames*, each of which displays a different document. The left frame is the same in both Figure 6.5a and 6.5b, and it contains a series of hyperlinks that are associated with other pages at the site. Click the About the Agency link in the left pane of Figure 6.5a, for example, and you display information about the agency in the right pane. Click the New York Weekend in Figure 6.5b, however, and you display a page describing a trip to New York City. Note, too, that each frame has its own vertical scroll bar. The scroll bars function independently of one another. Thus, you can click the vertical scroll bar in the left pane to view additional links, or you can click the scroll bar in the right pane to see additional information about the agency itself.

The document in the right pane of Figure 6.5a also illustrates the use of a **bookmark** or link to a location within the same document. A bookmark is a useful navigation aid in long documents, as it lets you jump from one place to another (within the same document) without having to manually scroll through the document. Thus, you can click the link that says, "click here for travel tips", and the browser will scroll automatically to the paragraph about travel tips and display that entire paragraph on the screen.

Creation of the Web site will require that you develop separate documents for the agency, as well as for each destination. There is no shortcut because the content of every site is unique and must be created specifically for that site. You can, however, use the **Web Page Wizard** to simplify the creation of the site itself as shown in Figure 6.6. The wizard asks you a series of questions, then it creates the site based on the answers you supply. The wizard takes care of the navigation and design. The content is up to you.

The wizard begins by asking for the name of the site and its location (Figure 6.6a). It's easiest to specify a local folder, such as the World Wide Travel folder on drive C, and then upload the entire folder to the Web server once the site is complete. Next you choose the means of navigation through the site (Figure 6.6b). Vertical frames are the most common, but you can also choose horizontal frames or a separate window for each document.

The essence of the wizard, however, is in the specification of the pages as shown in Figures 6.6c and 6.6d. The wizard suggests three pages initially (Figure 6.6c), but you can add additional pages or remove the suggested pages. The wizard gives you the opportunity to rename any of the pages and/or change the order in which they will appear in Figure 6.6d.

You choose the theme for your site in Figure 6.6e. The wizard then has all of the information it needs, and it creates the default page in Figure 6.6f. This page is much simpler than the completed site we saw in Figure 6.5. Nevertheless, the wizard has proved invaluable because it has created the site and established the navigation. It's now your task to complete and/or enhance the individual pages. The ease with which you create the documents in Figure 6.6 depends on your proficiency in Microsoft Word. Even if you have only limited experience with Microsoft Word, however, our instructions are sufficiently detailed that you should be able to complete our next exercise with little difficulty.

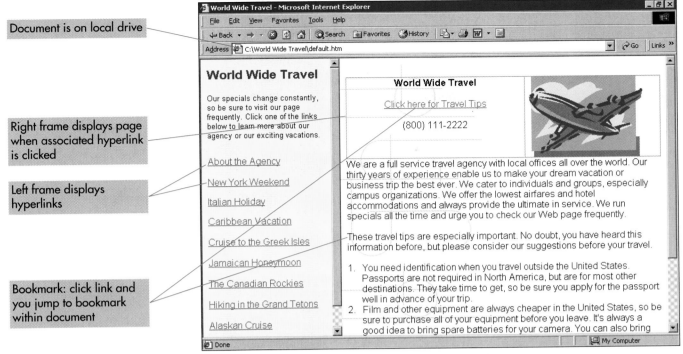

(a) About the Agency Page

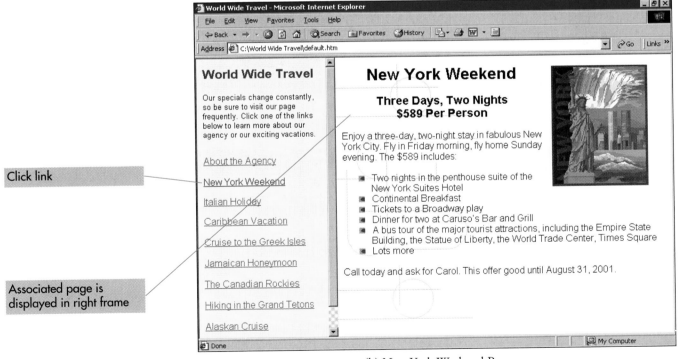

(b) New York Weekend Page

FIGURE 6.5 *A Web Site*

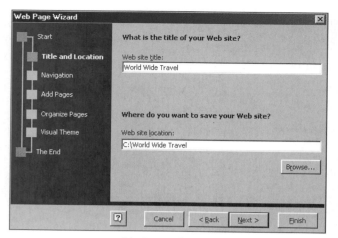
(a) Create the Site

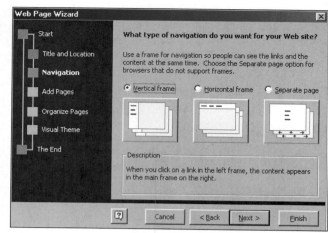
(b) Choose the Navigation

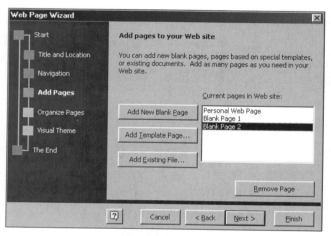
(c) Add the Pages

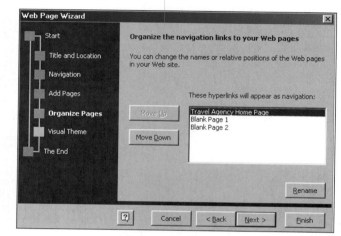
(d) Organize the Pages

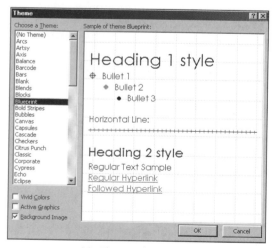
(e) Choose the Theme

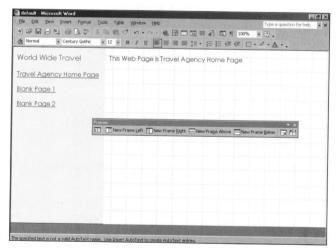
(f) The Initial Site

FIGURE 6.6 *The Web Page Wizard*

Hands-on Exercise 3

Creating a Web Site

Objective To use the Web Page Wizard to create a Web site; to facilitate navigation within a document by creating a bookmark. Use Figure 6.7.

Step 1: **Start the Web Page Wizard**

- Start Word. Pull down the **File menu** and click the **New command** to display the task pane. (If a new document is already open, pull down the **View menu** and click the **Task Pane command**.)
- Click the link to **General templates** to open the Templates dialog box. Click the **Web pages** tab, then double click the **Web Page Wizard** icon to start the wizard. Click **Next**.
- Enter **World Wide Travel** as the title of the Web site. Choose a separate folder to hold all of the documents for the site. We suggest **C:\World Wide Travel** as shown in Figure 6.7a. Click **Next**.
- The option button for **Vertical frame** as the means of navigation is already selected. Click **Next**.

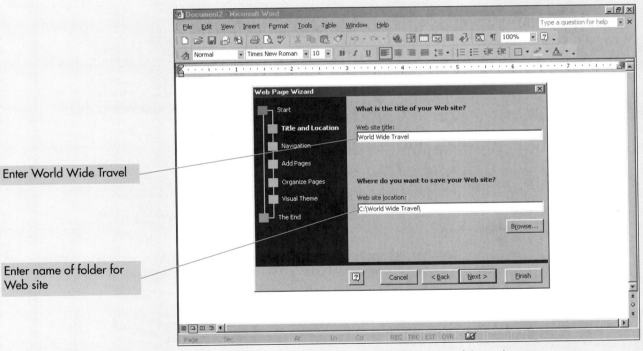

(a) Start the Web Page Wizard (step 1)

Figure 6.7 *Hands-on Exercise 3*

WIZARDS AND TEMPLATES

Office XP includes wizards and templates for a variety of documents. A template is a partially completed document that contains formatting, text, and/or graphics. A wizard introduces additional flexibility by first asking you a series of questions, then creating a template based on your answers.

Step 2: **Specify the Pages**

➤ The Web Page Wizard creates a site with three pages: Personal Web Page, Blank Page 1, and Blank Page 2. You can, however, add, remove, and/or rename these pages as appropriate for your site.

➤ Select (click) **Personal Web page** and click the **Remove Page button**. Click the **Add New Blank Page button** to add Blank Page 3. Click **Next** to display the screen in Figure 6.7b.

➤ Select **Blank Page 1** and click the **Rename button** (or simply double click **Blank Page 1**) to display the Rename Hyperlink dialog box. Enter **About the Agency** and click **OK**. Rename Blank Page 2 and Blank Page 3 to **New York Weekend** and **Italian Holiday**, respectively.

➤ Check that the order of pages is what you intend (About the Agency, New York Weekend, and Italian Holiday). Click **Next**.

➤ Click the **Browse Themes button** and choose a suitable theme. We chose **Capsules**. Click **OK**. Click **Next**. Click **Finish**.

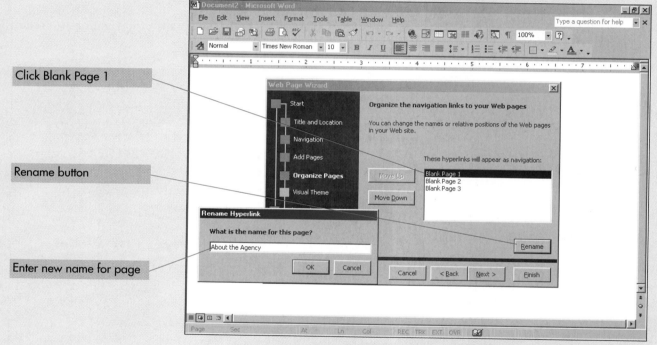

(b) Specify the Pages (step 2)

FIGURE 6.7 *Hands-on Exercise 3 (continued)*

CHANGING THE WEB SITE

The Web Page Wizard is intended to get you up and running as quickly as possible. It takes you through all of the steps to create a Web site, and it builds the appropriate links to all of the pages on that site. You will, however, need to modify the site after it has been created by adding text, adding new links, and/or deleting or modifying existing links.

Step 3: **Modify the Default Page**

➤ You should see the default page for the World Wide Travel site as shown in Figure 6.7c. Our figure, however, already reflects the changes that you will make to the default page. Close the Frames toolbar.

➤ Click and drag to select **World Wide Travel**. Click the **Bold button**. Click to the right of the text to deselect it. Press the **enter key** twice. Enter the text shown in Figure 6.7c.

➤ Click the **Save button** to save the changes to the default page. Pull down the **File menu** and click the **Close command** to close the default page but remain in Microsoft Word.

➤ We will reopen the default page in Internet Explorer later in the exercise to view the page as it will appear to others.

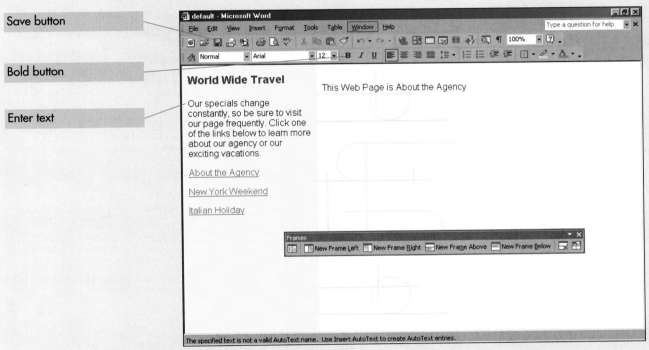

(c) Modify the Default Page (step 3)

FIGURE 6.7 *Hands-on Exercise 3 (continued)*

OBTAINING A PASSPORT

You can't obtain a passport online, but you can get all of the information you need. Go to travel.state.gov, the home page of the Bureau of Consular Affairs in the U.S. Department of State, then scroll down the page until you can click the link to Passport Information. You will be able to download an actual passport application with detailed instructions including a list of the documents you need to supply. You can also access a nationwide list of where to apply. See practice exercise 3 at the end of the chapter.

Step 4: **Modify the Agency Page**

➤ You should still be in Microsoft Word. Click the **Open button** on the Standard toolbar to display the Open dialog box. Click the **down arrow** on the Look in text box, then double click the **World Wide Travel folder** on drive C (the location you specified when you created the Web site).

➤ Double click the **About the Agency document** that currently consists of a single line of text. Click at the end of this sentence and press the **enter key**.

➤ Pull down the **Insert menu** and click the **File command** to display the Insert File dialog box. Change to the **Exploring Word folder** and insert the **Text for Travel document**.

➤ Your document should match the document in Figure 6.7d. Click and drag to select the original sentence as shown in the figure. Press the **Del key** to delete this sentence. Delete the blank line as well. Save the document.

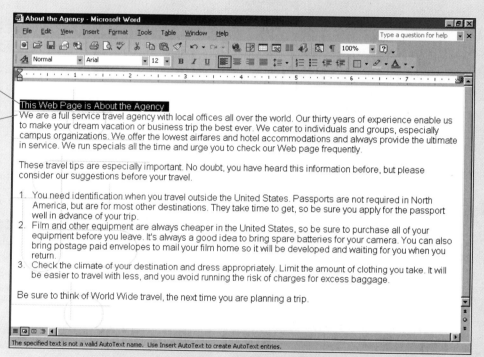

(d) Modify the Agency Page (step 4)

FIGURE 6.7 *Hands-on Exercise 3 (continued)*

THE WORLD WIDE TRAVEL FOLDER

A Web site consists of multiple Web pages, each of which is saved as a separate document. Each document in turn may contain graphical elements that are also saved separately. Thus the World Wide Travel folder contains an About the Agency document that in turn references graphic elements in its own folder. The other pages (e.g., Italian Holiday and New York Weekend) have their own folders. The World Wide Travel folder also contains additional documents such as a default page and a TOC (table of contents) frame that is displayed in the left pane.

Step 5: **Test the Navigation**

- Start your Web browser. Pull down the **File menu** and click the **Open command** to display the Open dialog box. Click the **Browse button**, then select the drive (e.g., drive C) and folder (e.g., World Wide Travel).
- Select (click) the **default** document, click the **Open command button**, then click **OK** to display the document in Figure 6.7e. You should see the Web page that was created earlier, except that you are viewing it in your browser rather than Microsoft Word. The Address bar reflects the local address of the document (C:\World Wide Travel\default.htm).
- Click the link to **New York Weekend** to display this page in the right pane. The page is not yet complete, but the link works properly. Click the link to **Italian Holiday** to display this page. Again, the page is not complete, but the link works properly.
- Click the link to **About the Agency** to access this page.

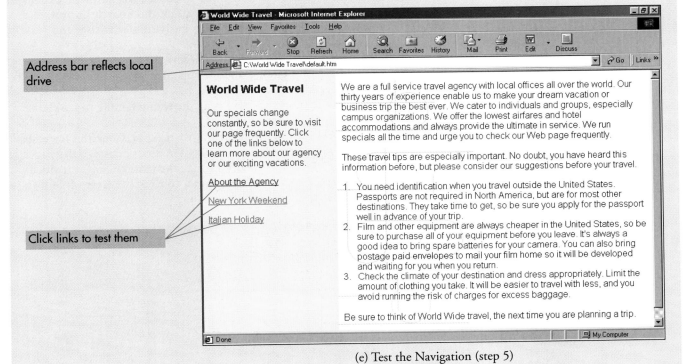

(e) Test the Navigation (step 5)

FIGURE 6.7 *Hands-on Exercise 3 (continued)*

UNDER CONSTRUCTION

Use prototyping to let the end user experience the "look and feel" of a site before the site has been completed. The user sees the opening document, and is provided with a set of links to partially completed documents. This provides a sense of the eventual site even though the latter is far from finished. Prototyping also provides valuable feedback to the developer, who is able to make the necessary adjustments before any extensive work has been done. See practice exercise 4 at the end of the chapter.

Step 6: **Insert the Clip Art**

- Click the **Word button** on the Windows taskbar. The About the Agency document should still be open. Press **Ctrl+Home** to move to the beginning of the document. Press the **enter key** to add a blank line.
- Click the **Insert Table button** on the Standard toolbar, then click and drag to select a one-by-two grid. Release the mouse. Click in the **left cell**. Type **World Wide Travel**.
- Press the **enter key** twice. Type the sentence **Click here for Travel Tips** (we will create the link later), press the **enter key** twice, and enter the agency's phone number, **(800) 111-2222**.
- Select all three lines and click the **Center button**. Choose a suitable point size. We suggest 18 point for the first line and 12 point for the other text.
- Click in the **right pane**. Pull down the **Insert menu**, click **Picture**, then click **Clip Art** to display the Insert Clip Art task pane in Figure 6.7f.
- Click in the **Search text box** and type **airplane** to search for all pictures that have been catalogued to describe this attribute. Click the **Search button**. The search begins and the various pictures appear individually within the task pane.
- Point to the image you want, click the **down arrow** that appears, then click **Insert** to insert the clip art. The picture should appear in the document. Close the task pane.
- Click the picture to select it, then click and drag the sizing handle to make the picture smaller.
- Click the **Internet Explorer button** on the Windows taskbar. Click the **Refresh button** on the Internet Explorer toolbar to display the new version of the page.
- If necessary, return to the Word document to resize the picture. Save the document, then return to Internet Explorer. Remember to click the **Refresh button** in Internet Explorer to see the most recent version.

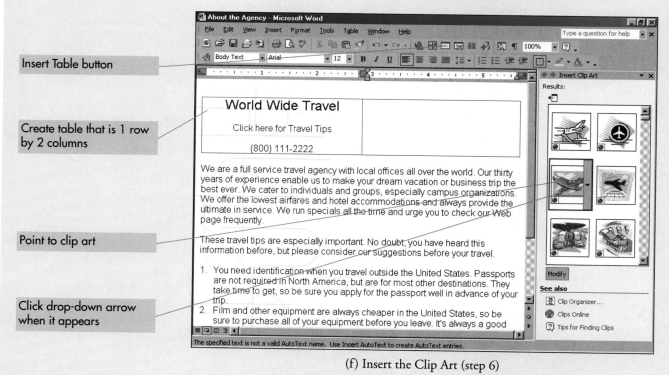

(f) Insert the Clip Art (step 6)

FIGURE 6.7 *Hands-on Exercise 3 (continued)*

Step 7: **Insert a Bookmark**

- Return to the Agency page in Microsoft Word to create a bookmark.
- To create the bookmark:
  - Click at the beginning of the second paragraph. Pull down the **Insert menu** and click **Bookmark** to display the Bookmark dialog box.
  - Enter **TravelTips** (spaces are not allowed) as the name of the bookmark, then click the **Add button** to add the bookmark and close the dialog box.
- To create the link to the bookmark:
  - Click and drag the text **Click here for Travel Tips** as shown in Figure 6.7g, then click the **Insert Hyperlink button** to display the dialog box.
  - Click the icon for **Place in This Document**, click the **plus sign** next to Bookmarks, then click **TravelTips**. Click **OK**.
- The sentence, Click here for Travel Tips, should appear as underlined text to indicate that it is now a hyperlink. Save the document.

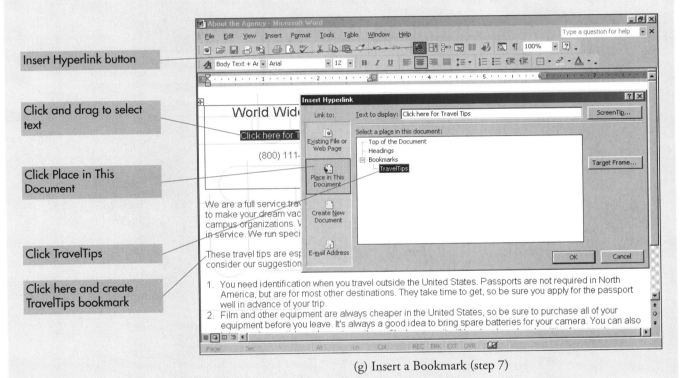

(g) Insert a Bookmark (step 7)

FIGURE 6.7 *Hands-on Exercise 3 (continued)*

THE TOP OF DOCUMENT BOOKMARK

Simplify the navigation within a long page with a link to the top of the document. Press Ctrl+End to move to the bottom of the document (one of several places where you can insert this link), then click the Insert Hyperlink button to display the Insert Hyperlink dialog box. Click the icon for Place in This Document, click Top of the Document from the list of bookmarks (Word creates this bookmark automatically), then click OK. You will see the underlined text, Top of Document, as a hyperlink.

Step 8: **View the Completed Page**

➤ Click the **Internet Explorer button** on the Windows taskbar. Click the **Refresh button** to view the completed document as shown in Figure 6.7h.

➤ If necessary, click the link to **About the Agency** to display the completed Agency page. There is a scroll bar in the right frame, because you cannot see the entire agency page at one time. Note, however, that you do not see a scroll bar in the left frame because this page can be seen in its entirety.

➤ Click in the **right pane**. Pull down the **File menu** and click the **Print command** to display the Print dialog box in Figure 6.7h. Click the option button to print **Only the selected frame** but check the box to **Print all linked documents** (which effectively prints every document in the site).

➤ Click **OK**, then submit the printed pages to your instructor as proof that you did this exercise.

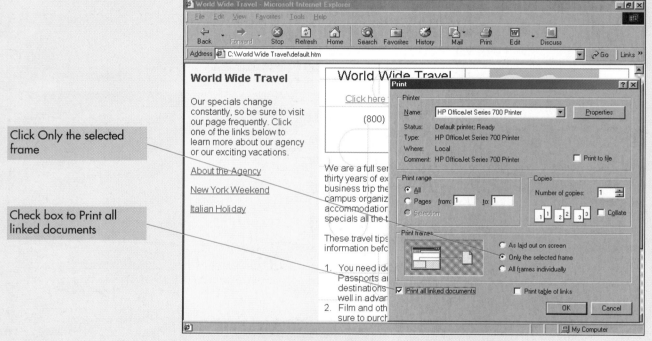

(h) View the Completed Page (step 8)

FIGURE 6.7 *Hands-on Exercise 3 (continued)*

CHANGE THE FONT SIZE

Internet Explorer enables you to view and/or print a page in one of five font settings (smallest, small, medium, large, and largest). Pull down the View menu, click the Text Size command, then choose the desired font size. The setting pertains to both the displayed page as well as the printed page.

SUMMARY

All Web documents are written in HyperText Markup Language (HTML), a language that consists of codes (or tags) that format a document for display on the World Wide Web. The easiest way to create an HTML document is through Microsoft Word. You start Word in the usual fashion and enter the text of the document with basic formatting. Then you pull down the File menu and click the Save As Web Page command.

Microsoft Word does the rest, generating the HTML tags that are needed to create the document. The resulting document can be modified with respect to its content and/or appearance just like an ordinary Word document. The Insert Hyperlink command is used to link a document to other pages. Graphics may come from a variety of sources and are inserted into a document through the Insert Picture command. The Format Theme command applies a professional design to the document. The Web page may be subsequently opened in Microsoft Word for additional editing and/or it can be opened in Internet Explorer for viewing.

After a Web document has been created, it can be placed on a server or local area network so that other people will be able to access it. This, in turn, requires you to check with your professor or system administrator to obtain the necessary username and password. Once you have this information, you use the FTP protocol to upload your page to the Web server. (FTP can also be used to upload duplicate copies of important files to a remote site to provide backup for these files.) There is an FTP capability within Office XP, but it is easier to use a standalone program. (Many sites on the Web provide a shareware version of FTP.) Even if your page is not placed on the Web, you can still view it locally on your PC through a Web browser.

The Web Page Wizard simplifies the creation of a multipage site. The wizard asks you a series of questions, then it creates the site based on the answers you supply. The opening page is divided into horizontal or vertical frames that provide links to subsidiary pages. Every page has a consistent format according to the theme that you selected. Additional pages can be added at any time. Existing pages can be modified or deleted. In short, the wizard takes care of the navigation and design. The content is up to you.

Prototyping can be used during the development process to provide the look and feel of the finished site. The navigation is complete, but the individual pages are still "under construction."

KEY TERMS

Bookmark (p. 274)
File Transfer Protocol (FTP) (p. 267)
Frame (p. 274)
Home page (p. 260)
Hyperlink (p. 260)
Hypertext Markup Language (HTML) (p. 257)

Insert Hyperlink command (p. 260)
Internet Explorer (p. 267)
Intranet (p. 267)
Prototyping (p. 281)
Save As Web Page command (p. 260)
Server (p. 267)
Tag (p. 258)

Telnet (p. 267)
Terminal session (p. 267)
Theme (p. 260)
Web Page Wizard (p. 274)
Web site (p. 274)

MULTIPLE CHOICE

1. Which of the following requires you to enter HTML tags explicitly in order to create a Web document?
   (a) A text editor such as the Notepad accessory
   (b) Microsoft Word
   (c) Both (a) and (b)
   (d) Neither (a) nor (b)

2. What is the easiest way to switch back and forth between Word and Internet Explorer, given that both are open?
   (a) Click the appropriate button on the Windows taskbar
   (b) Click the Start button, click Programs, then choose the appropriate program
   (c) Minimize all applications to display the Windows desktop, then double click the icon for the appropriate application
   (d) All of the above are equally convenient

3. When should you click the Refresh button on the Internet Explorer toolbar?
   (a) Whenever you visit a new Web site
   (b) Whenever you return to a Web site within a session
   (c) Whenever you view a document on a corporate Intranet
   (d) Whenever you return to a document that has changed during the session

4. How do you view the HTML tags for a Web document from Internet Explorer?
   (a) Pull down the View menu and select the Source command
   (b) Pull down the File menu, click the Save As command, and specify HTML as the file type
   (c) Click the Web Page Preview button on the Standard toolbar
   (d) All of the above

5. Internet Explorer can display an HTML page that is stored on:
   (a) A local area network
   (b) A Web server
   (c) Drive A or drive C of a standalone PC
   (d) All of the above

6. How do you save a Word document as a Web page?
   (a) Pull down the Tools menu and click the Convert to Web Page command
   (b) Pull down the File menu and click the Save As Web Page command
   (c) Both (a) and (b)
   (d) Neither (a) nor (b)

7. Which program transfers files between a PC and a remote computer?
   (a) Telnet
   (b) FTP
   (c) Homer
   (d) PTF

8. Which of the following requires an Internet connection?
   (a) Using Internet Explorer to view a document that is stored locally
   (b) Using Internet Explorer to view the Microsoft home page
   (c) Both (a) and (b)
   (d) Neither (a) nor (b)

9. Which of the following requires an Internet connection?
   (a) Telnet
   (b) FTP
   (c) Both (a) and (b)
   (d) Neither (a) nor (b)

10. Assume that you have an account on the server, www.myserver.edu, under the username, jdoe. What is the most likely Web address to view your home page?
    (a) www.myserver.edu
    (b) www.jdoe.edu
    (c) www.myserver.edu/~jdoe
    (d) www.myserver.edu.jdoe.html

11. The Insert Hyperlink command can reference:
    (a) An e-mail address
    (b) A bookmark
    (c) A Web page
    (d) All of the above

12. The Format Theme command:
    (a) Is required in order to save a Word document as a Web page
    (b) Applies a uniform design to the links and other elements within a document
    (c) Both (a) and (b)
    (d) Neither (a) nor (b)

13. The Web Page Wizard creates a default Web site and enables you to specify:
    (a) The means of navigation such as vertical or horizontal frames
    (b) The number of pages (links) that are found on the default page
    (c) The theme (design) of the Web site
    (d) All of the above

14. The Web Page Wizard creates a site with a Personal Web page, Blank Page 1, and Blank Page 2, but you can
    (a) Delete any of these pages
    (b) Add new pages to those it creates for you
    (c) Rename any pages it supplies
    (d) All of the above

15. Assume that the Web Page Wizard was used to create a site called Personal Computer Store, and further, that the site contains links to four separate pages, each of which contains one or more graphics. Which of the following is true?
    (a) You can expect to see a Personal Computer Store folder on your system
    (b) The Personal Computer Store folder will contain a separate document for each of the four pages
    (c) The Personal Computer Store folder will contain a separate folder for each of the four pages
    (d) All of the above

ANSWERS

1. a	6. b	11. d
2. a	7. b	12. b
3. d	8. b	13. d
4. a	9. c	14. d
5. d	10. c	15. d

PRACTICE WITH WORD

1. **Web Page Review:** The document in *Chapter 6 Practice 1* provides a quick review of the process to create a Web page and upload to a server. Your assignment is to open the document shown in Figure 6.8 and fill in the blanks. Use boldface and italics to highlight your answers, then print the completed document for your instructor. This is a valuable exercise to review the material from the chapter.

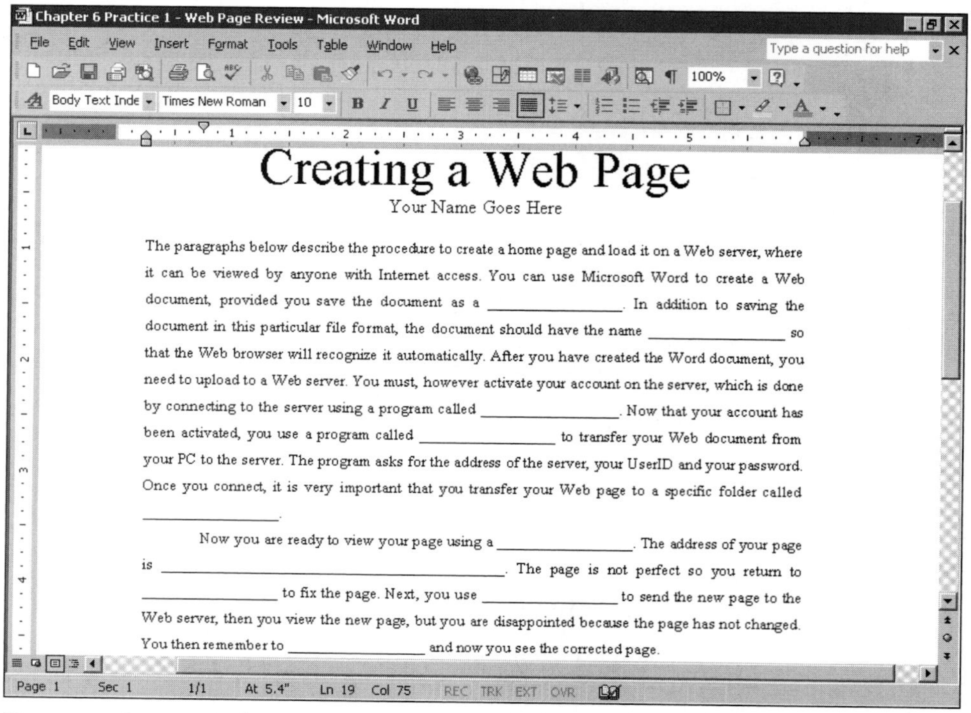

   FIGURE 6.8 *Web Page Review (Exercise 1)*

   BUILDS ON
   HANDS-ON
   EXERCISE 3
   PAGES 285–292

2. **New York Weekend:** Create the World Wide Travel Web site as described in the third hands-on exercise. Test the site to be sure that the navigation works. The About the Agency page is complete, but the New York Weekend and Italian Holiday documents exist only as one-sentence documents. Open either document and complete the page. Our suggestion for the New York weekend is shown in Figure 6.9, but you need not use our design. Print the additional page(s) and submit them to your instructor as proof you did this exercise.

   Use FTP to upload the Travel Web site to a Web server, provided you have this capability. Be sure to transfer the complete contents of the World Wide Travel folder from your local machine to the public_html folder on the server. What is the Web address of your site?

   BUILDS ON
   HANDS-ON
   EXERCISE 3
   PAGES 285–292

3. **Adding an External Link:** Open the default.htm document in the World Wide Travel folder, then add the link to Passport information to display the State Department page (travel.state.gov) shown in Figure 6.10. This site contains a wealth of information for the would-be traveler, including information on how to obtain a passport. Print the revised page for your instructor.

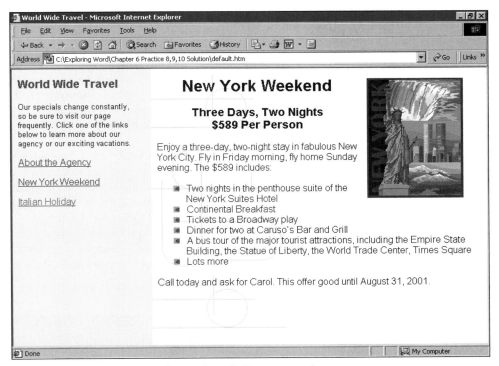

FIGURE 6.9 *New York Weekend (Exercise 2)*

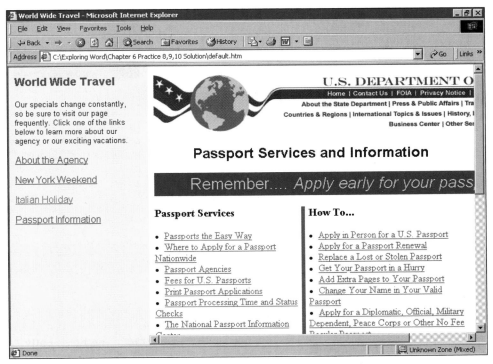

FIGURE 6.10 *Adding an External Link (Exercise 3)*

BUILDS ON
PRACTICE
EXERCISE 2
PAGE 288

4. Under Construction: The creation of a Web site is an interactive process between the client who is paying for the site and the developer. It's important for the client to get the "look and feel" of the site during the early stages of development, so that any errors can be corrected as soon as possible. One way to accomplish this goal is to implement the navigation for the complete site, prior to creating all of its content. This is done through an "Under Construction" (or prototype) page such as the document in Figure 6.11. The client obtains the look and feel of the eventual site and can communicate any changes immediately.
   a. Create an "Under Construction" page for the World Wide Travel agency. You can use our design or you can create your own.
   b. Expand the default.htm document from the previous exercise to include the additional vacations shown in Figure 6.11. Each of the additional links is to display the "Under Construction" page.
   c. Print the revised page for your instructor as shown in Figure 6.11. In addition, write a short note describing how you will add content to the site as the additional vacations become available. Explain how to remove a link if it is no longer available.

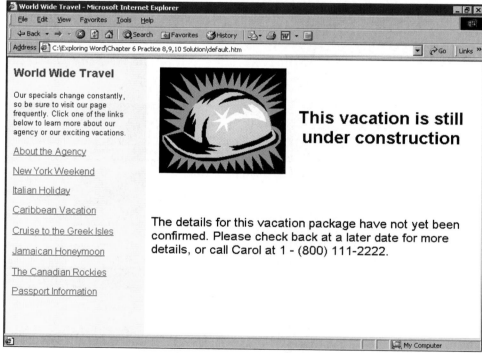

FIGURE 6.11   *Under Construction (Exercise 4)*

5. Frequently Asked Questions: Pull down the File menu, click the New command, and click the Web Pages tab to display a dialog box containing several templates for use as Web pages. Open the Frequently Asked Questions template, and use it to create a document with questions and answers about any subject that interests you. You could, for example, create a document with questions about travel, then add a link to this document to your World Wide Travel site. Print the completed document for your instructor as proof you did this exercise. Be sure to explore the bookmarks that are created automatically within the template. Include a short note explaining how to add new bookmarks and/or to remove existing bookmarks that are no longer relevant.

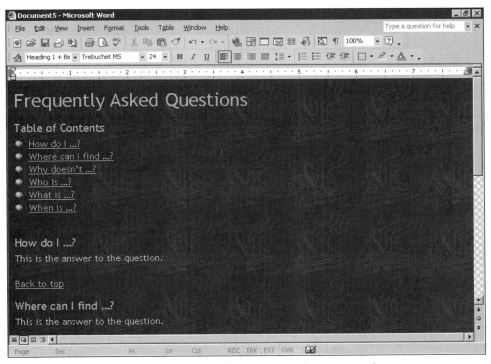

FIGURE 6.12   *Frequently Asked Questions (Exercise 5)*

6. **Personal Web Page:** Use the Personal Web Page template in Office XP to develop a page about yourself. The template in Figure 6.13 is intended to give you a quick start, but you are not obligated to use all of the headings. Choose an appropriate theme, then print the completed page for your instructor. Add a short note that compares the personal Web page in this exercise with the home page that you developed in the first hands-on exercise in the chapter.

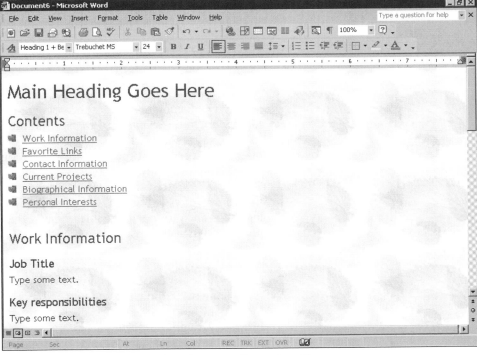

FIGURE 6.13   *Personal Web Page (Exercise 6)*

MICROSOFT WORD 2002

7. **A Commercial Web Site:** Use the Web Page Wizard to create a Web site for a hypothetical Computer Super Store as shown in Figure 6.14. Use the Web Page Wizard to create a site with horizontal or vertical frames. Choose any theme you like. Use the text in our document, *Chapter 6 Practice 7* for the home page. The subsidiary pages do not have to be completely developed; that is, the pages can be "under construction." The navigation, however, has to work. Print the completed home page and at least one subordinate page for your instructor.

   There is no requirement to upload the page to the Web, but it is worth doing if you have the capability. You will need additional information from your instructor about how to obtain an account on a Web server (if that is available), and further how to upload the Web page from your PC to the server.

FIGURE 6.14   *A Commercial Web Site (Exercise 7)*

8. **Milestones in Communications:** The use of hyperlinks enables the creation of interactive documents such as the document in Figure 6.15. Open the unformatted document in *Chapter 6 Practice 8* in the Exploring Word folder. Pull down the Format menu, click the AutoFormat command to display the AutoFormat dialog box, verify that the AutoFormat option is selected, and click OK.

   Modify the Body Text style and/or the Heading 1 style after the document has been formatted in any way that makes sense to you. The most important task, however, is to create the hyperlinks at the beginning of the document that let you branch to the various headings within the document. To create a hyperlink, pull down the Insert menu, click the Hyperlink command, then click the Place in this Document button to select from the bookmarks that are contained within the document. (Each heading in the document appears as a bookmark.)

   You also need to insert a hyperlink after each article to return to the top of the document. Follow the same procedure as before, but select the Top of the Document bookmark. The result is an interactive document that lets the user browse through the document in any sequence you choose. Complete the formatting, save the document as a Web page, then view it in a Web browser. Print the completed document for your instructor.

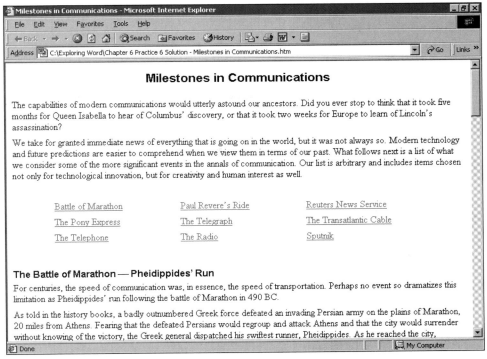

FIGURE 6.15   *Milestones in Communication (Exercise 8)*

9. **Meet Bob and Maryann:** Bob Grauer and Maryann Barber are full-time faculty at the University of Miami. The authors welcome you to visit their Web sites (www.bus.miami.edu/~rgrauer and www.bus.miami.edu/~mbarber) to view this semester's assignments. Maryann's site is shown in Figure 6.16 and should have a familiar look—it is based on the Web Page Wizard. Choose either author, look around, and summarize your findings in a brief note to your instructor.

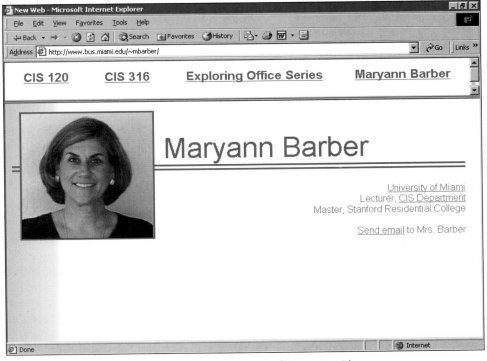

FIGURE 6.16   *Meet Bob and Maryann (Exercise 9)*

ON YOUR OWN

Designer Home Pages

Everyone has a personal list of favorite Web sites, but have you ever thought seriously about what makes an attractive Web page? Is an attractive page the same as a useful page? Try to develop a set of guidelines for a designer to follow as he or she creates a Web site, then incorporate these guidelines into a brief report for your instructor. Support your suggestions by referring to specific Web pages that you think qualify for your personal "Best (Worst) of the Web" award.

Register a Domain Name

InterNIC is the government agency responsible for assigning and maintaining domain names on the Web. A domain name is a mnemonic (easy-to-remember) name such as www.microsoft.com that corresponds to a numeric Internet Protocol (IP) address that represents the true location of the site. Use your favorite search engine to locate the InterNIC site, then search on the name for a hypothetical business and report your success to your instructor. Is the name of your potential business taken? If so, who owns the name and when was it registered? Remain at the site until you can find a name suitable for your business that is not yet taken. How much does it cost to register the name? Summarize your findings in a short note to your instructor.

Employment Opportunities

The Internet abounds with employment opportunities, help-wanted listings, and places to post your résumé. Your home page reflects your skills and experience to the entire world, and represents an incredible opportunity never before available to college students. You can encourage prospective employers to visit your home page, and make contact with hundreds more companies than would otherwise be possible. Update your home page to include a link to your résumé, and then surf the Net to find places to register it.

Front Page

Microsoft Word is an excellent way to begin creating Web documents. It is only a beginning, however, and there are many specialty programs with significantly more capability. One such product is Front Page, a product aimed at creating a Web site as opposed to isolated documents. Search the Microsoft Web site for information on Front Page, then summarize your findings in a short note to your instructor. Be sure to include information on capabilities that are included in Front Page that are not found in Word.

UNIX Permissions

As a developer, you need to be able to log on to your account, to add or modify documents. You also want others to be able to go to your URL to view the page, but you want to prevent the world at large from being able to modify the documents. This is controlled on the Web server through the operating system by setting appropriate permissions for each document. Return to the second hands-on exercise that described how to Telnet to a Web server and view the associated documents on that server. Look to the left of the listed files and note the various letters that indicate different levels of permission for different users.

CHAPTER 7

The Expert User: Workgroups, Forms, Master Documents, and Macros

OBJECTIVES

AFTER READING THIS CHAPTER YOU WILL BE ABLE TO:

1. Describe how to highlight editing changes in a document, and how to review, accept, or reject those changes.
2. Save multiple versions of a document and/or save a document with password protection.
3. Create and modify a form containing text fields, check boxes, and a drop-down list.
4. Create and modify a table; perform calculations within a table.
5. Sort the rows in a table in ascending or descending sequence according to the value of a specific column in the table.
6. Create a master document; add and/or modify subdocuments.
7. Explain how macros facilitate the execution of repetitive tasks; record and run a macro; view and edit the statements in a simple macro.
8. Use the Copy and Paste commands to duplicate an existing macro; modify the copied macro to create an entirely new macro.

OVERVIEW

This chapter introduces several capabilities that will make you a true expert in Microsoft Word. The features go beyond the needs of the typical student and extend to capabilities that you will appreciate in the workplace, as you work with others on a collaborative project. We begin with a discussion of workgroup editing, whereby suggested revisions from one or more individuals can be stored electronically within a document. This enables the original author to review each suggestion individually before it is incorporated into the document, and further, allows multiple people to work on a document in collaboration with one another.

The forms feature is covered as a means to facilitate data entry. Forms are ideal for documents that are used repetitively, where much of the text is constant but where there is variation in specific places (fields) within the document. The chapter also extends the earlier discussion on tables to include both sorting and calculations within a table, giving a Word document the power of a simple spreadsheet. We also describe the creation of a master document, a special type of structure that references one or more subdocuments, each of which is saved under a different name. A master document is useful when many individuals work on a common project, with each person assigned to a different task within the project.

The chapter ends with a discussion of macros, a technique that lets you automate the execution of any type of repetitive task. We create a simple macro to insert your name into a document, then we expand that macro to create a title page for any document. As always, the hands-on exercises enable you to implement the conceptual material at the computer.

WORKGROUPS

As a student, you have the final say in the content of your documents. In the workplace, however, it's common for several people to work on the same document. You may create the initial draft, then submit your work to a supervisor who suggests various changes and revisions. Word facilitates this process by enabling the revisions to be stored electronically within the document. The revisions can come from a single individual or from several persons working together on a project team or ***workgroup***.

Consider, for example, Figure 7.1, which displays two different versions of a document. Figure 7.1a contains the original document with suggested revisions, whereas Figure 7.1b shows the finished document after the changes have been made. The persons who entered the revisions could have made the changes directly in the document, but decided instead to give the original author the opportunity to accept or reject each change individually. The suggestions are entered into the Word document and appear on screen and/or the printed page, just as they might appear if they had been marked with pencil and paper.

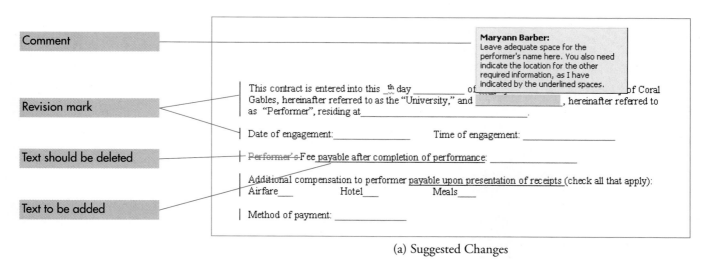

(a) Suggested Changes

FIGURE 7.1 *Workgroup Editing*

```
This contract is entered into this __th day _____ of 20__, by and between the University of
Coral Gables, hereinafter referred to as the "University," and _____,
hereinafter referred to as "Performer", residing at _____.

Date of engagement: _____        Time of engagement: _____

Fee payable after completion of performance: _____

Additional compensation to performer payable upon presentation of receipts (check all that apply):
    Airfare ___        Hotel ___        Meals ___

Method of payment: _____
```

(b) Revised Document

FIGURE 7.1 *Workgroup Editing (continued)*

The notation is simple and intuitive. A ***revision mark*** (a vertical line outside the left margin) signifies a change (an addition or deletion) has been made at that point in the document. A line through existing text indicates that the text should be deleted, whereas text that is underlined is to be added. The suggestions of multiple reviewers appear in different colors, with each reviewer assigned a different color. Yellow highlighting denotes a comment indicating that the reviewer has added a descriptive note without making a specific change. The comment appears on the screen when the cursor is moved over the highlighted text. (Comments can be printed at the end of a document.)

The review process is straightforward. The initial document is sent for review to one or more individuals, who enter their changes through tools on the ***Reviewing toolbar*** or through the ***Track Changes command*** in the Tools menu. The author of the original document receives the corrected document, then uses the ***Accept and Review Changes command*** to review the document and implement the suggested changes.

Versions

The Save command is one of the most basic in Microsoft Office. Each time you execute the command, the contents in memory are saved to disk under the designated filename, and the previous contents of the file are erased. What if, however, you wanted to retain the previous version of the file in addition to the current version that was just saved? You could use the Save As command to create a second file. It's easier to use the ***Versions command*** in the File menu because it lets you save multiple versions of a document in a single file.

The existence of multiple versions is transparent in that the latest version is opened automatically when you open the file at the start of a session. You can, however, review previous versions to see the changes that were made. Word displays the date and time each version was saved as well as the name of the person who saved each version.

Word provides two different levels of ***password protection*** in conjunction with saving a document. You can establish one password to open the document and a different password to modify it. A password can contain any combination of letters, numbers, and symbols, and can be 15 characters long. Passwords are case-sensitive.

FORMS

Forms are ubiquitous in the workplace and our society. You complete a form, for example, when you apply for a job or open any type of account. The form may be electronic and completed online, or it may exist as a printed document. All forms, however, are designed for some type of data entry. Microsoft Word lets you create a special type of document called a *form*, which allows the user to enter data in specific places, but precludes editing the document in any other way. The process requires you to create the form and save it to disk, where it serves as a template for future documents. Then, when you need to enter data for a specific document, you open the original form, enter the data, and save the completed form as a new document.

Figure 7.2 displays a "forms" version of the document shown earlier in Figure 7.1. The form does not contain specific data, but it does contain the text of a document (a contract in this example) that is to be completed by the user. It also contains shaded entries, or *fields*, that represent the locations where the user enters the data. To complete the form, the user presses the Tab key to go from one field to the next and enters data as appropriate. Then, when all fields have been entered, the form is printed to produce the finished document (a contract for a specific event). The data that was entered into the various fields appears as regular text.

The form is created as a regular document with the various fields added through tools on the **Forms toolbar**. Word enables you to create three types of fields—text boxes, check boxes, and drop-down list boxes. A *text field* is the most common and is used to enter any type of text. The length of a text field can be set exactly; for example, to two positions for the day in the first line of the document. The length can also be left unspecified, in which case the field will expand to the exact number of positions that are required as the data is entered. A *check box*, as the name implies, consists of a box, which is checked or not. A *drop-down list box* enables the user to choose from one of several existing entries.

After the form is created, it is protected to prevent further modification other than data entry. Our next exercise has you open an existing document, review changes to that document as suggested by members of a workgroup, accept the changes as appropriate, then convert the revised document into a form for data entry.

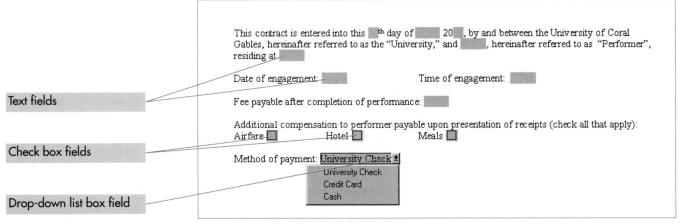

FIGURE 7.2 *A Blank Form*

Hands-on Exercise 1

Workgroups and Forms

Objective To review the editing comments within a document; to create a form containing text fields, check boxes, and a drop-down list.

Step 1: **Display the Forms and Reviewing Toolbars**

- Start Word. If Word is already open, pull down the **File menu** and click the **Close command** to close any open documents.
- Point to any visible toolbar, click the **right mouse button**, then click the **Customize command** to display the Customize dialog box as shown in Figure 7.3a.
- If necessary, click the **Toolbars tab** in the Customize dialog box. The boxes for the Standard and Formatting toolbars should be checked.
- Check the boxes to display the **Forms** and **Reviewing toolbars** as shown in Figure 7.3a. Click the **Close button** to close the Customize dialog box.

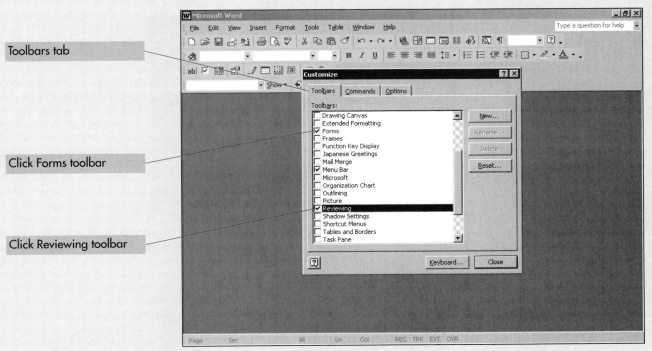

(a) Display the Forms and Reviewing Toolbars (step 1)

FIGURE 7.3 *Hands-on Exercise 1*

DOCKED VERSUS FLOATING TOOLBARS

A toolbar is either docked along an edge of a window or floating within the window. To move a docked toolbar, click and drag the move handle (the vertical line that appears at the left of the toolbar) to a new position. To move a floating toolbar, click and drag its title bar—if you drag a floating toolbar to the edge of the window, it becomes a docked toolbar and vice versa. You can also change the shape of a floating toolbar by dragging any border in the direction you want to go.

Step 2: **Highlight the Changes**

➤ Open the document called **Contract** in the **Exploring Word folder** as shown in Figure 7.3b. Save the document as **Modified Contract**.
➤ Pull down the **Tools menu**, click (or point to) the **Track Changes command**.
➤ The Track Changes command functions as a toggle switch; that is, execute the command, and the tracking is in effect. Execute the command a second time, and the tracking is off. You can track changes in one of three ways:
- Pull down the **Tools menu** and click the **Track Changes command**.
- Double click the **TRK indicator** on the status bar.
- Click the **Track Changes button** on the Reviewing toolbar.
➤ Tracking is in effect if you see the TRK indicator on the status bar.
➤ Press **Ctrl+Home** to move to the beginning of the document. Press the **Del key** four times to delete the word "The" and the blank space that follows. You will see an indication in the right margin that the text was deleted.
➤ Move to the end of the address (immediately after the zip code). Press the **space bar** three or four times, then enter the phone number, **(305) 111-2222**. The new text is underlined.

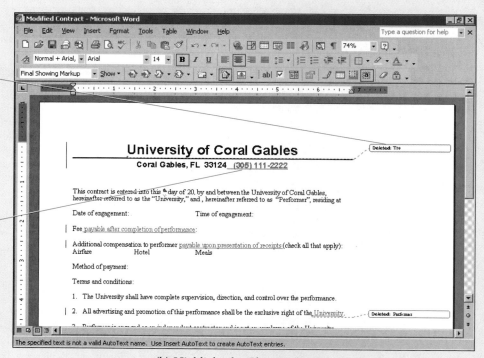

(b) Highlight the Changes (step 2)

FIGURE 7.3 *Hands-on Exercise 1 (continued)*

CHANGE THE EDITING MARKS

Red is the default color used to indicate changes to a document. Text that is added to a document is underlined in red, whereas text that is deleted is shown with a line through the deleted portion. You can, however, change either the color or the editing marks. Pull down the Tools menu, click the Track Changes command, click Highlight Changes to display the Highlight Changes dialog box, then click the Options buttons to display the Track Changes dialog box. Enter your editing preferences and click OK.

Step 3: **Accept or Reject Changes**

- Press **Ctrl+Home** to move to the beginning of the document, then click the **Next button** on the Reviewing toolbar to move to the first change, which is your deletion of the word "the".
- Click the **Accept Change button** to accept the change. Click the **Next button** to move to the next change, where you will review the next change.
- You can continue to review changes individually, or you can accept all of the changes as written. Click the **down arrow** on the Accept Change button and click **Accept All Changes in Document** as shown in Figure 7.3c.
- Save the document.

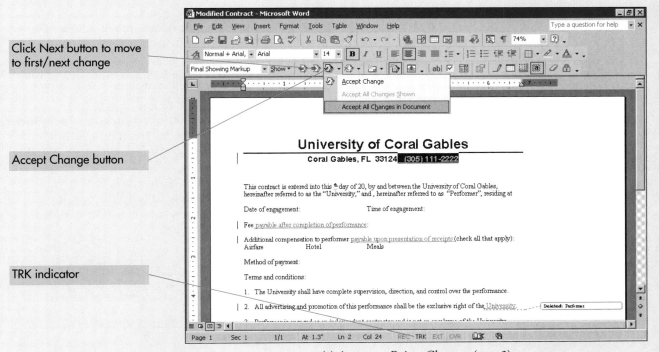

(c) Accept or Reject Changes (step 3)

FIGURE 7.3 *Hands-on Exercise 1 (continued)*

INSERT COMMENTS INTO A DOCUMENT

Add comments to a document to remind yourself (or a reviewer) of action that needs to be taken. Click in the document where you want the comment to appear, then pull down the Insert menu and click the Comment command to open the Comments window. Enter the text of the comment, then close the Comments window. The word containing the insertion point is highlighted in yellow to indicate that a comment has been added. Point to the highlighted entry, and the text of the comment is displayed in a ScreenTip. Edit or delete existing comments by right clicking the comment, then choosing the Edit Comment or Delete Comment command.

Step 4: **Create the Text and Check Box Fields**

- Click the **Track Changes button** on the Reviewing toolbar to stop tracking changes, which removes the TRK indicator from the status bar. Click the button a second time and tracking is again in effect. (You can also double click the **TRK indicator** on the status bar to toggle tracking on or off.)
- Move to the first line of text in the contract, then click to the right of the space following the second occurrence of the word "this".
- Click the **Text Form Field button** on the Forms toolbar to create a text field as shown in Figure 7.3d. The field should appear in the document as a shaded entry (see boxed tip). Do not worry, however, about the length of this field as we adjust it shortly via the Text Form Field Options dialog box (which is not yet visible).
- Click after the word **of** on the same line and insert a second text field followed by a blank space. Insert the six additional text fields as shown in Figure 7.3d. Add blank spaces as needed before each field.
- Click immediately after the word **Airfare**. Add a blank, then click the **Check Box Form Field** to create a check box as shown in the figure. Create additional check boxes after the words **Hotel** and **Meals**.
- Click in the first text field (after the word *this*), then click the **Form Field Options button** on the Forms toolbar to display the Text Form Field Options dialog box. Click the **down arrow** in the Type list box and choose **Number**. Enter **2** in the Maximum Length box.
- Click **OK** to accept these settings and close the dialog box. The length of the form field changes automatically to two positions. Change the options for the Year (Number, 2 positions) and Date of Engagement fields (Date, MMMM d, yyyy format) in similar fashion. Save the document.

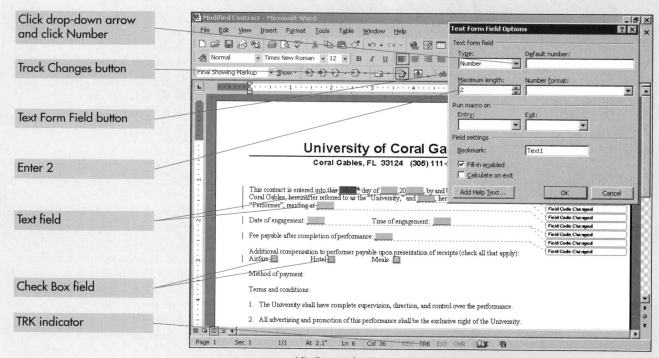

(d) Create the Text and Check Box Fields (step 4)

FIGURE 7.3 *Hands-on Exercise 1 (continued)*

Step 5: **Add the Drop-Down List Box**

- Double click the **TRK indicator** on the status bar to stop tracking changes. (The indicator should be dim after double clicking.)
- Click the **down arrow** on the Accept Changes button on the Reviewing toolbar. Click **Accept All Changes in Document**.
- Click in the document after the words **Method of Payment**, then click the **Drop-down Form Field button** to create a drop-down list box. **Double click** the newly created field to display the dialog box in Figure 7.3e.
- Click in the Drop-down Item text box, type **University Check**, and click the **Add button** to move this entry to the Items in drop-down list box. Type **Credit Card** and click the **Add button**. Type **Cash**, then click the **Add button** to complete the entries for the drop-down list box.
- Click **OK** to accept the settings and close the dialog box. Save the document.

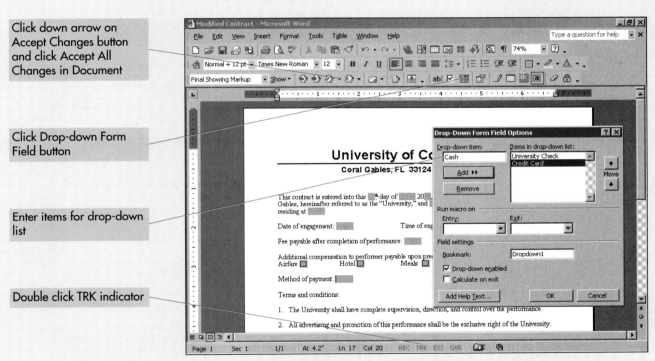

(e) Add the Drop-down List Box (step 5)

FIGURE 7.3 *Hands-on Exercise 1 (continued)*

FIELD CODES VERSUS FIELD RESULTS

All fields are displayed in a document in one of two formats, as a field code or as a field result. A field code appears in braces and indicates instructions to insert variable data when the document is printed; a field result displays the information as it will appear in the printed document. (The field results of a form field are blank until the data is entered into a form.) You can toggle the display between the field code and field result by selecting the field and pressing Shift+F9 during editing.

Step 6: **Save a New Version**

- Proofread the document to be sure that it is correct. Once you are satisfied with the finished document, click the **Protect Form button** on the Forms toolbar to prevent further changes to the form. (You can still enter data into the fields on the form, as we will do in the next step.)
- Pull down the **File menu** and click the **Versions command** to display the Versions dialog box for this document. There is currently one previous version, the one created by Robert Grauer on August 28, 1998.
- Click the **Save Now button** to display the Save Version dialog box in Figure 7.3f. Enter the text of a comment you want to associate with this version. The author's name will be different on your screen and will reflect the person who registered the version of Microsoft Word you are using.
- Click **OK** to save the version and close the dialog box.

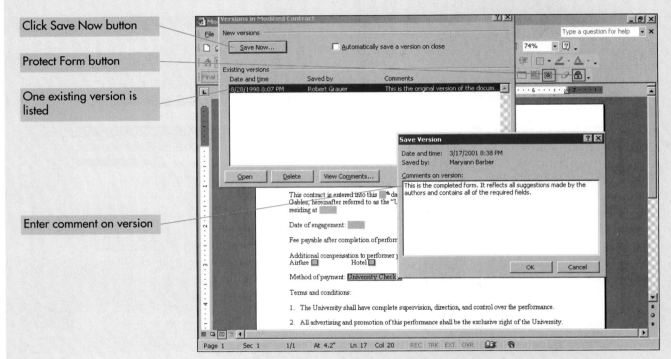

(f) Save a New Version (step 6)

FIGURE 7.3 *Hands-on Exercise 1 (continued)*

CREATE A BACKUP COPY

Microsoft Word enables you to automatically create a backup copy of a document in conjunction with the Save command. Pull down the Tools menu, click the Options button, click the Save tab, then check the box to Always create backup copy. The next time you save the file, the previously saved version is renamed "Backup of document", after which the document in memory is saved as the current version. In other words, the disk will contain the two most recent versions of the document.

Step 7: **Fill In the Form**

- Be sure that the form is protected; that is, that all buttons are dim on the Forms toolbar except for the Protect Form and Form Field Shading buttons. Press **Ctrl+Home** to move to the first field.
- Enter today's date, press the **Tab key** (to move to the next field), enter today's month, press the **Tab key**, and enter the year.
- Continue to press the **Tab key** to complete the form. Enter your name as the performer. Press the **space bar** on the keyboard to check or clear the various check boxes. Check the boxes for airfare, hotel, and meals, and enter a fee of $1,000. Click the **down arrow** on the Method of Payment list box and choose **University Check**.
- Your completed form should be similar to our form as shown in Figure 7.3g. You can make changes to the text of the contract by unprotecting the form. Do not, however, click the Protect Form button after data has been entered or you will lose the data.

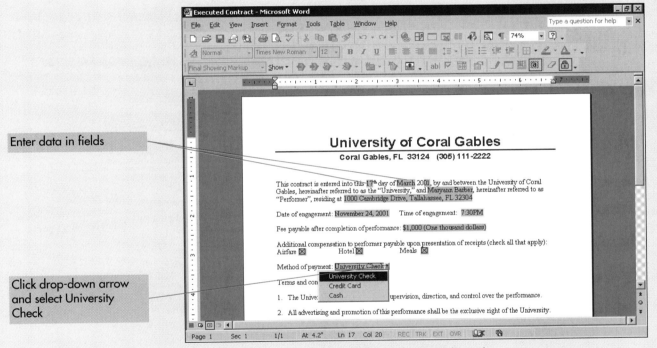

(g) Fill In the Form (step 7)

FIGURE 7.3 *Hands-on Exercise 1 (continued)*

PROTECTING AND UNPROTECTING A FORM

The Protect Form button toggles protection on and off. Click the button once, and the form is protected; data can be entered into the various fields, but the form itself cannot be modified. Click the button a second time, and the form is unprotected and can be fully modified. Be careful, however, about unprotecting a form once data has been entered. That action will not create a problem in and of itself, but protecting a form a second time (after the data was previously entered) will reset all of its fields.

Step 8: **Password Protect the Executed Contract**

- Pull down the **File menu**, click the **Save As command** to display the Save As dialog box, then type **Executed Contract** as the filename.
- Click the **drop-down arrow** next to the Tools button and click the **General Options command** to display the Save dialog box in Figure 7.3h. Click in the **Password to open** text box and enter **password** (the password is case-sensitive) as the password. Click **OK**.
- A Confirm Password dialog box will open, asking you to reenter the password and warning you not to forget the password; once a document is protected by a password, it cannot be opened without that password. Reenter the password and click **OK** to establish the password.
- Click **Save** to save the document and close the Save As dialog box. Exit Word if you do not want to continue with the next exercise at this time.

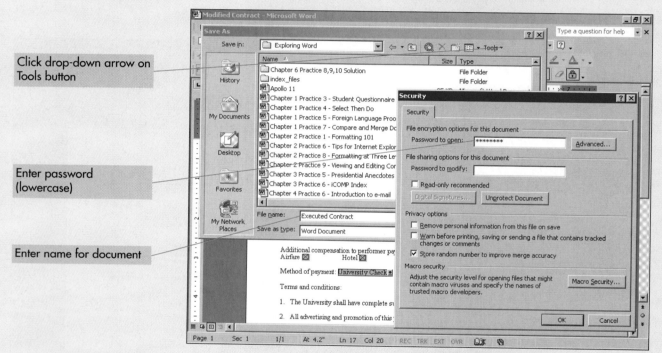

(h) Password Protect the Executed Contract (step 8)

FIGURE 7.3 *Hands-on Exercise 1 (continued)*

AUTHENTICATE YOUR DOCUMENTS

What if you sent a contract or other important document to a third party and the document was intercepted and altered en route? Or more likely, what if someone sent a forged document to a third party in your name? You can avoid both situations by using a digital signature to authenticate your correspondence. A digital signature is an electronic stamp of authenticity that confirms the origin and status of an e-mail attachment. You can obtain a digital signature from a variety of sources, then use Word to apply that signature to any document. See exercise 6 at the end of the chapter.

TABLE MATH

Tables were introduced in an earlier chapter and provide an easy way to arrange text, numbers, and/or graphics within a document. This section extends that discussion to include calculations within a table, giving a Word document the power of a simple spreadsheet. We also describe how to *sort* the rows within a table in a different sequence, according to the entries in a specific column of the table.

We begin by reviewing a few basic concepts. The rows and columns in a table intersect to form *cells*, each of which can contain text, numbers, and/or graphics. Text is entered into each cell individually, enabling you to add, delete, or format text in one cell without affecting the text in other cells. The rows within a table can be different heights, and each row may contain a different number of columns.

The commands in the *Tables menu* or the *Tables and Borders toolbar* operate on one or more cells. The Insert and Delete commands add new rows or columns, or delete existing rows or columns, respectively. Other commands shade and/or border selected cells or the entire table. You can also select multiple cells and merge them into a single cell. All of this was presented earlier, and should be familiar.

Figure 7.4 displays a table of expenses that is associated with the performer's contract. The table also illustrates two additional capabilities that are associated with a table. First, you can sort the rows in a table to display the data in different sequences as shown in Figures 7.4a and 7.4b. Both figures display the same 6×4 table (six rows and four columns). The first row in each figure is a header row and contains the field names for each column. The next four rows contain data for a specific expense, while the last row displays the total for all expenses

Figure 7.4a lists the expenses in alphabetical order—airfare, hotel, meals, and performance fee. Figure 7.4b, however, lists the expenses in *descending* (high to low) *sequence* according to the amount. Thus the performance fee (the largest expense) is listed first, and the meals (the smallest expense) appear last. Note, too, that the sort has been done in such a way as to affect only the four middle rows; that is, the header and total rows have not moved. This is accomplished according to the select-then-do methodology that is used for many operations in Microsoft Word. You select the rows that are to be sorted, then you execute the command (the Sort command in the Tables menu in this example).

Figure 7.4c displays the same table as in Figure 7.4b, albeit in a different format that displays the field codes rather than the field results. The entries consist of formulas that were entered into the table to perform a calculation. The entries are similar to those in a spreadsheet. Thus, the rows in the table are numbered from one to six while the columns are labeled from A to D. The row and column labels do not appear in the table per se, but are used to enter the formulas.

The intersection of a row and column forms a cell. Cell D4, for example, contains the entry to compute the total hotel expense by multiplying the number of days (in cell B4) by the per diem amount (in cell C4). In similar fashion, the entry in cell D5 computes the total expense for meals by multiplying the values in cells B5 and C5, respectively. The formula is not entered (typed) into the cell explicitly, but is created through the Formula command in the Tables menu.

Figure 7.4d is a slight variation of Figure 7.4c in which the field codes for the hotel and meals have been toggled off to display the calculated values as opposed to the field codes. The cells are shaded, however, to emphasize that these cells contain formulas (fields), as opposed to numerical values. (The shading is controlled by the Options command in the Tools menu. The *field codes* are toggled on and off by selecting the formula and pressing the Shift+F9 key or by right clicking the entry and selecting the Toggle Field Codes command.)

The formula in cell D6 has a different syntax and sums the value of all cells directly above it. You do not need to know the syntax since Word provides a dialog box which supplies the entry for you. It's easy, as you shall see in our next hands-on exercise.

Header row →

Expenses are in alphabetical order →

Totals are displayed →

Expense	Number of Days	Per Diem Amount	Amount
Airfare			$349.00
Hotel	2	$129.99	$259.98
Meals	2	$75.00	$150.00
Performance Fee			$1000.00
Total			**$1758.98**

(a) Expenses (Alphabetical Order by Expense)

Descending order by amount →

Expense	Number of Days	Per Diem Amount	Amount
Performance Fee			$1000.00
Airfare			$349.00
Hotel	2	$129.99	$259.98
Meals	2	$75.00	$150.00
Total			**$1758.98**

(b) Expenses (Descending Order by Amount)

Column labels →
Row labels →
Field codes →

	A	B	C	D
1	Expense	Number of Days	Per Diem Amount	Amount
2	Performance Fee			$1000.00
3	Airfare			$349.00
4	Hotel	2	$129.99	{=b4*c4}
5	Meals	2	$75.00	{=b5*c5}
6	Total			{=SUM(ABOVE)}

(c) Field codes

Shading indicates a cell formula →

	A	B	C	D
1	Expense	Number of Days	Per Diem Amount	Amount
2	Performance Fee			$1000.00
3	Airfare			$349.00
4	Hotel	2	$129.99	$259.98
5	Meals	2	$75.00	$150.00
6	Total			$1758.98

(d) Field Codes (Toggles and Shading)

Figure 7.4 *Sorting and Table Math*

CHAPTER 7: THE EXPERT USER

HANDS-ON EXERCISE 2

TABLE MATH

Objective To open a password-protected document and remove the password protection; to create a table containing various cell formulas. Use Figure 7.5.

Step 1: **Open the Document**

➤ Open the **Executed Contract** in the **Exploring Word folder** from the first exercise. You will be prompted for a password as shown in Figure 7.5a.

➤ Type **password** (in lowercase) since this was the password that was specified when you saved the document originally.

➤ Pull down the **File menu** and click the **Save As command** to display the Save As dialog box. Click the **Tools button**. Click **Security Options**.

➤ Click and drag to select the existing password (which appears as a string of eight asterisks). Press the **Del key** to remove the password. Click **OK** to close the Save dialog box. Click the **Save command button** to close the Save As dialog box. The document is no longer password protected.

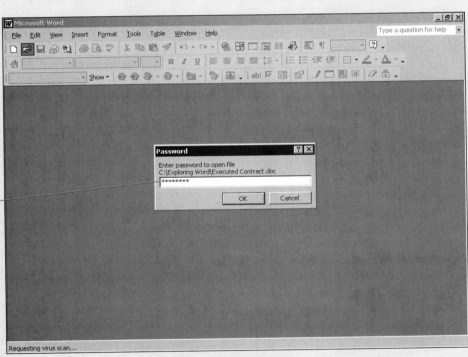

Enter password (lowercase)

(a) Open the Document (step 1)

FIGURE 7.5 *Hands-on Exercise 2*

CHANGE THE DEFAULT FOLDER

The default folder is the folder where Word saves and retrieves documents unless it is otherwise instructed. To change the default folder, pull down the Tools menu, click Options, click the File Locations tab, click Documents, and click the Modify command button. Enter the name of the new folder (for example, C:\Exploring Word), click OK, then click the Close button.

Step 2: **Review the Contract**

➤ You should see the executed contract from the previous exercise. Click the **Protect Form button** on the Forms toolbar to unprotect the document so that its context can be modified.

➤ Do *not* click the Protect Form button a second time or else the data will disappear. (You can quit the document without saving the changes, as described in the boxed tip below.)

➤ Point to any toolbar, and click the **right mouse button** to display the list of toolbars shown in Figure 7.5b. Click the **Forms toolbar** to toggle the toolbar off (the check will disappear). Right click any toolbar a second time, and toggle the Reviewing toolbar off as well.

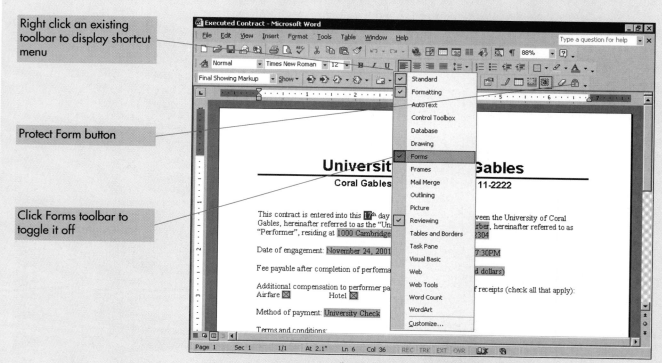

(b) Review the Contract (step 2)

FIGURE 7.5   *Hands-on Exercise 2 (continued)*

QUIT WITHOUT SAVING

There will be times when you do not want to save the changes to a document—for example, when you have edited it beyond recognition and wish you had never started. Pull down the File menu and click the Close command, then click No in response to the message asking whether you want to save the changes to the document. Pull down the File menu and reopen the file (it should be the first file in the list of most recently edited documents), then start over from the beginning.

Step 3: **Create the Table**

> ➤ Press **Ctrl+End** to move to the end of the contract, then press **Ctrl+Enter** to create a page break. You should be at the top of page two of the document.
> ➤ Press the **enter key** three times and then enter **Summary of Expenses** in **24-point Arial bold** as shown in Figure 7.5c. Center the text. Press **enter** twice to add a blank line under the heading.
> ➤ Change to **12-point Times New Roman**. Click the **Insert Table button** on the Standard toolbar to display a grid, then drag the mouse across and down the grid to create a 6 × 4 table (six rows and four columns). Release the mouse to create the table.
> ➤ Enter data into the table as shown in Figure 7.5c. You can format the column headings by selecting multiple cells, then clicking the **Center button** on the Standard toolbar. In similar fashion, you can right justify the numerical data by selecting the cells and clicking the **Align Right button**.
> ➤ Save the document.

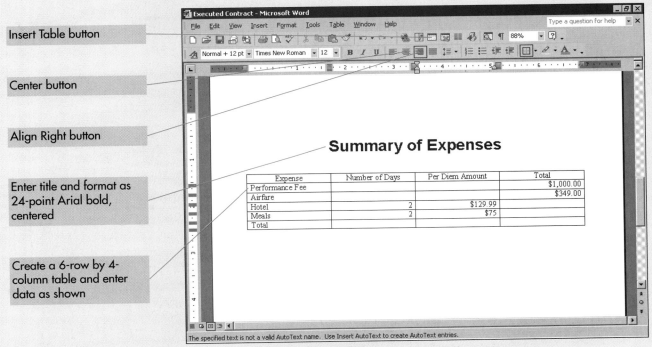

(c) Create the Table (step 3)

FIGURE 7.5 *Hands-on Exercise 2 (continued)*

TABS AND TABLES

The Tab key functions differently in a table than in a regular document. Press the Tab key to move to the next cell in the current row (or to the first cell in the next row if you are at the end of a row). Press Tab when you are in the last cell of a table to add a new blank row to the bottom of the table. Press Shift+Tab to move to the previous cell in the current row (or to the last cell in the previous row). You must press Ctrl+Tab to insert a regular tab character within a cell.

Step 4: **Sort the Table**

> ➤ Click and drag to select the entire table except for the last row. Pull down the **Table menu** and click the **Sort command** to display the Sort dialog box in Figure 7.5d.
> ➤ Click the **drop-down arrow** in the Sort by list box and select **Expense** (the column heading for the first column). The **Ascending option button** is selected by default.
> ➤ Verify that the option button to include a Header row is selected. Click **OK**. The entries in the table are rearranged alphabetically according to the entry in the Expenses column. The Total row remains at the bottom of the table since it was not included in the selected rows for the Sort command.
> ➤ Save the document.

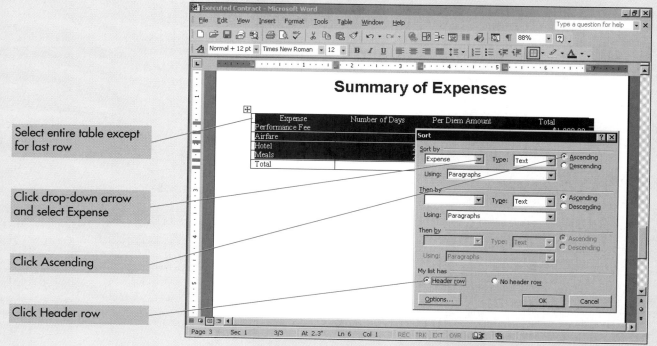

(d) Sort the Table (step 4)

FIGURE 7.5 *Hands-on Exercise 2 (continued)*

THE HEADER ROW

The first row in a table is known as the header row and contains the column names (headings) that describe the value in each column of the table. The header row is typically included in the range selected for the sort so that the Sort by list box displays the column names. The header row must remain at the top of the table, however, and thus it is important that the option button that indicates a header row be selected. In similar fashion, the last row typically contains the totals and should remain as the bottom row of the table. Hence it (the total row) is not included in the rows that are selected for sorting.

Step 5: **Enter the Formulas for Row Totals**

> ➤ Click in **cell D3** (the cell in the fourth column and third row). Pull down the **Table menu** and click the **Formula command** to display the Formula dialog box.
> ➤ Click and drag to select the =SUM(ABOVE) function, which is entered by default. Type **=b3*c3** as shown in Figure 7.5e to compute the total hotel expense. The total is computed by multiplying the number of days (in cell B3) by the per diem amount (in cell C3). Click **OK**. You should see $259.98 in cell D3.
> ➤ Click in **cell D4** and repeat the procedure to enter the formula **=b4*c4** to compute the total expense for meals. You should see $150.00 (two days at $75.00 per day). Save the document.

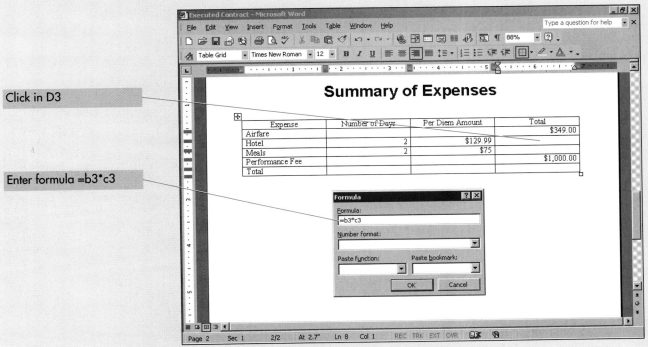

(e) Enter the Formulas for Row Totals (step 5)

FIGURE 7.5 *Hands-on Exercise 2 (continued)*

IT'S NOT EXCEL

Your opinion of table math within Microsoft Word depends on what you know about a spreadsheet. If you have never used Excel, then you will find table math to be very useful, especially when simple calculations are necessary within a Word document. If, on the other hand, you know Excel, you will find table math to be rather limited; for example, you cannot copy a formula from one cell to another, but must enter it explicitly in every cell. Nevertheless, the feature enables simple calculations to be performed entirely within Word, without having to link an Excel worksheet to a Word document.

Step 6: **Enter the SUM(ABOVE) Formula**

➤ Click in **cell D6** (the cell in row 6, column 4), which is to contain the total of all expenses. Pull down the **Table menu** and click the **Formula command** to display the Formula dialog box in Figure 7.5f.

➤ The =SUM(ABOVE) function is entered by default. Click **OK** to accept the formula and close the dialog box. You should see $1,758.98 (the sum of the cells in the last column) displayed in the selected cell.

➤ Select the formula and press **Shift+F9** to display the code {=SUM (ABOVE)}. Press **Shift+F9** a second time to display the field value ($1,758.98).

➤ Click in **cell D2** (the cell containing the airfare). Replace $349 with **$549.00** and press the **Tab key** to move out of the cell. The total expenses are *not* yet updated in cell D6.

➤ Point to **cell D6**, click the **right mouse button** to display a context-sensitive menu, and click the **Update Field command**. Cell D6 displays $1,958.98, the correct total for all expenses.

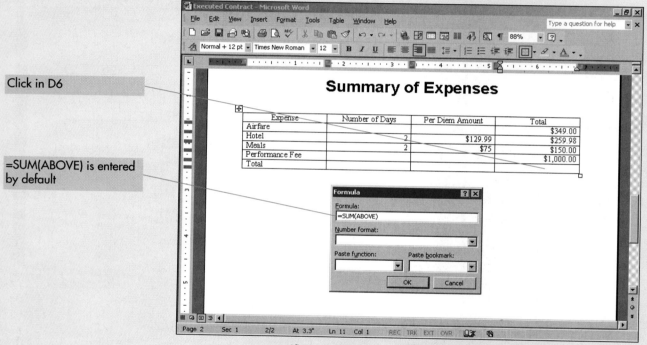

(f) Enter the SUM(ABOVE) Formula (step 6)

FIGURE 7.5   *Hands-on Exercise 2 (continued)*

FORMATTING A CALCULATED VALUE

Word does its best to format a calculation according to the way you want it. You can, however, change the default format by clicking the down arrow on the Number format list box and choosing a different format. You can also enter a format directly in the Number format text box. To display a dollar sign and comma without a decimal point, enter $#,##0 into the text box. You can use trial and error to experiment with other formats.

Step 7: **Print the Completed Contract**

- Zoom to two pages to preview the completed document. The first page contains the text of the executed contract that was completed in the previous exercise. The second page contains the table of expenses from this contract.
- Pull down the **File menu** and click the **Print command** to display the Print dialog box in Figure 7.5g. Click the **Options command button** to display the second Print dialog box.
- Check the box to include **Field codes** with the document. Click **OK** to close that dialog box, then click **OK** to print the document.
- Repeat the process to print the document a second time, but this time with field values, rather than field codes. Thus, pull down the **File menu**, click the **Print command** to display the Print dialog box, and click the **Options command button** to display a second Print dialog box.
- Clear the box to include **Field codes** with the document. Click **OK** to close that dialog box, then click **OK** to print the document.
- Exit Word if you do not want to continue with the next exercise at this time. Click **Yes** if asked to save the changes.

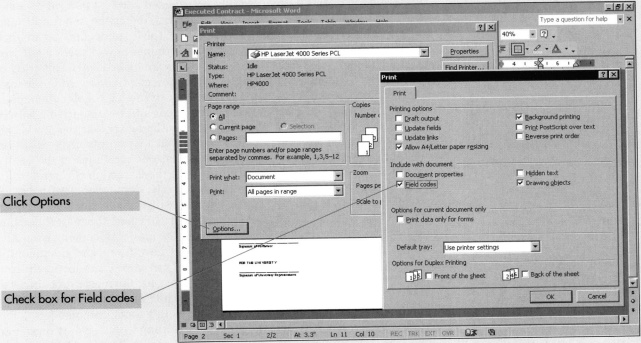

(g) Print the Completed Contract (step 7)

FIGURE 7.5 *Hands-on Exercise 2 (continued)*

DOCUMENT PROPERTIES

Prove to your instructor how hard you've worked by printing various statistics about your document, including the number of revisions and the total editing time. Pull down the File menu, click the Print command to display the Print dialog box, click the drop-down arrow in the Print What list box, select Document properties, then click OK.

MASTER DOCUMENTS

A ***master document*** is composed of multiple ***subdocuments***, each of which is stored as a separate file. The advantage of the master document is that you can work with several smaller documents, as opposed to a single large document. Thus, you edit the subdocuments individually and more efficiently than if they were all part of the same document. You can create a master document to hold the chapters of a book, where each chapter is stored as a subdocument. You can also use a master document to hold multiple documents created by others, such as a group project, where each member of the group is responsible for a section of the document.

Figure 7.6 displays a master document with five subdocuments. The subdocuments are collapsed in Figure 7.6a and expanded in Figure 7.6b. (The ***Outlining toolbar*** contains the Collapse and Expand Subdocument buttons, as well as other tools associated with master documents.) The collapsed structure in Figure 7.6a enables you to see at a glance the subdocuments that comprise the master document. You can insert additional subdocuments and/or remove existing subdocuments from the master document. Deleting a subdocument from within a master document does *not* delete the subdocument from disk.

The expanded structure in Figure 7.6b enables you to view and/or edit the contents of the subdocuments. Look carefully, however, at the first two subdocuments in Figure 7.6b. A padlock appears to the left of the first line in the first subdocument, whereas it is absent from the second subdocument. These subdocuments are locked and unlocked, respectively, and the distinction determines how changes made within the master document are saved. (All subdocuments are locked when collapsed as in Figure 7.6a.) Changes made to the locked subdocument will be saved in the master document, but not in the subdocument. Changes to the unlocked subdocument, however, will be saved in both the master document and the underlying subdocument. (The Lock Subdocuments button on the Outlining toolbar toggles between locked and unlocked subdocuments.) Either approach is acceptable. You just need to understand the difference, as you may want to use one technique or the other.

Regardless of how you edit the subdocuments, the attraction of a master document is the ability to work with multiple subdocuments simultaneously. The subdocuments are created independently of one another, with each subdocument stored in its own file. Then, when all of the subdocuments are finished, the master document is created and the subdocuments are inserted into the master document, from where they are easily accessed. Inserting page numbers into the master document, for example, causes the numbers to run consecutively from one subdocument to the next. You can also create a table of contents or index for the master document that will reflect the entries in all of the subdocuments. And finally you can print all of the subdocuments from within the master document with a single command.

Alternatively, you can reverse the process by starting with an empty master document and using it as the basis to create the subdocuments. This is ideal for organizing a group project in school or at work, the chapters in a book, or the sections in a report. Start with a new document, enter the topics assigned to each group member as headings within the master document, then use the ***Create Subdocument command*** to create subdocuments based on those headings. Saving the master document will automatically save each subdocument in its own file. This is the approach that we will follow in our next hands-on exercise.

The exercise also illustrates the ***Create New Folder command*** that lets you create a new folder on your hard drive (or floppy disk) from within Microsoft Word, as opposed to using Windows Explorer. The new folder can then be used to store the master document and all of its subdocuments in a single location apart from any other documents.

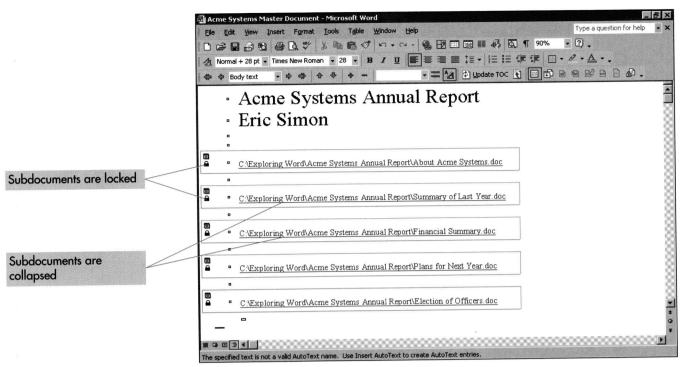

(a) Collapsed Subdocuments

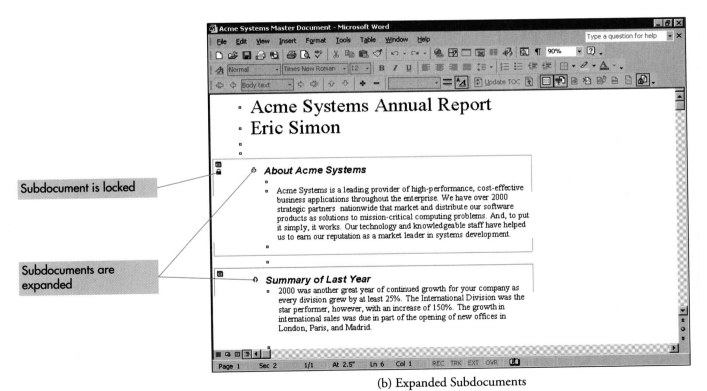

(b) Expanded Subdocuments

FIGURE 7.6   *A Master Document*

Hands-on Exercise 3

Master Documents

Objective To create a master document and various subdocuments; to create a new folder from within the Save As dialog box in Microsoft Word. Use Figure 7.7 as a guide in the exercise.

Step 1: **Create a New Folder**

- Start Word. If necessary, click the **New button** on the Standard toolbar to begin a new document. Enter the text of the document in Figure 7.7a in **12-point Times New Roman**.
- Press **Ctrl+Home** to move to the beginning of the document. Pull down the **Style list box** on the Formatting toolbar, then select **Heading 2** as the style for the document title.
- Click the **Save button** to display the Save As dialog box. If necessary, click the **drop-down arrow** on the Save in list box to select the **Exploring Word folder** you have used throughout the text.
- Click the **Create New Folder button** to display the New Folder dialog box. Type **Acme Systems Annual Report** as the name of the new folder. Click **OK** to create the folder and close the New Folder dialog box.
- The Save in list box indicates that the Acme Systems Annual Report folder is the current folder. The name of the document, **About Acme Systems**, is entered by default (since this text appears at the beginning of the document).
- Click the **Save button** to save the document and close the Save As dialog box.

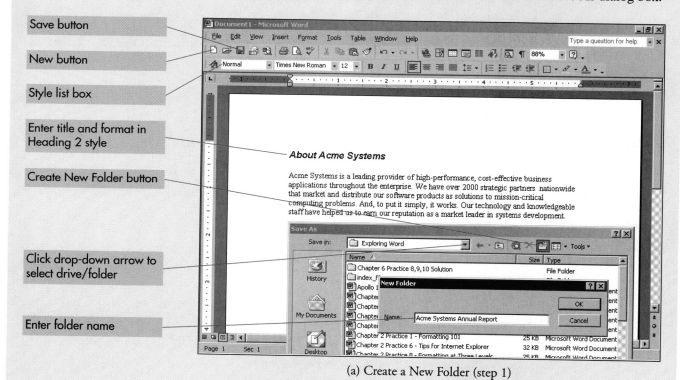

(a) Create a New Folder (step 1)

FIGURE 7.7 *Hands-on Exercise 3*

Step 2: **Create the Master Document**

- Click the **New button** on the Standard toolbar. Enter **Acme Systems Annual Report** as the first line of the document. Enter your name under the title.
- Press the **enter key** twice to leave a blank line or two after your name before the first subdocument. Type **Summary of Last Year** in the default typeface and size. Press **enter**. Enter the remaining topics, **Financial Summary**, **Plans for Next Year**, and **Election of Officers**.
- Change the format of the title and your name to **28-pt Times New Roman**.
- Pull down the **View menu** and click **Outline** to change to the Outline view. Click and drag to select the four headings as shown in Figure 7.7b. Click the **drop-down arrow** on the Style list box and select **Heading 2**.
- Be sure that all four headings are still selected. Press **Create Subdocument button**. Each heading expands automatically into a subdocument.

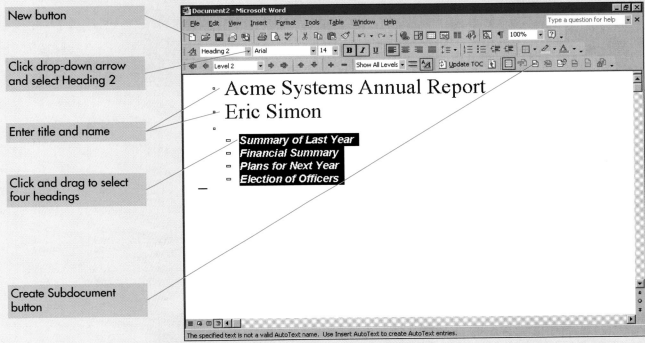

(b) Create the Master Document (step 2)

FIGURE 7.7 *Hands-on Exercise 3 (continued)*

THE CREATE SUBDOCUMENT BUTTON

You can enter subdocuments into a master document in one of two ways, through the Insert Subdocuments button if the subdocuments already exist, or through the Create Subdocument button to create the subdocuments from within the master document. Start a new document, enter the title of each subdocument on a line by itself that is formatted in a heading style, then click the Create Subdocument button to create the subdocuments. Save the master document. The subdocuments are saved automatically as individual files in the same folder.

Step 3: **Save the Documents**

- Click the **Save button** to display the Save As dialog box in Figure 7.7c. If necessary, click the **drop-down arrow** on the Save In list box to select the **Acme Systems Annual Report folder** that was created in step 1. You should see the About Acme Systems document in this folder.
- Enter **Acme Systems Master Document** in the File name list box, then click the **Save button** within the Save As dialog box to save the master document (which automatically saves the subdocuments in the same folder).
- Press the **Collapse Subdocuments button** to collapse the subdocuments. You will see the name of each subdocument as it appears on disk, with the drive and folder information. Press the **Expand Subdocuments button**, and the subdocuments are reopened within the master document.

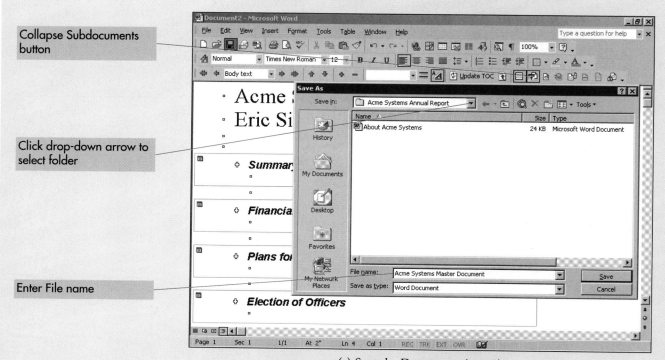

(c) Save the Documents (step 3)

FIGURE 7.7   *Hands-on Exercise 3 (continued)*

HELP WITH TOOLBAR BUTTONS

The Outlining toolbar is displayed automatically in the Outline view and suppressed otherwise. As with every toolbar you can point to any button to see a ToolTip with the name of the button. You can also press Shift+F1 to change the mouse pointer to a large arrow next to a question mark, then click any button to learn more about its function. The Outlining toolbar contains buttons that pertain specifically to master documents such as buttons to expand and collapse subdocuments, or insert and remove subdocuments. The Outlining toolbar also contains buttons to promote and demote items, to display or suppress formatting, and/or to collapse and expand the outline.

Step 4: **Insert a Subdocument**

- Click below your name, but above the first subdocument. Click the **Insert Subdocument button** to display the Insert Subdocument dialog box in Figure 7.7d. If necessary, click the **drop-down arrow** on the Look in list box to change to the Acme Systems Annual Report folder.
- There are six documents, which include the About Acme Systems document from step 1, the Acme Systems Master document that you just saved, and the four subdocuments that were created automatically in conjunction with the master document.
- Select the **About Acme Systems** document, then click the **Open button** to insert this document into the master document. Save the master document.

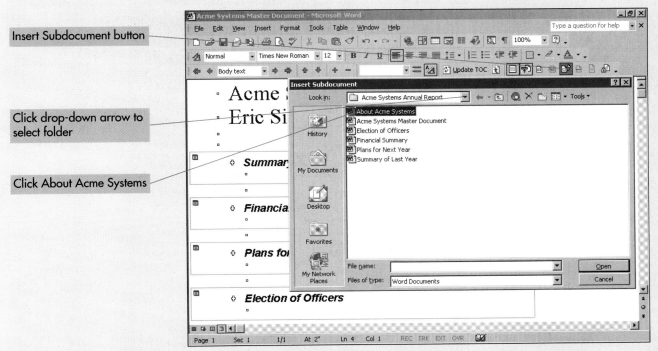

(d) Insert a Subdocument (step 4)

FIGURE 7.7 *Hands-on Exercise 3 (continued)*

CHANGE THE VIEW

The Outline view is used to create and/or modify a master document through insertion, repositioning, or deletion of its subdocuments. You can also modify the text of a subdocument within the Outline view and/or implement formatting changes at the character level such as a change in font, type size, or style. More sophisticated formatting, however, such as changes in alignment, indentation, or line spacing has to be implemented in the Normal or Print Layout views.

Step 5: **Modify a Subdocument**

> ➤ Click within the second subdocument, which will summarize the activities of last year. (The text of the document has not yet been entered.)
> ➤ Click the **Lock Document button** on the Outlining toolbar to display the padlock for this document. Click the **Lock Document button** a second time, which unlocks the document.
> ➤ Enter the text of the document as shown in Figure 7.7e, then click the **Save button** to save the changes to the master document. Be sure the subdocument is unlocked so that the changes you have made will be reflected in the subdocument file as well.

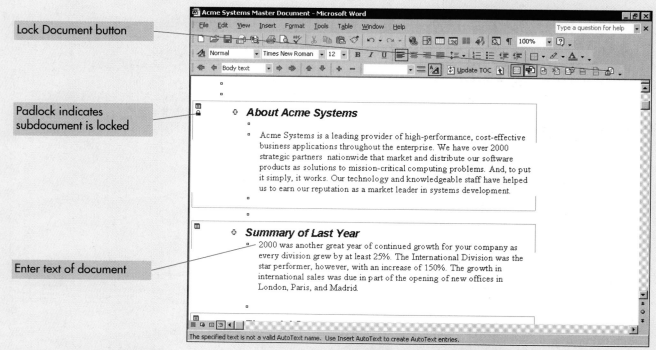

(e) Modify a Subdocument (step 5)

FIGURE 7.7 *Hands-on Exercise 3 (continued)*

OPEN THE SUBDOCUMENT

You can edit the text of a subdocument from within a master document, but it is often more convenient to open the subdocument when the editing is extensive. You can open a subdocument in one of two ways, by double clicking the document icon in the Outline view when the master document is expanded, or by clicking the hyperlink to the document when the Master Document is collapsed. Either way, the subdocument opens in its own window. Enter the changes into the subdocument, then save the subdocument and close its window to return to the master document, which now reflects the modified subdocument.

Step 6: **Print the Completed Document**

- Click the **Collapse Subdocuments button** to collapse the subdocuments as shown in Figure 7.7f. Click **OK** if asked to save the changes in the master document.
- Click the **Print button** on the Standard toolbar to print the document. Click **No** when asked whether to open the subdocuments before printing. The entire document appears on a single page. The text of the subdocuments is not printed, only the address of the documents.
- Click the **Print button** a second time, but click **Yes** when asked whether to open the subdocuments before printing.
- Submit both versions of the printed document to your instructor as proof that you did this exercise. Exit Word if you do not want to continue with the next exercise at this time.

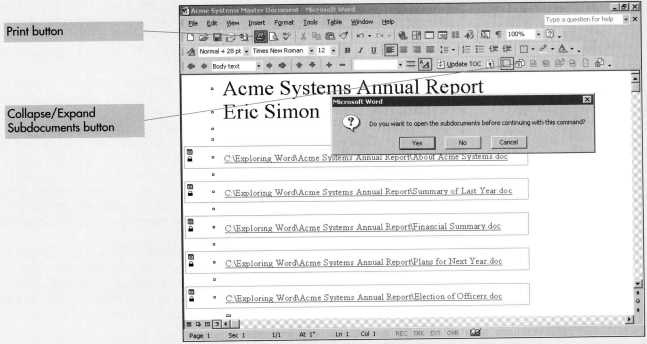

(f) Print the Completed Document (step 6)

FIGURE 7.7 *Hands-on Exercise 3 (continued)*

THE DOCUMENT MAP

The Document Map is one of our favorite features when working with large documents. Be sure that the master document is expanded to display the text of the subdocuments, then click the Document Map button on the Standard toolbar to divide the screen into two panes. The headings in a document are displayed in the left pane, and the text of the document is visible in the right pane. To go to a specific point in a document, click its heading in the left pane, and the insertion point is moved automatically to that point in the document, which is visible in the right pane. Click the Document Map button a second time to turn the feature off.

INTRODUCTION TO MACROS

Have you ever pulled down the same menus and clicked the same sequence of commands over and over? Easy as the commands may be to execute, it is still burdensome to continually repeat the same mouse clicks or keystrokes. If you can think of any task that you do repeatedly, whether in one document or in a series of documents, you are a perfect candidate to use macros.

A *macro* is a set of instructions (that is, a program) that executes a specific task. It is written in *Visual Basic for Applications (VBA)*, a programming language that is built into Microsoft Office. Fortunately, however, you don't have to be a programmer to use VBA. Instead, you use the *macro recorder* within Word to record your actions, which are then translated automatically into VBA. You get results that are immediately usable, and you can learn a good deal about VBA through observation.

Figure 7.8 illustrates a simple macro to enter your name, date, and class into a Word document. We don't expect you to be able to write the VBA code by yourself, but, as indicated, you don't have to. You just invoke the macro recorder and let it create the VBA statements for you. It is important, however, for you to understand the individual statements so that you can modify them as necessary. Do not be concerned with the precise syntax of every statement, but try instead to get an overall appreciation of what the statements do.

Every macro begins and ends with a Sub and End Sub statement, respectively. These statements identify the macro and convert it to a VBA *procedure*. The *Sub statement* contains the name of the macro, such as NameAndCourse in Figure 7.8. (Spaces are not allowed in a macro name.) The *End Sub statement* is always the last statement in a VBA procedure. Sub and End Sub are Visual Basic key words and appear in blue.

The next several statements begin with an apostrophe, appear in green, and are known as *comments*. Comments provide information about the procedure, but do not affect its execution. The comments are inserted automatically by the macro recorder and include the name of the macro, the date it was recorded, and the author. Additional comments can be inserted at any time.

Every other statement in the procedure corresponds directly to a command that was executed in Microsoft Word. It doesn't matter how the commands were executed—whether from a pull-down menu, toolbar, or keyboard shortcut, because the end results, the VBA statements that are generated by the commands, are the same. In this example, the user began by changing the font and font size, and these com-

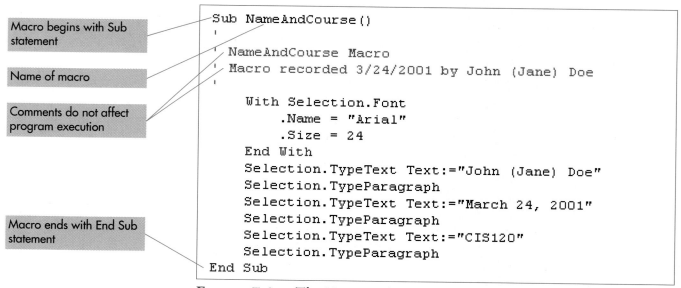

FIGURE 7.8 *The NameAndCourse Macro*

mands were converted by the macro recorder to the VBA statements that specify Arial and 24-point, respectively. Next, the user entered his name and pressed the enter key to begin a new paragraph. Again, the macro recorder converts these actions to the equivalent VBA statements. The user entered the date, pressed the enter key, entered the class, and pressed the enter key. Each of these actions resulted in an additional VBA statement.

You do not have to write VBA statements from scratch, but you should understand their function once they have been recorded. You can also edit the statements after they have been recorded, to change the selected text and/or its appearance. It's easy, for example, to change the procedure to include your name instead of John Doe. All changes to a macro are done through the Visual Basic Editor.

The Visual Basic Editor

Figure 7.9a displays the NameAndCourse macro as it appears within the ***Visual Basic Editor (VBE)***. The Visual Basic Editor is a separate application (as can be determined from its button on the taskbar in Figure 7.9), and it is accessible from any application in Office XP. The left side of the VBE window displays the ***Project Explorer***, which is similar in concept and appearance to the Windows Explorer. Macros are stored by default in the Normal template, which is available to all Word documents. The VBA code is stored in the NewMacros module. (A ***module*** contains one or more procedures.)

The macros for the selected module (NewMacros in Figure 7.9) appear in the ***Code window*** in the right pane. (Additional macros, if any, are separated from one another by a horizontal line.) The VBA statements are identical to what we described earlier. The difference between Figure 7.8 and 7.9a is that the latter shows the macro within the Visual Basic Editor.

Figure 7.9b displays the TitlePage macro, which is built from the NameAndCourse macro. The new macro (a VBA procedure) is more complicated than its predecessor. "Complicated" is an intimidating word, however, and we prefer to use "powerful" instead. In essence, the TitlePage procedure moves the insertion point to the beginning of a Word document, inserts three blank lines at the beginning of the document, then enters three additional lines that center the student's name, date, and course in 24-point Arial. The last statement creates a page break within the document so that the title appears on a page by itself. The macro recorder created these statements for us, as we executed the corresponding actions from within Word.

Note, too, that the TitlePage macro changed the way in which the date is entered to make the macro more general. The NameAndCourse macro in Figure 7.9a specified a date (March 24, 2001). The TitlePage macro, however, uses the VBA InsertDateTime command to insert the current date. We did not know the syntax of this statement, but we didn't have to. Instead we pulled down the Insert menu from within Word, and chose the Date and Time command. The macro recorder kept track of our actions and created the appropriate VBA statement for us. In similar fashion, the macro recorder kept track of our actions when we moved to the beginning of the document and when we inserted a page break.

A SENSE OF FAMILIARITY

Visual Basic for Applications has the basic capabilities found in any other programming language. If you have programmed before, whether in Pascal, C, or even COBOL, you will find all of the logic structures you are used to. These include the Do While and Do Until statements, the If-Then-Else statement for decision making, nested If statements, a Case statement, and calls to subprograms.

MICROSOFT WORD 2002

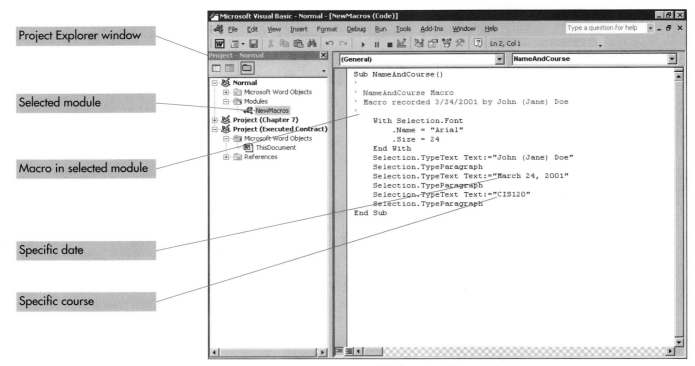

(a) NameAndCourse Macro

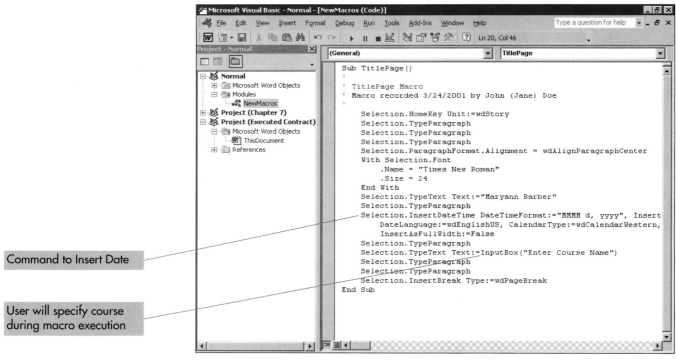

(b) TitlePage Macro

FIGURE 7.9   *The Visual Basic Editor*

HANDS-ON EXERCISE 4

INTRODUCTION TO MACROS

Objective To record, run, view, and edit simple macros; to run a macro from an existing Word document via a keyboard shortcut. Use Figure 7.10.

Step 1: **Create a Macro**

- Start Word. Open a new document if one is not already open.
- Pull down the **Tools menu**, click (or point to) the **Macro command**, then click **Record New Macro** to display the Record Macro dialog box in Figure 7.10a.
- Enter **NameAndCourse** as the name of the macro. (Do not leave any spaces.) If necessary, change the description to include your name.
- Click **Yes** if asked whether you want to replace the existing macro. (The existing macro may have been created by another student or if you previously attempted the exercise. Either way, you want to replace the existing macro.)
- Click **OK** to begin recording the macro. The mouse pointer changes to include a recording icon, and the Stop Recording toolbar is displayed.

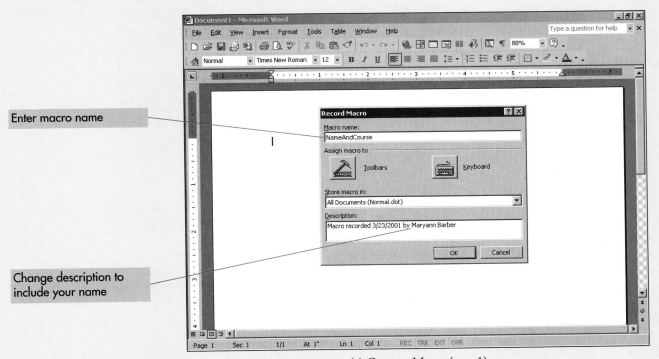

(a) Create a Macro (step 1)

FIGURE 7.10 *Hands-on Exercise 4*

MACRO NAMES

Macro names are not allowed to contain spaces or punctuation except for the underscore character. To create a macro name containing more than one word, capitalize the first letter of each word and/or use the underscore character; for example, NameAndCourse or Name_And_Course.

MICROSOFT WORD 2002 327

Step 2: **Record the Macro**

- The first task is to change the font. Normally we would change the font by using the Font list box on the Formatting toolbar, but there appears to be a bug in the macro recorder in that the font change is not recorded from the list box. Thus, we changed the font via the Font list box.
- Pull down the **Format menu** and click the **Font command** to display the Font dialog box in Figure 7.10b. Select **14-point Arial**. Click **OK** to accept the setting and close the dialog box.
- Pull down the **Insert menu** and click the **Date and Time command** to display the Date and Time dialog box in Figure 7.10b. Choose the format of the date that you prefer. Check the box to **Update Automatically**, then click **OK** to accept the settings and close the dialog box. Press the **enter key**.
- Enter the course you are taking this semester. Press the **enter key** a final time. Click the **Stop Recording button** to end the macro.

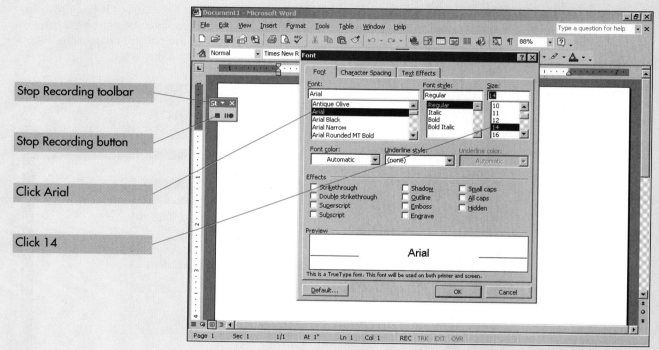

(b) Record the Macro (step 2)

FIGURE 7.10 *Hands-on Exercise 4 (continued)*

THE INSERT DATE COMMAND

A date is inserted into a document in one of two ways—as a field that is updated automatically to reflect the current date or as a specific value (the date and time on which the command is executed). The determination of which way the date is entered depends on whether the Update Automatically check box is checked or cleared, respectively. Be sure to choose the option that reflects your requirements.

Step 3: **Test the Macro**

- Click and drag to select your name, date, and class, then press the **Del key** to erase this information from the document.
- Pull down the **Tools menu**. Click **Macro**, then click the **Macros . . . command** to display the Macros dialog box in Figure 7.10c. Select **NameAndCourse** (the macro you just recorded) and click **Run**.
- Your name and class information should appear in the document. The typeface is 14-point Arial, which corresponds to your selection when you recorded the macro initially. Do not be dismayed if the macro did not work properly as we show you how to correct it in the next several steps.
- Press the **enter key** a few times. Press **Alt+F8** (a keyboard shortcut) to display the Macros dialog box.
- Double click the **NameAndCourse** macro to execute the macro. Your name and class information is entered a second time.

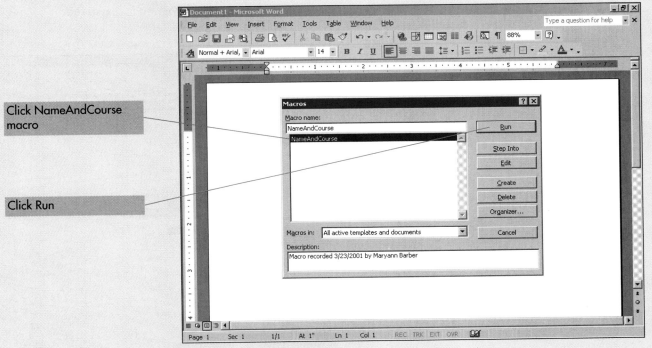

(c) Test the Macro (step 3)

FIGURE 7.10 *Hands-on Exercise 4 (continued)*

KEYBOARD SHORTCUTS

Take advantage of built-in shortcuts to facilitate the creation and testing of a macro. Press Alt+F11 to toggle between the VBA editor and the Word document. Use the Alt+F8 shortcut to display the Macros dialog box, then double click a macro to run it. You can also assign your own keyboard shortcut to a macro, as will be shown later in the exercise.

Step 4: **View the Macro**

➤ Pull down the **Tools menu**, click the **Macro command**, then click **Visual Basic Editor** (or press **Alt+F11**) to open the Visual Basic Editor. Maximize the VBE window. If necessary, pull down the **View menu** and click **Project Explorer** to open the Project window in the left pane. Close the Properties window if it is open.

➤ There is currently one project open, Document1, corresponding to the Word document on which you are working. Click the **plus sign** next to the Normal folder to expand that folder. Click the **plus sign** next to the **Modules folder** (within the Normal folder), then click **NewMacros**.

➤ Pull down the **View menu**, and click **Code** to open the Code window in the right pane. If necessary, click the **Maximize Button** in the Code window.

➤ Your screen should be similar to the one in Figure 7.10d except that it will reflect your name within the macro. The name in the comment statement may be different, however (especially if you are doing the exercise at school), as it corresponds to the person in whose name the program is registered.

➤ Delete the superfluous statements.

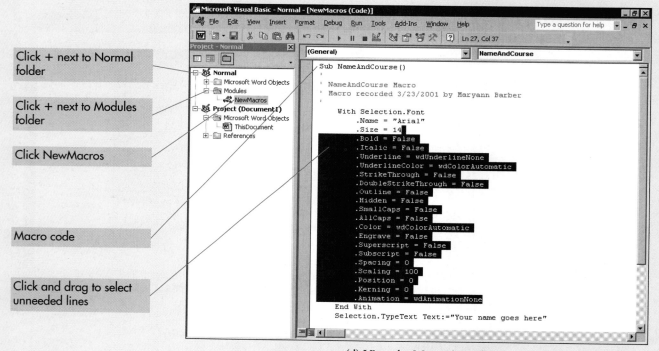

(d) View the Macro (step 4)

FIGURE 7.10 *Hands-on Exercise 4 (continued)*

RED, GREEN, AND BLUE

Visual Basic automatically assigns different colors to different types of statements (or a portion of those statements). Comments appear in green and are nonexecutable (i.e., they do not affect the outcome of a macro). Any statement containing a syntax error appears in red. Key words such as Sub and End Sub, With and End With, and True and False, appear in blue.

Step 5: **Edit the Macro**

- If necessary, change the name in the comment statement to reflect your name. The macro will run identically regardless of the changes in the comments. Changes to the statements within the macro, however, affect its execution.
- Click and drag to select the existing font name, **Arial**, then enter **Times New Roman** as shown in Figure 7.10e. Be sure that the **Times New Roman** appears within quotation marks. Change the font size to **24**.
- Click and drag to select the name of the course, which is "CIS120" in our example. Type **InputBox("Enter Course Name")** to replace the selected text.
- Note that as you enter the Visual Basic key word, InputBox, a prompt (containing the correct syntax) is displayed on the screen as shown in Figure 7.10e.
- Ignore the prompt and keep typing to complete the entry. Be sure you enter a closing parenthesis. Click the **Save button**.

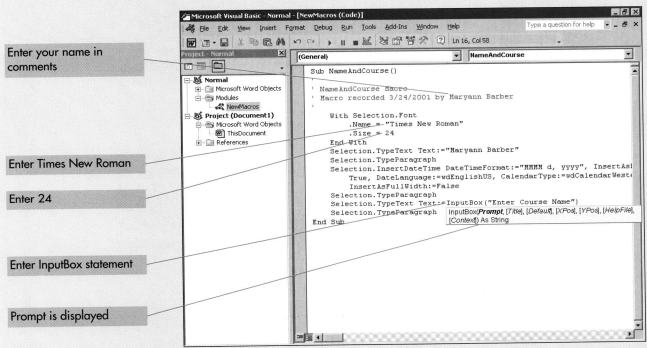

(e) Edit the Macro (step 5)

FIGURE 7.10 *Hands-on Exercise 4 (continued)*

COPY, RENAME, AND DELETE MACRO

You can copy a macro, rename it, then use the duplicate macro as the basis of a new macro. Click and drag to select the entire macro, click the Copy button, click after the End Sub statement, and click the Paste button to copy the macro. Click and drag to select the macro name in the Sub statement, type a new name, and you have a new (duplicate) macro. To delete a macro, click and drag to select the entire macro and press the Del key.

Step 6: **Test the Revised Macro**

➤ Press **Alt+F11** to toggle back to the Word document (or click the **Word button** on the taskbar). **Delete any text that is in the document**. If necessary, press **Ctrl+Home** to move to the beginning of the Word document.

➤ Press the **Alt+F8 key** to display the Macros dialog box, then double click the **NameAndCourse macro**. The macro enters your name and date, then displays the input dialog box shown in Figure 7.10f.

➤ Enter any appropriate course and click **OK** (or press the **enter key**). You should see your name, today's date, and the course you entered in 24-point Times New Roman type.

➤ Press **Alt+F11** to return to the Visual Basic Editor if the macro does not work as intended. Correct your macro so that its statements match those in step 5.

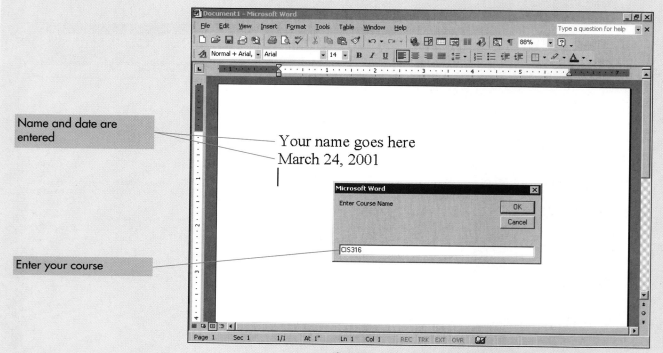

(f) Test the Revised Macro (step 6)

FIGURE 7.10 *Hands-on Exercise 4 (continued)*

HELP FOR VISUAL BASIC

Click within any Visual Basic key word, then press the F1 key for context-sensitive help. You will see a help screen containing a description of the statement, its syntax, key elements, and several examples. You can print the help screen by clicking the Options command button and selecting Print. (If you do not see the help screens, ask your instructor to install Visual Basic Help.)

Step 7: **Record the TitlePage Macro**

➤ If necessary, return to Word and delete the existing text in the document. Pull down the **Tools menu**. Click the **Macro command**, then click **Record New Macro** from the cascaded menu. You will see the Record Macro dialog box as described earlier.

➤ Enter **TitlePage** as the name of the macro. Do not leave any spaces in the macro name. Click the **Keyboard button** in the Record Macro dialog box to display the Customize Keyboard dialog box in Figure 7.10g. The insertion point is positioned in the Press New Shortcut Key text box.

➤ Press **Ctrl+T** to enter this keystroke combination as the new shortcut; note, however, that this shortcut is currently assigned to the Hanging Indent command:

- Click the **Assign button** if you do not use the Hanging Indent shortcut,
- *Or,* choose a different shortcut for the macro (or omit the shortcut altogether) if you are already using Ctrl+T for the Hanging Indent command.

➤ Close the Customize Keyboard dialog box.

➤ You are back in your document and can begin recording your macro:

- Press **Ctrl+Home** to move to the beginning of the document.
- Press the **enter key** three times to insert three blank lines.
- Click the **Center button** to center the text that will be subsequently typed.
- Press the **enter key** to create an additional blank line.
- Press **Ctrl+Enter** to create a page break.

➤ Click the **Stop Recording button** to end the macro.

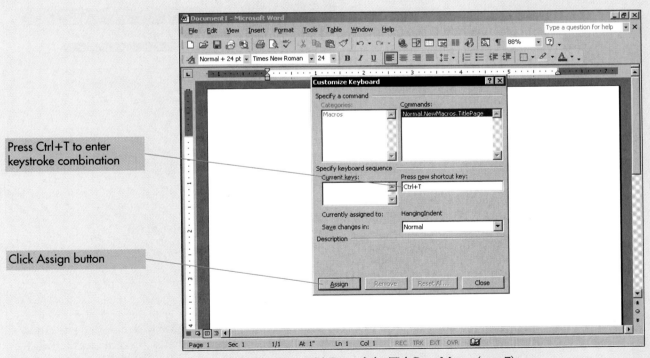

(g) Record the TitlePage Macro (step 7)

FIGURE 7.10 *Hands-on Exercise 4 (continued)*

Step 8: **Complete the TitlePage Macro**

> ➤ Press **Alt+F11** to return to the Visual Basic Editor. You should see two macros, NameAndCourse and TitlePage. Click and drag to select the statements in the **NameAndCourse macro** as shown in Figure 7.10h. Do not select the End Sub statement.
> ➤ Click the **Copy button** on the Standard toolbar (or use the **Ctrl+C** shortcut) to copy these statements to the clipboard.
> ➤ Move to the TitlePage macro and click after the VBA statement to center a paragraph. Press **enter** to start a new line. Click the **Paste button** on the Standard toolbar (or use the **Ctrl+V** shortcut) to paste the statements from the NameAndCourse macro into the TitlePage macro.
> ➤ You can see the completed macro by looking at Figure 7.10j, the screen in step 10. Click the **Save button** to save your macros.
> ➤ Press **Alt+F11** to return to Word. Pull down the File menu and click the **Close command** to close the document you were using to create the macros in this exercise. There is no need to save that document.

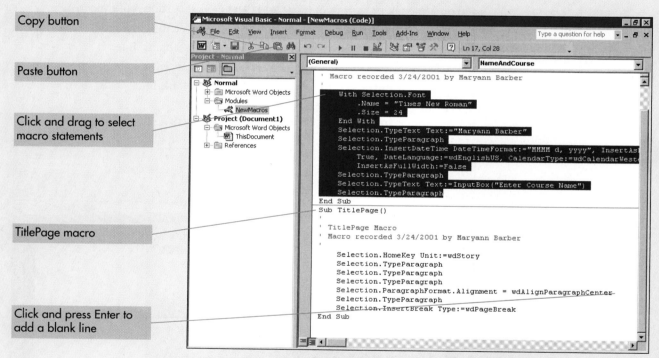

(h) Complete the TitlePage Macro (step 8)

FIGURE 7.10   *Hands-on Exercise 4 (continued)*

THE PAGE BORDER COMMAND

Add interest to a title page with a border. Click anywhere on the page, pull down the Format menu, click the Borders and Shading command, then click the Page Border tab in the Borders and Shading dialog box. You can choose a box, shadow, or 3-D style in similar fashion to placing a border around a paragraph. You can also click the drop-down arrow on the Art list box to create a border consisting of a repeating clip art image.

Step 9: **Test the TitlePage Macro**

- Open the completed Word document (Executed Contract) from the second hands-on exercise.
- Pull down the **Tools menu** and click the **Unprotect command**. Click anywhere in the document, then press **Ctrl+T** to execute the TitlePage macro.
- Your name and date should appear, after which you will be prompted for your course. Enter the course you are taking, and the macro will complete the title page.
- Pull down the **View menu** and change to the **Print Layout view**. Pull down the **View menu** a second time, click the **Zoom command**, click the option button for **Many Pages**, then click and drag the monitor to display three pages.
- You should see the executed contract with a title page as shown in Figure 7.10i. Print this document for your instructor. Save the document.

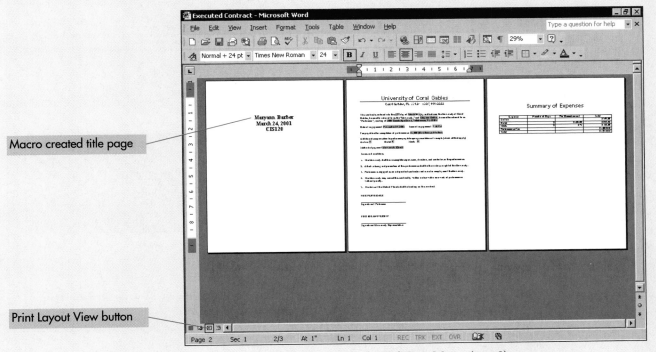

(i) Test the TitlePage Macro (step 9)

FIGURE 7.10 *Hands-on Exercise 4 (continued)*

TROUBLESHOOTING

If the shortcut keys do not work, it is probably because they were not defined properly. Pull down the View menu, click Toolbars, click Customize, then click the Keyboard command button to display the Customize Keyboard dialog box. Drag the scroll box in the Categories list box until you can select the Macros category. Select (click) the macro that is to receive the shortcut and click in the Press New Shortcut Key text box. Enter the desired shortcut, click the Assign button to assign the shortcut, then click the Close button to close the dialog box.

Step 10: **Print the Module**

- Press **Alt+F11** to return to the Visual Basic Editor. Delete the second **Selection.TypeParagraph** line, as it is unnecessary.
- Pull down the **File menu**. Click **Print** to display the Print dialog box in Figure 7.10j. Click the option button to print the current module. Click **OK**. Submit the listing of the current module, which contains the procedures for both macros, to your instructor as proof you did this exercise.
- Delete all of the macros you have created in this exercise if you are not working on your own machine. Pull down the **File menu**. Click the **Close and Return to Word command**.
- Exit Word. The Title Page macro will be waiting for you the next time you use Microsoft Word provided you did the exercise on your own computer.

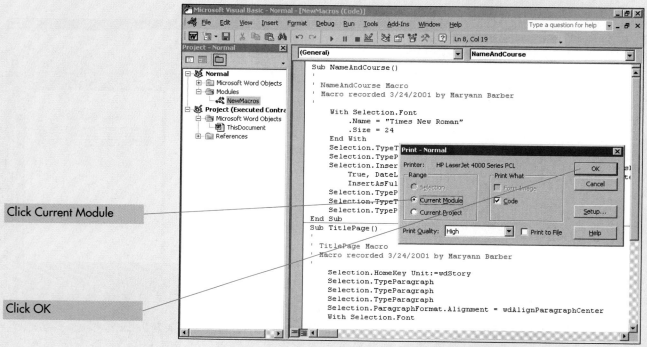

(j) Print the Module (step 10)

FIGURE 7.10   *Hands-on Exercise 4 (continued)*

INVEST IN MACROS

Creating a macro takes time, but that time can be viewed as an investment, because a well-designed macro will simplify the creation of subsequent documents. A macro is recorded once, tested and corrected as necessary, then run (executed) many times. It is stored by default in the Normal template, where it is available to every Word document. Yes, it takes time to create a meaningful macro, but once that's done, it is only a keystroke away.

SUMMARY

Multiple persons within a workgroup can review a document and have their revisions stored electronically within that document. The changes are entered via various tools on the Reviewing toolbar. A red line through existing text indicates that the text should be deleted, whereas text that is underlined is to be added. Yellow highlighting indicates a comment where the reviewer has added a descriptive note without making a specific change.

A form facilitates data entry when the document is made available to multiple individuals via a network. It is created as a regular document with the various fields added through tools on the Forms toolbar. Word enables you to create three types of fields—text boxes, check boxes, and drop-down list boxes. After the form is created, it is protected to prevent further modification other than data entry.

The rows in a table can be sorted to display the data in ascending or descending sequence, according to the values in one or more columns in the table. Sorting is accomplished by selecting the rows within the table that are to be sorted, then executing the Sort command in the Tables menu. Calculations can be performed within a table using the Formula command in the Tables menu.

A master document consists of multiple subdocuments, each of which is stored as a separate file. It is especially useful for very large documents such as a book or dissertation, which can be divided into smaller, more manageable documents. The attraction of a master document is that you can work with multiple subdocuments simultaneously.

A macro is a set of instructions that automates a repetitive task. It is in essence a program, and its instructions are written in Visual Basic for Applications (VBA), a programming language. A macro is created initially through the macro recorder in Microsoft Word, which records your commands and generates the corresponding VBA statements. Once a macro has been created, it can be edited manually by inserting, deleting, or changing its statements. A macro is run (executed) by the Run command in the Tools menu or more easily through a keyboard shortcut.

KEY TERMS

Accept and Review Changes command (p. 297)
Cells (p. 307)
Check box (p. 298)
Code window (p. 325)
Comment (p. 324)
Create New Folder command (p. 316)
Create Subdocument command (p. 316)
Descending sequence (p. 307)
Drop-down list box (p. 298)
End Sub statement (p. 324)
Field (p. 298)
Field codes (p. 307)

Form (p. 298)
Forms toolbar (p. 298)
Header row (p. 312)
InputBox function (p. 331)
Insert Date command (p. 328)
Keyboard shortcut (p. 329)
Macro (p. 324)
Macro recorder (p. 324)
Master document (p. 316)
Module (p. 325)
Outlining toolbar (p. 316)
Password protection (p. 297)
Procedure (p. 324)
Project Explorer (p. 325)
Reviewing toolbar (p. 297)

Revision mark (p. 297)
Shortcut key (p. 329)
Sort (p. 307)
Sub statement (p. 324)
Subdocument (p. 316)
Tables menu (p. 307)
Tables and Borders toolbar (p. 307)
Text field (p. 298)
Track Changes command (p. 297)
Versions command (p. 297)
Visual Basic for Applications (VBA) (p. 324)
Visual Basic Editor (VBE) (p. 325)
Workgroup (p. 296)

MULTIPLE CHOICE

1. Which of the following is a true statement regarding password protection?
   (a) All documents are automatically saved with a default password
   (b) The password is case-sensitive
   (c) A password cannot be changed once it has been implemented
   (d) All of the above

2. Which statement describes the way revisions are marked in a document?
   (a) A red line appears through text that is to be deleted
   (b) A red underline appears beneath text that is to be added
   (c) Yellow highlighting indicates a comment, where the user has made a suggestion, but has not indicated the actual revision in the document
   (d) All of the above

3. Which of the following types of fields *cannot* be inserted into a form?
   (a) Check boxes
   (b) Text fields
   (c) A drop-down list
   (d) Radio buttons

4. Which of the following is true about a protected form?
   (a) Data can be entered into the form
   (b) The text of the form cannot be modified
   (c) Both (a) and (b)
   (d) Neither (a) nor (b)

5. Which of the following describes the function of the Form Field Shading button on the Forms toolbar?
   (a) Clicking the button shades every field in the form
   (b) Clicking the button shades every field in the form and also prevents further modification to the form
   (c) Clicking the button removes the shading from every field
   (d) Clicking the button toggles the shading on or off

6. You have created a table containing numerical values and have entered the SUM(ABOVE) function at the bottom of a column. You then delete one of the rows included in the sum. Which of the following is true?
   (a) The row cannot be deleted because it contains a cell that is included in the sum function
   (b) The sum is updated automatically
   (c) The sum cannot be updated unless the Form Protect button is toggled off
   (d) The sum will be updated provided you right click the cell and select the Update field command

7. Which of the following is suitable for use as a master document?
   (a) An in-depth proposal that contains component documents
   (b) A lengthy newsletter with stories submitted by several people
   (c) A book
   (d) All of the above

8. Which of the following is true regarding changes made to a subdocument from within a master document?
   (a) The changes will be saved in the master document only
   (b) The changes will be saved in both the master document and the subdocument provided the subdocument is unlocked
   (c) The changes will be saved in both the master document and the subdocument provided the subdocument is locked
   (d) Changes cannot be made to a subdocument from within a master document

9. What happens if you click inside a subdocument, then click the Lock button on the Outlining toolbar?
   (a) The subdocument is locked
   (b) The subdocument is unlocked
   (c) The subdocument is locked or unlocked depending on its status prior to clicking the button
   (d) All editing to the subdocument is disabled

10. Which of the following describes the storage of a master document and the associated subdocuments?
    (a) Each document is saved under its own name as a separate file
    (b) All of the subdocuments must be stored in the same folder
    (c) Both (a) and (b)
    (d) Neither (a) nor (b)

11. Which of the following best describes the recording and execution of a macro?
    (a) A macro is recorded once and executed once
    (b) A macro is recorded once and executed many times
    (c) A macro is recorded many times and executed once
    (d) A macro is recorded many times and executed many times

12. Which of the following is true regarding comments in Visual Basic?
    (a) A comment is not executable; that is, its inclusion or omission does not affect the outcome of a macro
    (b) A comment begins with an apostrophe
    (c) Both (a) and (b)
    (d) Neither (a) nor (b)

13. Which commands are used to copy an existing macro so that it can become the basis of a new macro?
    (a) Copy command
    (b) Paste command
    (c) Both (a) and (b)
    (d) Neither (a) nor (b)

14. What is the default location for a macro created in Microsoft Word?
    (a) In the Normal template where it is available to every Word document
    (b) In the document in which it was created where it is available only to that document
    (c) In the Macros folder on your hard drive
    (d) In the Office folder on your hard drive

15. Which of the following correctly matches the shortcut to the associated task?
    (a) Alt+F11 toggles between Word and the Visual Basic Editor
    (b) Alt+F8 displays the Macros dialog box
    (c) Both (a) and (b)
    (d) Neither (a) nor (b)

ANSWERS

1. b
2. d
3. d
4. c
5. d
6. d
7. d
8. b
9. c
10. a
11. b
12. c
13. c
14. a
15. c

PRACTICE WITH WORD

1. **Reviewing a Document:** Open the *Chapter 7 Practice 1* document shown in Figure 7.11, then revise that document by incorporating all of the suggested revisions. Delete the existing comment, which appears as a ScreenTip in Figure 7.11, then insert your own comment indicating that you have completed the necessary revisions. Click the Show button on the Reviewing toolbar to display the Reviewing Pane as shown in the figure. Note that the document contains revisions from multiple reviewers. Look closely and you will see that the comments for each reviewer appear in a different color. (You can also click the Show button, then click the Reviewers command to toggle the comments from individual reviewers on or off.)

    Save the revised document as its own version within the Chapter 7 Practice 1 document. Print both versions of the document, the original and the one you created, and include the associated comments and properties for each version. These elements can be included with the printed document by clicking the Options button within the Print command.

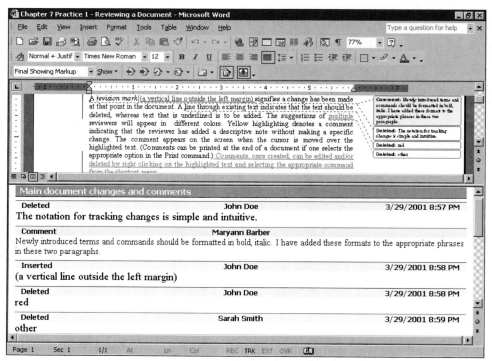

FIGURE 7.11 *Reviewing a Document (Exercise 1)*

2. **Route a Document for Review:** Do the first two hands-on exercises as described in the text, then send a copy of the executed contract to your instructor as shown in Figure 7.12. You can use the Send To Mail Recipient button on the Reviewing toolbar to start your e-mail program, which in turn attaches the document automatically. Alternatively, you can start your e-mail program independently, then use the Insert Attachment command to select the appropriate document.

    Either way, you will be sending an attached file to an e-mail recipient. You can include multiple attachments in the same message and/or send attachments of any file type. It's faster and cheaper than sending an overnight package.

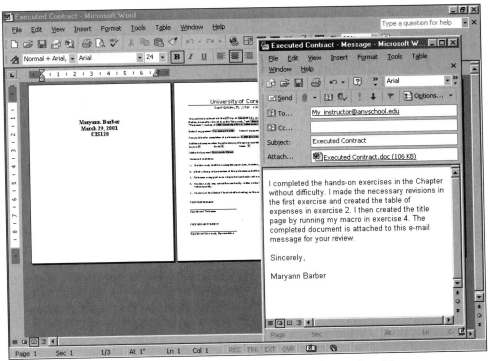

FIGURE 7.12   *Route a Document for Review (Exercise 2)*

3. Table Math: Figure 7.13 displays two versions of a table. Figure 7.13a shows the original table prior to any modification, whereas Figure 7.13b displays the table at the end of the exercise.
   a. Open the *Chapter 7 Practice 3* file in the Exploring Windows folder.
   b. Click in the cell containing "Enter your name" and enter your last name. (White was the person added in our exercise.)
   c. Sort the table so that the names appear in alphabetical order. (By coincidence the names in our example are in the same sequence after sorting.)
   d. Enter the appropriate formula for each person to compute the gain in sales.
   e. Enter the appropriate formulas in the total row to complete the totals as shown in Figure 7.13b.
   f. Add a short memo to your instructor indicating that you have completed the table. Print the memo twice, once with displayed values as shown in Figure 7.13b, then a second time to show the cell formulas.

Sales Person	Last Year	This Year	Gain
Brown	200	225	
Jones	200	300	
Smith	125	140	
Your Name Goes Here	100	450	
Total			

(a) Original Table (as it exists on disk)

Sales Person	Last Year	This Year	Gain
Brown	200	225	**25**
Jones	200	300	**100**
Smith	125	140	**15**
White	100	450	**350**
Total	**625**	**1115**	**490**

(b) Completed Table

FIGURE 7.13   *Table Math (Exercise 3)*

BUILDS ON

HANDS-ON
EXERCISE 3
PAGES 318–323

4. The Master Document: Do the third hands-on exercise as described in the chapter, then modify the master document as follows:
   a. Complete the subdocument, "Election of Officers" in which you propose nominations for the Board of Directors. Nominate yourself as the CEO and various classmates to fill the other positions.
   b. Delete the subdocument, "Plans for Next Year".
   c. Modify the Financial Summary document to include the table of fiscal results shown in Figure 7.14.
   d. Print the completed master document for your instructor. Print the document twice, in both the expanded and the collapsed format.

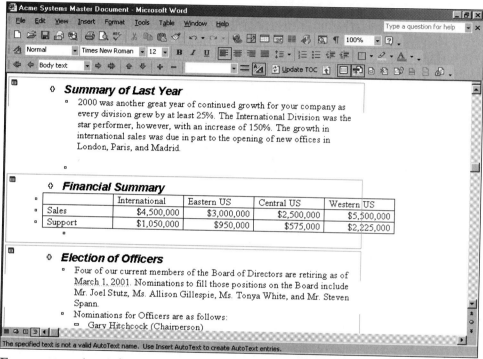

FIGURE 7.14 *The Master Document (Exercise 4)*

5. Customizing Toolbars and Menus: The chapter illustrated different ways to execute a macro—from the Tools menu in Word, by pressing Alt+F8 to display the Macros dialog box, and via a keyboard shortcut. You can also add a customized button to a toolbar (or a command to a menu) as shown in Figure 7.15. Do the following:
   a. Pull down the View menu, click Toolbars, then click Customize to display the Customize dialog box in Figure 7.15. Click the Commands tab, click the down arrow in the Categories list box until you see the Macros category, then click and drag the TitlePage macro from the Commands area in the Customize dialog box (it is not visible in Figure 7.15) to an existing toolbar. (You can add the macro to a menu by dragging the macro to the menu and placing it wherever you like within the commands in that menu.)
   b. Click the Modify Selection command button, click the Change Button Image command, then click the image you want for the toolbar button. Click the Modify Selection command button a second time, then click the Default style option to display just the image on the button as opposed to the image and the text. Click the Close button to accept the settings.
   c. Open an existing Word document, then click the button to test it. Experiment with other options in the Customize dialog box, then summarize your findings in a brief note to your instructor.

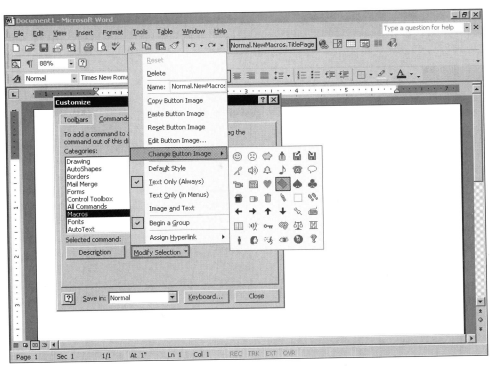

FIGURE 7.15 *Customizing a Toolbar (Exercise 5)*

6. Authenticating a Document: Open any document, pull down the Tools menu, click the Options command, then click the Security tab to display the dialog box in Figure 7.16. Click the Digital Signatures command button to display the associated dialog box, then click the Add button to authenticate the open document. Note, however, that you can add a signature only if you have applied for a digital certificate. Summarize your thoughts on authentication for your instructor.

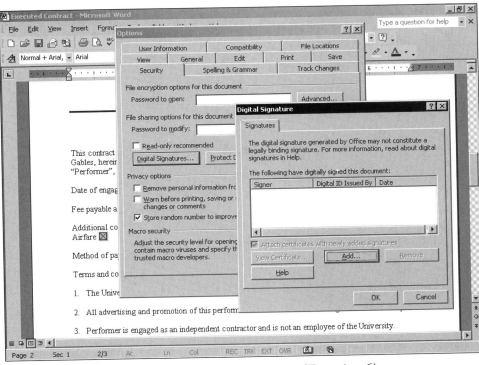

FIGURE 7.16 *Authenticating a Document (Exercise 6)*

7. Debugging a Macro: A "bug" is a mistake in a computer program; hence, "debugging" refers to the process of correcting a programming error. One useful tool for debugging a macro is the STEP Into command, in which you execute a macro one statement at a time as shown in Figure 7.17:
   a. Open any Word document, then press Alt+F11 to open the VBE window. Click the Close button in the left pane to close the Project window within the Visual Basic Editor. The Code window expands to take the entire Visual Basic Editor window.
   b. Point to an empty area on the Windows taskbar, then click the right mouse button to display a shortcut menu. Click Tile Windows Vertically to tile the open windows (Word and the Visual Basic Editor). Your desktop should be similar to Figure 7.16. It doesn't matter if the document is in the left or right window. (If additional windows are open on the desktop, minimize the other windows, then repeat the previous step to tile the open windows.)
   c. Click in the Visual Basic Editor window, then click anywhere within the TitlePage macro. Pull down the Debug menu and click the STEP Into command (or press the F8 key) to enter the macro. The Sub statement is highlighted. Press the F8 key a second time to move to the first executable statement (the comments are skipped). The statement is selected (highlighted), but it has not yet been executed. Press the F8 key again to execute this statement and move to the next statement.
   d. Continue to press the F8 key to execute the statements in the macro one at a time. You can see the effect of each statement as it is executed in the Word window.
   e. Do you think this procedure is useful in finding any bugs that might exist? Summarize the steps in debugging a macro in a short note to your instructor.

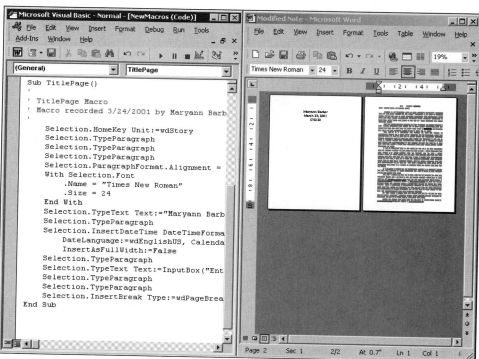

FIGURE 7.17   *Debugging a Macro (Exercise 7)*

ON YOUR OWN

Customize the Toolbar

The Create Envelope button is a perfect addition to the Standard toolbar if you print envelopes frequently. Pull down the Tools menu, click Customize, then click the Toolbars tab in the dialog box. If necessary, click the arrow in the Categories list box, select Tools, then drag the Create Envelope button to the Standard toolbar. Close the Customize dialog box. The Create Envelope button appears on the Standard toolbar and can be used the next time you need to create an envelope or label.

¿Cómo Está Usted?

The Insert Symbol command can be used to insert foreign characters into a document, but this technique is too slow if you use these characters with any frequency. Alternatively, you can use the predefined shortcut keys; for example, Ctrl+' followed by the letter "a" will insert á into a document. It's cumbersome, however, to remember the apostrophe, and we find it easier to create a macro and assign the shortcut Ctrl+A. Parallel macros can be developed for the other vowels or special characters, such as Ctrl+q for ¿. Try creating the appropriate macros, then summarize the utility of this technique in a short note to your instructor. Be sure to address the issue of using shortcuts that are already assigned to other macros; Ctrl+A, for example, is assigned to the Select All command by default.

Object Linking and Embedding

Table math is fine for simple calculations, but it is exceedingly limited when compared to Microsoft Excel. Thus, if you know Excel, you would be wise to explore the ability to link or embed an Excel workbook into a Word document. What is the difference between linking and embedding? Can the same workbook be associated with multiple Word documents? Does Object Linking and Embedding (OLE) pertain to applications other than Word and Excel? Use Chapter 5 as a starting point to learn about OLE, then summarize your findings in a short note to your instructor.

File Management in Microsoft Office

Most newcomers to Microsoft Office take the Open and Save As dialog boxes for granted. Look closely, however, and you will discover that these dialog boxes provide access to virtually all of the file management capabilities in Windows. You can create a folder, as was done in the chapter. You can also search for specific documents or create entries in the Favorites list. Write a short note to your instructor that summarizes the file management capabilities within the Open and Save dialog boxes. What additional capabilities are available through Windows Explorer that are not found within these dialog boxes?

Macros in Microsoft Excel

Do you use Microsoft Excel on a regular basis? Are there certain tasks that you do repeatedly, whether in the same document or in a series of different documents? If so, you would do well to explore the macro capabilities within Microsoft Excel. Does Excel have a macro recorder? Does it convert its macros to VBA? Can you modify an Excel macro after it has been created? Summarize the creation and execution of macros in Excel as compared to Word in a short note to your instructor.

MULTIPLE CHOICE

1. Which of the following is a true statement regarding password protection?
   (a) All documents are automatically saved with a default password
   (b) The password is case-sensitive
   (c) A password cannot be changed once it has been implemented
   (d) All of the above

2. Which statement describes the way revisions are marked in a document?
   (a) A red line appears through text that is to be deleted
   (b) A red underline appears beneath text that is to be added
   (c) Yellow highlighting indicates a comment, where the user has made a suggestion, but has not indicated the actual revision in the document
   (d) All of the above

3. Which of the following types of fields *cannot* be inserted into a form?
   (a) Check boxes
   (b) Text fields
   (c) A drop-down list
   (d) Radio buttons

4. Which of the following is true about a protected form?
   (a) Data can be entered into the form
   (b) The text of the form cannot be modified
   (c) Both (a) and (b)
   (d) Neither (a) nor (b)

5. Which of the following describes the function of the Form Field Shading button on the Forms toolbar?
   (a) Clicking the button shades every field in the form
   (b) Clicking the button shades every field in the form and also prevents further modification to the form
   (c) Clicking the button removes the shading from every field
   (d) Clicking the button toggles the shading on or off

6. You have created a table containing numerical values and have entered the SUM(ABOVE) function at the bottom of a column. You then delete one of the rows included in the sum. Which of the following is true?
   (a) The row cannot be deleted because it contains a cell that is included in the sum function
   (b) The sum is updated automatically
   (c) The sum cannot be updated unless the Form Protect button is toggled off
   (d) The sum will be updated provided you right click the cell and select the Update field command

7. Which of the following is suitable for use as a master document?
   (a) An in-depth proposal that contains component documents
   (b) A lengthy newsletter with stories submitted by several people
   (c) A book
   (d) All of the above

8. Which of the following is true regarding changes made to a subdocument from within a master document?
   (a) The changes will be saved in the master document only
   (b) The changes will be saved in both the master document and the subdocument provided the subdocument is unlocked
   (c) The changes will be saved in both the master document and the subdocument provided the subdocument is locked
   (d) Changes cannot be made to a subdocument from within a master document

APPENDIX A

Toolbars

OVERVIEW

Microsoft Word has multiple toolbars that provide access to commonly used commands. The toolbars are displayed in Figure A.1 and are listed here for convenience. They are: the Standard, Formatting, 3-D Settings, AutoText, Control Toolbox, Database, Diagram, Drawing, Drawing Canvas, Equation Editor, Extended Formatting, Forms, Frames, Function Key Display, Header/Footer, Japanese Greeting, Mail Merge, Microsoft, Organization Chart, Outlining, Picture, Reviewing, Shadow Settings, Shortcut Menus, Tables and Borders, Visual Basic, Web, Web Tools, Word Count, and WordArt. The Standard and Formatting toolbars are displayed by default and appear on the same row immediately below the menu bar. The other predefined toolbars are displayed (hidden) at the discretion of the user.

The buttons on the toolbars are intended to indicate their functions. Clicking the Printer button (the sixth button from the left on the Standard toolbar), for example, executes the Print command. If you are unsure of the purpose of any toolbar button, point to it, and a ScreenTip will appear that displays its name.

- To separate the Standard and Formatting toolbars and simultaneously display all of the buttons for each toolbar, pull down the Tools menu, click the Customize command, click the Options tab, then clear the check box that has the toolbars share one row. Alternatively, the toolbars appear on the same row so that only a limited number of buttons are visible on each toolbar and hence you may need to click the double arrow (More Buttons) tool at the end of the toolbar to view additional buttons. Additional buttons will be added to either toolbar as you use the associated feature, and conversely, buttons will be removed from the toolbar if the feature is not used.
- To display or hide a toolbar, pull down the View menu and click the Toolbars command. Select (deselect) the toolbar(s) that you want to display (hide). The selected toolbar(s) will be displayed in the same position as when last displayed. You may also point to any toolbar and click with the right mouse button to bring up a shortcut menu, after which you can select the toolbar to be displayed (hidden).

- To change the size of the buttons, suppress the display of the ScreenTips, or display the associated shortcut key (if available), pull down the View menu, click Toolbars, and click Customize to display the Customize dialog box. If necessary, click the Options tab, then select (deselect) the appropriate check box. Alternatively, you can right click on any toolbar, click the Customize command from the context-sensitive menu, then select (deselect) the appropriate check box from within the Options tab in the Customize dialog box.
- Toolbars are either docked (along the edge of the window) or floating (in their own window). A toolbar moved to the edge of the window will dock along that edge. A toolbar moved anywhere else in the window will float in its own window. Docked toolbars are one tool wide (high), whereas floating toolbars can be resized by clicking and dragging a border or corner as you would with any window.
  - To move a docked toolbar, click anywhere in the gray background area and drag the toolbar to its new location. You can also click and drag the move handle (the vertical line) at the left of the toolbar
  - To move a floating toolbar, drag its title bar to its new location.
- To customize one or more toolbars, display the toolbar(s) on the screen. Then pull down the View menu, click Toolbars, and click Customize to display the Customize dialog box. Alternatively, you can click on any toolbar with the right mouse button and select Customize from the shortcut menu.
  - To move a button, drag the button to its new location on that toolbar or any other displayed toolbar.
  - To copy a button, press the Ctrl key as you drag the button to its new location on that toolbar or any other displayed toolbar.
  - To delete a button, drag the button off the toolbar and release the mouse button.
  - To add a button, click the Commands tab in the Customize dialog box, select the category from the Categories list box that contains the button you want to add, then drag the button to the desired location on the toolbar. (To see a description of a tool's function prior to adding it to a toolbar, select the tool, then click the Description command button.)
  - To restore a predefined toolbar to its default appearance, click the Toolbars tab, select the desired toolbar, and click the Reset command button.
- Buttons can also be moved, copied, or deleted without displaying the Customize dialog box.
  - To move a button, press the Alt key as you drag the button.
  - To copy a button, press the Alt and Ctrl keys as you drag the button.
  - To delete a button, press the Alt key as you drag the button off the toolbar.
- To create your own toolbar, pull down the View menu, click Toolbars, click Customize, click the Toolbars tab, then click the New command button. Alternatively, you can click on any toolbar with the right mouse button, select Customize from the shortcut menu, click the Toolbars tab, and then click the New command button.
  - Enter a name for the toolbar in the dialog box that follows. The name can be any length and can contain spaces. Click OK.
  - The new toolbar will appear on the screen. Initially it will be big enough to hold only one button, but you can add, move, and delete buttons following the same procedures as for an existing toolbar. The toolbar will automatically size itself as new buttons are added and deleted.
  - To delete a custom toolbar, pull down the View menu, click Toolbars, click Customize, and click the Toolbars tab. *Verify that the custom toolbar to be deleted is the only one selected (highlighted).* Click the Delete command button. Click OK to confirm the deletion.

MICROSOFT WORD 2002 TOOLBARS

Standard Toolbar

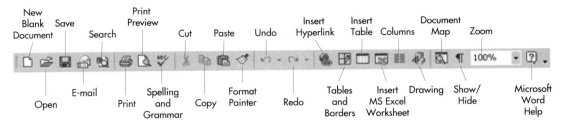

Formatting Toolbar

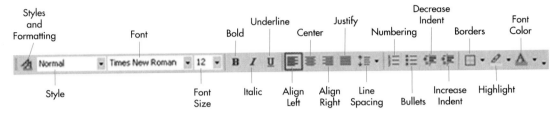

3-D Settings Toolbar

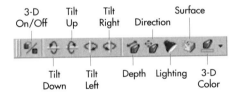

AutoText Toolbar

Control Toolbox

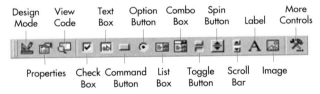

Database Toolbar

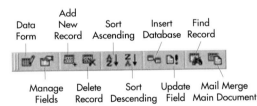

Diagram Toolbar

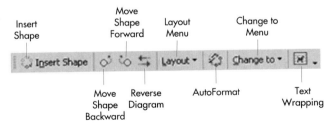

Drawing Toolbar

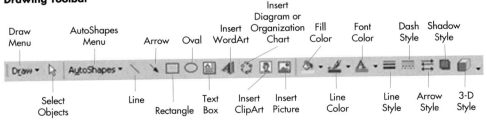

FIGURE A.1   *Toolbars*

Drawing Canvas Toolbar

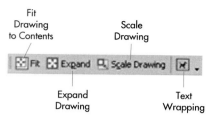

Extended Formatting Toolbar

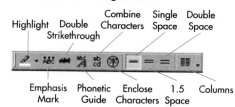

Equation Editor Toolbar

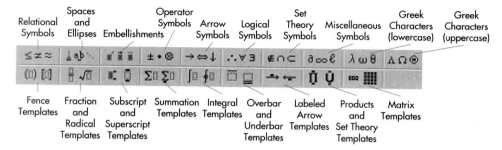

Forms Toolbar

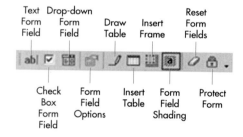

Function Key Display Toolbar

Frames Toolbar

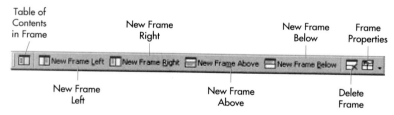

Header/Footer Toolbar

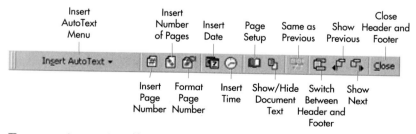

FIGURE A.1   *Toolbars (continued)*

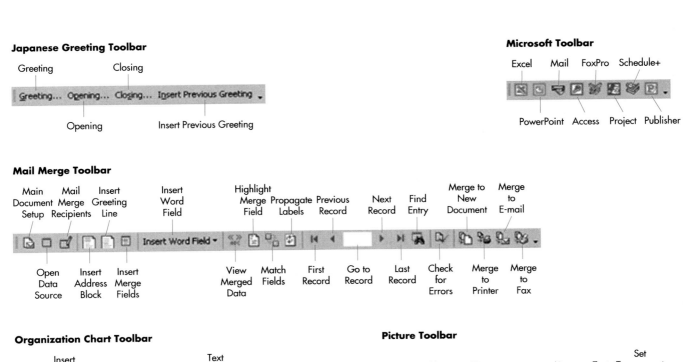
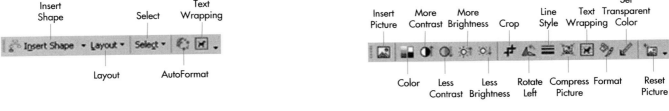
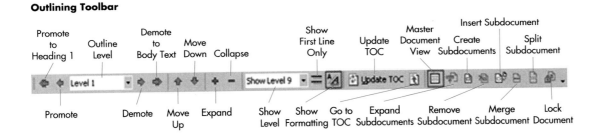
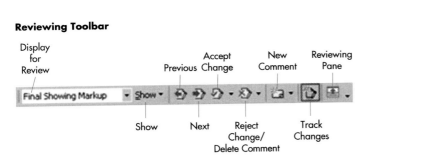

FIGURE A.1 *Toolbars (continued)*

MICROSOFT WORD 2002

Shadow Settings Toolbar

Shortcut Menus Toolbar

Visual Basic Toolbar

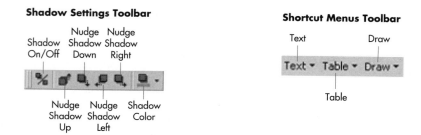
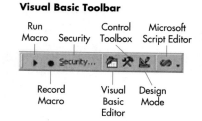

Tables and Borders Toolbar

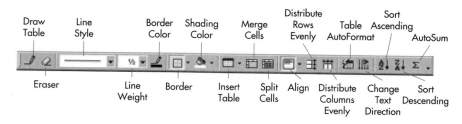

Web Toolbar

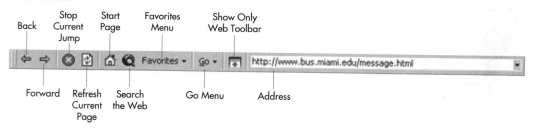

Web Tools Toolbar

Word Count Toolbar

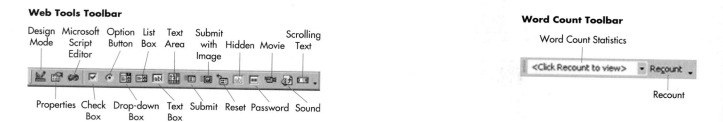

WordArt Toolbar

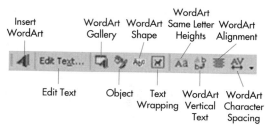

FIGURE A.1   *Toolbars (continued)*

A VBA Primer: Extending Microsoft® Office XP

OBJECTIVES

AFTER READING THIS SUPPLEMENT YOU WILL BE ABLE TO:

1. Describe the relationship of VBA to Microsoft Office XP; explain how to open the VBA editor within an Office application.
2. Distinguish between key words, statements, procedures, and modules; use the Office Assistant to obtain detailed information about any VBA statement.
3. Explain how to create, edit, and run a VBA procedure; explain how the Quick Info and Complete Word tools facilitate VBA coding.
4. Explain how to continue a VBA statement from one line to the next; add and remove comments from a procedure.
5. Distinguish between the MsgBox and InputBox statements; describe at least two arguments for each statement.
6. Explain how to debug a procedure by stepping through its statements; describe the role of the Local and Immediate windows in debugging.
7. Use the If . . . Then . . . Else statement to implement a decision; explain the advantage of the Case statement over multiple ElseIf clauses.
8. Create a custom toolbar with buttons corresponding to the VBA procedures you have developed.
9. Describe several statements used to implement a loop; explain the difference between placing a condition at the beginning or end of a loop.
10. Distinguish between event-driven and traditional programming; create event procedures associated with opening and closing an Excel workbook and with an Access database.

OVERVIEW

Visual Basic for Applications (VBA) is a powerful programming language that is accessible from all major applications in Microsoft Office XP. You do not have to know VBA in order to use Office effectively, but even a basic understanding will help you to create more powerful documents. Indeed, you may already have been exposed to VBA through the creation of simple macros in Word or Excel. A *macro* is a set of instructions (i.e., a program) that simplifies the execution of repetitive tasks. It is created through the *macro recorder* that captures commands as they are executed, then converts those commands to a VBA program. (The macro recorder is present in Word, Excel, and PowerPoint, but not in Access.) You can create and execute macros without ever looking at the underlying VBA, but you gain an appreciation for the language when you do.

The macro recorder is limited, however, in that it captures only commands, mouse clicks, and/or keystrokes. As you will see, VBA is much more than just recorded keystrokes. It is a language unto itself, and thus, it contains all of the statements you would expect to find in any programming language. This lets you enhance the functionality of any macro by adding extra statements as necessary—for example, an InputBox function to accept data from the user, followed by an If . . . Then . . . Else statement to take different actions based on the information supplied by the user.

This supplement presents the rudiments of VBA and is suitable for use with any Office application. We begin by describing the VBA Editor and how to create, edit, and run simple procedures. The examples are completely general and demonstrate the basic capabilities of VBA that are found in any programming language. We illustrate the MsgBox statement to display output to the user and the InputBox function to accept input from the user. We describe the For . . . Next statement to implement a loop and the If . . . Then . . . Else and Case statements for decision making. We also describe several debugging techniques to help you correct the errors that invariably occur. The last two exercises introduce the concept of event-driven programming, in which a procedure is executed in response to an action taken by the user. The material here is application-specific in conjunction with Excel and Access, but it can be easily extended to Word or PowerPoint.

One last point before we begin is that this supplement assumes no previous knowledge on the part of the reader. It is suitable for someone who has never been exposed to a programming language or written an Office macro. If, on the other hand, you have a background in programming or macros, you will readily appreciate the power inherent in VBA. VBA is an incredibly rich language that can be daunting to the novice. Stick with us, however, and we will show you that it is a flexible and powerful tool with consistent rules that can be easily understood and applied. You will be pleased at what you will be able to accomplish.

INTRODUCTION TO VBA

VBA is a programming language, and like any other programming language its programs (or procedures, as they are called) are made up of individual statements. Each *statement* accomplishes a specific task such as displaying a message to the user or accepting input from the user. Statements are grouped into *procedures*, and procedures, in turn, are grouped into *modules*. Every VBA procedure is classified as either public or private. A *private procedure* is accessible only from within the module in which it is contained. A *public procedure*, on the other hand, can be accessed from any module.

The statement, however, is the basic unit of the language. Our approach throughout this supplement will be to present individual statements, then to develop simple procedures using those statements in a hands-on exercise. As you read the discussion, you will see that every statement has a precise *syntax* that describes how the statement is to be used. The syntax also determines the *arguments* (or parameters) associated with that statement, and whether those arguments are required or optional.

THE MSGBOX STATEMENT

The ***MsgBox statement*** displays information to the user. It is one of the most basic statements in VBA, but we use it to illustrate several concepts in VBA programming. Figure 1a contains a simple procedure called MsgBoxExamples, consisting of four individual MsgBox statements. All procedures begin with a ***procedure header*** and end with the **End Sub statement**.

The MsgBox statement has one required argument, which is the message (or prompt) that is displayed to the user. All other arguments are optional, but if they are used, they must be entered in a specified sequence. The simplest form of the MsgBox statement is shown in example 1, which specifies a single argument that contains the text (or prompt) to be displayed. The resulting message box is shown in Figure 1b. The message is displayed to the user, who responds accordingly, in this case by clicking the OK button.

Example 2 extends the MsgBox statement to include a second parameter that displays an icon within the resulting dialog box as shown in Figure 1c. The type of icon is determined by a VBA ***intrinsic*** (or predefined) ***constant*** such as vbExclamation, which displays an exclamation point in a yellow triangle. VBA has many such constants that enable you to simplify your code, while at the same time achieving some impressive results.

Example 3 uses a different intrinsic constant, vbInformation, to display a different icon. It also extends the MsgBox statement to include a third parameter that is displayed on the title bar of the resulting dialog box. Look closely, for example, at Figures 1c and 1d, whose title bars contain "Microsoft Excel" and "Grauer/Barber", respectively. The first is the default entry (given that we are executing the procedure from within Microsoft Excel). You can, however, give your procedures a customized look by displaying your own text in the title bar.

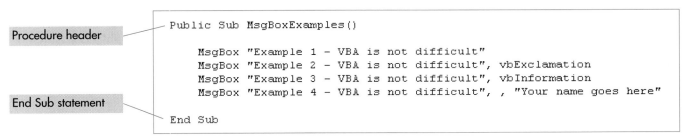

(a) VBA Code

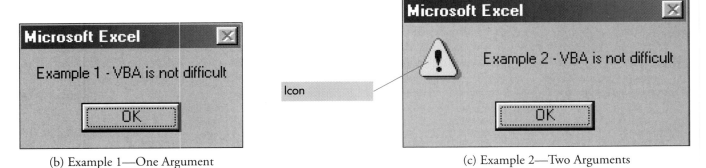

(b) Example 1—One Argument (c) Example 2—Two Arguments

FIGURE 1 *The MsgBox Statement*

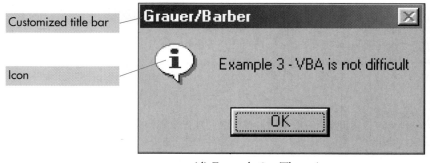
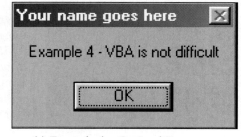

Customized title bar

Icon

(d) Example 3—Three Arguments   (e) Example 4—Omitted Parameter

FIGURE 1 *The MsgBox Statement (continued)*

Example 4 omits the second parameter (the icon), but includes the third parameter (the entry for the title bar). The parameters are positional, however, and thus the MsgBox statement contains two commas after the message to indicate that the second parameter has been omitted.

THE INPUTBOX FUNCTION

The MsgBox statement displays a prompt to the user, but what if you want the user to respond to the prompt by entering a value such as his or her name? This is accomplished using the ***InputBox function***. Note the subtle change in terminology in that we refer to the InputBox *function*, but the MsgBox *statement*. That is because a function returns a value, in this case the user's name, which is subsequently used in the procedure. In other words, the InputBox function asks the user for information, then it stores that information (the value returned by the user) for use in the procedure.

Figure 2 displays a procedure that prompts the user for a first and last name, after which it displays the information using the MsgBox statement. (The Dim statement at the beginning of the procedure is explained shortly.) Let's look at the first InputBox function, and the associated dialog box in Figure 2b. The InputBox function displays a prompt on the screen, the user enters a value ("Bob" in this example), and that value is stored in the variable that appears to the left of the equal sign (strFirstName). The concept of a variable is critical to every programming language. Simply stated, a ***variable*** is a named storage location that contains data that can be modified during program execution.

The MsgBox statement then uses the value of strFirstName to greet the user by name as shown in Figure 2c. This statement also introduces the ampersand to ***concatenate*** (join together) two different character strings, the literal "Good morning", followed by the value within the variable strFirstName.

The second InputBox function prompts the user for his or her last name. In addition, it uses a second argument to customize the contents of the title bar (VBA Primer in this example) as can be seen in Figure 2d. Finally, the MsgBox statement in Figure 2e displays both the first and last name through concatenation of multiple strings. This statement also uses the ***underscore*** to continue a statement from one line to the next.

VBA is not difficult, and you can use the MsgBox statement and InputBox function in conjunction with one another as the basis for several meaningful procedures. You will get a chance to practice in the hands-on exercise that follows shortly.

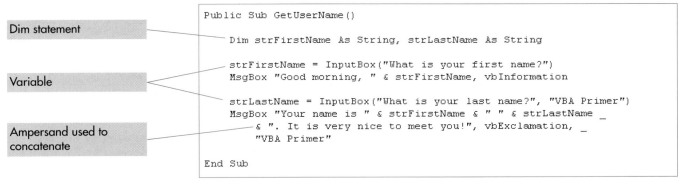

Dim statement

Variable

Ampersand used to concatenate

(a) VBA Code

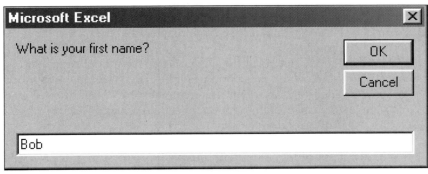

(b) InputBox

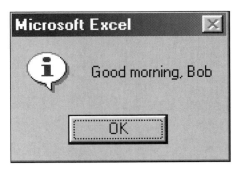

(c) Concatenation

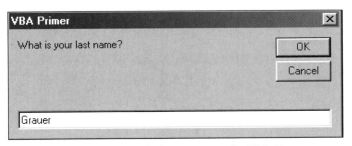

(d) InputBox includes Argument for Title Bar

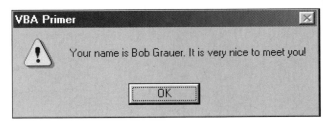

(e) Concatenation and Continuation

FIGURE 2   *The InputBox Statement*

Declaring Variables

Every variable must be declared (defined) before it can be used. This is accomplished through the ***Dim*** (short for Dimension) ***statement*** that appears at the beginning of a procedure. The Dim statement indicates the name of the variable and its type (for example, whether it will hold characters or numbers), which in turn reserves the appropriate amount of memory for that variable.

A variable name must begin with a letter and cannot exceed 255 characters. It can contain letters, numbers, and various special characters such as an underscore, but it cannot contain a space or the special symbols !, @, &, $, or #. Variable names typically begin with a prefix to indicate the type of data that is stored within the variable such as "str" for a character string or "int" for integers. The use of a prefix is optional with respect to the rules of VBA, but it is followed almost universally.

EXTENDING MICROSOFT OFFICE XP   5

THE VBA EDITOR

All VBA procedures are created using the ***Visual Basic Editor*** as shown in Figure 3. You may already be familiar with the editor, perhaps in conjunction with creating and/or editing macros in Word or Excel, or event procedures in Microsoft Access. Let's take a moment, however, to review its essential components.

The left side of the editor displays the ***Project Explorer***, which is similar in concept and appearance to the Windows Explorer, except that it displays the objects associated with the open document. If, for example, you are working in Excel, you will see the various sheets in a workbook, whereas in an Access database you will see forms and reports.

The VBA statements for the selected module (Module1 in Figure 3) appear in the code window in the right pane. The module, in turn, contains declarations and procedures that are separated by horizontal lines. There are two procedures, MsgBoxExamples and GetUserName, each of which was explained previously. A ***comment*** (nonexecutable) statement has been added to each procedure and appears in green. It is the apostrophe at the beginning of the line, rather than the color, that denotes a comment.

The ***Declarations section*** appears at the beginning of the module and contains a single statement, ***Option Explicit***. This option requires every variable in a procedure to be explicitly defined (e.g., in a Dim statement) before it can be used elsewhere in the module. It is an important option and should appear in every module you write (see exercise 5 at the end of the chapter).

The remainder of the window should look reasonably familiar in that it is similar to any other Office application. The title bar appears at the top of the window and identifies the application (Microsoft Visual Basic) and the current document (VBA Examples.xls). The right side of the title bar contains the Minimize, Restore, and Close buttons. A menu bar appears under the title bar. Toolbars are displayed under the menu bar. Commands are executed by pulling down the appropriate menu, via buttons on the toolbar, or by keyboard shortcuts.

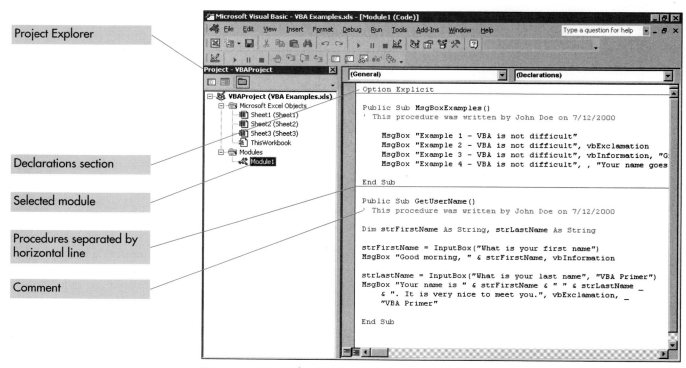

FIGURE 3 *The VBA Editor*

A VBA PRIMER

HANDS-ON EXERCISE 1

INTRODUCTION TO VBA

Objective   To create and test VBA procedures using the MsgBox and InputBox statements. Use Figure 4 as a guide in the exercise. You can do the exercise in any Office application.

Step 1a: **Start Microsoft Excel**

> ➤ We suggest you do the exercise in either Excel or Access (although you could use Word or PowerPoint just as easily). Go to step 1b for Access.
> ➤ Start **Microsoft Excel** and open a new workbook. Pull down the **File menu** and click the **Save command** (or click the **Save button** on the Standard toolbar) to display the Save As dialog box. Choose an appropriate drive and folder, then save the workbook as **VBA Examples**.
> ➤ Pull down the **Tools menu**, click the **Macro command**, then click the **Visual Basic Editor command** as shown in Figure 4a. Go to step 2.

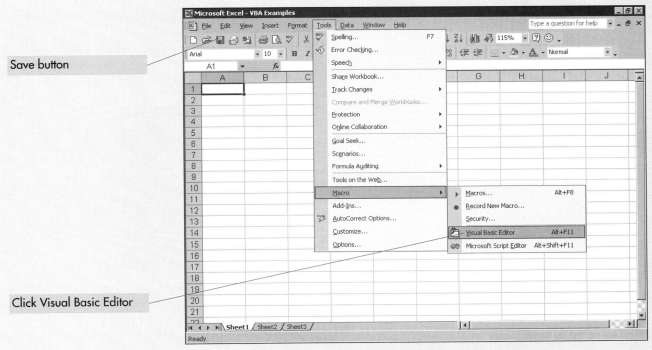

(a) Start Microsoft Excel (step 1a)

FIGURE 4   *Hands-on Exercise 1*

Step 1b: **Start Microsoft Access**

> ➤ Start **Microsoft Access** and choose the option to create a **Blank Access database**. Save the database as **VBA Examples**.
> ➤ Pull down the **Tools menu**, click the **Macro command**, then click the **Visual Basic Editor command**. (You can also use the **Alt+F11** keyboard shortcut to open the VBA editor without going through the Tools menu.)

Step 2: **Insert a Module**

- You should see a window similar to Figure 4b, but Module1 is not yet visible. Close the Properties window if it appears.
- If necessary, pull down the **View menu** and click **Project Explorer** to display the Project Explorer pane at the left of the window. Our figure shows Excel objects, but you will see the "same" window in Microsoft Access.
- Pull down the **Insert menu** and click **Module** to insert Module1 into the current project. The name of the module, Module1 in this example, appears in the Project Explorer pane.
- The Option Explicit statement may be entered automatically, but if not, click in the code window and type the statement **Option Explicit**.
- Pull down the **Insert menu** a second time, but this time select **Procedure** to display the Add Procedure dialog box in Figure 4b. Click in the **Name** text box and enter **MsgBoxExamples** as the name of the procedure. (Spaces are not allowed in a procedure name.)
- Click the option buttons for a **Sub procedure** and for **Public scope**. Click **OK**. The sub procedure should appear within the module and consist of the Sub and End Sub statements.

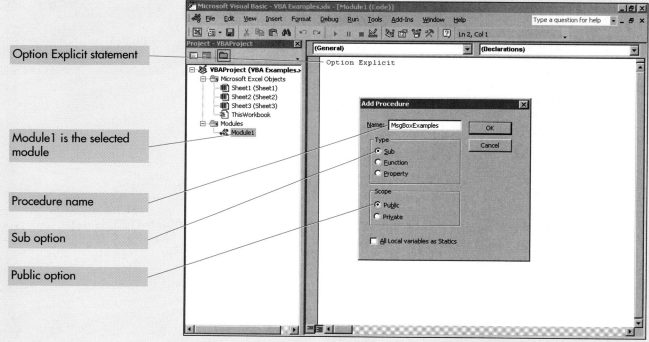

(b) Insert a Module (step 2)

FIGURE 4   *Hands-on Exercise 1 (continued)*

OPTION EXPLICIT

We say more about this important statement later on, but for now be sure that it appears in every module. See exercise 5 at the end of the chapter.

---

8   A VBA PRIMER

Step 3: **The MsgBox Statement**

- The insertion point (the flashing cursor) appears below the first statement. Press the **Tab key** to indent, type the key word **MsgBox**, then press the **space bar**. VBA responds with Quick Info that displays the syntax of the statement as shown in Figure 4c.
- Type a **quotation mark** to begin the literal, enter the text of your message, **This is my first VBA procedure**, then type the closing **quotation mark**.
- Click the **Run Sub button** on the Standard toolbar (or pull down the **Run menu** and click the **Run Sub command**) to execute the procedure. You should see a dialog box, containing the text you entered, within the Excel workbook (or other Office document) on which you are working.
- After you have read the message, click **OK** to return to the VBA Editor.

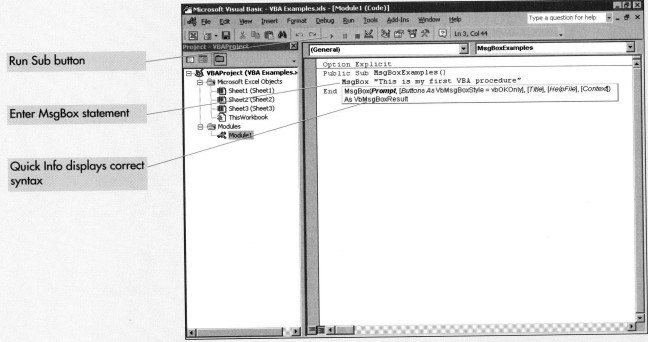

(c) The MsgBox Statement (step 3)

FIGURE 4 *Hands-on Exercise 1 (continued)*

QUICK INFO—HELP WITH VBA SYNTAX

Press the space bar after entering the name of a statement (e.g., MsgBox), and VBA responds with a Quick Info box that displays the syntax of the statement. You see the arguments in the statement and the order in which those arguments appear. Any argument in brackets is optional. If you do not see this information, pull down the Tools menu, click the Options command, then click the Editor tab. Check the box for Auto Quick Info and click OK.

Step 4: **Complete the Procedure**

➤ You should be back within the MsgBoxExamples procedure. If necessary, click at the end of the MsgBox statement, then press **enter** to begin a new line. Type **MsgBox** and press the **space bar** to begin entering the statement.

➤ The syntax of the MsgBox statement will appear on the screen. Type a **quotation mark** to begin the message, type **Add an icon** as the text of this message, then type the closing **quotation mark**. Type a **comma**, then press the **space bar** to enter the next parameter.

➤ VBA automatically displays a list of appropriate parameters, in this case a series of intrinsic constants that define the icon or command button that is to appear in the statement.

➤ You can type the first several letters (e.g., **vbi**, for vbInformation), then press the **space bar**, or you can use the **down arrow** to select **vbInformation** and then press the **space bar**. Either way you should complete the second MsgBox statement as shown in Figure 4d. Press **enter**.

➤ Enter the third MsgBox statement as shown in Figure 4d. Note the presence of the two consecutive commas to indicate that we omitted the second parameter within the MsgBox statement. Enter your name instead of John Doe where appropriate. Press **enter**.

➤ Enter the fourth (and last) MsgBox statement following our figure. Select **vbExclamation** as the second parameter, type a **comma**, then enter the text of the title bar, as you did for the previous statement.

➤ Click the **Save button** to save the changes to the module.

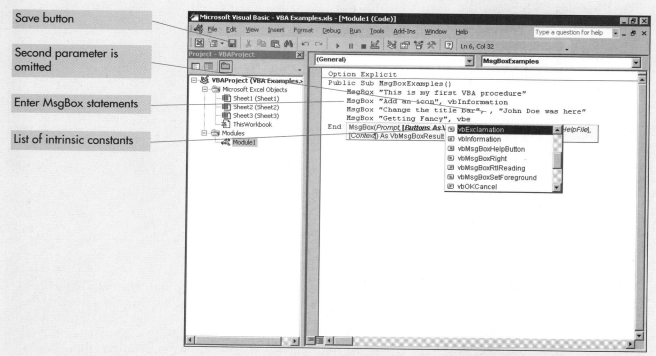

(d) Complete the Procedure (step 4)

FIGURE 4 *Hands-on Exercise 1 (continued)*

10 A VBA PRIMER

Step 5: **Test the Procedure**

➤ It's convenient if you can see the statements in the VBA procedure at the same time you see the output of those statements. Thus we suggest that you tile the VBA Editor and the associated Office application.
- Minimize all applications except the VBA Editor and the Office application (e.g., Excel).
- Right click the taskbar and click **Tile Windows Horizontally** to tile the windows as shown in Figure 4e. (It does not matter which window is on top. (If you see more than these two windows, minimize the other open window, then right click the taskbar and retile the windows.)
- Click anywhere in the VBA procedure, then click the **Run Sub button** on the Standard toolbar.
- The four messages will be displayed one after the other. Click **OK** after each message.

➤ Maximize the VBA window to continue working.

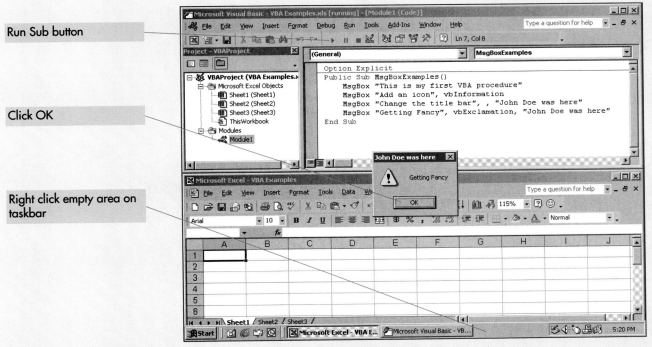

(e) Test the Procedure (step 5)

FIGURE 4   *Hands-on Exercise 1 (continued)*

HIDE THE WINDOWS TASKBAR

You can hide the Windows taskbar to gain additional space on the desktop. Right click any empty area of the taskbar to display a context-sensitive menu, click Properties to display the Taskbar properties dialog box, and if necessary click the Taskbar Options tab. Check the box to Auto Hide the taskbar, then click OK. The taskbar disappears from the screen but will reappear as you point to the bottom edge of the desktop.

Step 6: **Comments and Corrections**

➤ All VBA procedures should be documented with the author's name, date, and other comments as necessary to explain the procedure. Click after the procedure header. Press the **enter key** to leave a blank line.

➤ Press **enter** a second time. Type an **apostrophe** to begin the comment, then enter a descriptive statement similar to Figure 4f. Press **enter** when you have completed the comment. The line turns green to indicate it is a comment.

➤ The best time to experiment with debugging is when you know your procedure is correct. Go to the last MsgBox statement and delete the quotation mark in front of your name. Move to the end of the line and press **enter**.

➤ You should see the error message in Figure 4f. Unfortunately, the message is not as explicit as it could be; VBA cannot tell that you left out a quotation mark, but it does detect an error in syntax.

➤ Click **OK** in response to the error. Click the **Undo button** twice, to restore the quotation mark, which in turn corrects the statement.

➤ Click the **Save button** to save the changes to the module.

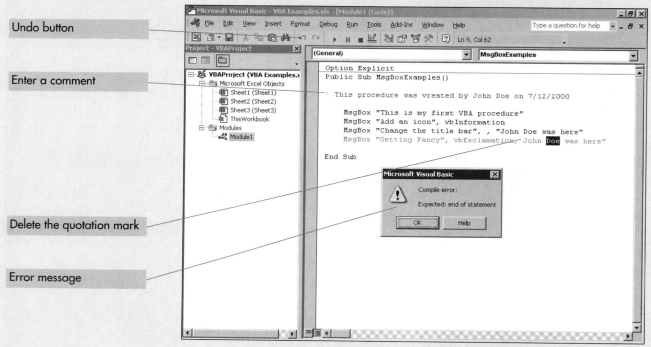

(f) Comments and Corrections (step 6)

FIGURE 4   *Hands-on Exercise 1 (continued)*

RED, GREEN, AND BLUE

Visual Basic for Applications uses different colors for different types of statements (or a portion of those statements). Any statement containing a syntax error appears in red. Comments appear in green. Key words, such as Sub and End Sub, appear in blue.

Step 7: **Create a Second Procedure**

- Pull down the **Insert menu** and click **Procedure** to display the Add Procedure dialog box. Enter **InputBoxExamples** as the name of the procedure. (Spaces are not allowed in a procedure name.)
- Click the option buttons for a **Sub procedure** and for **Public scope**. Click **OK**. The new sub procedure will appear within the existing module below the existing MsgBoxExamples procedure.
- Enter the statements in the procedure as they appear in Figure 4g. Be sure to type a space between the ampersand and the underscore in the second MsgBox statement. Click the **Save button** to save the procedure before testing it.
- You can display the output of the procedure directly in the VBA window if you minimize the Excel window. Thus, **right click** the Excel button on the taskbar to display a context-sensitive menu, then click the **Minimize command**. There is no visible change on your monitor.
- Click the **Run Sub button** to test the procedure. This time you see the Input box displayed on top of the VBA window because the Excel window has been minimized.
- Enter your first name in response to the initial prompt, then click **OK**. Click **OK** when you see the message box that says "Hello".
- Enter your last name in response to the second prompt and click **OK**. You should see a message box similar to the one in Figure 4g. Click **OK**.
- Return to the VBA procedure to correct any mistakes that might occur. Save the module.

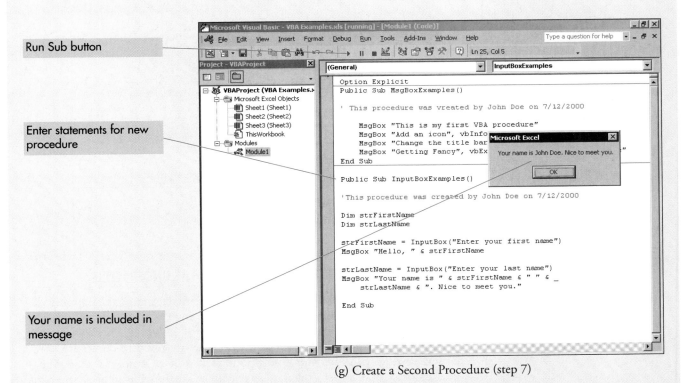

(g) Create a Second Procedure (step 7)

FIGURE 4 *Hands-on Exercise 1 (continued)*

EXTENDING MICROSOFT OFFICE XP

Step 8: **Create a Public Constant**

- Click after the Options Explicit statement and press **enter** to move to a new line. Type the statement to define the constant, **ApplicationTitle**, as shown in Figure 4h, and press **enter**.
- Click anywhere in the MsgBoxExamples procedure, then change the third argument in the last MsgBox statement to ApplicationTitle. Make the four modifications in the InputBoxExamples procedure as shown in Figure 4h.
- Click anywhere in the InputBoxExamples procedure, then click the **Run Sub button** to test the procedure. The title bar of each dialog box will contain a descriptive title corresponding to the value of the ApplicationTitle constant.
- Change the value of the ApplicationTitle constant in the General Declarations section, then rerun the InputBoxExamples procedure. The title of every dialog box changes to reflect the new value. Save the procedure.

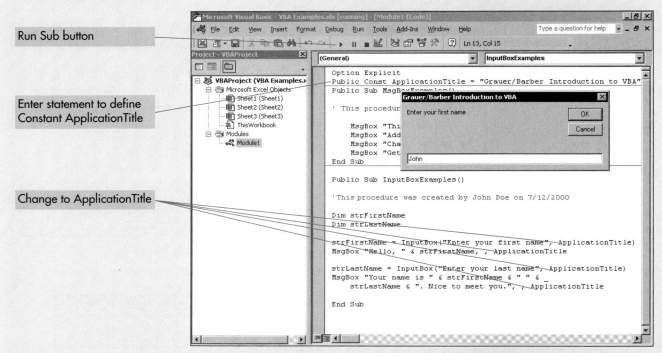

(h) Create a Public Constant (step 8)

FIGURE 4 *Hands-on Exercise 1 (continued)*

CONTINUING A VBA STATEMENT—THE & AND THE UNDERSCORE

A VBA statement can be continued from one line to the next by typing a space at the end of the line to be continued, typing the underscore character, then continuing on the next line. You may not, however, break a line in the middle of a literal (character string). Thus, you need to complete the character string with a closing quotation mark, add an ampersand (as the concatenation operator to display this string with the character string on the next line), then leave a space followed by the underscore to indicate continuation.

Step 9: **Help with VBA**

- You should be in the VBA editor. If necessary, pull down the **Help menu** and click **Microsoft Visual Basic Help** (or press the **F1 key**) to display the Office Assistant.
- Click the **Assistant**, type **InputBox**, then click the **Search button** in the Assistant's balloon for a list of topics pertaining to this entry. Click the first entry, **InputBox function**, to display the Help window.
- Click the **down arrow** on the Options button in the Help window, then click **Show tabs** to expand the Help window to include the Contents, Answer Wizard, and Index tabs as shown in Figure 4i.
- Take a minute to explore the information that is available. The Office Assistant functions identically in VBA as it does in all other Office applications. Close the Help window.
- Pull down the **File menu** and click the **Close command** (or click the **Close button** on the VBA title bar) to close the VBA window and return to the application. Click **Yes** if asked whether to save the changes to Module1.
- You should be back in the Excel (or Access) application window. Close the Office application if you do not want to continue with the next hands-on exercise at this time.
- Congratulations! You have just completed your first VBA procedure. Remember to use Help anytime you have a question.

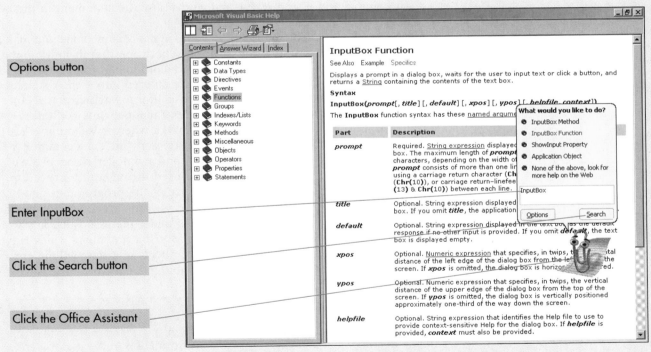

(i) Help with VBA (step 9)

FIGURE 4   *Hands-on Exercise 1 (continued)*

IF...THEN...ELSE STATEMENT

The ability to make decisions within a program, and then execute alternative sets of statements based on the results of those decisions, is crucial to any programming language. This is typically accomplished through an *If statement*, which evaluates a condition as either true or false, then branches accordingly. The If statement is not used in isolation, however, but is incorporated into a procedure to accomplish a specific task as shown in Figure 5a. This procedure contains two separate If statements, and the results are displayed in the message boxes shown in the remainder of the figure.

The InputBox statement associated with Figure 5b prompts the user for the name of his or her instructor, then it stores the answer in the variable strInstructorName. The subsequent If statement then compares the user's answer to the literal "Grauer". If the condition is true (i.e., Grauer was entered into the input box), then the message in Figure 5c is displayed. If, however, the user entered any other value, then the condition is evaluated as false, the MsgBox is not displayed, and processing continues with the next statement in the procedure.

The second If statement includes an optional *Else clause*. Again, the user is asked for a value, and the response is compared to the number 50. If the condition is true (i.e., the value of intUserStates equals 50), the message in Figure 5d is displayed to indicate that the response is correct. If, however, the condition is false (i.e., the user entered a number other than 50), the user sees the message in Figure 5e. Either way, true or false, processing continues with the next statement in the procedure. That's it—it's simple and it's powerful, and we will use the statement in the next hands-on exercise.

You can learn a good deal about VBA by looking at existing code and making inferences. Consider, for example, the difference between literals and numbers. *Literals* (also known as *character strings*) are stored differently from numbers, and this is manifested in the way that comparisons are entered into a VBA statement. Look closely at the condition that references a literal (strInstructorName = "Grauer") compared to the condition that includes a number (intUserStates = 50). The literal ("Grauer") is enclosed in quotation marks, whereas the number (50) is not. (The prefix used in front of each variable, "str" and "int", is a common VBA convention to indicate the variable type—a string and an integer, respectively.)

Note, too, that indentation and spacing are used throughout a procedure to make it easier to read. This is for the convenience of the programmer and not a requirement for VBA. The If, Else, and End If key words are aligned under one another, with the subsequent statements indented under the associated key word. We also indent a continued statement, such as a MsgBox statement, which is typically coded over multiple lines. Blank lines can be added anywhere within a procedure to separate blocks of statements from one another.

THE MSGBOX FUNCTION—YES OR NO

A simple MsgBox statement merely displays information to the user. MsgBox can also be used as a function, however, to accept information from the user such as clicking a Yes or No button, then combined with an If statement to take different actions based on the user's input. In essence, you enclose the arguments of the MsgBox function in parentheses (similar to what is done with the InputBox function), then test for the user response using the intrinsic constants vbYes and vbNo. See exercise 10 at the end of the chapter.

```
Public Sub IfThenElseExamples()

    Dim intUserStates As Integer
    Dim strInstructorName As String

    strInstructorName = InputBox("What is the last name of your instructor?")
    If strInstructorName = "Grauer" Then
        MsgBox "I hope you are enjoying the class", vbInformation
    End If

    intUserStates = InputBox("How many states are in the United States?")
    If intUserStates = 50 Then
        MsgBox "Correct. You know your geography!", vbExclamation
    Else
        MsgBox "Incorrect. You need to study geography!", vbExclamation
    End If

End Sub
```

- MsgBox executed if condition is true
- Executed if condition is true
- Optional Else clause
- Executed if condition is false

(a) VBA Code

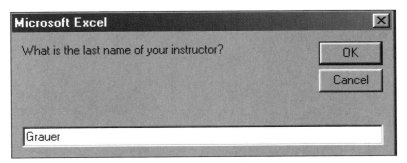

(b) Input Box Prompts for User Response

(c) Condition Is True

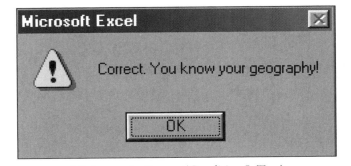

(d) Answer Is Correct (Condition Is True)

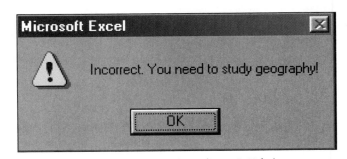

(e) Answer Is Wrong (Condition Is False)

FIGURE 5 *The If Statement*

CASE STATEMENT

The If statement is ideal for testing simple conditions and taking one of two actions. Although it can be extended to include additional actions by including one or more ElseIf clauses (If ... Then ... ElseIf ... ElseIf ...), this type of construction is often difficult to follow. Hence, the *Case statement* is used when multiple branches are possible.

The procedure in Figure 6a accepts a student's GPA, then displays one of several messages, depending on the value of the GPA. The individual cases are evaluated in sequence. Thus, we check first to see if the GPA is greater than or equal to 3.9, then 3.75, then 3.5, and so on. If none of the cases is true, the statement following the Else clause is executed.

Note, too, the format of the comparison in that numbers (such as 3.9 or 3.75) are not enclosed in quotation marks because the associated variable (sngUserGPA) was declared as numeric. If, however, we had been evaluating a string variable (such as, strUserMajor), quotation marks would have been required around the literal values (e.g., Case Is = "Business", Case Is = "Liberal Arts", and so on.) The distinction between numeric and character (string) variables is important.

```
Public Sub CaseExample()

    Dim sngUserGPA As Single

    sngUserGPA = InputBox("What is your GPA?")
    Select Case sngUserGPA
        Case Is >= 3.9
            MsgBox "Congratulations! You are graduating Summa Cum Laude!"
        Case Is >= 3.75
            MsgBox "Well Done! You are graduating Magna Cum Laude!"
        Case Is >= 3.5
            MsgBox "Congratulations! You are graduating Cum Laude!"
        Case Is >= 1.8
            MsgBox "You made it"
        Case Else
            MsgBox "Check the schedule for Summer School"
    End Select

End Sub
```

Numbers are not enclosed in quotes

Executed if none of the cases is true

(a) VBA Code

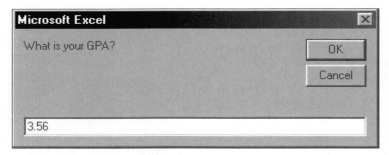

(b) Enter the GPA

(c) Third Option Is Selected

FIGURE 6 *The Case Statement*

CUSTOM TOOLBARS

A VBA procedure can be executed in several different ways. It can be run from the Visual Basic Editor, by pulling down the Run menu, clicking the Run Sub button on the Standard toolbar, or using the F5 function key. It can also be run from within the Office application (Word, Excel, or PowerPoint, but not Access), by pulling down the Tools menu, clicking the Macro command, then choosing the name of the macro that corresponds to the name of the procedure.

Perhaps the best way, however, is to create a ***custom toolbar*** that is displayed within the application as shown in Figure 7. The toolbar has its own name (Bob's Toolbar), yet it functions identically to any other Office toolbar. You have your choice of displaying buttons only, text only, or both buttons and text. Our toolbar provides access to four commands, each corresponding to a procedure that was discussed earlier. Click the Case Example button, for example, and the associated procedure is executed, starting with the InputBox statement asking for the user's GPA.

A custom toolbar is created via the Toolbars command within the View menu. The new toolbar is initially big enough to hold only a single button, but you can add, move, and delete buttons following the same procedure as for any other Office toolbar. You can add any command at all to the toolbar; that is, you can add existing commands from within the Office application, or you can add commands that correspond to VBA procedures that you have created. Remember, too, that you can add more buttons to existing office toolbars.

Once the toolbar has been created, it is displayed or hidden just like any other Office toolbar. It can also be docked along any edge of the application window or left floating as shown in Figure 7. It's fun, it's easy, and as you may have guessed, it's time for the next hands-on exercise.

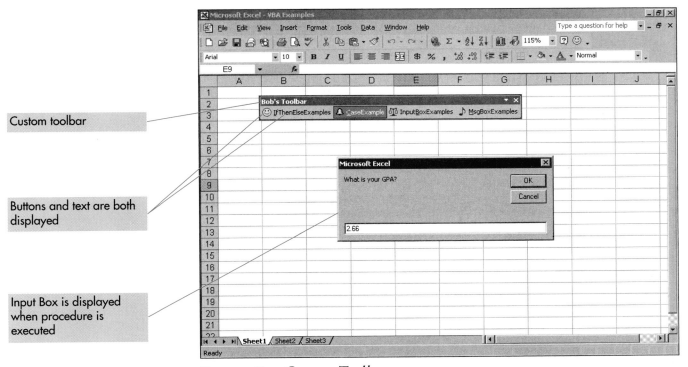

FIGURE 7 *Custom Toolbars*

HANDS-ON EXERCISE 2

DECISION MAKING

Objective To create procedures with If . . . Then . . . Else and Case statements, then create a custom toolbar to execute those procedures. Use Figure 8 as a guide in the exercise.

Step 1: **Open the Office Document**

➤ Open the **VBA Examples workbook** or Access database from the previous exercise. The procedure differs slightly, depending on whether you are using Access or Excel. In Access, you simply open the database. In Excel, however, you will be warned that the workbook contains a macro as shown in Figure 8a. Click the button to **Enable Macros**.

➤ Pull down the **Tools menu**, click the **Macro command**, then click the **Visual Basic Editor command**. You can also use the **Alt+F11** keyboard shortcut to open the VBA Editor without going through the Tools menu.

Click Enable Macros

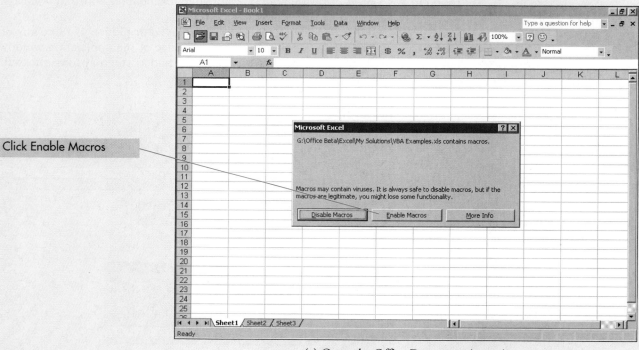

(a) Open the Office Document (step 1)

FIGURE 8 *Hands-on Exercise 2*

MACRO VIRUSES AND VBA PROCEDURES

An Excel macro is always associated with a VBA procedure. Thus, whenever Excel detects a procedure within a workbook, it warns you that the workbook contains a macro, which in turn may carry a macro virus. If you are confident the workbook is safe, click the button to Enable macros; otherwise open the workbook with the macros disabled.

Step 2: **Insert a New Procedure**

- You should be in the Visual Basic Editor as shown in Figure 8b. If necessary, double click **Module1** in the Explorer Window to open this module. Pull down the **Insert menu** and click the **Procedure command** to display the Add Procedure dialog box.
- Click in the **Name** text box and enter **IfThenElseExamples** as the name of the procedure. Click the option buttons for a **Sub procedure** and for **Public scope**. Click **OK**. The sub procedure should appear within the module and consist of the Sub and End Sub statements.
- Click within the newly created procedure, then click the **Procedure View button** at the bottom of the window. The display changes to show just the current procedure.
- Click the **Save button** to save the module with the new procedure.

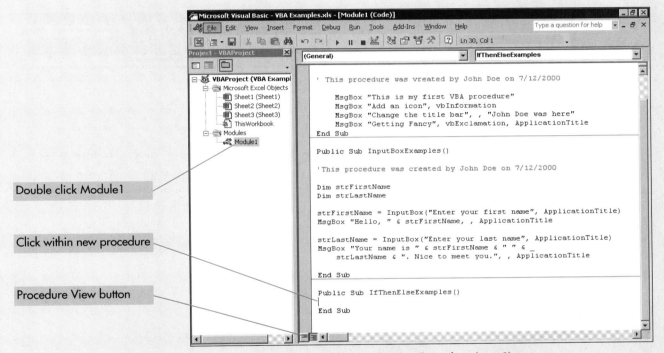

(b) Insert a New Procedure (step 2)

FIGURE 8 *Hands-on Exercise 2 (continued)*

PROCEDURE VIEW VERSUS FULL MODULE VIEW

The procedures within a module can be displayed individually, or alternatively, multiple procedures can be viewed simultaneously. To go from one view to the other, click the Procedure View button at the bottom of the window to display just the procedure you are working on, or click the Full Module View button to display multiple procedures. You can press Ctrl+PgDn and Ctrl+PgUp to move between procedures in either view.

Step 3: **Create the If...Then...Else Procedure**

➤ Enter the IfThenElseExamples procedure as it appears in Figure 8c, but use your instructor's name instead of Bob's. Note the following:
- The Dim statements at the beginning of the procedure are required to define the two variables that are used elsewhere in the procedure.
- The syntax of the comparison is different for string variables versus numeric variables. String variables require quotation marks around the comparison value (e.g., strInstructorName = "Grauer"). Numeric variables (e.g., intUserStates = 50) do not.
- Indentation and blank lines are used within a procedure to make the code easier to read, as distinct from a VBA requirement. Press the **Tab key** to indent one level to the right.

➤ Save the procedure.

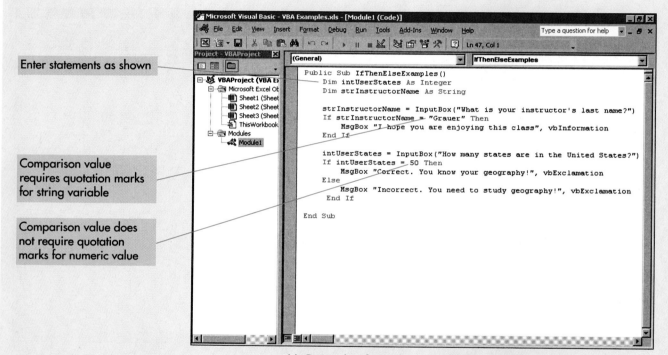

(c) Create the If...Then...Else Procedure (step 3)

FIGURE 8 *Hands-on Exercise 2 (continued)*

THE COMPLETE WORD TOOL

It's easy to misspell a variable name within a procedure, which is why the Complete Word tool is so useful. Type the first several characters in a variable name (e.g., "intU" or "strI" in the current procedure), then press Ctrl+Space. VBA will complete the variable for you, if you have already entered a sufficient number of letters for a unique reference. Alternatively, it will display all of the elements that begin with the letters you have entered. Use the down arrow to scroll through the list until you find the item, then press the space bar to complete the entry.

Step 4: **Test the Procedure**

- The best way to test a procedure is to display its output directly in the VBA window (without having to switch back and forth between that and the application window). Thus, right click the Excel button on the taskbar to display a context-sensitive menu, then click the **Minimize command**.
- There is no visible change on your monitor. Click anywhere within the procedure, then click the **Run Sub button**. You should see the dialog box in Figure 8d.
- Enter your instructor's name, exactly as it was spelled within the VBA procedure. Click **OK**. You should see a second message box that hopes you are enjoying the class. This box will be displayed only if you spell the instructor's name correctly. Click **OK**.
- You should see a second input box that asks how many states are in the United States. Enter **50** and click **OK**. You should see a message indicating that you know your geography. Click **OK** to close the dialog box.
- Click the **Run Sub button** a second time, but enter a different set of values in response to the prompts. Misspell your instructor's name, and you will not see the associated message box.
- Enter any number other than 50, and you will be told to study geography. Continue to test the procedure until you are satisfied it works under all conditions.

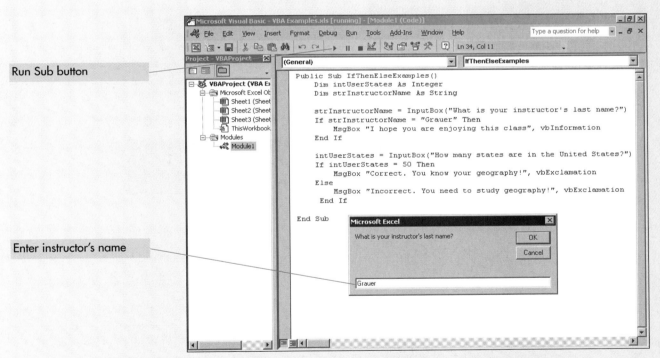

(d) Test the Procedure (step 4)

FIGURE 8 *Hands-on Exercise 2 (continued)*

EXTENDING MICROSOFT OFFICE XP

Step 5: **Create and Test the CaseExample Procedure**

➤ Pull down the **Insert menu** and create a new procedure called **CaseExample**, then enter the statements exactly as they appear in Figure 8e. Note:
- The variable sngUserGPA is declared to be a single-precision floating-point number (as distinct from the integer type that was used previously). A floating-point number is required in order to maintain a decimal point.
- You may use any editing technique with which you are comfortable. You could, for example, enter the first case, copy it four times in the procedure, then modify the copied text as necessary.
- The use of indentation and blank lines is for the convenience of the programmer and not a requirement of VBA.

➤ Click the **Run Sub button**, then test the procedure. Be sure to test it under all conditions; that is, you need to run it several times and enter a different GPA each time to be sure that all of the cases are working correctly.

➤ Save the procedure.

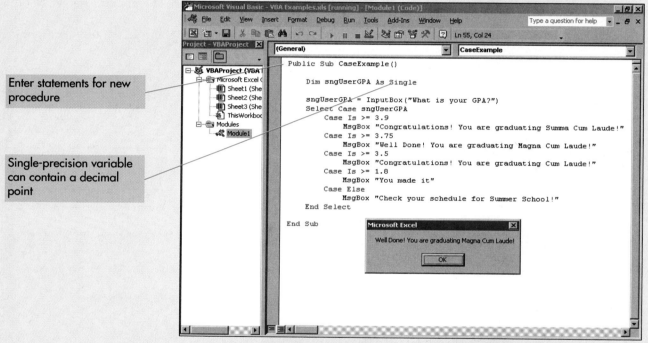

(e) Create and Test the CaseExample Procedure (step 5)

FIGURE 8 *Hands-on Exercise 2 (continued)*

RELATIONAL OPERATORS

The condition portion of an If or Case statement uses one of several relational operators. These include =, <, and > for equal to, less than, or greater than, respectively. You can also use >=, <=, or <> for greater than or equal to, less than or equal to, or not equal. This is basic, but very important, information if you are to code these statements correctly.

Step 6: **Create a Custom Toolbar**

- Click the **Excel** (or **Access**) **button** to display the associated application window. Pull down the **View menu**, click (or point to) the **Toolbars command**, then click **Customize** to display the Customize dialog box in Figure 8f. (Bob's toolbar is not yet visible.) Click the **Toolbars tab**.
- Click the **New button** to display the New Toolbar dialog box. Enter the name of your toolbar—e.g., **Bob's toolbar**—then click **OK** to create the toolbar and close the dialog box.
- Your toolbar should appear on the screen, but it does not yet contain any buttons. If necessary, click and drag the title bar of your toolbar to move the toolbar within the application window.
- Toggle the check box that appears next to your toolbar within the Customize dialog box on and off to display or hide your toolbar. Leave the box checked to display the toolbar and continue with this exercise.

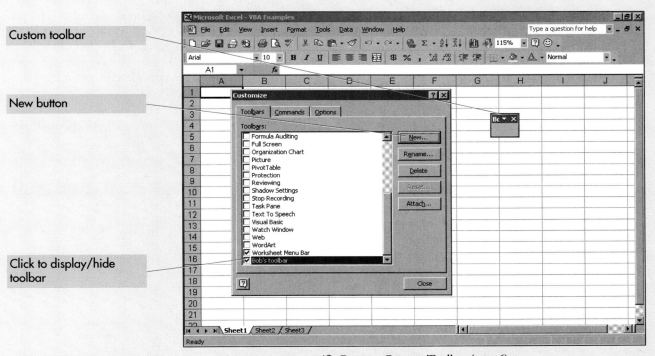

(f) Create a Custom Toolbar (step 6)

FIGURE 8   *Hands-on Exercise 2 (continued)*

FIXED VERSUS FLOATING TOOLBARS

A toolbar may be docked (fixed) along the edge of the application window, or it can be displayed as a floating toolbar anywhere within the window. You can switch back and forth by dragging the move handle of a docked toolbar to move the toolbar away from the edge. Conversely, you can drag the title bar of a floating toolbar to the edge of the window to dock the toolbar. You can also click and drag the border of a floating toolbar to change its size.

Step 7: **Add Buttons to the Toolbar**

➤ Click the **Commands tab** in the Customize dialog box, click the **down arrow** in the Categories list box, then scroll until you can select the **Macros category**. (If you are using Access and not Excel, you need to select the **File category**, then follow the steps as described in the boxed tip on the next page.)

➤ Click and drag the **Custom button** to your toolbar and release the mouse. A "happy face" button appears on the toolbar you just created. (You can remove a button from a toolbar by simply dragging the button from the toolbar.)

➤ Select the newly created button, then click the **Modify Selection command button** (or right click the button to display the context-sensitive menu) in Figure 8g. Change the button's properties as follows:
 • Click the **Assign Macro command** at the bottom of the menu to display the Assign Macro dialog box, then select the **IfThenElseExamples** macro (procedure) to assign it to the button. Click **OK**.
 • Click the **Modify Selection button** a second time.
 • Click in the **Name Textbox** and enter an appropriate name for the button, such as **IfThenElseExamples**.
 • Click the **Modify Selection button** a third time, then click **Text Only (Always)** to display text rather than an image.

➤ Close the Customize dialog box when you have completed the toolbar. Save the workbook.

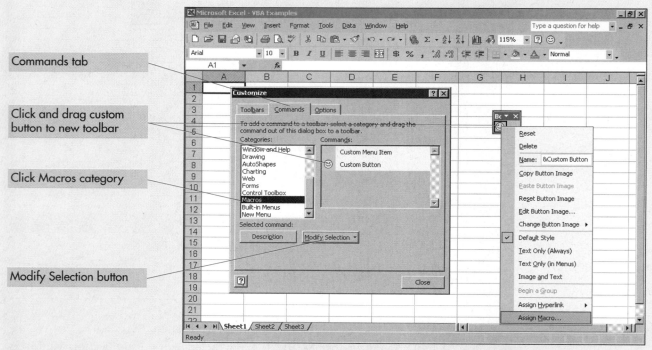

(g) Add Buttons to the Toolbar (step 7)

FIGURE 8 *Hands-on Exercise 2 (continued)*

Step 8: **Test the Custom Toolbar**

- Click any command on your toolbar as shown in Figure 8h. We clicked the **InputBoxExamples button**, which in turn executed the InputBoxExamples procedure that was created in the first exercise.
- Enter the appropriate information in any input boxes that are displayed. Click **OK**. Close your toolbar when you have completed testing it.
- If this is not your own machine, you should delete your toolbar as a courtesy to the next student. Pull down the **View menu**, click the **Toolbars command**, click **Customize** to display the Customize dialog box, then click the **Toolbars tab**. Select (highlight) the toolbar, then click the **Delete button** in the Customize dialog box. Click **OK** to delete the button. Close the dialog box.
- Exit Office if you do not want to continue with the next exercise.

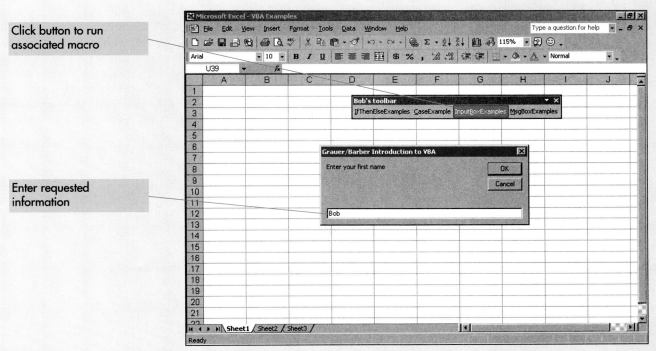

(h) Test the Custom Toolbar (step 8)

FIGURE 8 *Hands-on Exercise 2 (continued)*

ACCESS IS DIFFERENT

The procedure to create a custom toolbar in Access is different from the procedure in Excel. Select the File category within the Customize dialog box, then click and drag the Custom command to the newly created toolbar. Select the command on the toolbar, then click the Modify Selection command button in the dialog box. Click Properties, click the On Action text box, then type the name of the procedure you want to run in the format, =procedurename(). Close the dialog boxes, then press Alt+F11 to return to the VBA Editor. Change the key word "Sub" that identifies the procedure to "Function". Return to the database window, then test the newly created toolbar.

FOR . . . NEXT STATEMENT

The *For . . . Next statement* executes all statements between the words For and Next a specified number of times, using a counter to keep track of the number of times the statements are executed. The simplest form of the statement, For intCounter = 1 To N, executes the statements within the loop N times.

The procedure in Figure 9 contains two For . . . Next statements that sum the numbers from 1 to 10, counting by one and two, respectively. The Dim statements at the beginning of the procedure declare two variables, intSumofNumbers to hold the sum and intCounter to hold the value of the counter. The sum is initialized to zero immediately before the first loop. The statements in the loop are then executed 10 times, each time incrementing the sum by the value of the counter. The result (the sum of the numbers from 1 to 10) is displayed after the loop in Figure 9b.

The second For . . . Next statement increments the counter by two rather than by one. (The increment or step is assumed to be one unless a different value is specified.) The sum of the numbers is reset to zero prior to entering the second loop, the loop is entered, and the counter is initialized to the starting value of one. Each subsequent time through the loop, however, the counter is incremented by two. Each time the value of the counter is compared to the ending value, until it (the counter) exceeds the ending value, at which point the For . . . Next statement is complete. Thus the second loop will be executed for values of 1, 3, 5, 7, and 9. After the fifth time through the loop, the counter is incremented to 11, which is greater than the ending value of 10, and the loop is terminated.

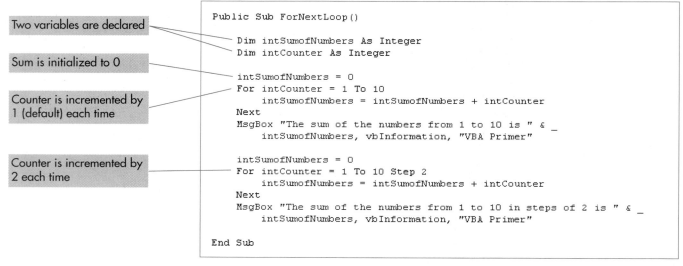

(a) VBA Code

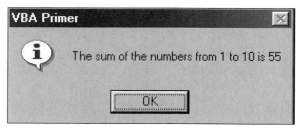

(b) In Increments of 1

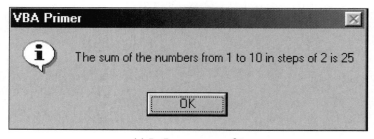

(c) In Increments of 2

FIGURE 9 *For . . . Next Loops*

DO LOOPS

The For...Next statement is ideal when you know in advance how many times you want to go through a loop. There are many instances, however, when the number of times through the loop is indeterminate. You could, for example, give a user multiple chances to enter a password or answer a question. This type of logic is implemented through a Do loop. You can repeat the loop as long as a condition is true (Do While), or until a condition becomes true (Do Until). The choice depends on how you want to state the condition.

Regardless of which key word you choose, Do While or Do Until, two formats are available. The difference is subtle and depends on whether the key word (While or Until) appears at the beginning or end of the loop. Our discussion will use the Do Until statement, but the Do While statement works in similar fashion.

Look closely at the procedure in Figure 10a, which contains two different loops. In the first example, the Until condition appears at the end of the loop, which means the statements in the loop are executed, and then the condition is tested. This ensures that the statements in the loop will be executed at least once. The second loop, however, places the Until condition at the beginning of the loop, so that it (the condition) is tested prior to the loop being executed. Thus, if the condition is satisfied initially, the second loop will never be executed. In other words, there are two distinct statements **Do...Loop Until** and **Do Until...Loop**. The first statement executes the loop, then tests the condition. The second statement tests the condition, then enters the loop.

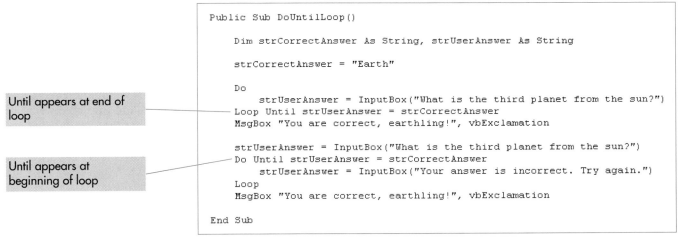

(a) VBA Code

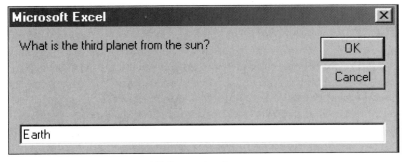

(b) Input the Answer

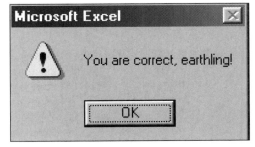

(c) Correct Response

FIGURE 10   *Do Until Loops*

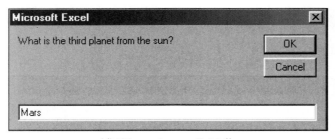

(d) Wrong Answer Initially

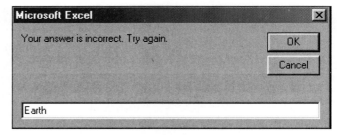

(e) Second Chance

FIGURE 10   *Do Until Loops (continued)*

It's tricky, but stay with us. In the first example, the user is asked the question within the loop, and the loop is executed repeatedly until the user gives the correct answer. In the second example, the user is asked the question outside of the loop, and the loop is bypassed if the user answers it correctly. The latter is the preferred logic because it enables us to phrase the question differently, before and during the loop. Look carefully at the difference between the InputBox statements and see how the question changes within the second loop.

DEBUGGING

As you learn more about VBA and develop more powerful procedures, you are more likely to make mistakes. The process of finding and correcting errors within a procedure is known as ***debugging*** and it is an integral part of programming. Do not be discouraged if you make mistakes. Everyone does. The important thing is how quickly you are able to find and correct the errors that invariably occur. We begin our discussion of debugging by describing two types of errors, ***compilation errors*** and ***execution*** (or ***run-time***) ***errors***.

A compilation error is simply an error in VBA syntax. (Compilation is the process of translating a VBA procedure to machine language, and thus a compilation error occurs when the VBA Editor is unable to convert a statement to machine language.) Compilation errors occur for many reasons, such as misspelling a key word, omitting a comma, and so on. VBA recognizes the error before the procedure is run and displays the invalid statement in red together with an associated error message. The programmer corrects the error and then reruns the procedure.

Execution errors are caused by errors in logic and are more difficult to detect because they occur without any error message. VBA, or for that matter any other programming language, does what you tell it to do, which is not necessarily what you want it to do. If, for example, you were to compute the sales tax of an item by multiplying the price by 60% rather than 6%, VBA will perform the calculation and simply display the wrong answer. It is up to you to realize that the results of the procedure are incorrect, and you will need to examine its statements and correct the mistake.

So how do you detect an execution error? In essence, you must decide what the expected output of your procedure should be, then you compare the actual results of the procedure to the intended result. If the results are different, an error has occurred, and you have to examine the logic in the procedure to find the error. You may see the mistake immediately (e.g., using 60% rather than 6% in the previous example), or you may have to examine the code more closely. And as you might expect, VBA has a variety of tools to help you in the debugging process. These tools are accessed from the ***Debug toolbar*** or the ***Debug menu*** as shown in Figure 11.

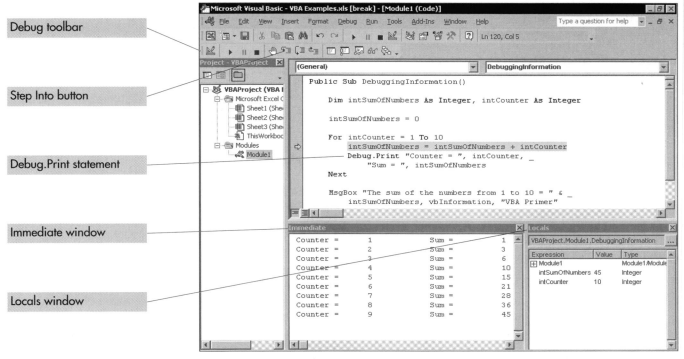

FIGURE 11 *Debugging*

The procedure in Figure 11 is a simple For . . . Next loop to sum the integers from 1 to 10. The procedure is correct as written, but we have introduced several debugging techniques into the figure. The most basic technique is to step through the statements in the procedure one at a time to see the sequence in which the statements are executed. Click the ***Step Into button*** on the Debug toolbar to enter (step into) the procedure, then continue to click the button to move through the procedure. Each time you click the button, the statement that is about to be executed is highlighted.

Another useful technique is to display the values of selected variables as they change during execution. This is accomplished through the ***Debug.Print statement*** that displays the values in the ***Immediate window***. The Debug.Print statement is placed within the For . . . Next loop so that you can see how the counter and the associated sum change during execution.

As the figure now stands, we have gone through the loop nine times, and the sum of the numbers from 1 to 9 is 45. The Step Into button is in effect so that the statement to be executed next is highlighted. You can see that we are back at the top of the loop, where the counter has been incremented to 10, and further, that we are about to increment the sum.

The ***Locals window*** is similar in concept except that it displays only the current values of all the variables within the procedure. Unlike the Immediate window, which requires the insertion of Debug.Print statements into a procedure to have meaning, the Locals window displays its values automatically, without any effort on the part of the programmer, other than opening the window. All three techniques can be used individually, or in conjunction with one another, as the situation demands.

We believe that the best time to practice debugging is when you know there are no errors in your procedure. As you may have guessed, it's time for the next hands-on exercise.

Hands-on Exercise 3

Loops and Debugging

Objective To create a loop using the For . . . Next and Do Until statements; to open the Locals and Immediate windows and illustrate different techniques for debugging. Use Figure 12 as a guide in the exercise.

Step 1: **Insert a New Procedure**

➤ Open the **VBA Examples workbook** or the Access database from the previous exercise. Either way, pull down the **Tools menu**, click the **Macro command**, then click **Visual Basic Editor** (or use the **Alt+F11** keyboard shortcut) to start the VBA editor.

➤ If necessary, double click **Module1** within the Project Explorer window to open this module. Pull down the **Insert menu** and click the **Procedure command** to display the Add Procedure dialog box.

➤ Click in the **Name** text box and enter **ForNextLoop** as the name of the procedure. Click the option buttons for a **Sub procedure** and for **Public scope**. Click **OK**. The sub procedure should appear within the module and consist of the Sub and End Sub statements.

➤ Click the **Procedure View button** at the bottom of the window as shown in Figure 12a. The display changes to show just the current procedure, giving you more room in which to work.

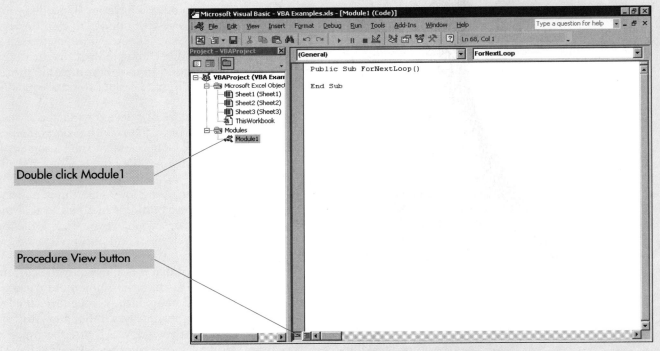

(a) Insert a New Procedure (step 1)

FIGURE 12 *Hands-on Exercise 3*

Step 2: **Test the For . . . Next Procedure**

➤ Enter the procedure exactly as it appears in Figure 12b. Note the following:
- A comment is added at the beginning of the procedure to identify the author and the date.
- Two variables are declared at the beginning of the procedure, one to hold the sum of the numbers and the other to serve as a counter.
- The sum of the numbers is initialized to zero. The For . . . Next loop varies the counter from 1 to 10.
- The statement within the For . . . Next loop increments the sum of the numbers by the current value of the counter. The equal sign is really a replacement operator; that is, replace the variable on the left (the sum of the numbers) by the expression on the right (the sum of the numbers plus the value of the counter.
- Indentation and spacing within a procedure are for the convenience of the programmer and not a requirement of VBA. We align the For and Next statements at the beginning and end of a loop, then indent all statements within a loop.
- The MsgBox statement displays the result and is continued over two lines.

➤ Click the **Save button** to save the module. Right click the **Excel button** on the Windows taskbar to display a context-sensitive menu, then click the **Minimize command**.

➤ Click the **Run Sub button** to test the procedure, which should display the MsgBox statement in Figure 12b. Correct any errors that may occur.

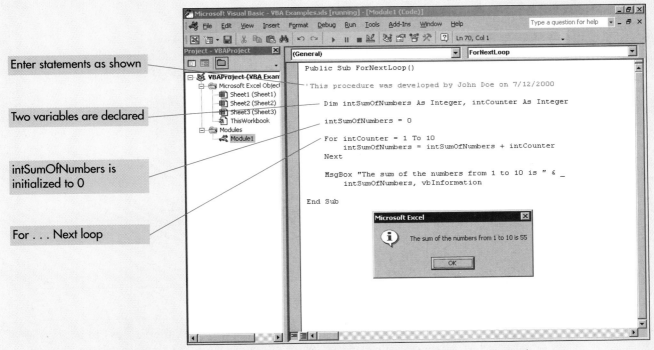

(b) Test the For . . . Next Procedure (step 2)

FIGURE 12 *Hands-on Exercise 3 (continued)*

Step 3: **Compilation Errors**

- The best time to practice debugging is when you know that the procedure is working properly. Accordingly, we will make some deliberate errors in our procedure to illustrate different debugging techniques.
- Pull down the **View menu**, click the **Toolbars command**, and (if necessary) toggle the Debug toolbar on, then dock it under the Standard toolbar.
- Click on the statement that initializes intSumOfNumbers to zero and delete the "s" at the end of the variable name. Click the **Run Sub button**.
- You will see the message in Figure 12c. Click **OK** to acknowledge the error, then click the **Undo button** to correct the error.
- The procedure header is highlighted, indicating that execution is temporarily suspended and that additional action is required from you to continue testing. Click the **Run Sub button** to retest the procedure.
- This time the procedure executes correctly and you see the MsgBox statement indicating that the sum of the numbers from 1 to 10 is 55. Click **OK**.

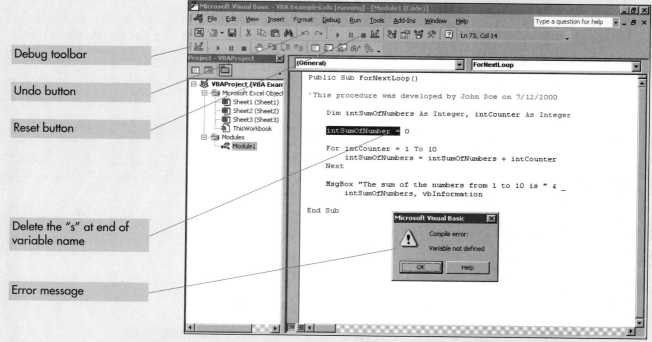

(c) Compilation Errors (step 3)

FIGURE 12   *Hands-on Exercise 3 (continued)*

USE HELP AS NECESSARY

Pull down the Help menu at any time (or press the F1 key) to access the VBA Help facility to explore at your leisure. You can also obtain context-sensitive help by clicking the Help button when it appears within a dialog box. Click the Help button in Figure 12c, for example, and you will be advised to correct the spelling of the variable.

Step 4: **Step Through a Procedure**

- Pull down the **View menu** a second time and click the **Locals Window command** (or click the **Locals Window button** on the Debug toolbar).
- If necessary, click and drag the top border of the Locals window to size the window appropriately as shown in Figure 12d.
- Click anywhere within the procedure. Pull down the **Debug menu** and click the **Step Into command** (or click the **Step Into button** on the Debug toolbar). The first statement (the procedure header) is highlighted, indicating that you are about to enter the procedure.
- Click the **Step Into button** (or use the **F8** keyboard shortcut) to step into the procedure and advance to the next executable statement. The statement that initializes intSumOfNumbers to zero is highlighted, indicating that this statement is about to be executed.
- Continue to press the **F8 key** to step through the procedure. Each time you execute a statement, you can see the values of intSumOfNumbers and intCounter change within the Locals window. (You can click the **Step Out button** at any time to end the procedure.)
- Correct errors as they occur. Click the **Reset button** on the Standard or Debug toolbars at any time to begin executing the procedure from the beginning.
- Eventually you exit from the loop, and the sum of the numbers (from 1 to 10) is displayed within a message box.
- Click **OK** to close the message box. Press the **F8 key** a final time, then close the Locals window.

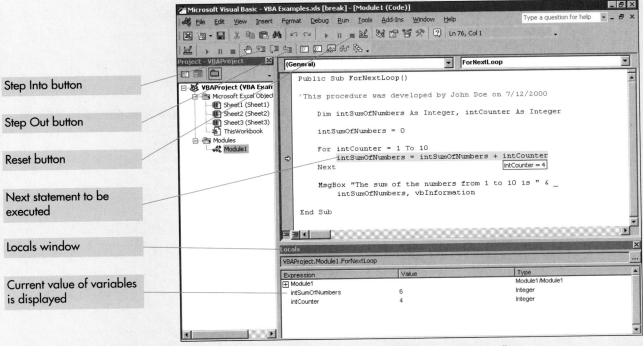

(d) Step Through a Procedure (step 4)

FIGURE 12 *Hands-on Exercise 3 (continued)*

EXTENDING MICROSOFT OFFICE XP

Step 5: **The Immediate Window**

- You should be back in the VBA window. Click immediately to the left of the Next statement and press **enter** to insert a blank line. Type the **Debug.Print** statement exactly as shown in Figure 12e. (Click **OK** if you see a message indicating that the procedure will be reset.)
- Pull down the **View menu** and click the **Immediate Window command** (or click the **Immediate Window button** on the Debug toolbar). The Immediate window should be empty, but if not, you can click and drag to select the contents, then press the Del key to clear the window.
- Click anywhere within the For ... Next procedure, then click the **Run Sub button** on the Debug toolbar to execute the procedure. You will see the familiar message box indicating that the sum of the numbers is 55. Click **OK**.
- You should see 10 lines within the Immediate window as shown in Figure 12e, corresponding to the values displayed by the Debug.Print statement as it was executed within the loop.
- Close the Immediate window.

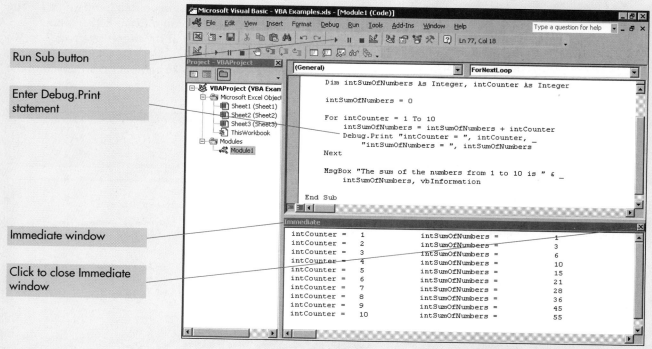

(e) The Immediate Window (step 5)

FIGURE 12   Hands-on Exercise 3 (continued)

INSTANT CALCULATOR

Use the Print method (action) in the Immediate window to use VBA as a calculator. Press Ctrl+G at any time to display the Immediate window. Click in the window, then type the statement Debug.Print, followed by your calculation, for example, Debug.Print 2+2, and press enter. The answer is displayed on the next line in the Immediate window.

Step 6: **A More General Procedure**

➤ Modify the existing procedure to make it more general; for example, to sum the values from any starting value to any ending value:
- Click at the end of the existing Dim statement to position the insertion point, press **enter** to create a new line, then add the second Dim statement as shown in Figure 12f.
- Click before the For statement, press **enter** to create a blank line, press **enter** a second time, then enter the two InputBox statements to ask the user for the beginning and ending value.
- Modify the For statement to execute from **intStart** to **intEnd** rather than from 1 to 10.
- Change the MsgBox statement to reflect the values of intStart and intEnd, and a customized title bar. Note the use of the ampersand and the underscore, to indicate concatenation and continuation, respectively.

➤ Click the **Save button** to save the module.

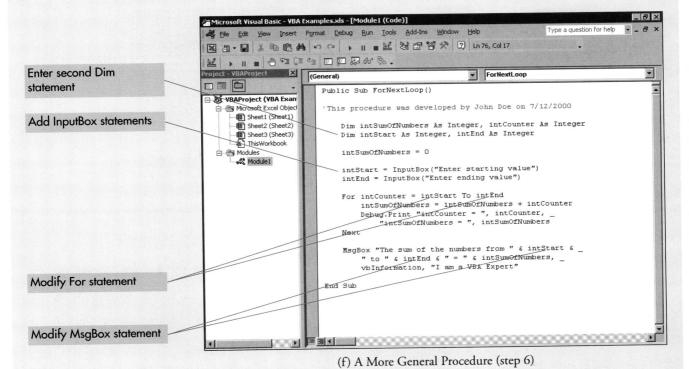

(f) A More General Procedure (step 6)

FIGURE 12 *Hands-on Exercise 3 (continued)*

> **USE WHAT YOU KNOW**
>
> Use the techniques acquired from other applications such as Microsoft Word to facilitate editing within the VBA window. Press the Ins key to toggle between the insert and overtype modes as you modify the statements within a VBA procedure. You can also cut, copy, and paste statements (or parts of statements) within a procedure and from one procedure to another. The Find and Replace commands are also useful.

Step 7: **Test the Procedure**

- Click the **Run Sub button** to test the procedure. You should be prompted for a beginning and an ending value. Enter any numbers you like, such as 10 and 20, respectively, to match the result in Figure 12g.
- The value displayed in the MsgBox statement should reflect the numbers you entered. For example, you will see a sum of 165 if you entered 10 and 20 as the starting and ending values.
- Look carefully at the message box that is displayed in Figure 12g. Its title bar displays the literal "I am a VBA expert", corresponding to the last argument in the MsgBox statement.
- Note, too, the spacing that appears within the message box, which includes spaces before and after each number. Look at your results and, if necessary, modify the MsgBox statement so that you have the same output. Click **OK**.
- Save the procedure.

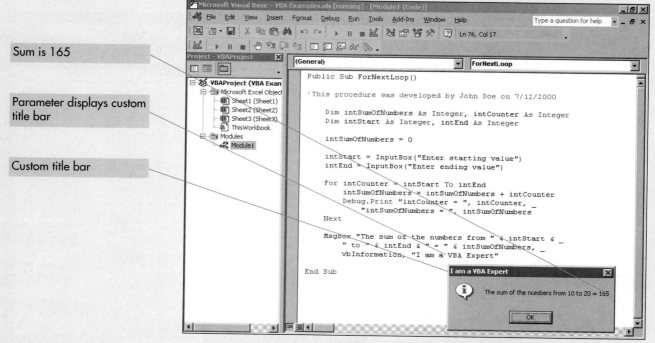

(g) Test the Procedure (step 7)

FIGURE 12 *Hands-on Exercise 3 (continued)*

CHANGE THE INCREMENT

The For...Next statement can be made more general by supplying an increment within the For statement. Try For intCount = 1 To 10 Step 2, or more generally, For intCount = intStart to intEnd Step intStepValue. "Step" is a Visual Basic key word and must be entered that way. intCount, intEnd, and intStepValue are user-defined variables. The variables must be defined at the beginning of a procedure and can be initialized by requesting values from the user through the InputBox statement.

---

38 A VBA PRIMER

Step 8: **Create a Do Until Loop**

➤ Pull down the **Insert menu** and click the **Procedure command** to insert a new procedure called **DoUntilLoop**. Enter the procedure as it appears in Figure 12h. Note the following:
  - Two string variables are declared to hold the correct answer and the user's response, respectively.
  - The variable strCorrectAnswer is set to "Earth", the correct answer for our question.
  - The initial InputBox function prompts the user to enter his/her response to the question. A second InputBox function appears in the loop that is executed if and only if the user enters the wrong answer.
  - The Until condition appears at the beginning of the loop, so that the loop is entered only if the user answers incorrectly. The loop executes repeatedly until the correct answer is supplied.
  - A message to the user is displayed at the end of the procedure after the correct answer has been entered.

➤ Click the **Run Sub button** to test the procedure. Enter the correct answer on your first attempt, and you will see that the loop is never entered.

➤ Rerun the procedure, answer incorrectly, then note that a second input box appears, telling you that your answer was incorrect.

➤ Save the procedure.

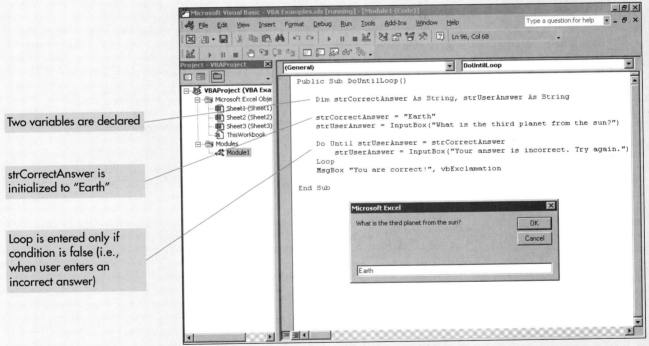

(h) Create a Do Until Loop (step 8)

FIGURE 12 *Hands-on Exercise 3 (continued)*

Step 9: **A More Powerful Procedure**

➤ Modify the procedure as shown in Figure 12i to include the statements to count and print the number of times the user takes to get the correct answer.
  • The variable intNumberOfAttempts is declared as an integer and is initialized to 1 after the user inputs his/her initial answer.
  • The Do loop is expanded to increment intNumberOfAttempts by 1 each time the loop is executed.
  • The MsgBox statement after the loop is expanded prints the number of attempts the user took to answer the question.
➤ Save the module, then click the **Run Sub button** to test the module. You should see a dialog box similar to the one in Figure 12i. Click **OK**.
➤ Pull down the **File menu** and click the **Print command** to display the Print dialog box. Click the option button to print the current module. Click **OK**.
➤ Exit Office if you do not want to continue at this time.

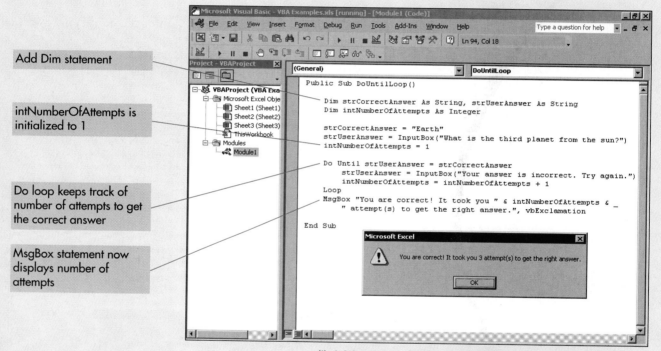

(i) A More Powerful Procedure (step 9)

FIGURE 12 *Hands-on Exercise 3 (continued)*

IT'S NOT EQUAL, BUT REPLACE

All programming languages use statements of the form N = N + 1, in which the equal sign does not mean equal in the literal sense; that is, N cannot equal N + 1. The equal sign is really a replacement operator. Thus, the expression on the right of the equal sign is evaluated, and that result replaces the value of the variable on the left. In other words, the statement N = N + 1 increments the value of N by one.

PUTTING VBA TO WORK (MICROSOFT EXCEL)

Our approach thus far has focused on VBA as an independent entity that can be run without specific reference to the applications in Microsoft Office. We have covered several individual statements, explained how to use the VBA editor to create and run procedures, and how to debug those procedures, if necessary. We hope you have found the material to be interesting, but you may be asking yourself, "What does this have to do with Microsoft Office?" In other words, how can you use your knowledge of VBA to enhance your ability in Microsoft Excel or Access? The answer is to create *event procedures* that run automatically in response to events within an Office application.

VBA is different from traditional programming languages in that it is event-driven. An *event* is defined as any action that is recognized by an application such as Excel or Access. Opening or closing an Excel workbook or an Access database is an event. Selecting a worksheet within a workbook is also an event, as is clicking on a command button on an Access form. To use VBA within Microsoft Office, you decide which events are significant, and what is to happen when those events occur. Then you develop the appropriate event procedures.

Consider, for example, Figure 13, which displays the results of two event procedures in conjunction with opening and closing an Excel workbook. (If you are using Microsoft Access instead of Excel, you can skip this discussion and the associated exercise, and move to the parallel material for Access that appears after the next hands-on exercise.) The procedure associated with Figure 13a displays a message that appears automatically after the user executes the command to close the associated workbook. The procedure is almost trivial to write, and consists of a single MsgBox statement. The effect of the procedure is quite significant, however, as it reminds the user to back up his or her work after closing the workbook. Nor does it matter how the user closes the workbook—whether by pulling down the menu or using a keyboard shortcut—because the procedure runs automatically in response to the Close Workbook event, regardless of how that event occurs.

The dialog box in Figure 13b prompts the user for a password and appears automatically when the user opens the workbook. The logic here is more sophisticated in that the underlying procedure contains an InputBox statement to request the password, a Do Until loop that is executed until the user enters the correct password or exceeds the allotted number of attempts, then additional logic to display the worksheet or terminate the application if the user fails to enter the proper password. The procedure is not difficult, however, and it builds on the VBA statements that were covered earlier.

The next hands-on exercise has you create the two event procedures that are associated with Figure 13. As you do the exercise, you will gain additional experience with VBA and an appreciation for the potential event procedures within Microsoft Office.

HIDING AND UNHIDING A WORKSHEET

Look carefully at the workbooks in Figures 13a and 13b. Both figures reference the identical workbook, Financial Consultant, as can be seen from the title bar. Look at the worksheet tabs, however, and note that two worksheets are visible in Figure 13a, whereas the Calculations worksheet is hidden in Figure 13b. This was accomplished in the Open workbook procedure and was implemented to hide the calculations from the user until the correct password was entered. See exercise 7 at the end of the chapter.

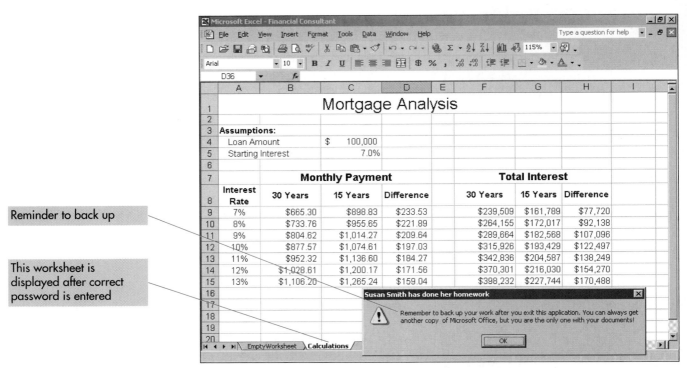

(a) Message to the User (Close Workbook event)

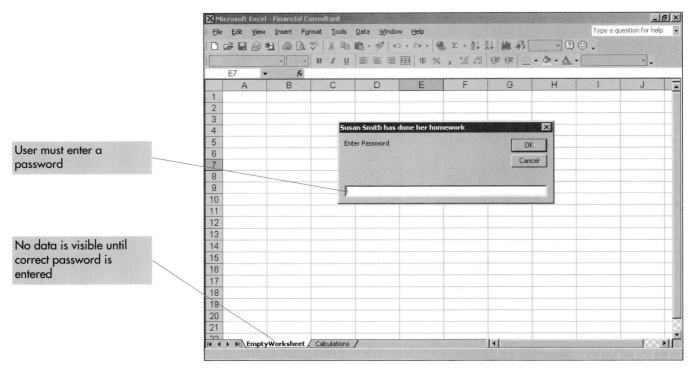

(b) Password Protection (Open Workbook event)

FIGURE 13 *Event-Driven Programming*

HANDS-ON EXERCISE 4

EVENT-DRIVEN PROGRAMMING (MICROSOFT EXCEL)

Objective To create an event procedure to implement password protection that is associated with opening an Excel workbook; to create a second event procedure that displays a message to the user upon closing the workbook. Use Figure 14 as a guide in the exercise.

Step 1: **Create the Close Workbook Procedure**

➤ Open the **VBA Examples workbook** you have used for the previous exercises and enable the macros. If you have been using Access rather than Excel, start Excel, open a new workbook, then save the workbook as **VBA Examples**.

➤ Pull down the **Tools menu**, click the **Macro command**, then click the **Visual Basic Editor command** (or use the **Alt+F11** keyboard shortcut).

➤ You should see the Project Explorer pane as shown in Figure 14a, but if not, pull down the **View menu** and click the **Project Explorer**. Double click **ThisWorkbook** to create a module for the workbook as a whole.

➤ Enter the **Option Explicit statement** if it is not there already, then press **enter** to create a new line. Type the statement to declare the variable, **ApplicationTitle**, using your name instead of Susan Smith.

➤ Click the **down arrow** in the Object list box and select **Workbook**, then click the **down arrow** in the Procedure list box and select the **BeforeClose event** to create the associated procedure. (If you choose a different event by mistake, click and drag to select the associated statements, then press the **Del key** to delete the procedure.)

➤ Enter the MsgBox statement as it appears in Figure 14a. Save the procedure.

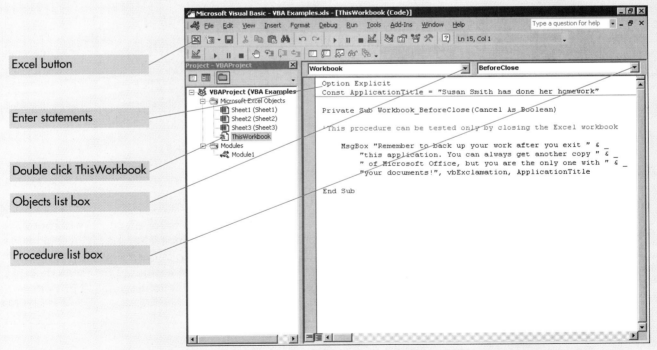

(a) Create the Close Workbook Procedure (step 1)

FIGURE 14 *Hands-on Exercise 4*

Step 2: **Test the Close Workbook Procedure**

- Click the **Excel button** on the Standard toolbar or on the Windows taskbar to view the Excel workbook. The workbook is not empty; that is, it does not contain any cell entries, but it does contain multiple VBA procedures.
- Pull down the **File menu** and click the **Close command**, which runs the procedure you just created and displays the dialog box in Figure 14b. Click **OK** after you have read the message, then click **Yes** if asked to save the workbook.
- Pull down the **File menu** and reopen the **VBA Examples workbook**, enabling the macros. Press **Alt+F11** to return to the VBA window to create an additional procedure.
- Double click **ThisWorkbook** from within the Projects Explorer pane to return to the BeforeClose procedure and make the necessary corrections, if any.
- Save the procedure.

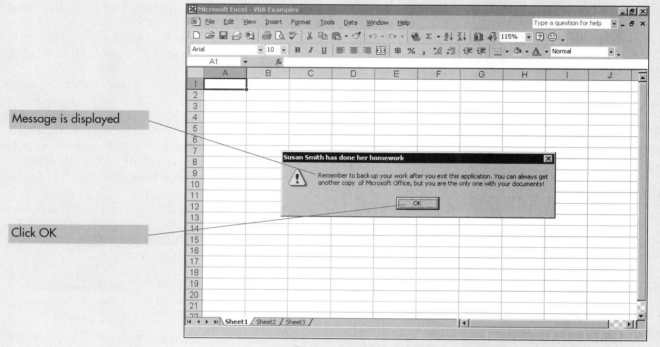

(b) Test the Close Workbook Procedure (step 2)

FIGURE 14   *Hands-on Exercise 4 (continued)*

THE MOST RECENTLY OPENED FILE LIST

One way to open a recently used workbook is to select the workbook directly from the File menu. Pull down the File menu, but instead of clicking the Open command, check to see if the workbook appears on the list of the most recently opened workbooks located at the bottom of the menu. If so, just click the workbook name, rather than having to make the appropriate selections through the Open dialog box.

Step 3: **Start the Open Workbook Event Procedure**

➤ Click the **Procedure View button** at the bottom of the Code window. Click the **down arrow** in the Procedure list box and select the **Open event** to create an event procedure.

➤ Enter the VBA statements as shown in Figure 14c. Note the following:
- Three variables are required for this procedure—the correct password, the password entered by the user, and the number of attempts.
- The user is prompted for the password, and the number of attempts is set to one. The user is given two additional attempts, if necessary, to get the password correct. The loop is bypassed, however, if the user supplies the correct password on the first attempt.

➤ Minimize Excel. Save the procedure, then click the **Run Sub button** to test it. Try different combinations in your testing; that is, enter the correct password on the first, second, and third attempts. The password is **case-sensitive**.

➤ Correct errors as they occur. Click the **Reset button** at any time to begin executing the procedure from the beginning. Save the procedure.

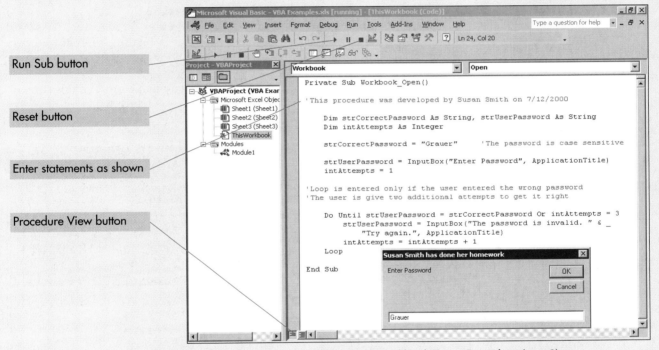

(c) Start the Open Workbook Event Procedure (step 3)

FIGURE 14 *Hands-on Exercise 4 (continued)*

THE OBJECT AND PROCEDURE BOXES

The Object box at the top of the code window displays the selected object such as an Excel workbook, whereas the Procedure box displays the name of the events appropriate to that object. Events that already have procedures appear in bold. Clicking an event that is not bold creates the procedure header and End Sub statements for that event.

Step 4: **Complete the Open Workbook Event Procedure**

➤ Enter the remaining statements in the procedure as shown in Figure 14d. Note the following:
- The If statement determines whether the user has entered the correct password and, if so, displays the appropriate message.
- If, however, the user fails to supply the correct password, a different message is displayed, and the workbook will close due to the **Workbooks.Close statement** within the procedure.
- As a precaution, put an apostrophe in front of the Workbooks.Close statement so that it is a comment, and thus it is not executed. Once you are sure that you can enter the correct password, you can remove the apostrophe and implement the password protection.

➤ Save the procedure, then click the **Run Sub button** to test it. Be sure that you can enter the correct password (**Grauer**), and that you realize the password is case-sensitive.

➤ Delete the apostrophe in front of the Workbooks.Close statement. The text of the statement changes from green to black to indicate that it is an executable statement rather than a comment. Save the procedure.

➤ Click the **Run Sub button** a second time, then enter an incorrect password three times in a row. You will see the dialog box in Figure 14d, followed by a message reminding you to back up your workbook, and then the workbook will close.

➤ The first message makes sense, the second does not make sense in this context. Thus, we need to modify the Close Workbook procedure when an incorrect password is entered.

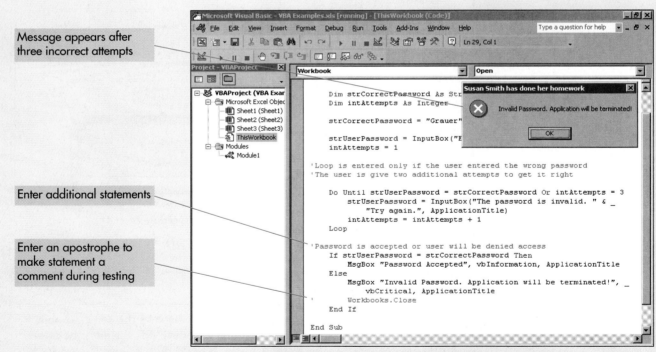

(d) Complete the Open Workbook Event Procedure (step 4)

FIGURE 14 *Hands-on Exercise 4 (continued)*

Step 5: **Modify the Before Close Event Procedure**

➤ Reopen the **VBA Examples workbook**. Click the button to **Enable Macros**.
➤ Enter the password, **Grauer** (the password is case-sensitive), press **enter**, then click **OK** when the password has been accepted.
➤ Press **Alt+F11** to reopen the VBA Editor, and (if necessary) double click **ThisWorkbook** within the list of Microsoft Excel objects.
➤ Click at the end of the line defining the ApplicationTitle constant, press **enter**, then enter the statement to define the **binNormalExit** variable as shown in Figure 14e. (The statement appears initially below the line ending the General Declarations section, but moves above the line when you press enter.)
➤ Modify the BeforeClose event procedure to include an If statement that tests the value of the binNormalExit variable as shown in Figure 14e. You must, however, set the value of this variable in the Open Workbook event procedure as described in step 6. Save the procedure.

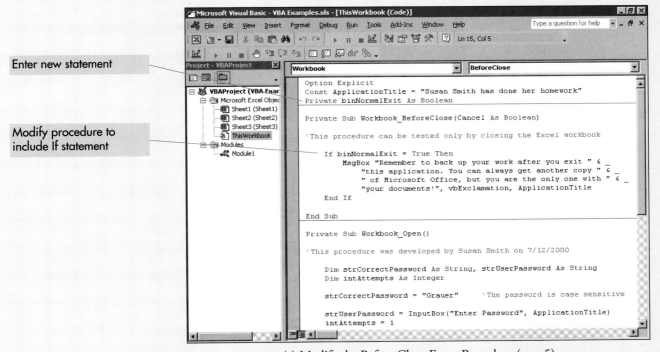

(e) Modify the Before Close Event Procedure (step 5)

FIGURE 14   *Hands-on Exercise 4 (continued)*

SETTING A SWITCH

The use of a switch (binNormalExit, in this example) to control an action within a procedure is a common programming technique. The switch is set to one of two values according to events that occur within the system, then the switch is subsequently tested and the appropriate action is taken. Here, the switch is set when the workbook is opened to indicate either a valid or invalid user. The switch is then tested prior to closing the workbook to determine whether to print the closing message.

Step 6: **Modify the Open Workbook Event Procedure**

➤ Scroll down to the Open Workbook event procedure, then modify the If statement to set the value of binNormExit as shown in Figure 14f:
- Take advantage of the Complete Word tool to enter the variable name. Type the first few letters, "binN", then press Ctrl+Space, and VBA will complete the variable name.
- The indentation within the statement is not a requirement of VBA per se, but is used to make the code easier to read. Blank lines are also added for this purpose.
- Comments appear throughout the procedure to explain its logic.
- Save the modified procedure.

➤ Click the **Run Sub button**, then enter an incorrect password three times in a row. Once again, you will see the dialog box indicating an invalid password, but this time you will not see the message reminding you to back up your workbook. The workbook closes as before.

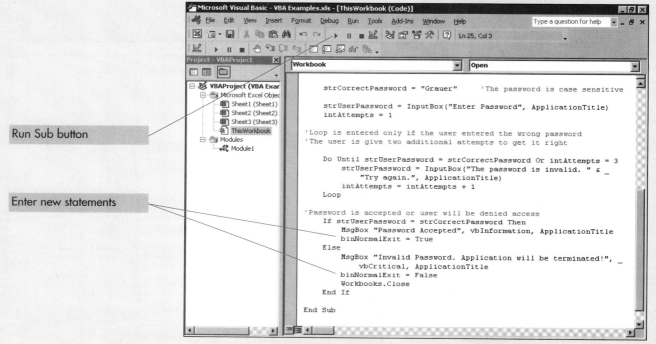

(f) Modify the Open Workbook Event Procedure (step 6)

FIGURE 14 *Hands-on Exercise 4 (continued)*

TEST UNDER ALL CONDITIONS

We cannot overemphasize the importance of thoroughly testing a procedure, and further, testing it under all conditions. VBA statements are powerful, but they are also complex, and a misplaced or omitted character can have dramatic consequences. Test every procedure completely at the time it is created, so that the logic of the procedure is fresh in your mind.

Step 7: **Open a Second Workbook**

- Reopen the **VBA Examples workbook**. Click the button to **Enable Macros**.
- Enter the password, **Grauer**, then press **enter**. Click **OK** when you see the second dialog box telling you that the password has been accepted.
- Pull down the **File menu** and click the **Open command** (or click the **Open button** on the Standard toolbar) and open a second workbook. We opened a workbook called **Financial Consultant**, but it does not matter which workbook you open.
- Pull down the **Window menu**, click the **Arrange command**, click the **Horizontal option button**, and click **OK** to tile the workbooks as shown in Figure 14g. The title bars show the names of the open workbooks.
- Pull down the **Tools menu**, click **Macro**, then click **Visual Basic Editor**.

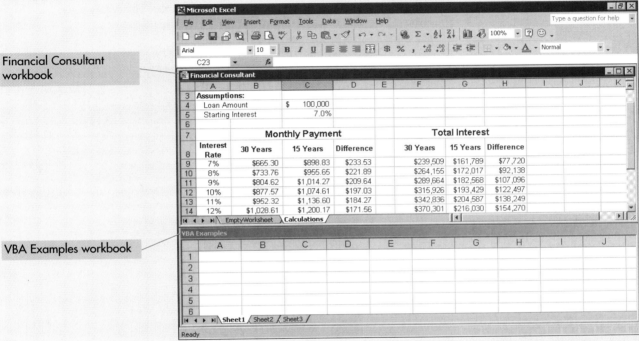

(g) Open a Second Workbook (step 7)

FIGURE 14 *Hands-on Exercise 4 (continued)*

THE COMPARISON IS CASE-SENSITIVE

Any literal comparison (e.g., strInstructorName = "Grauer") is case-sensitive, so that the user has to enter the correct name and case in order for the condition to be true. A response of "GRAUER" or "grauer", while containing the correct name, will be evaluated as false because the case does not match. You can, however, use the UCase (uppercase) function to convert the user's response to uppercase, and test accordingly. In other words, UCase(strInstructorName) = "GRAUER" will be evaluated as true if the user enters "Grauer" in any combination of upper or lowercase letters.

Step 8: **Copy the Procedure**

➤ You should be back in the Visual Basic Editor as shown in Figure 14h. Copy the procedures associated with the Open and Close Workbook events from the VBA Examples workbook to the other workbook, Financial Consultant.
- Double click **ThisWorkbook** within the list of Microsoft Excel objects under the VBA Examples workbook.
- Click and drag to select the definition of the ApplicationTitle constant in the General Declarations section plus the two procedures (to open and close the workbook) in their entirety.
- Click the **Copy button** on the Standard toolbar.
- If necessary, expand the Financial Consultant VBA Project, then double click **ThisWorkbook** with the list of Excel objects under the Financial Consultant workbook. Click underneath the **Option Explicit command**.
- Click the **Paste button** on the Standard toolbar. The VBA code should be copied into this module as shown in Figure 14h.

➤ Click the **Save button** to save the module.

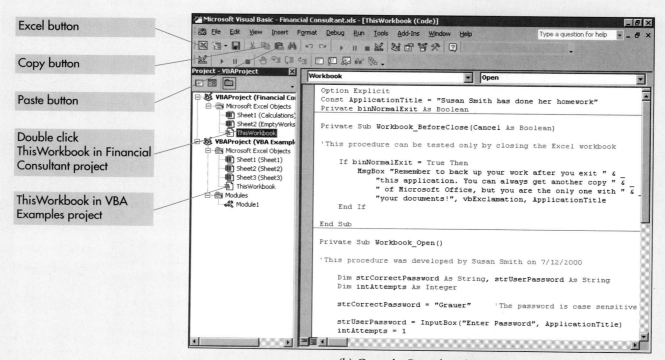

(h) Copy the Procedure (step 8)

FIGURE 14 *Hands-on Exercise 4 (continued)*

KEYBOARD SHORTCUTS—CUT, COPY, AND PASTE

Ctrl+X, Ctrl+C, and Ctrl+V are shortcuts to cut, copy, and paste, respectively, and apply to all applications in the Office suite as well as to Windows applications in general. (The shortcuts are easier to remember when you realize that the operative letters X, C, and V are next to each other at the bottom left side of the keyboard.)

Step 9: **Test the Procedure**

➤ Click the **Excel button** on the Standard toolbar within the VBA window (or click the **Excel button** on the Windows taskbar) to view the Excel workbook. Click in the window containing the Financial Consultant workbook (or whichever workbook you are using), then click the **Maximize button**.

➤ Pull down the **File menu** and click the **Close command**. (The dialog box in Figure 14i does not appear initially because the value of binNormalExit is not yet set; you have to open the workbook to set the switch.) Click **Yes** if asked whether to save the changes to the workbook.

➤ Pull down the **File menu** and reopen the workbook. Click the button to **Enable Macros**, then enter **Grauer** when prompted for the password. Click **OK** when the password has been accepted.

➤ Close this workbook, close the **VBA Examples workbook**, then pull down the **File menu** and click the **Exit command** to quit Excel.

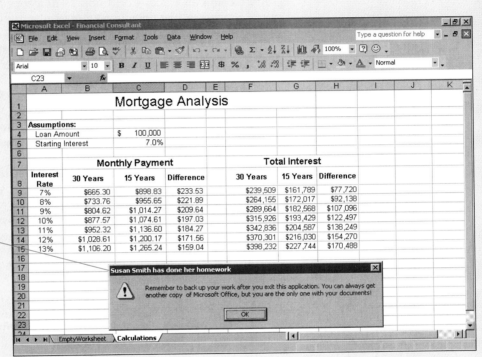

Title bar has been customized

(i) Test the Procedure (step 9)

FIGURE 14 *Hands-on Exercise 4 (continued)*

SCREEN CAPTURE

Prove to your instructor that you have completed the hands-on exercise correctly by capturing a screen, then pasting the screen into a Word document. Do the exercise until you come to the screen that you want to capture, then press the PrintScreen key at the top of the keyboard. Click the Start button to start Word and open a Word document, then pull down the Edit menu and click the Paste command to bring the captured screen into the Word document. See exercise 1 at the end of the chapter.

PUTTING VBA TO WORK (MICROSOFT ACCESS)

The same VBA procedure can be run from multiple applications in Microsoft Office, despite the fact that the applications are very different. The real power of VBA, however, is its ability to detect events that are unique to a specific application and to respond accordingly. An event is defined as any action that is recognized by an application. Opening or closing an Excel workbook or an Access database is an event. Selecting a worksheet within a workbook is also an event, as is clicking on a command button on an Access form. To use VBA within Microsoft Office, you decide which events are significant, and what is to happen when those events occur. Then you develop the appropriate *event procedures* that execute automatically when the event occurs.

Consider, for example, Figure 15, which displays the results of two event procedures in conjunction with opening and closing an Access database. (These are procedures similar to those we created in the preceding pages in conjunction with opening and closing an Excel workbook.) The procedure associated with Figure 15a displays a message that appears automatically after the user clicks the Switchboard button to exit the database. The procedure is almost trivial to write, and consists of a single MsgBox statement. The effect of the procedure is quite significant, however, as it reminds the user to back up his or her work. Indeed, you can never overemphasize the importance of adequate backup.

The dialog box in Figure 15b prompts the user for a password and appears automatically when the user opens the database. The logic here is more sophisticated in that the underlying procedure contains an InputBox statement to request the password, a Do Until loop that is executed until the user enters the correct password or exceeds the allotted number of attempts, then additional logic to display the switchboard or terminate the application if the user fails to enter the proper password. The procedure is not difficult, however, and it builds on the VBA statements that were covered earlier.

The next hands-on exercise has you create the event procedures that are associated with the database in Figure 15. The exercise references a switchboard, or user interface, that is created as a form within the database. The switchboard displays a menu that enables a nontechnical person to move easily from one object in the database (e.g., a form or report) to another.

The switchboard is created through a utility called the Switchboard Manager that prompts you for each item you want to add to the switchboard, and which action you want to be taken in conjunction with that menu item. You could do the exercise with any database, but we suggest you use the database we provide to access the switchboard that we created for you. The exercise begins, therefore, by having you download a data disk from our Web site.

EVENT-DRIVEN VERSUS TRADITIONAL PROGRAMMING

A traditional program is executed sequentially, beginning with the first line of code and continuing in order through the remainder of the program. It is the program, not the user, that determines the order in which the statements are executed. VBA, on the other hand, is event-driven, meaning that the order in which the procedures are executed depends on the events that occur. It is the user, rather than the program, that determines which events occur, and consequently which procedures are executed. Each application in Microsoft Office has a different set of objects and associated events that comprise the application's object model.

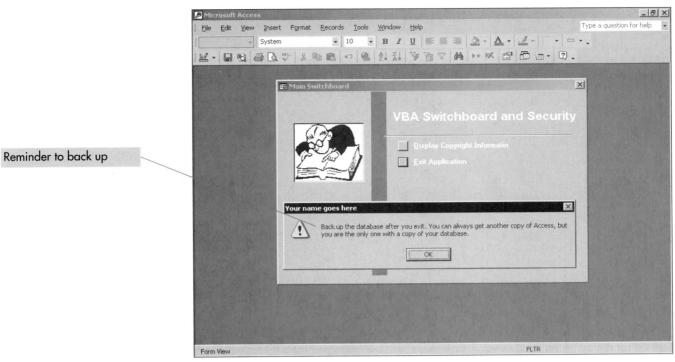

(a) Reminder to the User (Exit Application event)

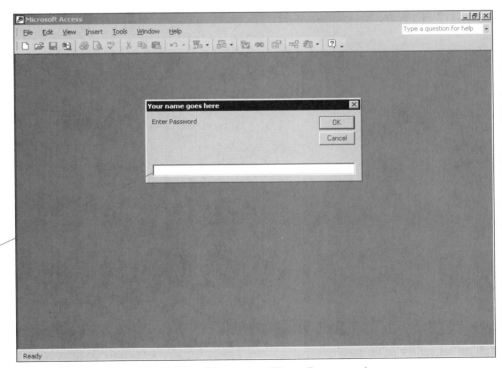

(b) Password Protection (Open Form event)

FIGURE 15 *Event-Driven Programming (Microsoft Access)*

HANDS-ON EXERCISE 5

EVENT-DRIVEN PROGRAMMING (MICROSOFT ACCESS)

Objective To implement password protection for an Access database; to create a second event procedure that displays a message to the user upon closing the database. Use Figure 16 as a guide in the exercise.

Step 1: **Open the Access Database**

➤ You can do this exercise with any database, but we suggest you use the database we have provided. Go to **www.prenhall.com/grauer**, click the **Office 2000 book**, click the **Student Resources tab**, then click the link to download the data disk.

➤ Scroll until you can select the disk for the **VBA Primer**. Download the file to the Windows desktop, then double click the file once it has been downloaded to your PC.

➤ Double click the file and follow the onscreen instructions to expand the self-extracting file that contains the database.

➤ Go to the newly created **Exploring VBA folder** and open the **VBA Switchboard and Security database** as shown in Figure 16a.

➤ Pull down the **Tools menu**, click the **Macro command**, then click the **Visual Basic Editor command**. Maximize the VBA Editor window.

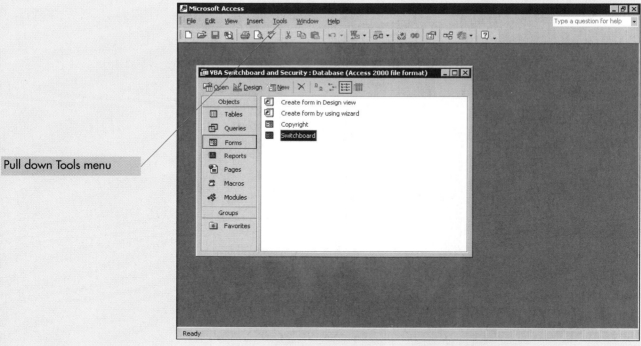

Pull down Tools menu

(a) Open the Access Database (step 1)

FIGURE 16   *Hands-on Exercise 5*

54   A VBA PRIMER

Step 2: **Create the ExitDatabase Procedure**

- Pull down the **Insert menu** and click **Module** to insert Module1. Complete the **General Declarations section** by adding your name to the definition of the ApplicationTitle constant as shown in Figure 16b.
- Pull down the **Insert menu** and click **Procedure** to insert a new procedure called **ExitDatabase**. Click the option buttons for a **Sub procedure** and for **Public scope**. Click **OK**.
- Complete the ExitDatabase procedure by entering the **MsgBox** and **DoCmd.Quit** statements. The DoCmd.Quit statement will close Access, but it is entered initially as a comment by beginning the line with an apostrophe.
- Click anywhere in the procedure, then click the **Run Sub button** to test the procedure. Correct any errors that occur, then when the MsgBox displays correctly, **delete the apostrophe** in front of the DoCmd.Quit statement.
- Save the module. The next time you execute the procedure, you should see the message box you just created, and then Access will be terminated.

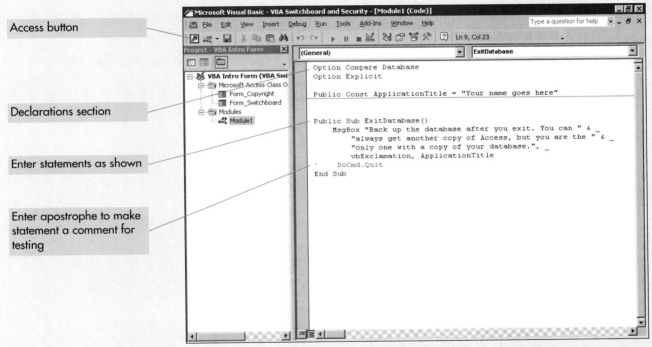

(b) Create the ExitDatabase Procedure (step 2)

FIGURE 16 *Hands-on Exercise 5 (continued)*

CREATE A PUBLIC CONSTANT

Give your application a customized look by adding your name or other identifying message to the title bar of the message and/or input boxes that you use. You can add the information individually to each statement, but it is easier to declare a public constant from within a general module. That way, you can change the value of the constant in one place and have the change reflected automatically throughout your application.

Step 3: **Modify the Switchboard**

- Click the **Access button** on the Standard toolbar within the VBA window to switch to the Database window (or use the **F11** keyboard shortcut).
- Pull down the **Tools menu**, click the **Database Utilities command**, then choose **Switchboard Manager** to display the Switchboard Manager dialog box in Figure 16c.
- Click the **Edit button** to edit the Main Switchboard and display the Edit Switchboard Page dialog box. Select the **&Exit Application command** and click its **Edit button** to display the Edit Switchboard Item dialog box.
- Change the command to **Run Code**. Enter **ExitDatabase** in the Function Name text box. Click **OK**, then close the two other dialog boxes. The switchboard has been modified so that clicking the Exit button will run the VBA procedure you just created.

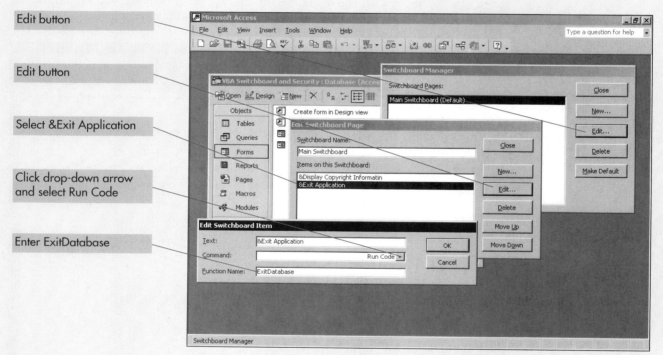

(c) Modify the Switchboard (step 3)

FIGURE 16 *Hands-on Exercise 5 (continued)*

CREATE A KEYBOARD SHORTCUT

The & has special significance when used within the name of an Access object because it creates a keyboard shortcut to that object. Enter "&Exit Application", for example, and the letter E (the letter immediately after the ampersand) will be underlined and appear as "E̲xit Application" on the switchboard. From there, you can execute the item by clicking its button, or you can use the Alt+E keyboard shortcut (where "E" is the underlined letter in the menu option).

Step 4: **Test the Switchboard**

➤ If necessary, click the **Forms tab** in the Database window. Double click the **Switchboard form** to open the switchboard as shown in Figure 16d. The switchboard contains two commands.

➤ Click the **Display Copyright Information command** to display a form that we use with all our databases. (You can open this form in Design view and modify the text to include your name, rather than ours. If you do, be sure to save the modified form, then close it.)

➤ Click the **Exit Application command** (or use the **Alt+E** keyboard shortcut). You should see the dialog box in Figure 16d, corresponding to the MsgBox statement you created earlier. Click **OK** to close the dialog box.

➤ Access itself will terminate because of the DoCmd.Quit statement within the ExitDatabase procedure. (If this does not happen, return to the VBA Editor and remove the apostrophe in front of the DoCmd statement.)

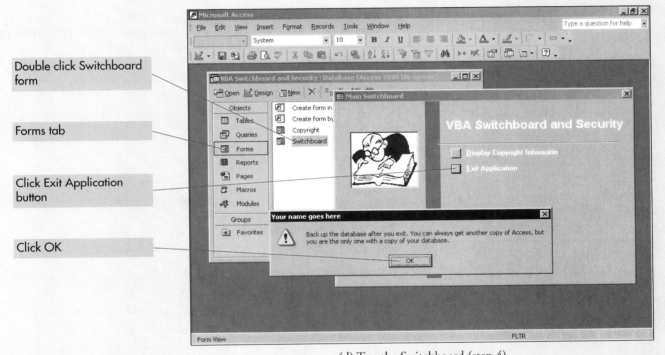

(d) Test the Switchboard (step 4)

FIGURE 16   *Hands-on Exercise 5 (continued)*

BACK UP IMPORTANT FILES

It's not a question of if it will happen, but when—hard disks die, files are lost, or viruses may infect a system. It has happened to us and it will happen to you, but you can prepare for the inevitable by creating adequate backup before the problem occurs. The essence of a backup strategy is to decide which files to back up, how often to do the backup, and where to keep the backup. Do it!

Step 5: **Complete the Open Form Event Procedure**

➤ Start Access and reopen the **VBA Switchboard and Security database**. Press **Alt+F11** to start the VBA Editor. Click the **plus sign** next to Microsoft Access Class objects, double click the module called **Form_Switchboard**, then look for the **Form_Open procedure** as shown in Figure 16e.

➤ The procedure was created automatically by the Switchboard Manager. You must, however, expand this procedure to include password protection. Note the following:
- Three variables are required—the correct password, the password entered by the user, and the number of attempts.
- The user is prompted for the password, and the number of attempts is set to one. The user is given two additional attempts, if necessary, to get the correct password.
- The If statement at the end of the loop determines whether the user has entered the correct password, and if so, it executes the original commands that are associated with the switchboard. If, however, the user fails to supply the correct password, an invalid password message is displayed and the **DoCmd.Quit** statement terminates the application.
- We suggest you place an **apostrophe** in front of the statement initially so that it becomes a comment, and thus it is not executed. Once you are sure that you can enter the correct password, you can remove the apostrophe and implement the password protection.

➤ Save the procedure. You cannot test this procedure from within the VBA window; you must cause the event to happen (i.e., open the form) for the procedure to execute. Click the **Access button** on the Standard toolbar to return to the Database window.

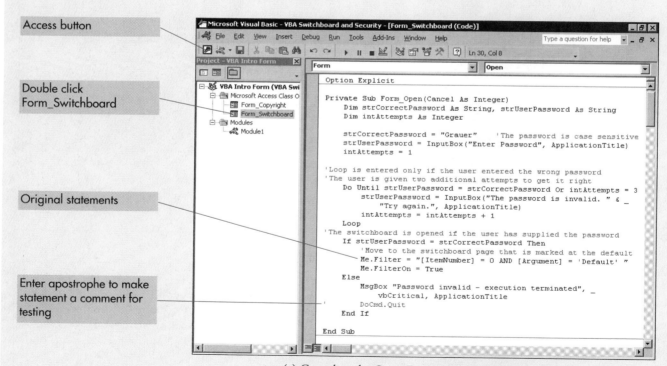

(e) Complete the Open Form Event Procedure (step 5)

FIGURE 16 *Hands-on Exercise 5 (continued)*

Step 6: **Test the Procedure**

- Close all open windows within the Access database except for the Database window. Click the **Forms tab**, then double click the **Switchboard Form**.
- You should be prompted for the password as shown in Figure 16f. The password (in our procedure) is **Grauer**.
- Test the procedure repeatedly to include all possibilities. Enter the correct password on the first, second, and third attempts to be sure that the procedure works as intended. Each time you enter the correct password, you will have to close the switchboard, then reopen it.
- Test the procedure one final time, by failing to enter the correct password. You will see a message box indicating that the password is invalid and that execution will be terminated. Termination will not take place, however, because the DoCmd.Quit statement is currently entered as a comment.
- Press **Alt+F11** to reopen the VBA Editor. Delete the apostrophe in front of the DoCmd.Quit statement. The text of the statement changes from green to black to indicate that it is an executable statement. Save the procedure.

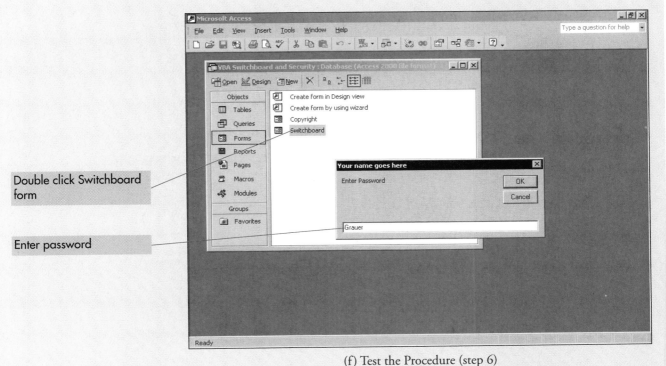

(f) Test the Procedure (step 6)

FIGURE 16 *Hands-on Exercise 5 (continued)*

TOGGLE COMMENTS ON AND OFF

Comments are used primarily to explain the purpose of VBA statements, but they can also be used to "comment out" code as distinct from deleting the statement altogether. Thus you can add or remove the apostrophe in front of the statement, to toggle the comment on or off.

Step 7: **Change the Startup Properties**

- Click the **Access button** on the VBA Standard toolbar to return to the Database window. Pull down the **Tools menu** and click **Startup** to display the Startup dialog box as shown in Figure 16g.
- Click in the **Application Title** text box and enter the title of the application, **VBA Switchboard and Security** in this example.
- Click the **drop-down arrow** in the Display Form/Page list box and select the **Switchboard form** as the form that will open automatically in conjunction with opening the database.
- Clear the check box to display the Database window. Click **OK** to accept the settings and close the dialog box. The next time you open the database, the switchboard should open automatically, which in turn triggers the Open Form event procedure that will prompt the user to enter a password.
- Close the Switchboard form.

(g) Change the Startup Properties (step 7)

FIGURE 16 *Hands-on Exercise 5 (continued)*

HIDE THE DATABASE WINDOW

Use the Startup property to hide the Database window from the novice user. You avoid confusion and you may prevent the novice from accidentally deleting objects in the database. Of course, anyone with some knowledge of Access can restore the Database window by pulling down the Window menu, clicking the Unhide command, then selecting the Database window from the associated dialog box. Nevertheless, hiding the Database window is a good beginning.

Step 8: **Test the Database**

➤ Close the database, then reopen the database to test the procedures we have created in this exercise. The sequence of events is as follows:
  • The database is loaded and the switchboard is opened but is not yet visible. The Open Form procedure for the switchboard is executed, and you are prompted for the password as shown in Figure 16h.
  • The password is entered correctly and the switchboard is displayed. The Database window is hidden, however, because the Startup Properties have been modified.
➤ Click the **Exit Application command** (or use the **Alt+E** keyboard shortcut). You will see the message box reminding you to back up the system, after which the database is closed and Access is terminated.
➤ Testing is complete and you can go on to add the other objects to your Access database. Congratulations on a job well done.

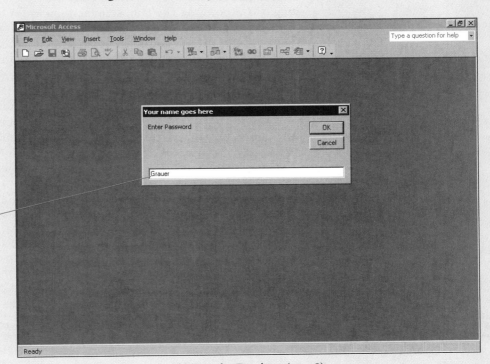

(h) Test the Database (step 8)

FIGURE 16   *Hands-on Exercise 5 (continued)*

HIDE MENUS AND TOOLBARS

You can use the Startup property to hide menus and/or toolbars from the user by clearing the respective check boxes. A word of caution, however—once the menus are hidden, it is difficult to get them back. Start Access, pull down the File menu, and click Open to display the Open dialog box, select the database to open, then press and hold the Shift key when you click the Open button. This powerful technique is not widely known.

SUMMARY

Visual Basic for Applications (VBA) is a powerful programming language that is accessible from all major applications in Microsoft Office XP. A VBA statement accomplishes a specific task such as displaying a message to the user or accepting input from the user. Statements are grouped into procedures, and procedures in turn are grouped into modules. Every procedure is classified as either private or public.

The MsgBox statement displays information to the user. It has one required argument, which is the message (or prompt) that is displayed to the user. The other two arguments, the icon that is to be displayed in the dialog box and the text of the title bar, are optional. The InputBox function displays a prompt to the user requesting information, then it stores that information (the value returned by the user) for use later in the procedure.

Every variable must be declared (defined) before it can be used. This is accomplished through the Dim (short for Dimension) statement that appears at the beginning of a procedure. The Dim statement indicates the name of the variable and its type (for example, whether it will hold a character string or an integer number), which in turn reserves the appropriate amount of memory for that variable.

The ability to make decisions within a procedure, then branch to alternative sets of statements is implemented through the If . . . Then . . . Else or Case statements. The Else clause is optional, but may be repeated multiple times within an If statement. The Case statement is preferable to an If statement with multiple Else clauses.

The For . . . Next statement (or For . . . Next loop as it is also called) executes all statements between the words For and Next a specified number of times, using a counter to keep track of the number of times the loop is executed. The Do . . . Loop Until and/or Do Until . . . Loop statements are used when the number of times through the loop is not known in advance.

VBA is different from traditional programming languages in that it is event-driven. An event is defined as any action that is recognized by an application, such as Excel or Access. Opening or closing an Excel workbook or an Access database is an event. Selecting a worksheet within a workbook is also an event, as is clicking on a command button on an Access form. To use VBA within Microsoft Office, you decide which events are significant, and what is to happen when those events occur. Then you develop the appropriate event procedures.

KEY TERMS

Argument (p. 2)
Case statement (p. 18)
Character string (p. 16)
Comment (p. 6)
Compilation error (p. 30)
Complete Word tool (p. 22)
Concatenate (p. 4)
Custom toolbar (p. 19)
Debug menu (p. 30)
Debug toolbar (p. 30)
Debug.Print statement (p. 31)
Debugging (p. 30)
Declarations section (p. 6)
Dim statement (p. 5)
Do Loops (p. 29)
Else clause (p. 16)
End Sub statement (p. 3)
Event (p. 41)

Event procedure (Access) (p. 52)
Event procedure (Excel) (p. 41)
Execution error (p. 30)
For . . . Next Statement (p. 28)
Full Module view (p. 21)
Help (p. 15)
If statement (p. 16)
Immediate window (p. 31)
InputBox function (p. 4)
Intrinsic constant (p. 3)
Literal (p. 16)
Locals window (p. 31)
Macro (p. 2)
Macro recorder (p. 2)
Module (p. 2)
MsgBox statement (p. 3)
Object box (p. 45)
Option Explicit (p. 6)

Private procedure (p. 2)
Procedure (p. 2)
Procedure box (p. 45)
Procedure header (p. 3)
Procedure view (p. 21)
Project Explorer (p. 6)
Public procedure (p. 2)
Quick Info (p. 9)
Run-time error (p. 30)
Statement (p. 2)
Step Into button (p. 31)
Syntax (p. 2)
Underscore (p. 4)
Variable (p. 4)
VBA (p. 1)
Visual Basic Editor (p. 6)
Visual Basic for Applications (p. 2)

MULTIPLE CHOICE

1. Which of the following applications in Office XP has access to VBA?
   (a) Word
   (b) Excel
   (c) Access
   (d) All of the above

2. Which of the following is a valid name for a VBA variable?
   (a) Public
   (b) Private
   (c) strUserFirstName
   (d) int Count Of Attempts

3. Which of the following is true about an If statement?
   (a) It evaluates a condition as either true or false, then executes the statement(s) following the keyword "Then" if the condition is true
   (b) It must contain the keyword Else
   (c) It must contain one or more ElseIf statements
   (d) All of the above

4. Which of the following lists the items from smallest to largest?
   (a) Module, procedure, statement
   (b) Statement, module, procedure
   (c) Statement, procedure, module
   (d) Procedure, module, statement

5. Given the statement, MsgBox "Welcome to VBA" , , "Bob was here", which of the following is true?
   (a) "Welcome to VBA" will be displayed within the resulting message box
   (b) "Welcome to VBA" will appear on the title bar of the displayed dialog box
   (c) The two adjacent commas will cause a compilation error
   (d) An informational icon will be displayed with the message

6. Where are the VBA procedures associated with an Office document stored?
   (a) In the same folder, but in a separate file
   (b) In the Office document itself
   (c) In a special VBA folder on drive C
   (d) In a special VBA folder on the local area network

7. The Debug.Print statement is associated with the:
   (a) Locals window
   (b) Immediate window
   (c) Project Explorer
   (d) Debug toolbar

8. Which of the following is the proper sequence of arguments for the MsgBox statement?
   (a) Text for the title bar, prompt, button
   (b) Prompt, button, text for the title bar
   (c) Prompt, text for the title bar, button
   (d) Button, prompt, text for the title bar

9. Which of the following is a true statement about Do loops?
   (a) Placing the Until clause at the beginning of the loop tests the condition prior to executing any statements in the loop
   (b) Placing the Until clause at the end of the loop executes the statements in the loop, then it tests the condition
   (c) Both (a) and (b)
   (d) Neither (a) nor (b)

10. Given the statement, For intCount = 1 to 10 Step 3, how many times will the statements in the loop be executed (assuming that there are no statements in the loop to terminate the execution)?
    (a) 10
    (b) 4
    (c) 3
    (d) Impossible to determine

11. Which of the following is a *false* statement?
    (a) A dash at the end of a line indicates continuation
    (b) An ampersand indicates concatenation
    (c) An apostrophe at the beginning of a line signifies a comment
    (d) A pair of quotation marks denotes a character string

12. What is the effect of deleting the apostrophe that appears at the beginning of a VBA statement?
    (a) A compilation error will occur
    (b) The statement is converted to a comment
    (c) The color of the statement will change from black to green
    (d) The statement is made executable

13. Which of the following If statements will display the indicated message if the user enters a response other than "Grauer" (assuming that "Grauer" is the correct password)?
    (a) If strUserResponse <> "Grauer" Then MsgBox "Wrong password"
    (b) If strUserResponse = "Grauer" Then MsgBox "Wrong password"
    (c) If strUserResponse > "Grauer" Then MsgBox "Wrong password"
    (d) If strUserResponse < "Grauer" Then MsgBox "Wrong password"

14. Which of the following will execute the statements in the loop at least once?
    (a) Do . . . Loop Until
    (b) Do Until ..... Loop
    (c) Both (a) and (b)
    (d) Neither (a) nor (b)

15. The copy and paste commands can be used to:
    (a) Copy statements within a procedure
    (b) Copy statements from a procedure in one module to a procedure in another module within the same document
    (c) Copy statements from a module in an Excel workbook to a module in an Access database
    (d) All of the above

ANSWERS

1. d   6. b   11. a
2. c   7. b   12. d
3. a   8. b   13. a
4. c   9. c   14. a
5. a   10. b   15. a

PRACTICE WITH VBA

BUILDS ON

HANDS-ON
EXERCISE 4
PAGES 43–51

1. **Screen Capture:** The ability to capture a screen, then print the captured screen as part of a document, is very useful. The Word document in Figure 17, for example, captures the Excel screen as a workbook is opened, with the dialog box in place.

   Open the completed workbook from the fourth hands-on exercise. Press the PrintScreen key when prompted for the password to copy the screen to the Windows clipboard, an area of memory that is accessible to any Windows application. Next, start (or switch to) a Word document, and then execute the Paste command in the Edit menu to paste the contents of the clipboard into the current document. That's all there is to it. The screen is now part of the Word document, where it can be moved and sized like any other Windows object. Print the Word document for your instructor.

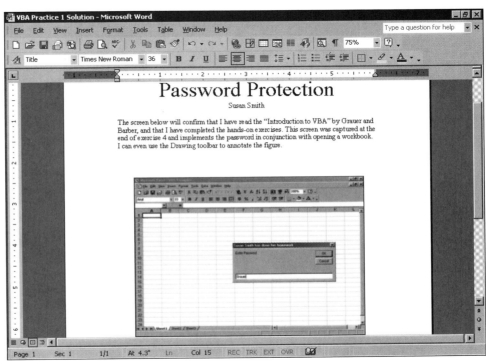

FIGURE 17   *Screen Capture (Exercise 1)*

2. **VBA in a Word Document:** Everything that you have learned with respect to creating VBA event procedures in Excel or Access is also applicable to Microsoft Word. Accordingly, start Microsoft Word and create the document in Figure 18. Read the document carefully, then create the Document_Close event procedure to display the indicated message box, adding your name to the title bar. Prove to your professor that you have completed this assignment by capturing the screen in Figure 18, as described in the previous exercise.

BUILDS ON

HANDS-ON
EXERCISE 4
PAGES 43–51

3. **The Before Print Event:** Open the Excel workbook that you used in the fourth hands-on exercise to create an event procedure associated with the Before_Print event. The procedure is to contain a MsgBox statement to remind the user to print a workbook with both displayed values and cell contents as shown in Figure 19.

   The easiest way to switch between the two views is to press Ctrl+~. (The tilde is located at the upper-left of the keyboard.) Prove to your professor that you have completed this assignment by capturing the screen in Figure 19.

EXTENDING MICROSOFT OFFICE XP

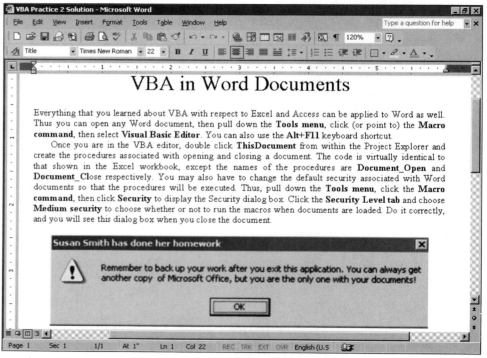

FIGURE 18 *VBA in a Word Document (Exercise 2)*

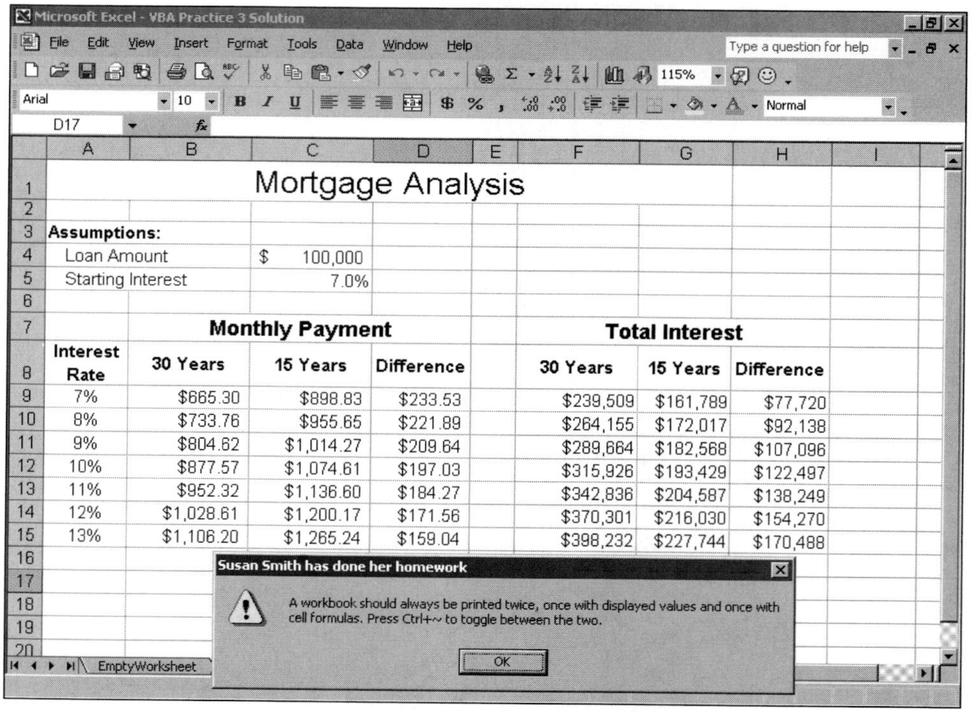

FIGURE 19 *The Before Print Event (Exercise 3)*

BUILDS ON

HANDS-ON
EXERCISE 5
PAGES 54–61

4. The On Click Event: Open the Access database that you used in the fifth hands-on exercise to create an event procedure associated with the On Click event for the indicated command button.
   a. Open the Copyright form from the switchboard, then change to the Design view. Modify the form so that it contains your name rather than ours.
   b. Right click the Technical Support button to display a context-sensitive menu, click Properties, click the Event tab, then select the On Click event. Click the Build button, select Code Builder as shown in Figure 20, and click OK.
   c. The VBA Editor will position you within the On Click event procedure for the command button. All you need to do is add a single MsgBox statement to identify yourself as "Tech Support". Be sure to include the parameter to display your name on the title bar of the message box.
   d. Save the form, then go to form view and click the button to view the message box you just created. Prove to your professor that you have completed this assignment by capturing the associated screen, as described in exercise 1.

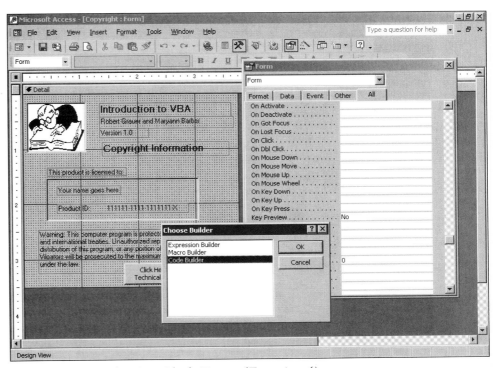

FIGURE 20   *The On Click Event (Exercise 4)*

5. The Option Explicit Statement: The Option Explicit statement should appear at the beginning of every module, but this is not a VBA requirement, only a suggestion from the authors. Omitting the statement can have serious consequences in that the results of a procedure are incorrect, a point that is illustrated in Figure 21.
   a. What is the answer that is displayed in the message box of Figure 21? What is the answer that should be displayed?
   b. Look at the statements within the For ... Next loop to see if you can detect the reason for the error. (*Hint:* Look closely at the variable names.)
   c. What does the Option Explicit statement do? How would including the statement in the procedure of Figure 21 help to ensure the correct result?
   d. Summarize your answers in a note to your instructor.

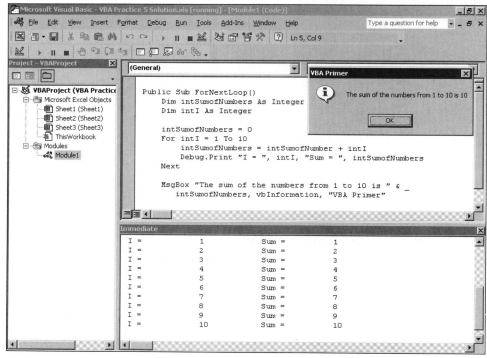

FIGURE 21 *The Option Explicit Statement (Exercise 5)*

6. String Processing: The procedure in Figure 22 illustrates various string processing functions to validate a user's e-mail address. Answer the following:
   a. What are the specific checks that are implemented to check the user's e-mail address? Are these checks reasonable?
   b. What does the VBA Len function do? What does the InStr function do? (Use the VBA Help menu to learn more about these functions.)
   c. What is the purpose of the variable binValidEmail within the procedure?

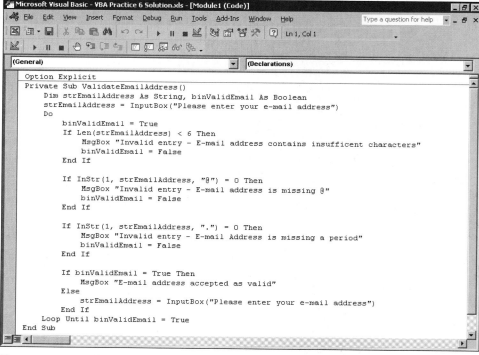

FIGURE 22 *String Processing (Exercise 6)*

BUILDS ON

HANDS-ON
EXERCISE 4
PAGES 43–51

7. **Hiding a Worksheet:** Figure 23 expands on the procedure to implement password protection in an Excel workbook by hiding the worksheet until the correct password has been entered.
   a. What additional statements have been added to the procedure in Figure 23 that were not present in Hands-on Exercise 4? What is the purpose of each statement?
   b. What statement could you add to the procedure to hide the empty worksheet after the correct password has been entered?
   c. Summarize your answers in a note to your instructor.

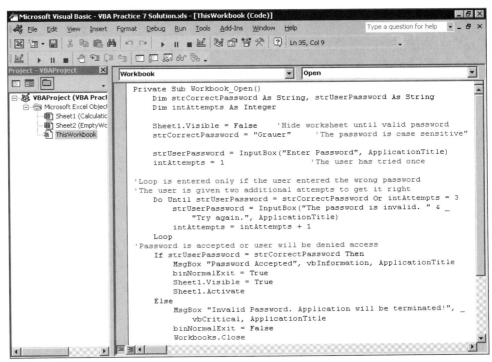

FIGURE 23 *Hiding a Worksheet (Exercise 7)*

8. **Help with VBA:** Help is just a mouse click away and it is invaluable. Use the Help facility to look up detailed information that expands a topic that was discussed in the chapter. The screen in Figure 24, for example, explains the integer data type and its use within a Dim statement. The information is quite detailed, but if you read carefully, you will generally find the answer. Print three different Help-screens for your instructor.

9. **Invoking a Procedure:** The same statement (or set of statements) is often executed from many places within a single procedure or from multiple procedures within an application. You can duplicate the code as necessary, but it is far more efficient to create a single procedure that contains the repeated statements, and then invoke that procedure. The advantage to this approach is that you have to write (or modify) the procedure only once.

   The module in Figure 25 illustrates how this is accomplished. The History-Quiz procedure asks the user multiple questions, then displays one of two messages, depending on whether the response is correct. These messages are contained in two separate procedures, then the appropriate procedure (CorrectAnswer or IncorrectAnswer) is called from within the HistoryQuiz procedure, depending on whether the user's answer is right or wrong. Create and test the module in Figure 25 to be sure you understand this technique.

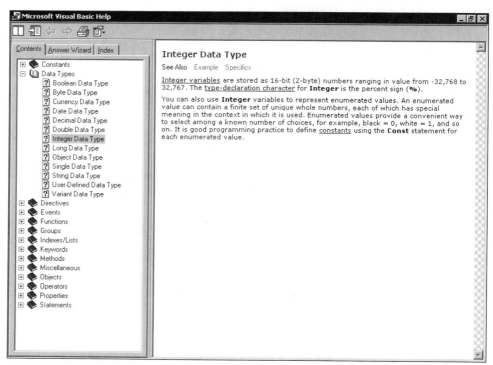

FIGURE 24 *Help with VBA (Exercise 8)*

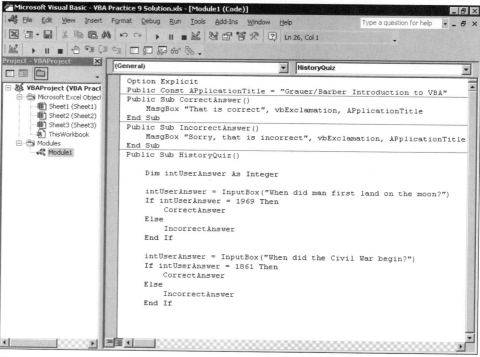

FIGURE 25 *Invoking a Procedure (Exercise 9)*

10. **The MsgBox Function:** The procedure in Figure 26 shows how the MsgBox statement can accept information from the user and branch accordingly. A simple MsgBox statement merely displays a message. If, however, you enclose the parameters of the MsgBox statement in parentheses, it becomes a function and returns a value (in this example, a mouse click indicating whether the user clicked yes or no). The use of parentheses requires that you include a second parameter such as vbYesNo to display the Yes and No command buttons. You then embed the MsgBox function within an If statement that tests for the intrinsic contstants, vbYes and vbNo, respectively.

    You can concatenate the vbYesNo intrinsic constant with another constant such as vbQuestion to display an icon next to the buttons as shown in Figure 26. You can also use other intrinsic constants such as vbOKCancel to display different sets of command buttons.

    Add the procedure in Figure 26 to the VBA Examples workbook (or Access database) that you created in the chapter. Print the procedure for your instructor.

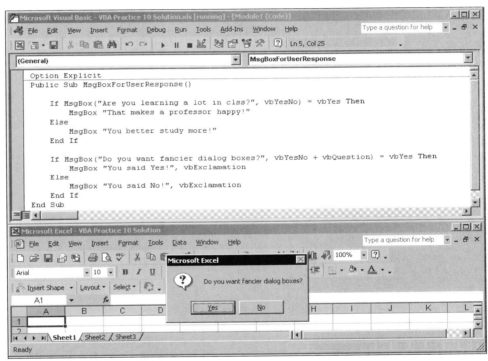

FIGURE 26   *The MsgBox Function (Exercise 10)*

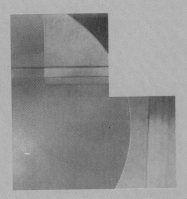

Essentials of Microsoft® Windows®

OBJECTIVES

AFTER READING THIS SUPPLEMENT YOU WILL BE ABLE TO:

1. Describe the objects on the Windows desktop; use the icons on the desktop to start the associated applications.
2. Explain the significance of the common user interface; identify the elements that are present in every window.
3. Explain in general terms the similarities and differences between various versions of Windows.
4. Use the Help command to learn about Windows.
5. Format a floppy disk.
6. Differentiate between a program file and a data file; explain the significance of the icons that appear next to a file in My Computer and Windows Explorer.
7. Explain how folders are used to organize the files on a disk; use the View menu and/or the Folder Options command to change the appearance of a folder.
8. Distinguish between My Computer and Windows Explorer with respect to viewing files and folders; explain the advantages of the hierarchical view available within Windows Explorer.
9. Use Internet Explorer to download a file; describe how to view a Web page from within Windows Explorer.
10. Copy and/or move a file from one folder to another; delete a file, then recover the deleted file from the Recycle Bin.

OVERVIEW

Microsoft® Windows is a computer program (actually many programs) that controls the operation of a computer and its peripherals. The Windows environment provides a common user interface and consistent command structure for every application. You have seen the interface many times, but do you really understand it? Can you move and copy files with confidence? Do you know how to back up the Excel spreadsheets, Access databases, and other Office documents that you work so hard to create? If not, now is the time to learn. This section is written for you, the computer novice, and it assumes no previous knowledge.

We begin with an introduction to the Windows desktop, the graphical user interface that enables you to work in intuitive fashion by pointing at icons and clicking the mouse. We identify the basic components of a window and describe how to execute commands and supply information through different elements in a dialog box. We introduce you to My Computer, an icon on the Windows desktop, and show you how to use My Computer to access the various components of your system. We also describe how to access the Help command.

The supplement concentrates, however, on disk and file management. We present the basic definitions of a file and a folder, then describe how to use My Computer to look for a specific file or folder. We introduce Windows Explorer, which provides a more efficient way of finding data on your system, then show you how to move or copy a file from one folder to another. We discuss other basic operations, such as renaming and deleting a file. We also describe how to recover a deleted file (if necessary) from the Recycle Bin.

There are also four hands-on exercises, which enable you to apply the conceptual discussion in the text at the computer. The exercises refer to a set of practice files (data disk) that we have created for you. You can obtain these files from our Web site (www.prenhall.com/grauer) or from a local area network if your professor has downloaded the files for you.

THE DESKTOP

Windows 95 was the first of the so-called "modern Windows" and was followed by Windows NT, Windows 98, Windows 2000, Windows Me (Millennium edition), and most recently, by Windows XP. Each of these systems is still in use. Windows 98 and its successor, Windows Me, are geared for the home user and provide extensive support for games and peripheral devices. Windows NT, and its successor Windows 2000, are aimed at the business user and provide increased security and reliability. Windows XP is the successor to all current breeds of Windows. It has a slightly different look, but maintains the conventions of its various predecessors. Hence we have called this module "Essentials of Microsoft Windows" and refer to Windows in a generic sense. (The screens were taken from Windows 2000 Professional, but could just as easily have been taken from other versions of the operating system.)

All versions of Windows create a working environment for your computer that parallels the working environment at home or in an office. You work at a desk. Windows operations take place on the *desktop* as shown in Figure 1. There are physical objects on a desk such as folders, a dictionary, a calculator, or a phone. The computer equivalents of those objects appear as icons (pictorial symbols) on the desktop. Each object on a real desk has attributes (properties) such as size, weight, and color. In similar fashion, Windows assigns properties to every object on its desktop. And just as you can move the objects on a real desk, you can rearrange the objects on the Windows desktop.

Figure 1a displays the typical desktop that appears when Windows is installed on a new computer. It has only a few objects and is similar to the desk in a new office, just after you move in. This desktop might have been taken from any of five systems—Windows 95, Windows NT, Windows 98, Windows 2000, or Windows Me—and is sometimes called "Classic Windows." The icons on this desktop are opened by double clicking. (It is possible to display an alternate desktop with underlined icons that are opened by single clicking, but that option is rarely used.) Figure 1b shows the new Windows XP desktop as it might appear on a home computer, where individual accounts are established for different users.

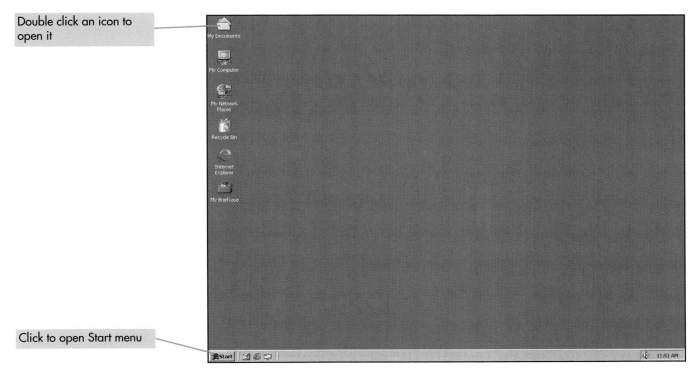

(a) Windows 95, Windows NT, Windows 98, Windows Me, and Windows 2000

(b) Windows XP

FIGURE 1 *The Different Faces of Windows*

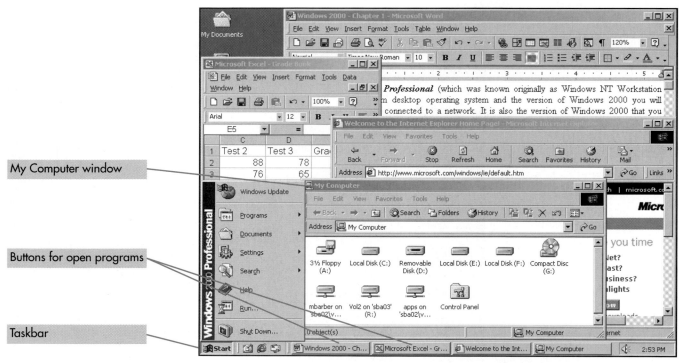

(c) A Working Desktop (all versions of Windows)

FIGURE 1 *The Different Faces of Windows (continued)*

Do not be concerned if your desktop is different from ours. Your real desk is arranged differently from those of your friends, just as your Windows desktop will also be different. Moreover, you are likely to work on different systems—at school, at work, or at home, and thus it is important that you recognize the common functionality that is present on all desktops. The ***Start button***, as its name suggests, is where you begin. Click the Start button and you see a menu that lets you start any program installed on your computer.

Look now at Figure 1c, which displays an entirely different desktop, one with four open windows that is similar to a desk in the middle of a working day. Each window in Figure 1c displays a program that is currently in use. The ability to run several programs at the same time is known as ***multitasking***, and it is a major benefit of the Windows environment. Multitasking enables you to run a word processor in one window, create a spreadsheet in a second window, surf the Internet in a third window, play a game in a fourth window, and so on. You can work in a program as long as you want, then change to a different program by clicking its window.

You can also change from one program to another by using the taskbar at the bottom of the desktop. The ***taskbar*** contains a button for each open program, and it enables you to switch back and forth between those programs by clicking the appropriate button. The taskbar in Figure 1a does not contain any buttons (other than the Start button) since there are no open applications. The taskbar in Figure 1c, however, contains four additional buttons, one for each open window.

The icons on the desktop are used to access programs or other functions. The ***My Computer*** icon is the most basic. It enables you to view the devices on your system, including the drives on a local area network to which you have direct access. Open My Computer in either Figure 1a or 1b, for example, and you see the objects in the My Computer window of Figure 1c. The contents of My Computer depend on the hardware of the specific computer system. Our system, for example, has one floppy drive, three local (hard or fixed) disks, a removable disk (an Iomega Zip drive), a CD-ROM, and access to various network drives. The My Computer window also contains the Control Panel folder that provides access to functions that control other elements of your computing environment. (These capabilities are not used by beginners, are generally "off limits" in a lab environment, and thus are not discussed further.)

The other icons on the desktop are also noteworthy. The ***My Documents*** folder is a convenient place in which to store the documents you create. ***My Network Places*** extends the view of your computer to include the other local area networks (if any) that your computer can access, provided you have a valid username and password. The ***Recycle Bin*** enables you to restore a file that was previously deleted. The Internet Explorer icon starts ***Internet Explorer***, the Web browser that is built into the Windows operating system.

THE DOJ (DEPARTMENT OF JUSTICE) VERSUS MICROSOFT

A simple icon is at the heart of the multibillion dollar lawsuit brought by 19 states against Microsoft. In short, Microsoft is accused of integrating its Internet Explorer browser into the Windows operating system with the goal of dominating the market and eliminating the competition. Is Internet Explorer built into every current version of Microsoft Windows? Yes. Can Netscape Navigator run without difficulty under every current version of Microsoft Windows? The answer is also yes. As of this writing the eventual outcome of the case against Microsoft has yet to be determined.

THE COMMON USER INTERFACE

All Windows applications share a ***common user interface*** and possess a consistent command structure. This means that every Windows application works essentially the same way, which provides a sense of familiarity from one application to the next. In other words, once you learn the basic concepts and techniques in one application, you can apply that knowledge to every other application. Consider, for example, Figure 2, which shows open windows for My Computer and My Network Places, and labels the essential elements in each.

The contents of the two windows are different, but each window has the same essential elements. The ***title bar*** appears at the top of each window and displays the name of the window, My Computer and My Network Places in Figure 2a and 2b, respectively. The icon at the extreme left of the title bar identifies the window and also provides access to a control menu with operations relevant to the window such as moving it or sizing it. The ***minimize button*** shrinks the window to a button on the taskbar, but leaves the window in memory. The ***maximize button*** enlarges the window so that it takes up the entire desktop. The ***restore button*** (not shown in either figure) appears instead of the maximize button after a window has been maximized, and restores the window to its previous size. The ***close button*** closes the window and removes it from memory and the desktop.

The ***menu bar*** appears immediately below the title bar and provides access to ***pull-down menus***. One or more ***toolbars*** appear below the menu bar and let you execute a command by clicking a button as opposed to pulling down a menu. The ***status bar*** at the bottom of the window displays information about the window as a whole or about a selected object within a window.

A vertical (or horizontal) ***scroll bar*** appears at the right (or bottom) border of a window when its contents are not completely visible and provides access to the unseen areas. A scroll bar does not appear in Figure 2a since all of the objects in the window are visible at the same time. A vertical scroll bar is found in Figure 2b, however, since there are other objects in the window.

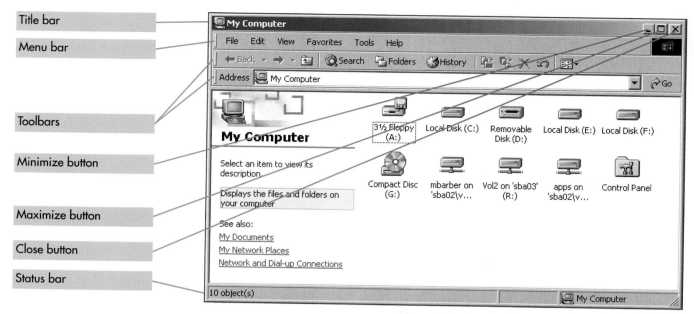

(a) My Computer

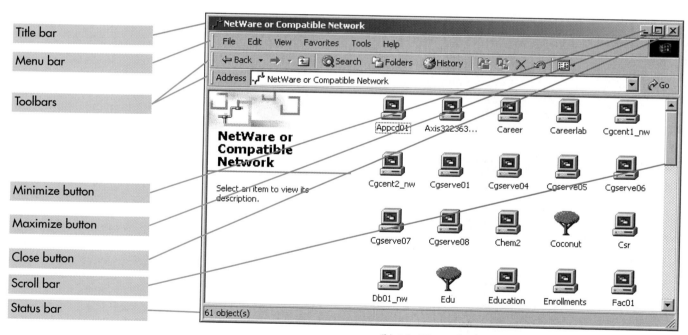

(b) My Network Places

FIGURE 2   *Anatomy of a Window*

Moving and Sizing a Window

A window can be sized or moved on the desktop through appropriate actions with the mouse. To *size a window*, point to any border (the mouse pointer changes to a double arrow), then drag the border in the direction you want to go—inward to shrink the window or outward to enlarge it. You can also drag a corner (instead of a border) to change both dimensions at the same time. To *move a window* while retaining its current size, click and drag the title bar to a new position on the desktop.

Pull-Down Menus

The menu bar provides access to *pull-down menus* that enable you to execute commands within an application (program). A pull-down menu is accessed by clicking the menu name or by pressing the Alt key plus the underlined letter in the menu name; for example, press Alt+V to pull down the View menu. (You may have to press the Alt key in order to see the underlines.) Three pull-down menus associated with My Computer are shown in Figure 3.

Commands within a menu are executed by clicking the command or by typing the underlined letter. Alternatively, you can bypass the menu entirely if you know the equivalent keystrokes shown to the right of the command in the menu (e.g., Ctrl+X, Ctrl+C, or Ctrl+V to cut, copy, or paste as shown within the Edit menu). A dimmed command (e.g., the Paste command in the Edit menu) means the command is not currently executable; some additional action has to be taken for the command to become available.

An ellipsis (. . .) following a command indicates that additional information is required to execute the command; for example, selection of the Format command in the File menu requires the user to specify additional information about the format-

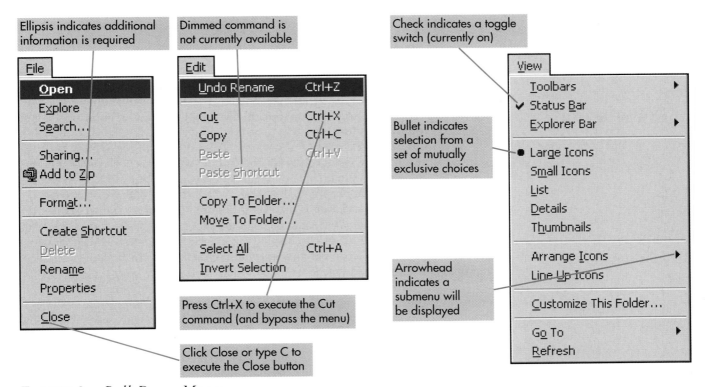

FIGURE 3   *Pull-Down Menus*

ESSENTIALS OF MICROSOFT WINDOWS

ting process. This information is entered into a dialog box (discussed in the next section), which appears immediately after the command has been selected.

A check next to a command indicates a toggle switch, whereby the command is either on or off. There is a check next to the Status Bar command in the View menu of Figure 3, which means the command is in effect (and thus the status bar will be displayed). Click the Status Bar command and the check disappears, which suppresses the display of the status bar. Click the command a second time and the check reappears, as does the status bar in the associated window.

A bullet next to an item (e.g., Large Icons in Figure 3) indicates a selection from a set of mutually exclusive choices. Click another option within the group (e.g., Small Icons) and the bullet will disappear from the previous selection (Large Icons) and appear next to the new selection (Small Icons).

An arrowhead after a command (e.g., the Arrange Icons command in the View menu) indicates that a submenu (also known as a cascaded menu) will be displayed with additional menu options.

Dialog Boxes

A ***dialog box*** appears when additional information is necessary to execute a command. Click the Print command in Internet Explorer, for example, and you are presented with the Print dialog box in Figure 4, requesting information about precisely what to print and how. The information is entered into the dialog box in different ways, depending on the type of information that is required. The tabs at the top of the dialog box provide access to different sets of options. The General and Paper tabs are selected in Figures 4a and 4b, respectively.

Option (Radio) buttons indicate mutually exclusive choices, one of which must be chosen, such as the page range in Figure 4a. You can print all pages, the selection (highlighted text), the current page, or a specific set of pages (such as pages 1–4), but you can choose one and only one option. Click a button to select an option, which automatically deselects the previously selected option.

A ***text box*** enters specific information such as the pages that will be printed in conjunction with selecting the radio button for pages. A flashing vertical bar (an I-beam) appears within the text box when the text box is active, to mark the insertion point for the text you will enter.

A ***spin button*** is another way to enter specific information such as the number of copies. Click the Up or Down arrow to increase or decrease the number of pages, respectively. You can also enter the information explicitly by typing it into a spin box, just as you would a text box.

Check boxes are used instead of option buttons if the choices are not mutually exclusive or if an option is not required. The Collate check box is checked in Figure 4a, whereas the Print to file box is not checked. Individual options are selected and cleared by clicking the appropriate check box, which toggles the box on and off.

A ***list box*** such as the Size is list box in Figure 4b displays some or all of the available choices, any one of which is selected by clicking the desired item. Just click the Down arrow on the list box to display the associated choices such as the paper source in Figure 4b. (A scroll bar appears within an open list box if all of the choices are not visible and provides access to the hidden choices.)

The ***Help button*** (a question mark at the right end of the title bar) provides help for any item in the dialog box. Click the button, then click the item in the dialog box for which you want additional information. The Close button (the X at the extreme right of the title bar) closes the dialog box without executing the command.

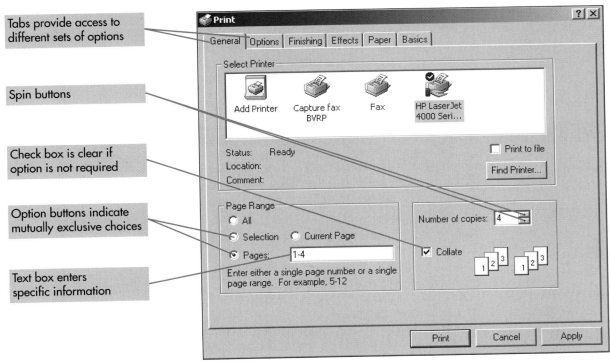

(a) General Tab

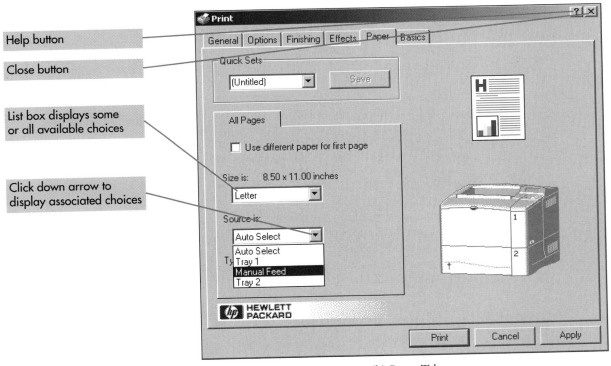

(b) Paper Tab

FIGURE 4 *Dialog Boxes*

ESSENTIALS OF MICROSOFT WINDOWS

All dialog boxes also contain one or more **command buttons**, the function of which is generally apparent from the button's name. The Print button, in Figure 4a, for example, initiates the printing process. The Cancel button does just the opposite, and ignores (cancels) any changes made to the settings, then closes the dialog box without further action. An ellipsis (three dots) on a command button indicates that additional information will be required if the button is selected.

THE MOUSE

The mouse is indispensable to Windows and is referenced continually in the hands-on exercises throughout the text. There are five basic operations with which you must become familiar:

- To *point* to an object, move the mouse pointer onto the object.
- To *click* an object, point to it, then press and release the left mouse button.
- To *right click* an object, point to the object, then press and release the right mouse button. Right clicking an object displays a context-sensitive menu with commands that pertain to the object.
- To *double click* an object, point to it and then quickly click the left button twice in succession.
- To *drag* an object, move the pointer to the object, then press and hold the left button while you move the mouse to a new position.

You may also encounter a mouse with a wheel between the left and right buttons that lets you scroll through a document by rotating the wheel forward or backward. The action of the wheel, however, may change, depending on the application in use. In any event, the mouse is a pointing device—move the mouse on your desk and the mouse pointer, typically a small arrowhead, moves on the monitor. The mouse pointer assumes different shapes according to the location of the pointer or the nature of the current action. You will see a double arrow when you change the size of a window, an I-beam as you insert text, a hand to jump from one help topic to the next, or a circle with a line through it to indicate that an attempted action is invalid.

The mouse pointer will also change to an hourglass to indicate Windows is processing your command, and that no further commands may be issued until the action is completed. The more powerful your computer, the less frequently the hourglass will appear.

The Mouse versus the Keyboard

Almost every command in Windows can be executed in different ways, using either the mouse or the keyboard. Most people start with the mouse and add keyboard shortcuts as they become more proficient. There is no right or wrong technique, just different techniques, and the one you choose depends entirely on personal preference in a specific situation. If, for example, your hands are already on the keyboard, it is faster to use the keyboard equivalent. Other times, your hand will be on the mouse and that will be the fastest way. Toolbars provide still other ways to execute common commands.

In the beginning, you may wonder why there are so many different ways to do the same thing, but you will eventually recognize the many options as part of Windows' charm. It is not necessary to memorize anything, nor should you even try; just be flexible and willing to experiment. The more you practice, the faster all of this will become second nature to you.

THE HELP COMMAND

All versions of Windows include extensive documentation with detailed information about virtually every function in Windows. It is accessed through the **Help command** on the Start menu, which provides different ways to search for information.

The **Contents tab** in Figure 5a is analogous to the table of contents in an ordinary book. The topics are listed in the left pane and the information for the selected topic is displayed in the right pane. The list of topics can be displayed in varying amounts of detail, by opening and closing the various book icons that appear. (The size of the left pane can be increased or decreased by dragging the border between the left and right pane in the appropriate direction.)

A closed book such as "Troubleshooting and Maintenance" indicates the presence of subtopics, which are displayed by opening (clicking) the book. An open book, on the other hand, such as "Internet, E-mail, and Communications," already displays its subtopics. Each subtopic is shown with one of two icons—a question mark to indicate "how to" information, or an open book to indicate conceptual information. Either way, you can click any subtopic in the left pane to view its contents in the right pane. Underlined entries in the right pane (e.g., Related Topics) indicate a hyperlink, which in turn displays additional information. Note, too, that you can print the information in the right pane by pulling down the Options menu and selecting the Print command.

The **Index tab** in Figure 5b is analogous to the index of an ordinary book. You enter the first several letters of the topic to look up, such as "floppy disk," choose a topic from the resulting list, and then click the Display button to view the information in the right pane. The underlined entries in the right pane represent hyperlinks, which you can click to display additional topics. And, as in the Contents window, you can print the information in the right pane by pulling down the Options menu and selecting the Print command.

The **Search tab** (not shown in Figure 5) displays a more extensive listing of entries than does the Index tab. It lets you enter a specific word or phrase and then it returns every topic containing that word or phrase.

The **Favorites tab** enables you to save the information within specified help topics as bookmarks, in order to return to those topics at a later date, as explained in the following hands-on exercise.

FORMATTING A FLOPPY DISK

You will soon begin to work on the computer, which means that you will be using various applications to create different types of documents. Each document is saved in its own file and stored on disk, either on a local disk (e.g., drive C) if you have your own computer, or on a floppy disk (drive A) if you are working in a computer lab at school.

All disks have to be formatted before they can hold data. The formatting process divides a disk into concentric circles called tracks, and then further divides each track into sectors. You don't have to worry about formatting a hard disk, as that is done at the factory prior to the machine being sold. You typically don't even have to format a floppy disk, since most floppies today are already formatted when you buy them. Nevertheless, it is very easy to format a floppy disk and it is a worthwhile exercise. Be aware, however, that formatting erases any data that was previously on a disk, so be careful not to format a disk with important data (e.g., one containing today's homework assignment). Formatting is accomplished through the **Format command**. The process is straightforward, as you will see in the hands-on exercise that follows.

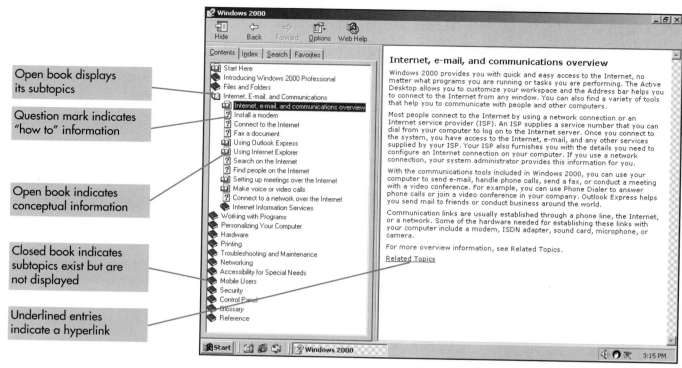

(a) Contents Tab

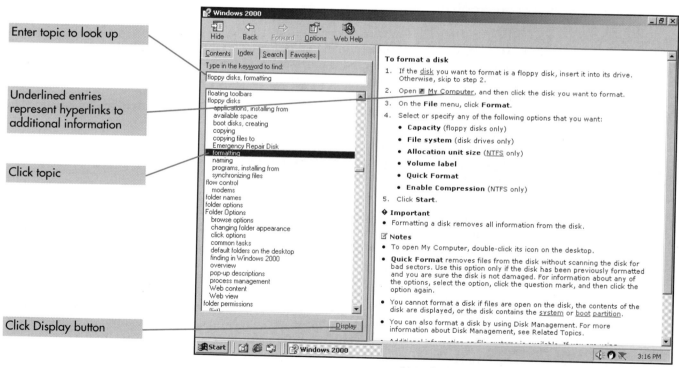

(b) Index Tab

FIGURE 5   *The Help Command*

HANDS-ON EXERCISE 1

WELCOME TO WINDOWS

Objective To turn on the computer, start Windows, and open My Computer; to move and size a window; to format a floppy disk and use the Help command. Use Figure 6 as a guide in the exercise.

Step 1: **Open My Computer**

> ➤ Start the computer by turning on the various switches appropriate to your system. Your system will take a minute or so to boot up, after which you may be asked for a **user name** and **password**.
> ➤ Enter this information, after which you should see the desktop in Figure 6a. It does not matter if you are using a different version of Windows.
> ➤ Close the Getting Started with Windows 2000 window if it appears. Do not be concerned if your desktop differs from ours.
> ➤ The way in which you open My Computer (single or double clicking) depends on the options in effect as described in step 2. Either way, however, you can **right click** the **My Computer icon** to display a context-sensitive menu, then click the **Open command**.
> ➤ The My Computer window will open on your desktop, but the contents of your window and/or its size and position will be different from ours. You are ready to go to work.

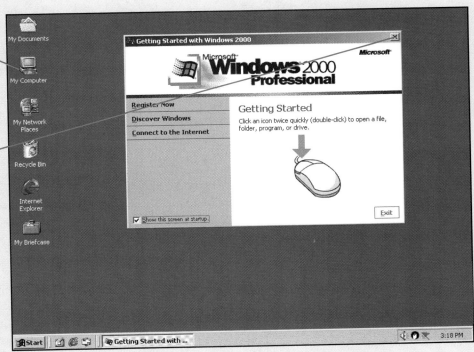

(a) Open My Computer (step 1)

FIGURE 6 *Hands-on Exercise 1*

Step 2: **Set the Folder Options**

➤ Pull down the **Tools menu** and click the **Folder Options command** to display the Folder Options dialog box. Click the **General tab**, then set the options as shown in Figure 6b. (Your network administrator may have disabled this command, in which case you will use the default settings.)
- The Active desktop enables you to display Web content directly on the desktop. We suggest that you disable this option initially.
- Enabling Web content in folders displays the template at the left side of the window. The Windows classic option does not contain this information.
- Opening each successive folder within the same window saves space on the desktop as you browse the system. We discuss this in detail later on.
- The choice between clicking underlined items and double clicking an icon (without the underline) is personal. We prefer to double click.

➤ Click **OK** to accept the settings and close the Folder Options dialog box. The My Computer window on your desktop should be similar to ours.

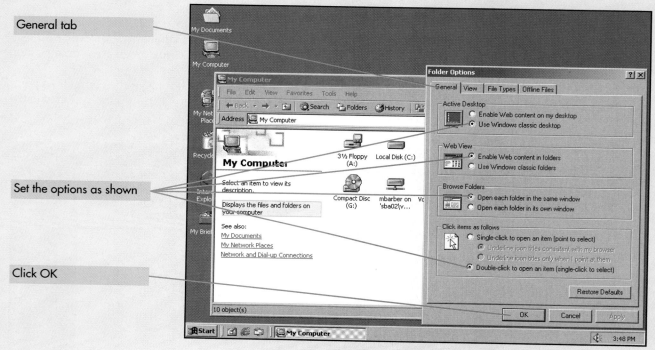

(b) Set the Folder Options (step 2)

FIGURE 6 *Hands-on Exercise 1 (continued)*

IT'S DIFFERENT IN WINDOWS 98

The Folder Options command is under the View menu in Windows 98, whereas it is found in the Tools menu in Windows 2000. Thus, to go from clicking to double clicking in Windows 98, pull down the View menu, click Folder Options, click the General tab, then choose Web style or Classic style, respectively. The procedure to display Web content in a folder is also different in Windows 98; you need to pull down the View menu and toggle the As Web Page command on.

Step 3: **Move and Size a Window**

- ➤ If necessary, pull down the **View menu** and click **Large Icons** so that your My Computer window more closely resembles the window in Figure 6c.
- ➤ Move and size the My Computer window on your desktop to match the display in Figure 6c.
  - To change the width or height of the window, click and drag a border (the mouse pointer changes to a double arrow) in the direction you want to go.
  - To change the width and height at the same time, click and drag a corner rather than a border.
  - To change the position of the window, click and drag the title bar.
- ➤ Click the **minimize button** to shrink the My Computer window to a button on the taskbar. My Computer is still active in memory, however. Click the **My Computer button** on the taskbar to reopen the window.
- ➤ Click the **maximize button** so that the My Computer window expands to fill the entire screen. Click the **restore button** (which replaces the maximize button and is not shown in Figure 6c) to return the window to its previous size.

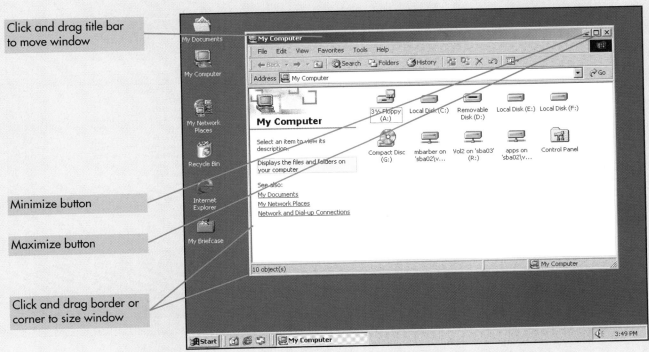

(c) Move and Size a Window (step 3)

FIGURE 6 *Hands-on Exercise 1 (continued)*

MINIMIZING VERSUS CLOSING AN APPLICATION

Minimizing an application leaves the application open in memory and available at the click of the taskbar button. Closing it, however, removes the application from memory, which also causes it to disappear from the taskbar. The advantage of minimizing an application is that you can return to the application immediately. The disadvantage is that leaving too many applications open simultaneously may degrade performance.

Step 4: **Use the Pull-Down Menus**

➤ Pull down the **View menu**, then click the **Toolbars command** to display a cascaded menu as shown in Figure 6d. If necessary, check the commands for the **Standard Buttons** and **Address Bar**, and clear the commands for Links and Radio.

➤ Pull down the **View menu** to make or verify the following selections. (You have to pull down the View menu each time you make an additional change.)
- The **Status Bar command** should be checked. The Status Bar command functions as a toggle switch. Click the command and the status bar is displayed; click the command a second time and the status bar disappears.)
- Click the **Details command** to change to this view. Notice that the different views are grouped within the menu and that only one view at a time can be selected.

➤ Pull down the **View menu** once again, click (or point to) the **Explorer Bar command**, and verify that none of the options is checked.

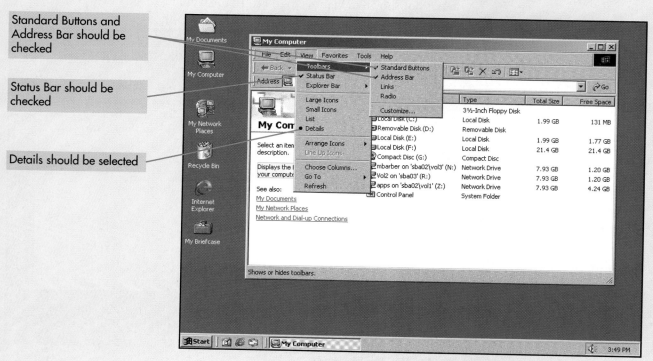

(d) Use the Pull-Down Menus (step 4)

FIGURE 6   *Hands-on Exercise 1 (continued)*

DESIGNATING THE DEVICES ON A SYSTEM

The first (usually only) floppy drive is always designated as drive A. (A second floppy drive, if it were present, would be drive B.) The first hard (local) disk on a system is always drive C, whether or not there are one or two floppy drives. Additional local drives, if any, a Zip (removable storage) drive, a network drive, and/or the CD-ROM are labeled from D on.

Step 5: **Format a Floppy Disk**

➤ Place a floppy disk in drive A. Select (click) drive A, then pull down the **File menu** and click the **Format command** to display the dialog box in Figure 6e.
- Set the **Capacity** to match the floppy disk you purchased (1.44MB for a high-density disk and 720KB for a double-density disk).
- Click the **Volume label text box** if it's empty or click and drag over the existing label. Enter a new label (containing up to 11 characters).
- You can check the **Quick Format box** if the disk has been previously formatted, as a convenient way to erase the contents of the disk.

➤ Click the **Start button**, then click **OK** after you have read the warning. The formatting process erases anything that is on the disk, so be sure that you do not need anything on the disk you are about to format.

➤ Click **OK** after the formatting is complete. Close the Format dialog box, then save the formatted disk for use with various exercises later in the text.

➤ Close the My Computer window.

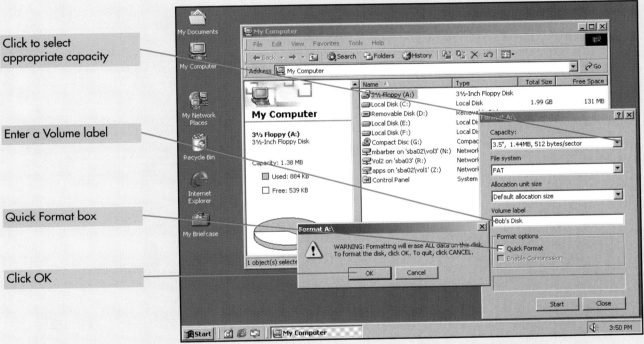

(e) Format a Floppy Disk (step 5)

FIGURE 6 *Hands-on Exercise 1 (continued)*

THE HELP BUTTON

The Help button (a question mark) appears in the title bar of almost every dialog box. Click the question mark, then click the item you want information about (which then appears in a pop-up window). To print the contents of the pop-up window, click the right mouse button inside the window, and click Print Topic. Click outside the pop-up window to close the window and continue working.

Step 6: **The Help Command**

- Click the **Start button** on the taskbar, then click the **Help command** to display the Help window in Figure 6f. Maximize the Help window.
- Click the **Contents tab**, then click a closed book such as **Hardware** to open the book and display the associated topics. Click any one of the displayed topics such as **Hardware overview** in Figure 6f.
- Pull down the **Options menu** and click the **Print command** to display the Print Topics dialog box. Click the option button to print the selected topic, click **OK**, then click the **Print button** in the resulting dialog box.
- Click the **Index tab**, type **format** (the first several letters in "Formatting disks," the topic you are searching for). Double click this topic within the list of index items. Pull down the **Options menu** and click the **Print command** to print this information as well.
- Submit the printed information to your instructor. Close the Help window.

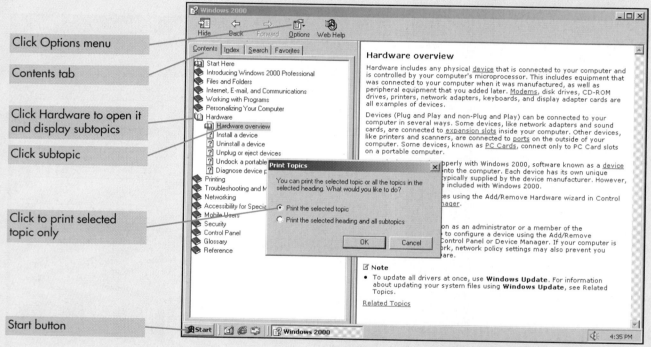

(f) The Help Command (step 6)

FIGURE 6 *Hands-on Exercise 1 (continued)*

THE FAVORITES TAB

Do you find yourself continually searching for the same Help topic? If so, you can make life a little easier by adding the topic to a list of favorite Help topics. Start Help, then use the Contents, Index, or Search tabs to locate the desired topic. Now click the Favorites tab in the Help window, then click the Add button to add the topic. You can return to the topic at any time by clicking the Favorites tab, then double clicking the bookmark to display the information.

Step 7: **Shut Down the Computer**

➤ It is very important that you shut down your computer properly as opposed to just turning off the power. This enables Windows to properly close all of its system files and to save any changes that were made during the session.

➤ Click the **Start button**, click the **Shut Down command** to display the Shut Down Windows dialog box in Figure 6g. Click the **drop-down arrow** to display the desired option:
- Logging off ends your session, but leaves the computer running at full power. This is the typical option you select in a laboratory setting.
- Shutting down the computer ends the session and also closes Windows so that you can safely turn the power off. (Some computers will automatically turn the power off for you if this option is selected.)
- Restarting the computer ends your sessions, then closes and restarts Windows to begin a new session.

➤ Welcome to Windows 2000!

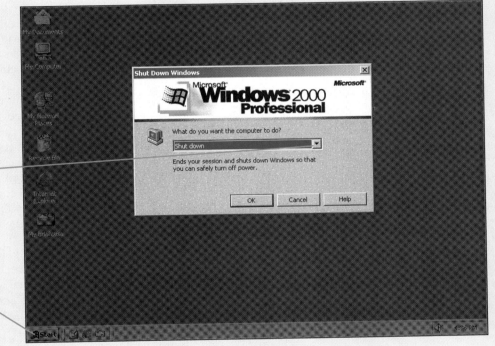

(g) Shut Down the Computer (step 7)

FIGURE 6   *Hands-on Exercise 1 (continued)*

THE TASK MANAGER

The Start button is the normal way to exit Windows. Occasionally, however, an application may "hang"—in which case you want to close the problem application but leave Windows open. Press Ctrl+Alt+Del to display the Windows Security dialog box, then click the Task Manager command button. Click the Applications tab, select the problem application, and click the End Task button.

FILES AND FOLDERS

A *file* is a set of instructions or data that has been given a name and stored on disk. There are two basic types of files, *program files* and *data files*. Microsoft Word and Microsoft Excel are examples of program files. The documents and workbooks created by these programs are examples of data files.

A *program file* is an executable file because it contains instructions that tell the computer what to do. A *data file* is not executable and can be used only in conjunction with a specific program. As a typical student, you execute (run) program files, then you use those programs to create and/or modify the associated data files.

Every file has a *file name* that identifies it to the operating system. The file name may contain up to 255 characters and may include spaces. (File names cannot contain the following characters: \, /, :, *, ?, ", <, >, or |. We suggest that you try to keep file names simple and restrict yourself to the use of letters, numbers, and spaces.) Long file names permit descriptive entries such as *Term Paper for Western Civilization* (as distinct from a more cryptic *TPWCIV* that was required under MS-DOS and Windows 3.1).

Files are stored in *folders* to better organize the hundreds (thousands, or tens of thousands) of files on a hard disk. A Windows folder is similar in concept to a manila folder in a filing cabinet into which you put one or more documents (files) that are somehow related to each other. An office worker stores his or her documents in manila folders. In Windows, you store your files (documents) in electronic folders on disk.

Folders are the keys to the Windows storage system. Some folders are created automatically; for example, the installation of a program such as Microsoft Office automatically creates one or more folders to hold the various program files. Other folders are created by the user to hold the documents he or she creates. You could, for example, create one folder for your word processing documents and a second folder for your spreadsheets. Alternatively, you can create a folder to hold all of your work for a specific class, which may contain a combination of word processing documents and spreadsheets. The choice is entirely up to you, and you can use any system that makes sense to you. Anything at all can go into a folder—program files, data files, even other folders.

Figure 7 displays the contents of a hypothetical Homework folder with six documents. Figure 7a enables Web content, and so we see the colorful logo at the left of the folder, together with links to My Documents, My Network Places, and My Computer. Figure 7b is displayed without the Web content, primarily to gain space within the window. The display or suppression of the Web content is determined by a setting in the Folder Options command.

Figures 7a and 7b are displayed in different views. Figure 7a uses the *Large Icons view*, whereas Figure 7b is displayed in the *Details view*, which shows additional information for each file. (Other possible views include Small Icons, List, and Thumbnail.) The file icon, whether large or small, indicates the *file type* or application that was used to create the file. The History of Computers file, for example, is a Microsoft Word document. The Grade Book is a Microsoft Excel workbook.

Regardless of the view and options in effect, the name of the folder (Homework) appears in the title bar next to the icon of an open folder. The minimize, maximize, and Close buttons appear at the right of the title bar. A menu bar with six pull-down menus appears below the title bar. The Standard Buttons toolbar appears below the menu, and the Address Bar (indicating the drive and folder) appears below the toolbar. A status bar appears at the bottom of both windows, indicating that the Homework folder contains six objects (documents) and that the total file size is 525KB.

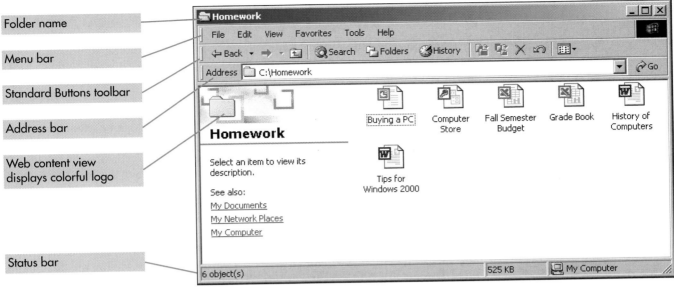

(a) Large Icons View with Web Content Enabled

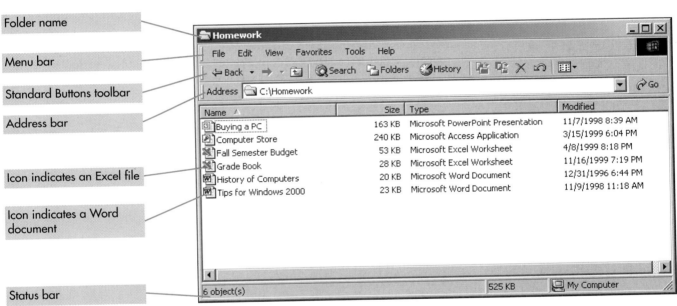

(b) Details View without Web Content

FIGURE 7 *The Homework Folder*

CHANGE THE VIEW

Look closely at the address bar in Figures 7a and 7b to see that both figures display the Homework folder on drive C, although the figures are very different in appearance. Figure 7a displays Web content to provide direct links to three other folders, and the contents of the Homework folder are displayed in the Large Icons view to save space. Figure 7b suppresses the Web content and uses the Details view to provide the maximum amount of information for each file in the Homework folder. You are free to choose whichever options you prefer.

MY COMPUTER

My Computer enables you to browse through the various drives and folders on a system in order to locate a document and go to work. Let's assume that you're looking for a document called "History of Computers" that you saved previously in the Homework folder on drive C. To get to this document, you would open My Computer, from where you would open drive C, open the Homework folder, and then open the document. It's a straightforward process that can be accomplished in two different ways, as shown in Figure 8.

The difference between the two figures is whether each drive or folder is opened in its own window, as shown in Figure 8a, or whether the same window is used for every folder, as in Figure 8b. (This is another option that is set through the Folder Options command.) In Figure 8a you begin by double clicking the My Computer icon on the desktop to open the My Computer window, which in turn displays the devices on your system. Next, you double click the icon for drive C to open a second window that displays the folders on drive C. From there, you double click the icon for the Homework folder to open a third window containing the documents in the Homework folder. Once in the Homework folder, you can double click the icon of an existing document, which starts the associated application and opens the document.

The process is identical in Figure 8b except that each object opens in the same window. The Back arrow on the Standard Buttons toolbar is meaningful in Figure 8b because you can click the button to return to the previous window (drive C), then click it again to go back to My Computer. Note, however, that the button is dimmed in all three windows in Figure 8a because there is no previous window, since each folder is opened in its own window.

THE EXPLORING OFFICE PRACTICE FILES

There is only one way to master the file operations inherent in Windows and that is to practice at the computer. To do so requires that you have a series of files with which to work. We have created these files for you, and we reference them in the next several hands-on exercises. Your instructor will make these files available to you in a variety of ways:

- The files can be downloaded from our Web site, assuming that you have access to the Internet and that you have a basic proficiency with Internet Explorer. Software and other files that are downloaded from the Internet are typically compressed (made smaller) to reduce the amount of time it takes to transmit the file. In essence, you will download a *compressed file* (which may contain multiple individual files) from our Web site and then uncompress the file onto a local drive as described in the next hands-on exercise.
- The files might be on a network drive, in which case you can use My Computer (or Windows Explorer, which is discussed later in the chapter) to copy the files from the network drive to a floppy disk. The procedure to do this is described in the third hands-on exercise.
- There may be an actual "data disk" in the computer lab. Go to the lab with a floppy disk, then use the Copy Disk command (on the File menu of My Computer) to duplicate the data disk and create a copy for yourself.

It doesn't matter how you obtain the practice files, only that you are able to do so. Indeed, you may want to try different techniques in order to gain additional practice with Windows.

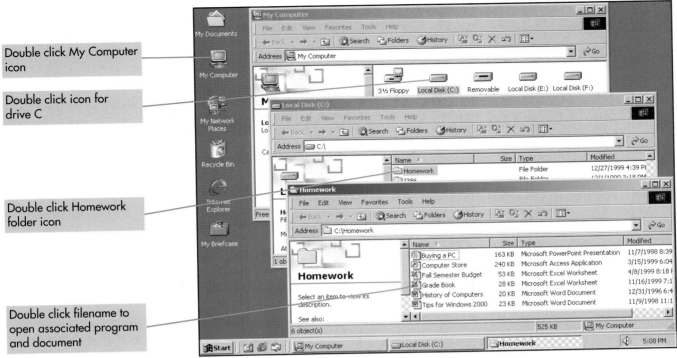

(a) Multiple Windows

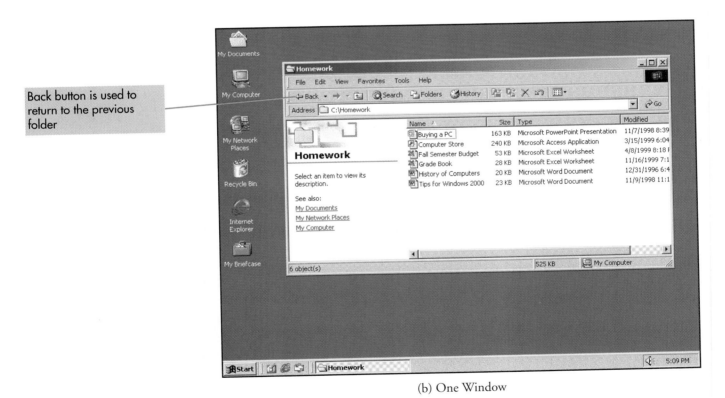

(b) One Window

FIGURE 8   *Browsing My Computer*

HANDS-ON EXERCISE 2

THE PRACTICE FILES VIA THE WEB

Objective To download a file from the Web. The exercise requires a formatted floppy disk and access to the Internet. Use Figure 9 as a guide in the exercise.

Step 1: **Start Internet Explorer**

➤ Start Internet Explorer, perhaps by double clicking the **Internet Explorer icon** on the desktop, or by clicking the **Start button**, clicking the **Programs command**, then locating the command to start the program. If necessary, click the **maximize button** so that Internet Explorer takes the entire desktop.

➤ Enter the address of the site you want to visit:
- Pull down the **File menu**, click the **Open command** to display the Open dialog box, and enter **www.prenhall.com/grauer** (the http:// is assumed). Click **OK**.
- *Or, c*lick in the **Address bar** below the toolbar, which automatically selects the current address (so that whatever you type replaces the current address). Enter the address of the site you want to visit, **www.prenhall.com/grauer** (the http:// is assumed). Press **enter**.

➤ You should see the *Exploring Office Series* home page as shown in Figure 9a. Click the book for **Office XP**, which takes you to the Office XP home page.

➤ Click the **Student Resources link** (at the top of the window) to go to the Student Resources page.

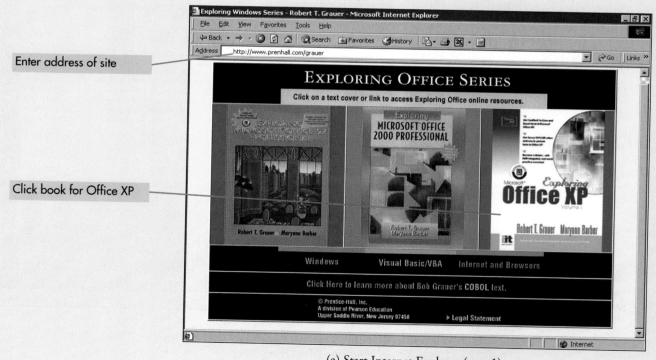

(a) Start Internet Explorer (step 1)

FIGURE 9   *Hands-on Exercise 2*

Step 2: **Download the Practice Files**

- Click the link to **Student Data Disk** (in the left frame), then scroll down the page until you can see **Essentials of Microsoft Windows**.
- Click the indicated link to download the practice files. The Save As dialog box is not yet visible.
- You will see the File Download dialog box asking what you want to do. The option button to save this program to disk is selected. Click **OK**. The Save As dialog box appears as shown in Figure 9b.
- Place a formatted floppy disk in drive A, click the **drop-down arrow** on the Save in list box, and select (click) **drive A**. Click **Save** to begin downloading the file.
- The File Download window will reappear on your screen and show you the status of the downloading operation. If necessary, click **OK** when you see the dialog box indicating that the download is complete.
- Close Internet Explorer.

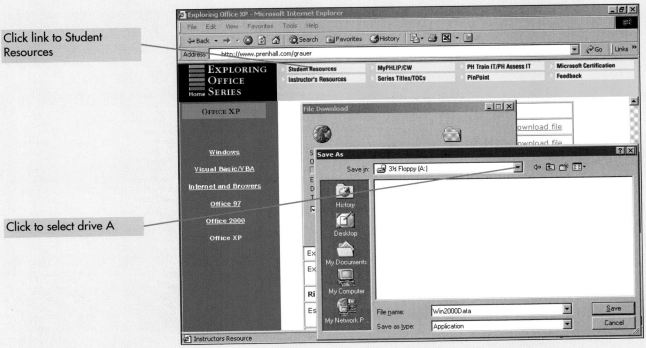

(b) Download the Practice Files (step 2)

FIGURE 9 *Hands-on Exercise 2 (continued)*

REMEMBER THE LOCATION

It's easy to download a file from the Web. The only tricky part, if any, is remembering where you have saved the file. This exercise is written for a laboratory setting, and thus we specified drive A as the destination, so that you will have the file on a floppy disk at the end of the exercise. If you have your own computer, however, it's faster to save the file to the desktop or in a temporary folder on drive C. Just remember where you save the file so that you can access it after it has been downloaded.

Step 3: **Open My Computer**

- ➤ Double click the My Computer icon on the desktop to open My Computer. If necessary, customize My Computer to match Figure 9c.
  - Pull down the **View menu** and change to the **Details view**.
  - Pull down the **View menu** a second time, click (or point to) the **Toolbars command**, then check the **Standard buttons** and **Address Bar** toolbars.
- ➤ Pull down the **Tools menu** and click the **Folder Options command** to verify the settings in effect so that your window matches ours. Be sure to **Enable Web content in folders** (in the Web View area), to **Open each folder in the same window** (in the Browse Folders area), and **Double Click to open an item** (in the Click Items area).
- ➤ Click the icon for **drive A** to select it. The description of drive A appears at the left of the window.
- ➤ Double click the icon for **drive A** to open this drive. The contents of the My Computer window are replaced by the contents of drive A.

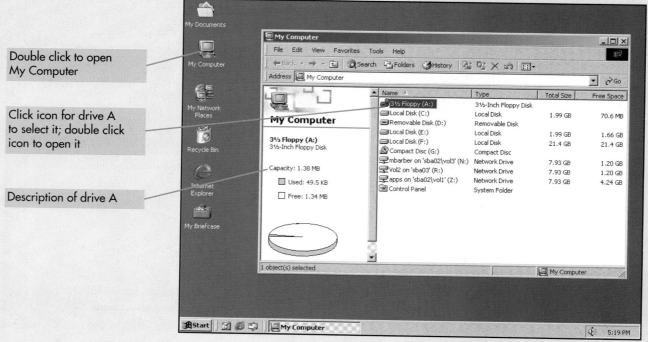

(c) Open My Computer (step 3)

FIGURE 9 *Hands-on Exercise 2 (continued)*

THE RIGHT MOUSE BUTTON

Point to any object on the Windows desktop or within an application window, then click the right mouse button to see a context-sensitive menu with commands pertaining to that object. You could, for example, right click the icon for drive A, then select the Open command from the resulting menu. The right mouse button is one of the most powerful Windows shortcuts and one of its best-kept secrets. Use it!

Step 4: **Install the Practice Files**

➤ You should see the contents of drive A as shown in Figure 9d. (If your desktop displays two windows rather than one, it is because you did not set the folder options correctly. Pull down the **Tools menu**, click the **Folder Options command**, and choose the option to **Open each folder in the same window**.)

➤ Double click the **Win2000data file** to install the data disk. You will see a dialog box thanking you for selecting the *Exploring Windows* series. Click **OK**.

- Check that the Unzip To Folder text box specifies **A:** to extract the files to the floppy disk. (You may enter a different drive and/or folder.)
- Click the **Unzip button** to extract the practice files and copy them onto the designated drive. Click **OK** after you see the message indicating that the files have been unzipped successfully. Close the WinZip dialog box.

➤ The practice files have been extracted to drive A and should appear in the Drive A window. If you do not see the files, pull down the **View menu** and click the **Refresh command**.

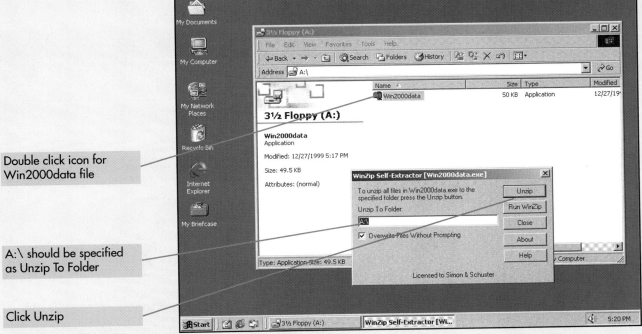

(d) Install the Practice Files (step 4)

FIGURE 9 *Hands-on Exercise 2 (continued)*

DOWNLOADING A FILE

Software and other files are typically compressed to reduce the amount of storage space the files require on disk and/or the time it takes to download the files. In essence, you download a compressed file (which may contain multiple individual files), then you uncompress (expand) the file on your local drive in order to access the individual files. After the file has been expanded, it is no longer needed and can be deleted.

Step 5: **Delete the Compressed File**

> ➤ If necessary, pull down the **View menu** and click **Details** to change to the Details view in Figure 9e. (If you do not see the descriptive information about drive A at the left of the window, pull down the **Tools menu**, click the **Folder Options command**, and click the option button to **Enable Web content in folders**.)
> ➤ You should see a total of six files in the Drive A window. Five of these are the practice files on the data disk. The sixth file is the original file that you downloaded earlier. This file is no longer necessary, since it has been already been expanded.
> ➤ Select (click) the **Win2000data file**. Pull down the **File menu** and click the **Delete command**, or click the **Delete button** on the toolbar. Pause for a moment to be sure you want to delete this file, then click **Yes** when asked to confirm the deletion as shown in Figure 9e.
> ➤ The Win2000Data file is permanently deleted from drive A. (Items deleted from a floppy disk or network drive are not sent to the Recycle Bin, and cannot be recovered.)

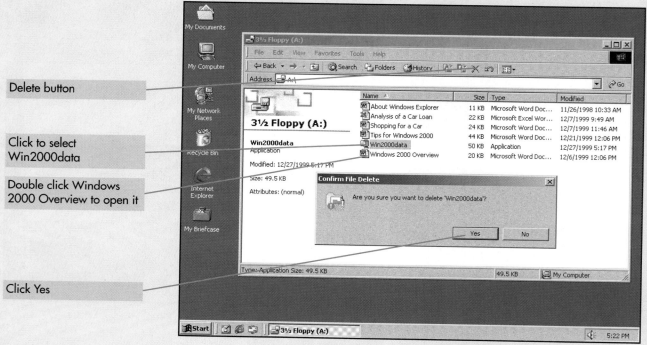

(e) Delete the Compressed File (step 5)

FIGURE 9   *Hands-on Exercise 2 (continued)*

SORT BY NAME, DATE, FILE TYPE, OR SIZE

Files can be displayed in ascending or descending sequence by name, date modified, file type, or size by clicking the appropriate column heading. Click Size, for example, to display files in the order of their size. Click the column heading a second time to reverse the sequence; that is, to switch from ascending to descending, and vice versa.

Step 6: **Modify a Document**

- Double click the **Windows 2000 Overview** document from within My Computer to open the document as shown in Figure 9f. (The document will open in the WordPad accessory if Microsoft Word is not installed on your machine.) If necessary, maximize the window for Microsoft Word.
- If necessary, click inside the document window, then press **Ctrl+End** to move to the end of the document. Add the sentence shown in Figure 9h followed by your name.
- Pull down the **File menu**, click **Print**, then click **OK** to print the document and prove to your instructor that you did the exercise.
- Pull down the **File menu** and click **Exit** to close the application. Click **Yes** when prompted to save the file.

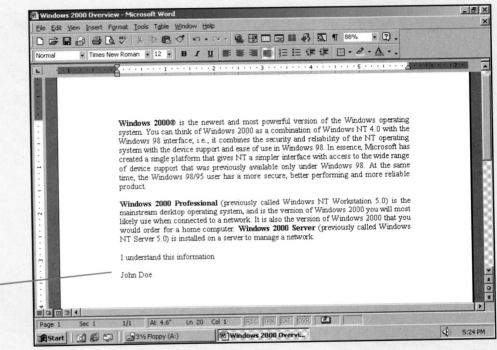

Add text and your name

(f) Modify a Document (step 6)

FIGURE 9 *Hands-on Exercise 2 (continued)*

THE DOCUMENT, NOT THE APPLICATION

All versions of Windows are document oriented, meaning that you are able to think in terms of the document rather than the application that created it. You can still open a document in traditional fashion by starting the application that created the document, then using the File Open command in that program to retrieve the document. It's often easier, however, to open the document from within My Computer (or Windows Explorer) by double clicking its icon. Windows then starts the application and opens the data file. In other words, you can open a document without explicitly starting the application.

Step 7: **Check Your Work**

➤ You should be back in the My Computer window as shown in Figure 9g. If necessary, click the **Views button** to change to the Details view.
➤ Look closely at the date and time that is displayed next to the Windows 2000 Overview document. It should show today's date and the current time (give or take a minute) since that is when the document was last modified.
➤ Look closely and see that Figure 9g also contains a sixth document, called "Backup of Windows 2000 Overview". This is a backup copy of the original document that will be created automatically by Microsoft Word if the appropriate options are in effect. (See the boxed tip below.)
➤ Exit Windows or, alternatively, continue with steps 8 and 9 to return to our Web site and explore additional resources.

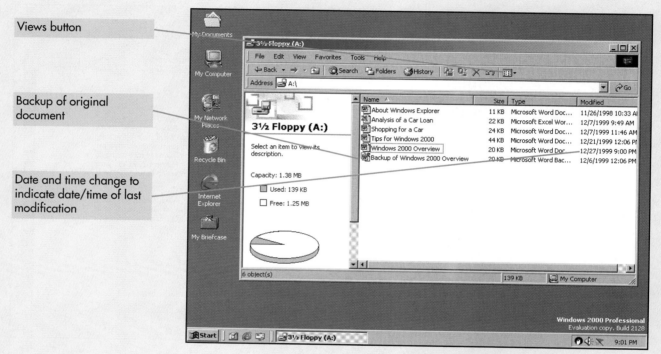

(g) Check Your Work (step 7)

FIGURE 9 *Hands-on Exercise 2 (continued)*

USE WORD TO CREATE A BACKUP COPY

Microsoft Word enables you to automatically create a backup copy of a document in conjunction with the Save command. The next time you are in Microsoft Word, pull down the Tools menu, click the Options command, click the Save tab, then check the box to always create a backup copy. Every time you save a file from this point on, the previously saved version is renamed "Backup of document," and the document in memory is saved as the current version. The disk will contain the two most recent versions of the document, enabling you to retrieve the previous version if necessary.

Step 8: **Download the PowerPoint Lecture**

➤ Restart Internet Explorer and connect to **www.prenhall.com/grauer**. Click the book for **Office XP**, click the link to **Student Resources**, then choose **PowerPoint Lectures** to display the screen in Figure 9h.

➤ Click the down arrow until you can click the link to the PowerPoint slides for **Essentials of Windows**. The File Download dialog box will appear with the option to save the file to disk selected by default. Click **OK**.

➤ Click the **drop-down arrow** on the Save in list box, and select **drive A**. Be sure that the floppy disk is still in drive A, then click **Save** to begin downloading the file. Click **OK** when you see the dialog box indicating that the download is complete.

➤ Click the taskbar button to return to the **My Computer window** for drive A. You should see all of the files that were previously on the floppy disk plus the file you just downloaded.

➤ Double click the **Win2000ppt file**, then follow the onscreen instructions to unzip the file to drive A.

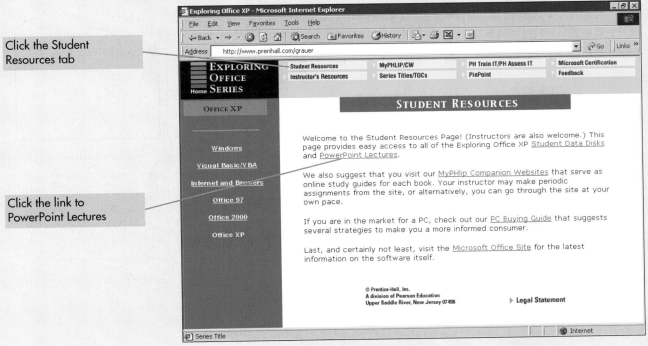

(h) Download the PowerPoint Lecture (step 8)

FIGURE 9 *Hands-on Exercise 2 (continued)*

THE MyPHLIP WEB SITE

The MyPHLIP (Prentice Hall Learning on the Internet Partnership) Web site is another resource that is available for the Exploring Office series. Click the MyPHLIP tab at the top of the screen, which takes you to www.prenhall.com/myphlip, where you will register and select the text you are using. See exercise 3 at the end of the chapter.

Step 9: **Show Time**

- Drive A should now contain a PowerPoint file in addition to the self-extracting file. (Pull down the **View menu** and click the **Refresh command** if you do not see the PowerPoint file.)
- Double click the PowerPoint file to open the presentation, then click the button to Enable Macros (if prompted). You should see the PowerPoint presentation in Figure 9i. (You must have PowerPoint installed on your computer in order to view the presentation.)
- Pull down the **View menu** and click **Slide Show** to begin the presentation, which is intended to review the material in this supplement. Click the left mouse button (or press the **PgDn key**) to move to the next slide.
- Click the left mouse button continually to move from one slide to the next. Close PowerPoint at the end of the presentation.
- Exit Windows if you do not want to continue with the next exercise at this time.

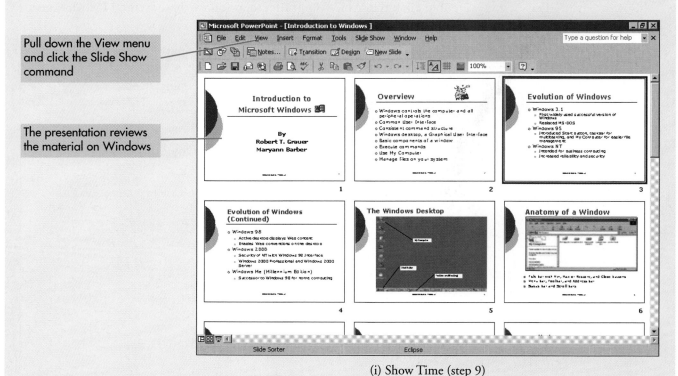

(i) Show Time (step 9)

FIGURE 9 *Hands-on Exercise 2 (continued)*

MISSING POWERPOINT—WHICH VERSION OF OFFICE DO YOU HAVE?

You may have installed Microsoft Office on your computer, but you may not have PowerPoint. That is because Microsoft has created several different versions of Microsoft Office, each with a different set of applications. Unfortunately, PowerPoint is not included in every configuration and may be missing from the suite that is shipped most frequently with new computers.

WINDOWS EXPLORER

Windows has two different programs to manage the files and folders on a system, My Computer and Windows Explorer. My Computer is intuitive, but less efficient, as you have to open each folder in succession. Windows Explorer is more sophisticated, as it provides a hierarchical view of the entire system in a single window. A beginner might prefer My Computer, whereas a more experienced user will most likely opt for Windows Explorer.

Assume, for example, that you are taking four classes this semester, and that you are using the computer in each course. You've created a separate folder to hold the work for each class and have stored the contents of all four folders on a single floppy disk. Assume further that you need to retrieve your third English assignment so that you can modify the assignment, then submit the revised version to your instructor. Figure 10 illustrates how Windows Explorer could be used to locate your assignment.

The Explorer window in Figure 10a is divided into two panes. The left pane contains a tree diagram (or hierarchical view) of the entire system showing all drives and, optionally, the folders in each drive. The right pane shows the contents of the active drive or folder. Only one object (a drive or folder) can be active in the left pane, and its contents are displayed automatically in the right pane.

Look carefully at the icon for the English folder in the left pane of Figure 10a. The folder is open, whereas the icon for every other folder is closed. The open folder indicates that the English folder is the active folder. (The name of the active folder also appears in the title bar of Windows Explorer and in the Address bar.) The contents of the active folder (three Word documents in this example) are displayed in the right pane. The right pane is displayed in Details view, but could just as easily have been displayed in another view (e.g., Large Icons).

As indicated, only one folder can be open (active) at a time in the left pane. Thus, to see the contents of a different folder such as Accounting, you would open (click on) the Accounting folder, which automatically closes the English folder. The contents of the Accounting folder would then appear in the right pane. You should organize your folders in ways that make sense to you, such as a separate folder for every class you are taking. You can also create folders within folders; for example, a correspondence folder may contain two folders of its own, one for business correspondence and one for personal letters.

Windows Explorer can also be used to display a Web page, as shown in Figure 10b. All you do is click the icon for Internet Explorer in the left pane to start the program and display its default home page. Alternatively, you can click in the Address bar and enter the address of any Web page directly; for example, click in the Address bar and type www.microsoft.com to display the home page for Microsoft. Once you are browsing pages on the Web, it's convenient to close the left pane so that the page takes the complete window. You can reopen the Folders window by pulling down the View menu, clicking the Explorer Bar command, and toggling Folders on.

THE SMART TOOLBAR

The toolbar in Windows Explorer recognizes whether you are viewing a Web page or a set of files and folders, and changes accordingly. The icons that are displayed when viewing a Web page are identical to those in Internet Explorer and include the Search, Favorites, and History buttons. The buttons that are displayed when viewing a file or folder include the Undo, Delete, and Views buttons that are used in file management.

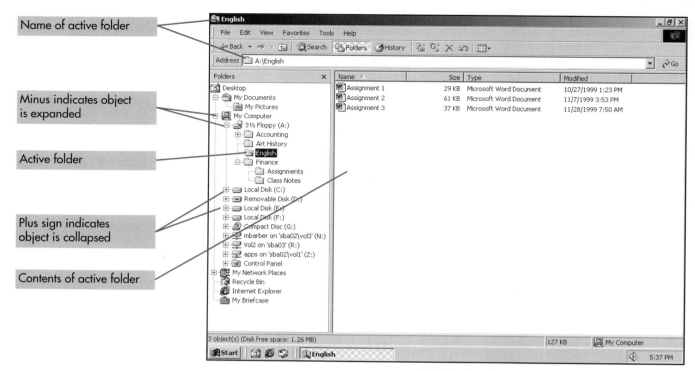

(a) Drive A

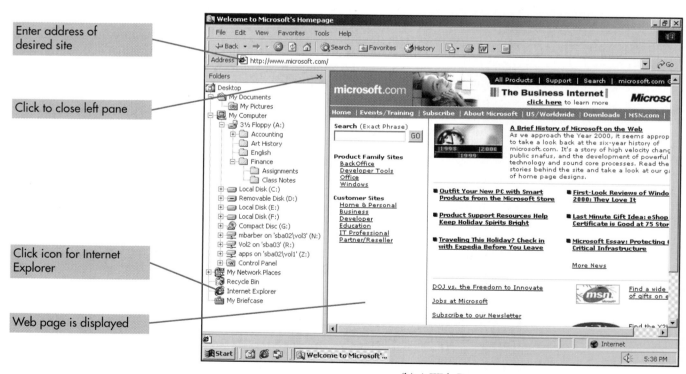

(b) A Web Page

FIGURE 10 *Windows Explorer*

Expanding and Collapsing a Drive or Folder

The tree diagram in Windows Explorer displays the devices on your system in hierarchical fashion. The desktop is always at the top of the hierarchy, and it contains icons such as My Computer, the Recycle Bin, Internet Explorer, and My Network Places. My Computer in turn contains the various drives that are accessible from your system, each of which contains folders, which in turn contain documents and/or additional folders. Each object may be expanded or collapsed by clicking the plus or minus sign, respectively. Click either sign to toggle to the other. Clicking a plus sign, for example, expands the drive, then displays a minus sign next to the drive to indicate that its subordinates are visible.

Look closely at the icon next to My Computer in either Figure 10a or 10b. It is a minus sign (as opposed to a plus sign) and it indicates that My Computer has been expanded to show the devices on the system. There is also a minus sign next to the icon for drive A to indicate that it too has been expanded to show the folders on the disk. Note, however, the plus sign next to drives C and D, indicating that these parts of the tree are currently collapsed and thus their subordinates (in this case, folders) are not visible.

Any folder may contain additional folders, and thus individual folders may also be expanded or collapsed. The minus sign next to the Finance folder, for example, indicates that the folder has been expanded and contains two additional folders, for Assignments and Class Notes, respectively. The plus sign next to the Accounting folder, however, indicates the opposite; that is, the folder is collapsed and its subordinate folders are not currently visible. A folder with neither a plus nor a minus sign, such as Art History, does not contain additional folders and cannot be expanded or collapsed.

The hierarchical view within Windows Explorer, and the ability to expand and collapse the various folders on a system, enables you to quickly locate a specific file or folder. If, for example, you want to see the contents of the Art History folder, you click its icon in the left pane, which automatically changes the display in the right pane to show the documents in that folder. Thus, Windows Explorer is ideal for moving or copying files from one folder or drive to another. You simply select (open) the folder that contains the files, use the scroll bar in the left pane (if necessary) so that the destination folder is visible, then click and drag the files from the right pane to the destination folder.

The Folder Options command functions identically in Windows Explorer and in My Computer. You can decide whether you want to single or double click the icons and/or whether to display Web content within a folder. You can also use the View menu to select the most appropriate view. Our preferences are to double click the icons, to omit Web content, and to use the Details view.

CONVERGENCE OF THE EXPLORERS

Windows Explorer and Internet Explorer are separate programs, but each includes some functionality of the other. You can use Windows Explorer to display a Web page by clicking the Internet Explorer icon within the tree structure in the left pane. Conversely, you can use Internet Explorer to display a local drive, document, or folder. Start Internet Explorer in the usual fashion, click in the Address bar, then enter the appropriate address, such as C:\ to display the contents of drive C.

HANDS-ON EXERCISE 3

THE PRACTICE FILES VIA A LOCAL AREA NETWORK

Objective To use Windows Explorer to copy the practice files from a network drive to a floppy disk. The exercise requires a formatted floppy disk and access to a local area network. Use Figure 11 as a guide in the exercise.

Step 1: **Start Windows Explorer**

➤ Click the **Start Button**, click **Programs**, click **Accessories**, then click **Windows Explorer**. Click the **maximize button** so that Windows Explorer takes the entire desktop as shown in Figure 11a. Do not be concerned if your desktop is different from ours.

➤ Make or verify the following selections using the **View menu**. You have to pull down the View menu each time you choose a different command.
  - The **Standard buttons** and **Address bar** toolbars should be selected.
  - The **Status Bar command** should be checked.
  - The **Details view** should be selected.

➤ Click (select) the **Desktop icon** in the left pane to display the contents of the desktop in the right pane. Your desktop may have different icons from ours, but your screen should almost match Figure 11a. We set additional options in the next step.

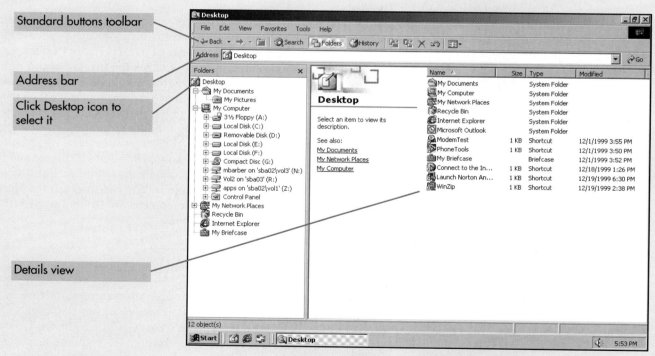

(a) Start Windows Explorer (step 1)

FIGURE 11 *Hands-on Exercise 3*

36 ESSENTIALS OF MICROSOFT WINDOWS

Step 2: **Change the Folder Options**

- Click the **minus** (or the **plus**) **sign** next to My Computer to collapse (or expand) My Computer and hide (or display) the objects it contains. Toggle the signs back and forth a few times for practice. End with a minus sign next to My Computer as shown in Figure 11b.
- Place a newly formatted floppy disk in drive A. Click the drive icon next to drive A to select the drive and display its contents in the right pane. The disk does not contain any files since zero bytes are used.
- Displaying Web content at the left of a folder (as is done in Figure 11b) is fine when a drive or folder does not contain a large number of files. It is generally a waste of space, however, and so we want to change the folder options.
- Pull down the **Tools menu** and click the **Folder Options command** to display the Folder Options dialog box in Figure 11a. Click the option to **Use Windows classic folders**. Click **OK**.

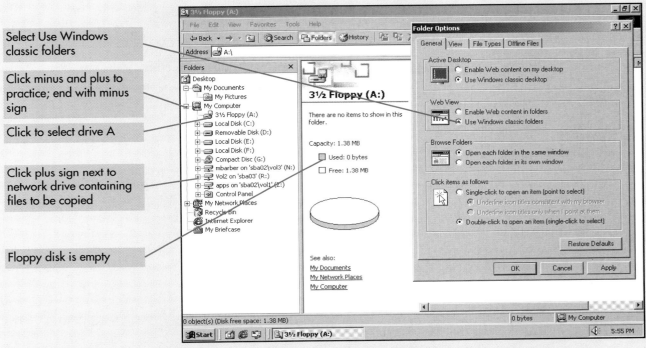

(b) Change the Folder Options (step 2)

FIGURE 11 *Hands-on Exercise 3 (continued)*

THE PLUS AND MINUS SIGN

Any drive, be it local or on the network, may be expanded or collapsed to display or hide its folders. A minus sign indicates that the drive has been expanded and that its folders are visible. A plus sign indicates the reverse; that is, the device is collapsed and its folders are not visible. Click either sign to toggle to the other. Clicking a plus sign, for example, expands the drive, then displays a minus sign next to the drive to indicate that the folders are visible. Clicking a minus sign has the reverse effect.

ESSENTIALS OF MICROSOFT WINDOWS   37

Step 3: **Select the Network Drive**

➤ Click the **plus sign** for the network drive that contains the files you are to copy (e.g., drive **R** in Figure 11c). Select (click) the **Exploring Windows 2000 folder** to open this folder.

➤ You may need to expand other folders on the network drive (such as the Datadisk folder on our network) as per instructions from your professor. Note the following:
  • The Exploring Windows 2000 folder is highlighted in the left pane, its icon is an open folder, and its contents are displayed in the right pane.
  • The status bar indicates that the folder contains five objects and the total file size is 119KB.

➤ Click the icon next to any other folder to select the folder, which in turn deselects the Exploring Windows 2000 folder. (Only one folder in the left pane is active at a time.) Reselect (click) the **Exploring Windows 2000 folder**.

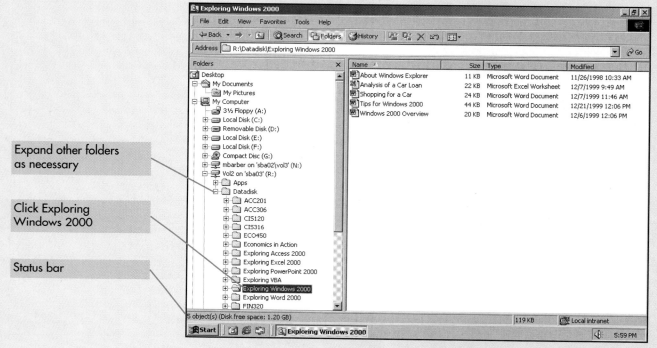

(c) Select the Network Drive (step 3)

FIGURE 11 *Hands-on Exercise 3 (continued)*

CUSTOMIZE WINDOWS EXPLORER

Increase or decrease the size of the left pane within Windows Explorer by dragging the vertical line separating the left and right panes in the appropriate direction. You can also drag the right border of the various column headings (Name, Size, Type, and Modified) in the right pane to increase or decrease the width of the column. And best of all, you can click any column heading to display the contents of the selected folder in sequence by that column. Click the heading a second time and the sequence changes from ascending to descending and vice versa.

Step 4: **Copy the Individual Files**

> Select (click) the file called **About Windows Explorer**, which highlights the file as shown in Figure 11d. Click and drag the selected file in the right pane to the **drive A icon** in the left pane:
> - You will see the ⊘ symbol as you drag the file until you reach a suitable destination (e.g., until you point to the icon for drive A). The ⊘ symbol will change to a plus sign when the icon for drive A is highlighted, indicating that the file can be copied successfully.
> - Release the mouse to complete the copy operation. You will see a pop-up window, which indicates the status of the copy operation.
>
> Select (click) the file **Tips for Windows 2000**, which automatically deselects the previously selected file. Copy the selected file to drive A by dragging its icon from the right pane to the drive A icon in the left pane.
>
> Copy the three remaining files to drive A as well. Select (click) drive **A** in the left pane, which in turn displays the contents of the floppy disk in the right pane. You should see the five files you have copied to drive A.

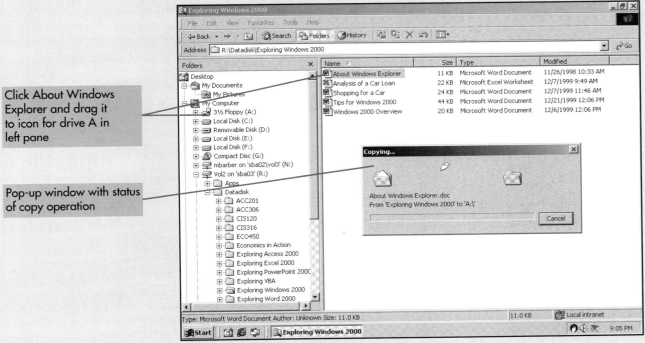

(d) Copy the Individual Files (step 4)

FIGURE 11 *Hands-on Exercise 3 (continued)*

SELECT MULTIPLE FILES

Selecting one file automatically deselects the previously selected file. You can, however, select multiple files by clicking the first file, then pressing and holding the Ctrl key as you click each additional file. Use the Shift key to select multiple files that are adjacent to one another by clicking the first file, then pressing and holding the Shift key as you click the last file.

ESSENTIALS OF MICROSOFT WINDOWS

Step 5: **Display a Web Page**

➤ This step requires an Internet connection. Click the **minus sign** next to the network drive to collapse that drive. Click the **minus sign** next to any other expanded drive so that the left pane is similar to Figure 11e.

➤ Click the **Internet Explorer icon** to start Internet Explorer and display the starting page for your configuration. The page you see will be different from ours, but you can click in the Address bar near the top of the window to enter the address of any Web site.

➤ Look closely at the icons on the toolbar, which have changed to reflect the tools associated with viewing a Web page. Click the **Back button** to return to drive A, the previously displayed item in Windows Explorer. The icons on the toolbar return to those associated with a folder.

➤ Close Windows Explorer. Shut down the computer if you do not want to continue with the next exercise at this time.

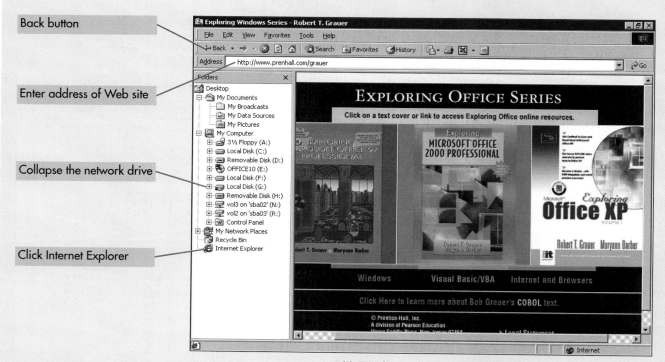

(e) Display a Web Page (step 5)

FIGURE 11   *Hands-on Exercise 3 (continued)*

SERVER NOT RESPONDING

Two things have to occur in order for Internet Explorer to display the requested document—it must locate the server on which the document is stored, and it must be able to connect to that computer. If you see a message similar to "Server too busy or not responding", it implies that Internet Explorer has located the server but was unable to connect because the site is busy or is temporarily down. Try to connect again, in a minute or so, or later in the day.

THE BASICS OF FILE MANAGEMENT

As you grow to depend on the computer, you will create a variety of files using applications such as Microsoft Word or Excel. Learning how to manage those files is one of the most important skills you can acquire. The previous hands-on exercises provided you with a set of files with which to practice. That way, when you have your own files you will be comfortable executing the various file management commands you will need on a daily basis. This section describes the basic file operations you will need, then presents another hands-on exercise in which you apply those commands.

Moving and Copying a File

The essence of file management is to move and copy a file or folder from one location to another. This can be done in different ways, most easily by clicking and dragging the file icon from the source drive or folder to the destination drive or folder, within Windows Explorer. There is one subtlety, however, in that the result of dragging a file (i.e., whether the file is moved or copied) depends on whether the source and destination are on the same or different drives. Dragging a file from one folder to another folder on the same drive moves the file. Dragging a file to a folder on a different drive copies the file. The same rules apply to dragging a folder, where the folder and every file in it are moved or copied as per the rules for an individual file.

This process is not as arbitrary as it may seem. Windows assumes that if you drag an object (a file or folder) to a different drive (e.g., from drive C to drive A), you want the object to appear in both places. Hence, the default action when you click and drag an object to a different drive is to copy the object. You can, however, override the default and move the object by pressing and holding the Shift key as you drag.

Windows also assumes that you do not want two copies of an object on the same drive, as that would result in wasted disk space. Thus, the default action when you click and drag an object to a different folder on the same drive is to move the object. You can override the default and copy the object by pressing and holding the Ctrl key as you drag. It's not as complicated as it sounds, and you get a chance to practice in the hands-on exercise, which follows shortly.

Deleting a File

The ***Delete command*** deletes (erases) a file from a disk. The command can be executed in different ways, most easily by selecting a file, then pressing the Del key. It's also comforting to know that you can usually recover a deleted file, because the file is not (initially) removed from the disk, but moved instead to the Recycle Bin, from where it can be restored to its original location. Unfortunately, files deleted from a floppy disk are not put in the Recycle Bin and hence cannot be recovered.

The ***Recycle Bin*** is a special folder that contains all files that were previously deleted from any hard disk on your system. Think of the Recycle Bin as similar to the wastebasket in your room. You throw out (delete) a report by tossing it into a wastebasket. The report is gone (deleted) from your desk, but you can still get it back by taking it out of the wastebasket as long as the basket wasn't emptied. The Recycle Bin works the same way. Files are not deleted from the hard disk per se, but moved instead to the Recycle Bin from where they can be restored to their original location.

The Recycle Bin will eventually run out of space, in which case the files that have been in the Recycle Bin the longest are permanently deleted to make room for additional files. Accordingly, once a file is removed from the Recycle Bin it can no longer be restored, as it has been physically deleted from the hard disk. Note, too, that the protection afforded by the Recycle Bin does not extend to files deleted from a floppy disk. Such files can be recovered, but only through utility programs outside of Windows 2000.

Renaming a File

Every file or folder is assigned a name at the time it is created, but you may want to change that name at some point in the future. Point to a file or a folder, click the right mouse button to display a menu with commands pertaining to the object, then click the **Rename command**. The name of the file or folder will be highlighted with the insertion point (a flashing vertical line) positioned at the end of the name. Enter a new name to replace the selected name, or click anywhere within the name to change the insertion point and edit the name.

Backup

It's not a question of if it will happen, but when—hard disks die, files are lost, or viruses may infect a system. It has happened to us and it will happen to you, but you can prepare for the inevitable by creating adequate backup *before* the problem occurs. The essence of a **backup strategy** is to decide which files to back up, how often to do the backup, and where to keep the backup. Once you decide on a strategy, follow it, and follow it faithfully!

Our strategy is very simple—back up what you can't afford to lose, do so on a daily basis, and store the backup away from your computer. You need not copy every file, every day. Instead, copy just the files that changed during the current session. Realize, too, that it is much more important to back up your data files than your program files. You can always reinstall the application from the original disks or CD, or if necessary, go to the vendor for another copy of an application. You, however, are the only one who has a copy of your term paper.

Write Protection

A floppy disk is normally **write-enabled** (the square hole is covered with the movable tab) so that you can change the contents of the disk. Thus, you can create (save) new files to a write-enabled disk and/or edit or delete existing files. Occasionally, however, you may want to **write-protect** a floppy disk (by sliding the tab to expose the square hole) so that its contents cannot be modified. This is typically done with a backup disk where you want to prevent the accidental deletion of a file and/or the threat of virus infection.

Our Next Exercise

Our next exercise begins with the floppy disk containing the five practice files in drive A. We ask you to create two folders on drive A (step 1) and to move the various files into these folders (step 2). Next, you copy a folder from drive A to the My Documents folder (step 3), modify one of the files in the My Documents folder (step 4), then copy the modified file back to drive A (step 5). We ask you to delete a file in step 6, then recover it from the Recycle Bin in step 7. We also show you how to write-protect a floppy disk in step 8. Let's get started.

HANDS-ON EXERCISE 4

FILE MANAGEMENT

Objective Use Windows Explorer to move, copy, and delete a file; recover a deleted file from the Recycle Bin; write-protect a floppy disk. Use Figure 12 as a guide in the exercise.

Step 1: **Create a New Folder**

> ➤ Start Windows Explorer, maximize its window, and if necessary, change to **Details view**. Place the floppy disk from Exercise 2 or 3 in drive A.
> ➤ Select (click) the icon for **drive A** in the left pane of the Explorer window. Drive A should contain the files shown in Figure 12a.
> ➤ You will create two folders on drive A, using two different techniques:
> - Point to a blank area anywhere in the **right pane**, click the **right mouse button** to display a context-sensitive menu, click (or point to) the **New command**, then click **Folder** as the type of object to create.
> - The icon for a new folder will appear with the name of the folder (New Folder) highlighted. Type **John Doe's Documents** (use your own name) to change the name of the folder. Press **Enter**.
> - Click the icon for **drive A** in the left pane. Pull down the **File menu**, click (or point to) the **New command**, and click **Folder** as the type of object to create. Type **Automobile** to change the name of the folder. Press **Enter**. The right pane should now contain five documents and two folders.
> ➤ Pull down the **View menu**. Click the **Arrange icons command**, then click the **By Name command** to display the folders in alphabetical order.

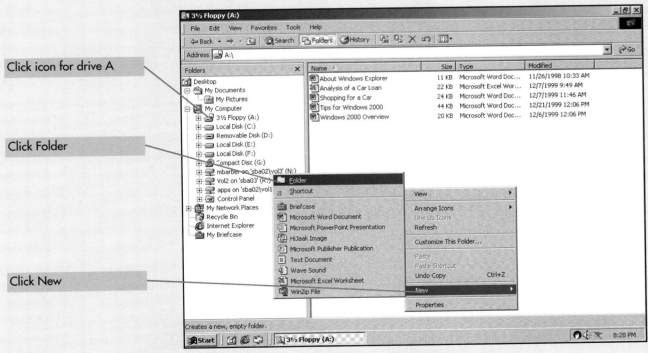

(a) Create a New Folder (step 1)

FIGURE 12 *Hands-on Exercise 4*

Step 2: **Move a File**

➤ Click the **plus sign** next to drive A to expand the drive as shown in Figure 12b. Note the following:
  - The left pane shows that drive A is selected. The right pane displays the contents of drive A (the selected object in the left pane).
  - There is a minus sign next to the icon for drive A in the left pane, indicating that it has been expanded and that its folders are visible. Thus, the folder names also appear under drive A in the left pane.
➤ Click and drag the icon for the file **About Windows Explorer** from the right pane, to the **John Doe's Documents folder** in the left pane, to move the file into that folder.
➤ Click and drag the **Tips for Windows 2000** and **Windows 2000 Overview** documents to the **John Doe's Documents folder** in similar fashion.
➤ Click the **John Doe's Documents folder** in the left pane to select the folder and display its contents in the right pane. You should see the three files that were just moved.
➤ Click the icon for **Drive A** in the left pane, then click and drag the remaining files, **Analysis of a Car** and **Shopping for a Car**, to the **Automobile folder**.

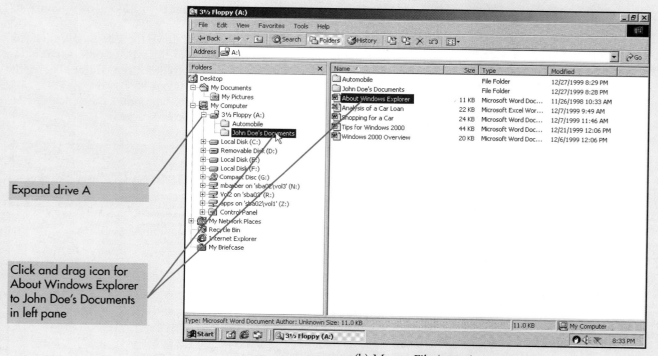

(b) Move a File (step 2)

FIGURE 12 *Hands-on Exercise 4*

RIGHT CLICK AND DRAG

Click and drag with the right mouse button to display a shortcut menu asking whether you want to copy or move the file. This simple tip can save you from making a careless (and potentially serious) error. Use it!

Step 3: **Copy a Folder**

➤ Point to **John Doe's Documents folder** in either pane, click the **right mouse button**, and drag the folder to the **My Documents folder** in the left pane, then release the mouse to display a shortcut menu. Click the **Copy Here command**.
- You may see a Copy files message box as the individual files within John Doe's folder are copied to the My Documents folder.
- If you see the Confirm Folder Replace dialog box, it means that you already copied the files or a previous student used the same folder when he or she did this exercise. Click the **Yes to All button** so that your files replace the previous versions in the My Documents folder.

➤ Click the **My Documents folder** in the left pane. Pull down the **View menu** and click the **Refresh command** (or press the **F5 key**) so that the tree structure shows the newly copied folder. (Please remember to delete John Doe's Documents folder at the end of the exercise.)

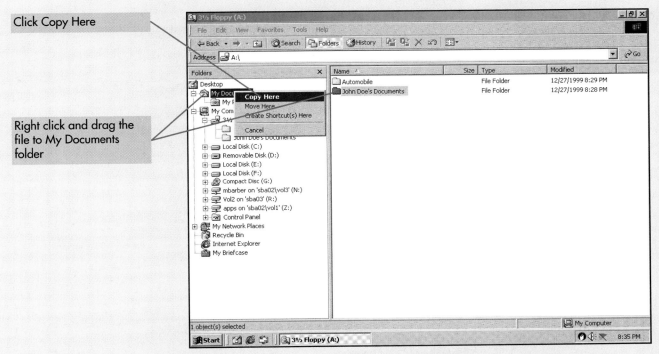

(c) Copy a Folder (step 3)

FIGURE 12   *Hands-on Exercise 4 (continued)*

THE MY DOCUMENTS FOLDER

The My Documents folder is created by default with the installation of Microsoft Windows. There is no requirement that you store your documents in this folder, but it is convenient, especially for beginners who may lack the confidence to create their own folders. The My Documents folder is also helpful in a laboratory environment where the network administrator may prevent you from modifying the desktop and/or from creating your own folders on drive C, in which case you will have to use the My Documents folder.

Step 4: **Modify a Document**

- Click **John Doe's Documents folder** within the My Documents folder to make it the active folder and to display its contents in the right pane. Change to the **Details view**.
- Double click the **About Windows Explorer** document to start Word and open the document. Do not be concerned if the size and/or position of the Microsoft Word window are different from ours.
- If necessary, click inside the document window, then press **Ctrl+End** to move to the end of the document. Add the sentence shown in Figure 12d.
- Pull down the **File menu** and click **Save** to save the modified file (or click the **Save button** on the Standard toolbar). Pull down the **File menu** and click **Exit**.
- Pull down the **View menu** in Windows Explorer and click **Refresh** (or press the **F5 key**) to update the contents of the right pane. The date and time associated with the About Windows Explorer file has been changed to indicate that the file has just been modified.

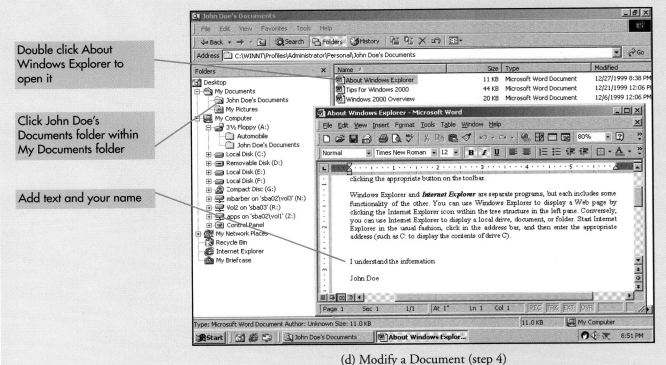

(d) Modify a Document (step 4)

FIGURE 12 *Hands-on Exercise 4 (continued)*

KEYBOARD SHORTCUTS

Ctrl+B, Ctrl+I, and Ctrl+U are shortcuts to boldface, italicize, and underline, respectively. Ctrl+X (the X is supposed to remind you of a pair of scissors), Ctrl+C, and Ctrl+V correspond to Cut, Copy, and Paste, respectively. Ctrl+Home and Ctrl+End move to the beginning or end of a document. These shortcuts are not unique to Microsoft Word, but are recognized in virtually every Windows application. See practice exercise 11 at the end of the chapter.

Step 5: **Copy (Back Up) a File**

➤ Verify that **John Doe's folder** within My Documents is the active folder, as denoted by the open folder icon. Click and drag the icon for the **About Windows Explorer** file from the right pane to John Doe's Documents folder on **Drive A** in the left pane.

➤ You will see the message in Figure 12e, indicating that the folder (on drive A) already contains a file called About Windows Explorer and asking whether you want to replace the existing file. Click **Yes** because you want to replace the previous version of the file on drive A with the updated version from the My Documents folder.

➤ You have just backed up the file; in other words, you have created a copy of the file on the disk in drive A. Thus, you can use the floppy disk to restore the file in the My Documents folder should anything happen to it.

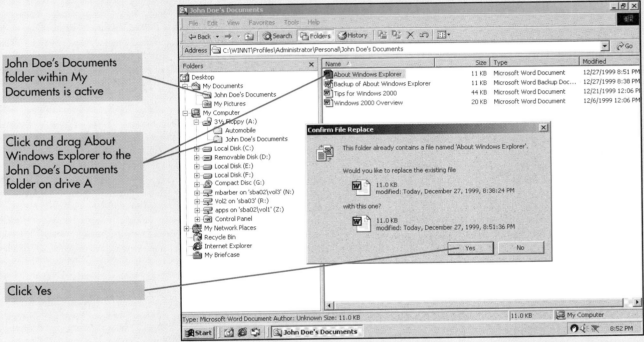

(e) Copy (Back Up) a File (step 5)

FIGURE 12 *Hands-on Exercise 4 (continued)*

FILE EXTENSIONS

Long-time DOS users remember a three-character extension at the end of a file name to indicate the file type; for example, DOC or XLS to indicate a Word document or Excel workbook, respectively. The extensions are displayed or hidden according to a setting in the Folder Options command. Pull down the Tools menu, click the Folder Options command to display the Folder Options dialog box, click the View tab, then check (or clear) the box to hide (or show) file extensions for known file types. Click OK to accept the setting and exit the dialog box.

Step 6: **Delete a Folder**

> ➤ Select (click) **John Doe's Documents folder** within the My Documents folder in the left pane. Pull down the **File menu** and click **Delete** (or press the **Del key**).
> ➤ You will see the dialog box in Figure 12f asking whether you are sure you want to delete the folder (i.e., send the folder and its contents to the Recycle Bin). Note the recycle logo within the box, which implies that you will be able to restore the file.
> ➤ Click **Yes** to delete the folder. The folder disappears from drive C. Pull down the **Edit menu**. Click **Undo Delete**. The deletion is cancelled and the folder reappears in the left pane. If you don't see the folder, pull down the **View menu** and click the **Refresh command**.

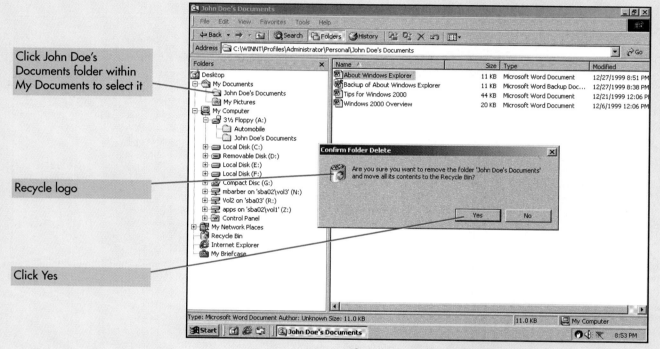

(f) Delete a Folder (step 6)

FIGURE 12 *Hands-on Exercise 4 (continued)*

THE UNDO COMMAND

The Undo command is present not only in application programs such as Word or Excel, but in Windows Explorer as well. You can use the Undo command to undelete a file provided you execute the command immediately (within a few commands) after the Delete command. To execute the Undo command, right-click anywhere in the right pane to display a shortcut menu, then select the Undo action. You can also pull down the Edit menu and click Undo to reverse (undo) the last command. Some operations cannot be undone (in which case the command will be dimmed), but Undo is always worth a try.

Step 7: **The Recycle Bin**

➤ Select John Doe's Documents folder within the My Documents folder in the left pane. Select (click) the **About Windows Explorer** file in the right pane. Press the **Del key**, then click **Yes**.

➤ Click the **Down arrow** in the vertical scroll bar in the left pane until you see the icon for the **Recycle Bin**. Click the icon to make the Recycle Bin the active folder and display its contents in the right pane.

➤ You will see a different set of files from those displayed in Figure 12g. Pull down the **View menu**, click (or point to) **Arrange icons**, then click **By Delete Date** to display the files in this sequence.

➤ Click in the **right pane**. Press **Ctrl+End** or scroll to the bottom of the window. Point to the **About Windows Explorer** file, click the **right mouse button** to display the shortcut menu in Figure 12g, then click **Restore**.

➤ The file disappears from the Recycle bin because it has been returned to John Doe's Documents folder.

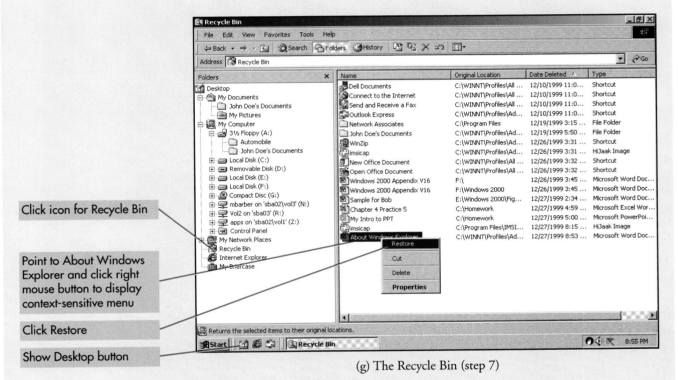

(g) The Recycle Bin (step 7)

FIGURE 12   *Hands-on Exercise 4 (continued)*

THE SHOW DESKTOP BUTTON

The Show Desktop button on the taskbar enables you to minimize all open windows with a single click. The button functions as a toggle switch. Click it once and all windows are minimized. Click it a second time and the open windows are restored to their positions on the desktop. If you do not see the Show Desktop button, right click a blank area of the taskbar to display a context-sensitive menu, click Toolbars, then check the Quick Launch toolbar.

Step 8: **Write-Protect a Floppy Disk**

- Remove the floppy disk from drive A, then move the built-in tab on the disk so that the square hole on the disk is open. Return the disk to the drive.
- If necessary, expand drive A in the left pane, select the **Automobile folder**, select the **Analysis of a Car Loan document** in the right pane, then press the **Del key**. Click **Yes** when asked whether to delete the file.
- You will see the message in Figure 12h indicating that the file cannot be deleted because the disk has been write-protected. Click **OK**. Remove the write-protection by moving the built-in tab to cover the square hole.
- Repeat the procedure to delete the **Analysis of a Car Loan document**. Click **Yes** in response to the confirmation message asking whether you want to delete the file.
- The file disappears from the right pane, indicating it has been deleted. The **Automobile folder** on drive A should contain only one file.
- Delete **John Doe's Documents folder** from My Documents as a courtesy to the next student. Exit Windows Explorer. Shut down the computer.

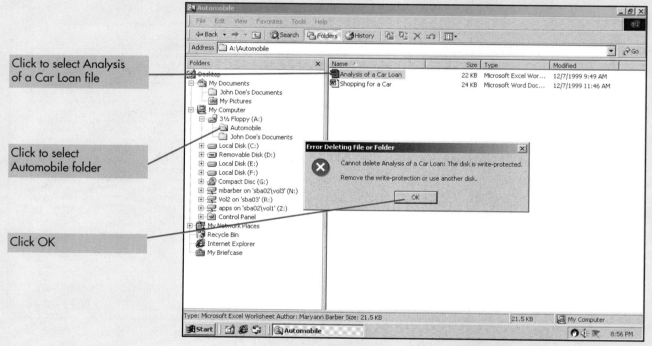

(h) Write-Protect a Floppy Disk (step 8)

FIGURE 12 *Hands-on Exercise 4 (continued)*

BACK UP IMPORTANT FILES

We cannot overemphasize the importance of adequate backup and urge you to copy your data files to floppy disks and store those disks away from your computer. You might also want to write-protect your backup disks so that you cannot accidentally erase a file. It takes only a few minutes, but you will thank us, when (not if) you lose an important file and don't have to wish you had another copy.

SUMMARY

Microsoft Windows controls the operation of a computer and its peripherals. Windows 98 and its successor, Windows Me, are geared for the home user and provide extensive support for games and peripheral devices. Windows NT and its successor, Windows 2000, are aimed at the business user and provide increased security and reliability. Windows XP replaces all current versions of Windows. All versions of Windows follow the same conventions and have the same basic interface.

All Windows operations take place on the desktop. Every window on the desktop contains the same basic elements, which include a title bar, a control-menu box, a minimize button, a maximize or restore button, and a close button. Other elements that may be present include a menu bar, vertical and/or horizontal scroll bars, a status bar, and various toolbars. All windows may be moved and sized. The Help command in the Start menu provides access to detailed information.

Multitasking is a major benefit of the Windows environment as it enables you to run several programs at the same time. The taskbar contains a button for each open program and enables you to switch back and forth between those programs by clicking the appropriate button.

A dialog box supplies information needed to execute a command. Option buttons indicate mutually exclusive choices, one of which must be chosen. Check boxes are used if the choices are not mutually exclusive or if an option is not required. A text box supplies descriptive information. A (drop-down or open) list box displays multiple choices, any of which may be selected. A tabbed dialog box provides access to multiple sets of options.

A floppy disk must be formatted before it can store data. Formatting is accomplished through the Format command within the My Computer window. My Computer enables you to browse the disk drives and other devices attached to your system. The contents of My Computer depend on the specific configuration.

A file is a set of data or set of instructions that has been given a name and stored on disk. There are two basic types of files, program files and data files. A program file is an executable file, whereas a data file can be used only in conjunction with a specific program. Every file has a file name and a file type. The file name can be up to 255 characters in length and may include spaces.

Files are stored in folders to better organize the hundreds (or thousands) of files on a disk. A folder may contain program files, data files, and/or other folders. There are two basic ways to search through the folders on your system, My Computer and Windows Explorer. My Computer is intuitive but less efficient than Windows Explorer, as you have to open each folder in succession. Windows Explorer is more sophisticated, as it provides a hierarchical view of the entire system.

Windows Explorer is divided into two panes. The left pane displays all of the devices and, optionally, the folders on each device. The right pane shows the contents of the active (open) drive or folder. Only one drive or folder can be active in the left pane. Any device, be it local or on the network, may be expanded or collapsed to display or hide its folders. A minus sign indicates that the drive has been expanded and that its folders are visible. A plus sign indicates that the device is collapsed and its folders are not visible.

The result of dragging a file (or folder) from one location to another depends on whether the source and destination folders are on the same or different drives. Dragging the file to a folder on the same drive moves the file. Dragging the file to a folder on a different drive copies the file. It's easier, therefore, to click and drag with the right mouse button to display a context-sensitive menu from which you can select the desired operation.

The Delete command deletes (removes) a file from a disk. If, however, the file was deleted from a local (fixed or hard) disk, it is not really gone, but moved instead to the Recycle Bin from where it can be subsequently recovered.

KEY TERMS

Backup strategy (p. 42)
Check box (p. 8)
Close button (p. 5)
Command button (p. 10)
Common user interface (p. 5)
Compressed file (p. 22)
Contents tab (p. 11)
Copy a file (p. 47)
Data file (p. 20)
Delete a file (p. 41)
Desktop (p. 2)
Details view (p. 20)
Dialog box (p. 8)
Favorites tab (p. 18)
File (p. 20)
Filename (p. 20)
File type (p. 20)
Folder (p. 20)
Folder Options command (p. 14)
Format command (p. 17)
Help command (p. 18)

Index tab (p. 14)
Internet Explorer (p. 40)
List box (p. 8)
Maximize button (p. 5)
Menu bar (p. 5)
Minimize button (p. 5)
Mouse operations (p. 10)
Move a file (p. 44)
Move a window (p. 15)
Multitasking (p. 4)
My Computer (p. 22)
My Documents folder (p. 45)
My Network Places (p. 5)
New command (p. 43)
Option button (p. 8)
Program file (p. 20)
Pull-down menu (p. 7)
Radio button (p. 8)
Recycle Bin (p. 49)
Rename command (p. 42)
Restore a file (p. 5)

Restore button (p. 5)
Scroll bar (p. 5)
Size a window (p. 15)
Spin button (p. 8)
Start button (p. 4)
Status bar (p. 5)
Taskbar (p. 4)
Text box (p. 8)
Task Manager (p. 19)
Title bar (p. 5)
Toolbar (p. 5)
Undo command (p. 48)
Windows 2000 (p. 2)
Windows 95 (p. 2)
Windows 98 (p. 2)
Windows Explorer (p. 33)
Windows Me (p. 2)
Windows NT (p. 2)
Windows XP (p. 2)
Write-enabled (p. 42)
Write-protected (p. 42)

MULTIPLE CHOICE

1. Which versions of the Windows operating system were intended for the home computer?
   (a) Windows NT and Windows 98
   (b) Windows NT and Windows XP
   (c) Windows NT and Windows 2000
   (d) Windows 98 and Windows Me

2. What happens if you click and drag a file from drive C to drive A?
   (a) The file is copied to drive A
   (b) The file is moved to drive A
   (c) A menu appears that allows you to choose between moving and copying
   (d) The file is sent to the recycle bin

3. Which of the following is *not* controlled by the Folder Options command?
   (a) Single or double clicking to open a desktop icon
   (b) The presence or absence of Web content within a folder
   (c) The view (e.g., using large or small icons) within My Computer
   (d) Using one or many windows when browsing My Computer

4. What is the significance of a faded (dimmed) command in a pull-down menu?
   (a) The command is not currently accessible
   (b) A dialog box will appear if the command is selected
   (c) A Help window will appear if the command is selected
   (d) There are no equivalent keystrokes for the particular command

5. Which of the following is true regarding a dialog box?
   (a) Option buttons indicate mutually exclusive choices
   (b) Check boxes imply that multiple options may be selected
   (c) Both (a) and (b)
   (d) Neither (a) nor (b)

6. Which of the following is the first step in sizing a window?
   (a) Point to the title bar
   (b) Pull down the View menu to display the toolbar
   (c) Point to any corner or border
   (d) Pull down the View menu and change to large icons

7. Which of the following is the first step in moving a window?
   (a) Point to the title bar
   (b) Pull down the View menu to display the toolbar
   (c) Point to any corner or border
   (d) Pull down the View menu and change to large icons

8. How do you exit from Windows?
   (a) Click the Start button, then click the Shut Down command
   (b) Right click the Start button, then click the Shut Down command
   (c) Click the End button, then click the Shut Down command
   (d) Right click the End button, then click the Shut Down command

9. Which button appears immediately after a window has been maximized?
   (a) The close button
   (b) The minimize button
   (c) The maximize button
   (d) The restore button

10. What happens to a window that has been minimized?
    (a) The window is still visible but it no longer has a minimize button
    (b) The window shrinks to a button on the taskbar
    (c) The window is closed and the application is removed from memory
    (d) The window is still open but the application is gone from memory

11. What is the significance of three dots next to a command in a pull-down menu?
    (a) The command is not currently accessible
    (b) A dialog box will appear if the command is selected
    (c) A Help window will appear if the command is selected
    (d) There are no equivalent keystrokes for the particular command

12. The Recycle Bin enables you to restore a file that was deleted from:
    (a) Drive A
    (b) Drive C
    (c) Both (a) and (b)
    (d) Neither (a) nor (b)

13. The left pane of Windows Explorer may contain:
    (a) One or more folders with a plus sign
    (b) One or more folders with a minus sign
    (c) Both (a) and (b)
    (d) Neither (a) nor (b)

14. Which of the following was suggested as essential to a backup strategy?
    (a) Back up all program files at the end of every session
    (b) Store backup files at another location
    (c) Both (a) and (b)
    (d) Neither (a) nor (b)

ANSWERS

1. d   5. c   9. d   13. c
2. a   6. c   10. b   14. b
3. c   7. a   11. b
4. a   8. a   12. b

PRACTICE WITH WINDOWS

1. **My Computer:** The document in Figure 13 is an effective way to show your instructor that you understand the My Computer window, and further that you have basic proficiency in Microsoft Word.
   a. Open My Computer to display the contents of your configuration. Pull down the View menu and switch to the Details view. Size the window as necessary. Press Alt + Print Screen to capture the copy of the My Computer window to the Windows clipboard. (The Print Screen key captures the entire screen. Using the Alt key, however, copies just the current window.)
   b. Click the Start menu, click Programs, then click Microsoft Word.
   c. Pull down the Edit menu. Click the Paste command to copy the contents of the clipboard to the document you are about to create. The My Computer window should be pasted into your document.
   d. Press Ctrl+End to move to the end of your document. Press Enter two or three times to leave blank lines as appropriate. Type a modified form of the memo in Figure 13 so that it conforms to your configuration.
   e. Finish the memo and sign your name. Pull down the File menu, click the Print command, then click OK in the dialog box to print the document.

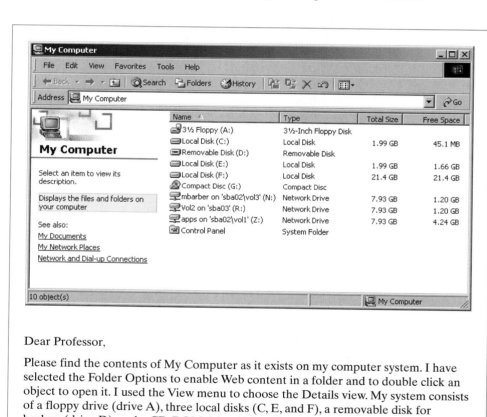

FIGURE 13   *My Computer (exercise 1)*

2. **Windows Explorer:** Prove to your instructor that you have completed the fourth hands-on exercise by creating a document similar to the one in Figure 14. Use the technique described in the previous problem to capture the screen and paste it into a Word document.

   Compare the documents in Figures 13 and 14 that show My Computer and Windows Explorer, respectively. My Computer is intuitive and preferred by beginners, but it is very limited when compared to Windows Explorer. The latter displays a hierarchical view of your system, showing the selected object in the left pane and the contents of the selected object in the right pane. We urge you, therefore, to become comfortable with Windows Explorer, as that will make you more productive.

Dear Professor,

This document shows that I have completed the last hands-on exercise and that I have a basic proficiency in Windows Explorer. I can now create a folder and move and copy files with confidence. These skills will become increasingly important as I use various applications throughout the semester.

I also understand the role of the Windows clipboard, an area of memory that is available to every application. I simply press the Print Screen key to capture a screen to the clipboard, and then I use the Paste button to copy the contents of the clipboard into the Word document.

Sincerely,

Alex Rogers

FIGURE 14 *Windows Explorer (exercise 2)*

3. **MyPHLIP Web Site:** Every text in the *Exploring Office XP* series has a corresponding MyPHLIP (Prentice Hall Learning on the Internet Partnership) Web site, where you will find a variety of student resources as well as online review questions for each chapter. Go to www.prenhall.com/myphlip and follow the instructions. The first time at the site you will be prompted to register by supplying your e-mail address and choosing a password. Next, you choose the discipline (CIS/MIS) and a book (e.g., *Exploring Microsoft Office XP, Volume I*), which in turn will take you to a page similar to Figure 15.

   Your professor will tell you whether he or she has created an online syllabus, in which case you should click the link to find your professor after adding the book. Either way, the next time you return to the site, you will be taken directly to your text. Select any chapter, click "Go", then use the review questions as directed.

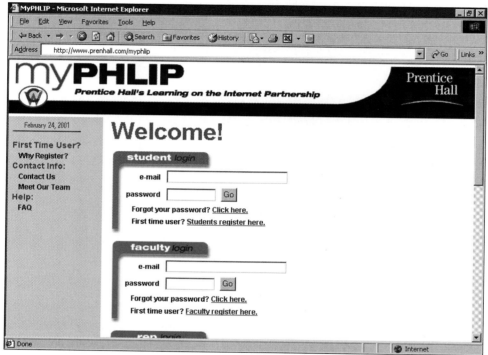

FIGURE 15 *MyPHLIP Web Site (Windows module) (exercise 3)*

4. **Organize Your Work:** A folder may contain documents, programs, or other folders. The My Classes folder in Figure 16, for example, contains five folders, one folder for each class you are taking this semester. Folders help you to organize your files, and you should become proficient in their use. The best way to practice with folders is on a floppy disk, as was done in Figure 16. Accordingly:
   a. Format a floppy disk or use the floppy disk you have been using throughout the chapter.
   b. Create a Correspondence folder. Create a Business and a Personal folder within the Correspondence folder.
   c. Create a My Courses folder. Within the My Courses folder create a separate folder for each course you are taking.
   d. Use the technique described in problems 1 and 2 to capture the screen in Figure 16 and incorporate it into a document. Add a short paragraph that describes the folders you have created, then submit the document.

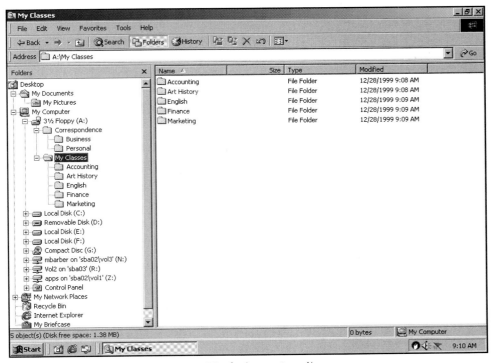

FIGURE 16 *Organize Your Work (exercise 4)*

5. **The Windows Web Site:** The Web is the best source for information on any application. Go to the Windows home page (www.microsoft.com/windows) as shown in Figure 17, then write a short note to your instructor summarizing the contents of that page and the associated links. Similar pages exist for all Microsoft applications such as www.microsoft.com/office for Microsoft Office.

6. **Implement a Screen Saver:** A screen saver is a delightful way to personalize your computer and a good way to practice with Microsoft Windows. This is typically not something you can do in a laboratory setting, but it is well worth doing on your own machine. Point to a blank area of the desktop, click the right mouse button to display a context-sensitive menu, then click the Properties command to open the Display Properties dialog box in Figure 18. Click the Screen Saver tab, click the Down arrow in the Screen Saver list box, and select Marquee Display. Click the Settings command button, enter the text and other options for your message, then click OK to close the Options dialog box. Click OK a second time to close the Display Properties dialog box.

7. **The Active Desktop:** The Active Desktop displays Web content directly on the desktop, then updates the information automatically according to a predefined schedule. You can, for example, display a stock ticker or scoreboard similar to what you see on television. You will need your own machine and an Internet connection to do this exercise, as it is unlikely that the network administrator will let you modify the desktop:
   a. Right click the Windows desktop, click Properties to show the Display Properties dialog box, then click the Web tab. Check the box to show Web content on the Active desktop.
   b. Click the New button, then click the Visit Gallery command button to go to the Active Desktop Gallery in Figure 19 on page 59. Choose any category, then follow the onscreen instructions to display the item on your desktop. We suggest you start with the stock ticker or sports scoreboard.
   c. Summarize your opinion of the active desktop in a short note to your instructor. Did the feature work as advertised? Is the information useful to you?

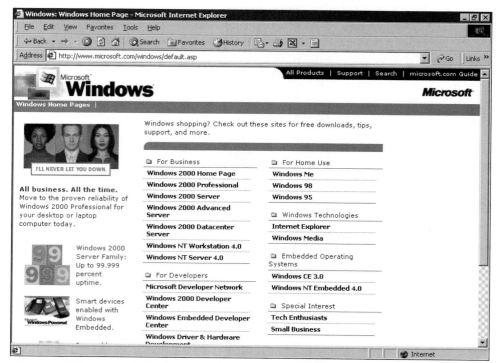

FIGURE 17   *The Windows Web Site (exercise 5)*

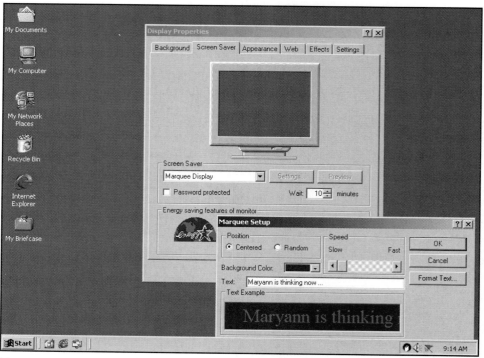

FIGURE 18   *Implement a Screen Saver (exercise 6)*

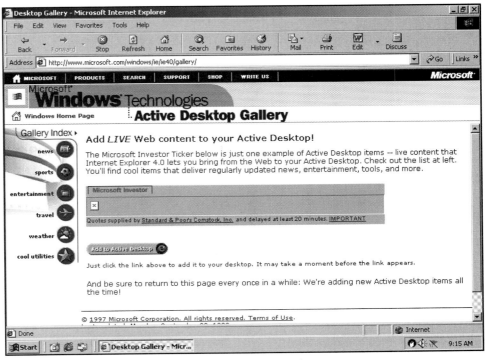

FIGURE 19   *The Active Desktop (exercise 7)*

8. The Control Panel: The Control Panel enables you to change the hardware or software settings on your system. You will not have access to the Control Panel in a lab environment, but you will need it at home if you change your configuration, perhaps by installing a new program. Click the Start button, click Settings, then select Control Panel to display the Control Panel window. Click the down arrow on the Views button to change to the Details view as shown in Figure 20. (The Control Panel can also be opened from My Computer.)

    Write a short report (two or three paragraphs is sufficient) that describes some of the capabilities within Control Panel. *Be careful about making changes, however, and be sure you understand the nature of the new settings before you accept any of the changes.*

9. Users and Passwords: Windows 2000 enables multiple users to log onto the same machine, each with his or her own user name and password. The desktop settings for each user are stored individually, so that all users have their own desktop. The administrator and default user is created when Windows 2000 is first installed, but new users can be added or removed at any time. Once again you will need your own machine:
    a. Click the Start button, click Settings, then click Control Panel to open the Control Panel window as shown in Figure 21. The Control Panel is a special folder that allows you to modify the hardware and/or software settings on your computer.
    b. Double click the Users and Passwords icon to display the dialog box in Figure 20. *Be very careful about removing a user or changing a password, because you might inadvertently deny yourself access to your computer.*
    c. Summarize the capabilities within the users and passwords dialog box in a short note to your instructor. Can you see how these principles apply to the network you use at school or work?

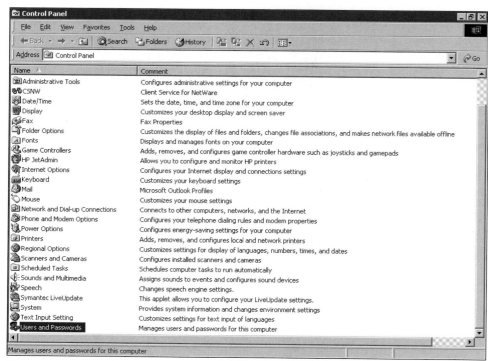

FIGURE 20 *The Control Panel (exercise 8)*

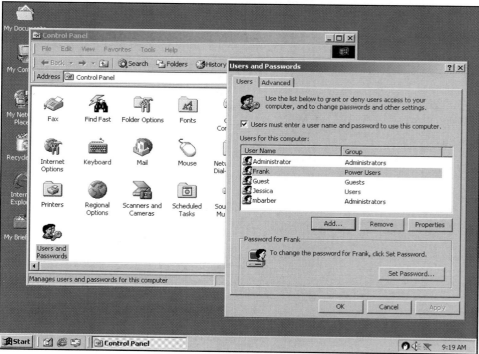

FIGURE 21 *Users and Passwords (exercise 9)*

10. The Fonts Folder: The Fonts folder within the Control Panel displays the names of the fonts available on a system and enables you to obtain a printed sample of any specific font. Click the Start button, click (or point to) the Settings command, click (or point to) Control Panel, then double click the Fonts icon to open the Fonts folder and display the fonts on your system.
    a. Double click any font to open a Fonts window as shown in Figure 22, then click the Print button to print a sample of the selected font.
    b. Open a different font. Print a sample page of this font as well.
    c. Locate the Wingdings font and print this page. Do you see any symbols you recognize? How do you insert these symbols into a document?
    d. How many fonts are there in your fonts Folder? Do some fonts appear to be redundant with others? How much storage space does a typical font require? Write the answers to these questions in a short paragraph.
    e. Start Word. Create a title page containing your name, class, date, and the title of this assignment (My Favorite Fonts). Center the title. Use boldface or italics as you see fit. Be sure to use a suitable type size.
    f. Staple the various pages together (the title page, the three font samples, and the answers to the questions in part d). Submit the assignment to your instructor.

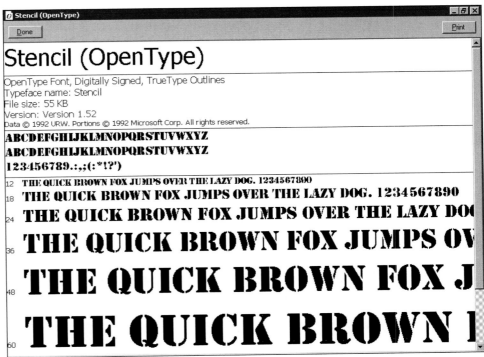

FIGURE 22   *The Fonts Folder (exercise 10)*

11. Keyboard Shortcuts: Microsoft Windows is a graphical user interface in which users "point and click" to execute commands. As you gain proficiency, however, you will find yourself gravitating toward various keyboard shortcuts as shown in Figures 23a and 23b. There is absolutely no need to memorize these shortcuts, nor should you even try. A few, however, have special appeal and everyone has his or her favorite. Use the Help menu to display this information, pick your three favorite shortcuts, and submit them to your instructor. Compare your selections with those of your classmates.

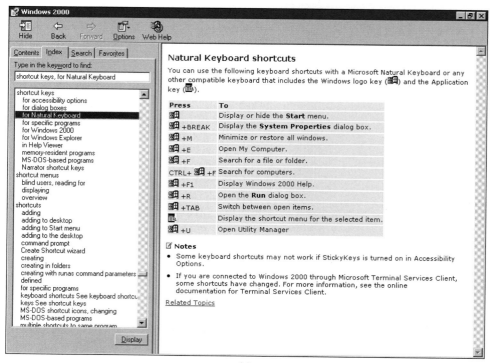

(a)

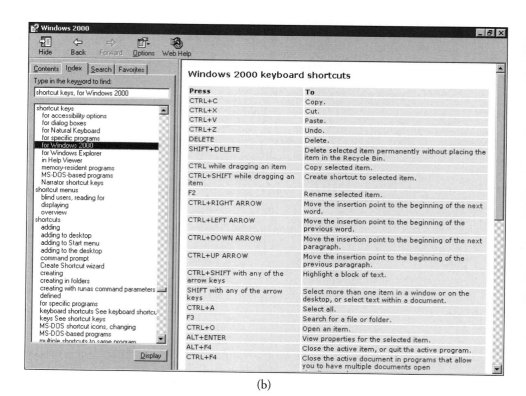

(b)

FIGURE 23 *Shortcut Keys for Natural Keyboard (Exercise 11)*

ON YOUR OWN

Planning for Disaster

Do you have a backup strategy? Do you even know what a backup strategy is? You had better learn, because sooner or later you will wish you had one. You will erase a file, be unable to read from a floppy disk, or worse yet suffer a hardware failure in which you are unable to access the hard drive. The problem always seems to occur the night before an assignment is due. The ultimate disaster is the disappearance of your computer, by theft or natural disaster. Describe, in 250 words or less, the backup strategy you plan to implement in conjunction with your work in this class.

Your First Consultant's Job

Go to a real installation such as a doctor's or attorney's office, the company where you work, or the computer lab at school. Determine the backup procedures that are in effect, then write a one-page report indicating whether the policy is adequate and, if necessary, offering suggestions for improvement. Your report should be addressed to the individual in charge of the business, and it should cover all aspects of the backup strategy; that is, which files are backed up and how often, and what software is used for the backup operation. Use appropriate emphasis (for example, bold italics) to identify any potential problems. This is a professional document (it is your first consultant's job), and its appearance should be perfect in every way.

File Compression

You've learned your lesson and have come to appreciate the importance of backing up all of your data files. The problem is that you work with large documents that exceed the 1.44MB capacity of a floppy disk. Accordingly, you might want to consider the acquisition of a file compression program to facilitate copying large documents to a floppy disk in order to transport your documents to and from school, home, or work. (A Zip file is different from a Zip drive. The latter is a hardware device, similar in concept to a large floppy disk, with a capacity of 100MB or 250MB.)

You can download an evaluation copy of the popular WinZip program at www.winzip.com. Investigate the subject of file compression and submit a summary of your findings to your instructor.

The Threat of Virus Infection

A computer virus is an actively infectious program that attaches itself to other programs and alters the way a computer works. Some viruses do nothing more than display an annoying message at an inopportune time. Most, however, are more harmful, and in the worst case, erase all files on the disk. Use your favorite search engine to research the subject of computer viruses in order to answer the following questions. When is a computer subject to infection by a virus? What precautions does your school or university take against the threat of virus infection in its computer lab? What precautions, if any, do you take at home? Can you feel confident that your machine will not be infected if you faithfully use a state-of-the-art antivirus program that was purchased in January 2001?

The Briefcase

It is becoming increasingly common for people to work on more than one machine. Students, for example, may alternate between machines at school and home. In similar fashion, an office worker may use a desktop and a laptop, or have a machine at work and at home. In every instance, you need to transfer files back and forth between the two machines. This can be done using the Copy command from within Windows Explorer. It can also be done via the Briefcase folder. Your instructor has asked you to look into the latter capability and to prepare a brief report describing its use. Do you recommend the Briefcase over a simple Copy command?

Cut, Copy, and Paste

The Cut, Copy, and Paste commands are used in conjunction with one another to move and copy data within a document, or from one Windows document to another. The commands can also be executed from within My Computer or Windows Explorer to move and copy files. You can use the standard Windows shortcuts of Ctrl+X, Ctrl+C, and Ctrl+V to cut, copy, and paste, respectively. You can also click the corresponding icons on the Standard Buttons toolbar within Windows Explorer or My Computer.

Experiment with this technique, then write a short note to your instructor that summarizes the various ways in which files can be moved or copied within Windows 2000.

Register Now

It is good practice to register every program you purchase, so that the vendor can notify you of new releases and/or other pertinent information. Windows provides an online capability whereby you can register via modem. To register your copy of Windows, click the Start button, click Programs, click Accessories, click Welcome to Windows, then click the Registration Wizard. Follow the directions that are displayed on the screen. (Registering a program does carry the risk of having unwanted sales messages sent to you by e-mail. At the Web site, look for a check box in which you choose whether to receive unsolicited e-mail.) You can do this exercise only if you are working on your own computer.

INDEX

A

Accept and Review Changes command, 297
Alignment, 77–78, 86
Application window, 5
Arial, 64, 66, 212
Ascending sequence, 312
AutoCorrect, 26, 37, 113
AutoFormat command, 113, 168, 174–175, 188
AutoMark, 195
Automatic replacement, 52
AutoSearch, 119
AutoShapes, 114, 234
AutoShapes toolbar, 234
AutoText, 26–27, 38, 167
AutoText toolbar, 38

B

Backspace key, 5
Backup (options for), 30–31, 312
Body Text style, 171–172, 177
Boldface, 72
Bookmark, 274
Borders and Shading command, 83, 168, 210, 219, 231
Browse object, 193
Bulleted list, 154–156, 210, 228
Bullets and Numbering command, 154

C

Calendar Wizard, 151
Case-insensitive search, 52
Case-sensitive search, 52
Cell, 162, 307
Character style, 171
   creation of, 181
Chart (linking of), 243–244
Check box
   in a form, 298, 302

Clip art, 109, 169, 210, 224–225, 229
   See also Media Gallery
Clipboard (*See* Office clipboard; Windows clipboard)
Close command, 8, 16
Code window, 325
Columns command, 84, 90–91, 213, 216, 227
   balancing of, 217
Comment, 63, 301
   in VBA, 324
Common user interface, 240
Compare and merge documents, 31, 46
Context-sensitive help button, 221
Copying text, 50
   with mouse, 62
   shortcut for, 61
Copyright, 117
Courier New, 64, 66
Create New Folder command, 316, 318
Create Subdocument command, 316, 319
Crop tool, 230
Custom dictionary, 24
Cut command, 50
   shortcut for, 61

D

Data disk (*See* Practice files)
Data points, 237
Data series, 237
Decrease Indent button, 159
Default folder, 309
Del key, 5
Delete rows (columns), 170
Deleting text, 5, 22
Descending sequence, 307
Desktop publishing, 210
Dialog box (shortcuts in), 74
Document Map, 179, 323
Document properties, 23, 315
Document window, 5
Drag and drop, 62

I1

Drawing canvas, 108
Drawing toolbar, 108, 226, 245
Drop-down list box
    in a form, 298, 302
Dropped capital letter, 210, 233

E

Editing marks, 300
E-mail (a Word document), 23
Embedded object, 237
Emphasis (in a document), 224–225
End Sub statement, 324
Endnote, 117
Envelope, 140, 149
Exit command, 8, 16

F

Fair use, 117
Field (in a form), 298
Field code, 303
    displaying of, 307
Field result, 303
File menu, 8–9
File name (rules for), 8
File Transfer Protocol (FTP), 267–269
File type, 8
Find command, 51
    special characters, 58
    with formatting, 71
First line indent, 77, 79
Folder (creation of), 124, 201–202
Font, 71, 212
Footer, 183, 192
Footnote, 117, 122
Foreign language proofing tools, 36, 44–45
Form, 298, 302–303
Form letter, 130
Format Font command, 65
Format Object command, 242
Format Painter, 73
Format Paragraph command, 81–82
Format Picture command, 110, 224–225, 263
Formatting properties, 72, 86, 176
Formatting toolbar, 5, 7
    separation of, 13, 20
Forms toolbar, 298, 299, 314
Frame, 274

G

Go To command, 51, 185, 189
Grammar check, 28–29, 35, 215
    foreign language tools for, 36
Grid (in document design), 222–224

H

Hanging indent, 77, 79
Hard page break, 69
Hard return, 2
    display of, 20
Header, 183, 192
Header and Footer command, 239
Header row (with Sort command), 312
Heading 1 style, 171–172, 178
Highlighting, 73, 115
Home page (creation of), 260
Horizontal ruler, 6
HTML document, 116, 126, 257–285
Hyperlink, 116, 123, 260, 264
    color of, 270
    editing of, 264
Hypertext Markup Language (*See* HTML)
Hyphenation, 81

I

Increase Indent button, 159
Indents, 77, 79, 87
Index, 185–186, 194–196
Index and Tables command, 185
InputBox function, 331
Ins key, 4, 21
Insert Bookmark command, 283
Insert Comment command, 301
Insert Date command, 133, 328
Insert Hyperlink command, 123, 260, 264
Insert mode, 4, 21
Insert Page Numbers command, 183
Insert Picture command, 109, 120, 169, 224, 263, 282
Insert Reference command, 117, 122
Insert rows (columns), 170
Insert Subdocument command, 321
Insert Symbol command, 106, 113
Insert Table command, 162
Insertion point, 2
Internet, 116
Internet Explorer, 267
Intranet, 117, 267
Italics, 72

K

Keyboard shortcut (with macros), 329

L

Landscape orientation, 69
Leader character, 80, 185
Left indent, 77, 79

Line spacing, 81, 86
Linked object, 237

M

Macro, 324–336
Macro recorder, 324
Mail merge, 130–141
Mail Merge toolbar, 137
Mail Merge Wizard, 134
Mailing label, 140, 150
Main document, 130
Margins (changing of), 214
Margins tab, 68–69, 74
Mark Index entry, 194
Master document, 316–323
Masthead, 210, 218
Media Gallery, 105, 110, 263
Merge fields, 130
Merging cells, 166
Microsoft Excel, 237–238, 240
Microsoft Word
    file types in, 34
    starting of, 12
    version of, 16
Microsoft WordArt, 107, 111–112, 235
Module, 325
Monospaced typeface, 65, 212
Moving text, 50
    with mouse, 62
Multitasking, 241

N

Nonbreaking hyphen, 81
Noncontiguous text (selection of), 88
Nonprinting symbols, 17
Normal style, 171
Normal view, 17–18, 55
Numbered list, 154–155, 157, 161, 210

O

Object linking and embedding (OLE), 237–246
Office Assistant, 12, 15
    hiding of, 20
Office clipboard, 50, 60
Open command, 8–9, 19
    previewing files, 32
Organization chart, 114, 151
Orphan, 81, 86
Outline, 154–155, 158–161, 173
Outline numbered list, 154–155
Outline view, 173, 179, 321
Outlining toolbar, 316
Overtype mode, 4, 21

P

Page Border command, 88, 334
Page break, 69
Page numbers, 183
Page Setup command, 67–69, 74, 76, 164, 191, 272
Paragraph (versus section), 84
Paragraph style, 171
    creation of, 180
Password protection, 297
    removing of, 309
Paste command, 50
    shortcut for, 61
Paste Link option, 241, 244
Paste options, 59
Paste Special command, 50, 59, 241–244
Picture
    cropping of, 121, 230
    downloading of, 119
    inserting of, 120
Places bar, 8
Point size, 65, 212
Portrait orientation, 69
Practice files (downloading of), 11
Print command, 8, 16
Print Layout view, 17–18, 55
Print Preview command, 63, 221
Print Preview toolbar, 221
Procedure, 324
Project Explorer, 325–326
Proportional typeface, 65, 212
Protect command, 305
Public domain, 117
Pull quote, 210, 232

R

Redo command, 22, 61
Replace command, 51, 58
    with formatting, 71
Résumé Wizard, 127–128
Reveal formatting, 72
Reverse, 210, 219
Reviewing toolbar, 297, 299
Revision mark, 297
Right indent, 77, 79
Round-trip HTML, 260
Ruler
    setting column width, 90, 217
    setting indents, 87

S

Sans serif typeface, 64, 212
Save As command, 30, 33
Save As Web Page command, 116, 124, 260, 262

Save command, 8–9, 14
ScreenTip, 5
Scrolling, 53, 57
Section, 84, 184
Section break, 84, 91, 190, 213, 217
Selecting text
   F8 key, 85
   noncontiguous, 88
   selection bar, 62
Selective replacement, 52
Select-then-do, 50
Serif typeface, 64, 212
Server, 267
Shortcut key (with macros), 329
Show/Hide ¶ button, 17, 20
Shrink-to-fit button, 221
Side-by-side column chart, 237–238
Sizing handle, 108, 110
Small caps, 65
Soft page break, 69
Soft return, 2
Sort command, 312
Source code (HTML), 259
Special characters (find and replace), 58
Special indent, 77, 79
Spell check, 24–25, 34, 215
   foreign language tools for, 36
Splitting cells, 166
Stacked column chart, 237
Standard toolbar, 5, 7
   separation of, 13, 20
Status bar, 6
Strikethrough, 65
Style, 171–172, 187, 220
Styles and Formatting command, 171
Sub statement, 324
Subdocument, 316–317, 320–322
Subscript, 65
Superscript, 65
Symbols (as clip art), 106, 145–146
Symbols font, 106

T

Table, 162–170
Table math, 307–308, 313–314
Table menu, 162, 307
Table of contents, 184–186, 188–189, 193
   updating of, 197
Tables and Borders toolbar, 231, 307
Tabs, 80
   in a table, 165, 311
Tag, 258
Task pane, 6
   reveal formatting, 72
Taskbar, 241

Telnet, 267, 273
Template, 127, 129, 138
Terminal session, 267
Text box, 235
Text field (in a form), 298, 302
Theme, 260, 265
Thesaurus, 27–28, 36
   foreign language tools for, 36
Times New Roman, 64, 66, 212
Tip of the Day, 15
Toggle switch, 3
Toolbar, 5, 7, 347–352
   docked versus floating, 299
   hiding or displaying, 17, 20
   separation of, 13
Tools menu, 20
Track Changes command, 297, 300
TRK indicator, 300
Troubleshooting, 17–18, 20
Type size, 65, 212
Type style, 65
Typeface, 212
Typography, 64–67, 212

U

Underlining, 72
Undo command, 22, 61
Unprotect command, 335

V

VBA, 324
Versions command, 297, 304
Vertical ruler, 6
View menu, 20, 55–56
Visual Basic Editor, 325–326
Visual Basic for Applications (*See* VBA)

W

Web Layout command, 261
Web page, 116
Web Page Wizard, 117, 274–278
Web site (creation of), 274–284
Webdings font, 106
Whole words only, 52
Widow, 81, 86
Wild card, 52
Windows clipboard, 50, 59
Windows desktop, 10
Wingdings font, 106, 113
Wizard, 127, 129
Word Count toolbar, 33

Word wrap, 2–3
WordArt (*See* Microsoft WordArt)
Workgroups, 296–297
Worksheet (linking of), 240–241
World Wide Web, 116
   searching of, 118

Z

Zoom command, 55, 89